Advanced Micro- and Nano-Manufacturing Technologies

Advanced Micro- and Nano-Manufacturing Technologies

Guest Editor

Kun Li

 Basel • Beijing • Wuhan • Barcelona • Belgrade • Novi Sad • Cluj • Manchester

Guest Editor
Kun Li
College of Mechanical and
Vehicle Engineering
Chongqing University
Chongqing
China

Editorial Office
MDPI AG
Grosspeteranlage 5
4052 Basel, Switzerland

This is a reprint of the Special Issue, published open access by the journal *Micromachines* (ISSN 2072-666X), freely accessible at: www.mdpi.com/journal/micromachines/special_issues/Advanced_AddictiveManufacturing.

For citation purposes, cite each article independently as indicated on the article page online and using the guide below:

Lastname, A.A.; Lastname, B.B. Article Title. *Journal Name* **Year**, *Volume Number*, Page Range.

ISBN 978-3-7258-3014-5 (Hbk)
ISBN 978-3-7258-3013-8 (PDF)
https://doi.org/10.3390/books978-3-7258-3013-8

© 2025 by the authors. Articles in this book are Open Access and distributed under the Creative Commons Attribution (CC BY) license. The book as a whole is distributed by MDPI under the terms and conditions of the Creative Commons Attribution-NonCommercial-NoDerivs (CC BY-NC-ND) license (https://creativecommons.org/licenses/by-nc-nd/4.0/).

Contents

About the Editor vii

Kun Li
Editorial for the Special Issue on Advanced Micro- and Nano-Manufacturing Technologies
Reprinted from: *Micromachines* 2024, 15, 1479, https://doi.org/10.3390/mi15121479 1

Hao Li, Shenghuan Zhang, Qiaoyu Chen, Zhaoyang Du, Xingyu Chen and Xiaodan Chen et al.
High-Quality Spherical Silver Alloy Powder for Laser Powder Bed Fusion Using Plasma Rotating Electrode Process
Reprinted from: *Micromachines* 2024, 15, 396, https://doi.org/10.3390/mi15030396 6

Chufan Chen, Mengyuan Sun, Leiyi Wang, Teng Huang and Min Xu
A Fast Transient Response Capacitor-Less LDO with Transient Enhancement Technology
Reprinted from: *Micromachines* 2024, 15, 299, https://doi.org/10.3390/mi15030299 22

Kai Li, Vyacheslav Trofimov, Changjun Han, Gaoling Hu, Zhi Dong and Yujin Zou et al.
The Printability, Microstructure, and Mechanical Properties of $Fe_{80-x}Mn_xCo_{10}Cr_{10}$ High-Entropy Alloys Fabricated by Laser Powder Bed Fusion Additive Manufacturing
Reprinted from: *Micromachines* 2024, 15, 123, https://doi.org/10.3390/mi15010123 34

Huibo Zhao, Yan Gu, Yuan Xi, Xingbao Fu, Yinghuan Gao and Jiali Wang et al.
Simulation and Experimental Study of Non-Resonant Vibration-Assisted Lapping of SiCp/Al
Reprinted from: *Micromachines* 2024, 15, 113, https://doi.org/10.3390/mi15010113 53

Bo Zhao, Xifeng Gao, Jiansheng Pan, Huan Liu and Pengyue Zhao
Investigation of Gallium Arsenide Deformation Anisotropy During Nanopolishing via Molecular Dynamics Simulation
Reprinted from: *Micromachines* 2024, 15, 110, https://doi.org/10.3390/mi15010110 75

Jeon-Woong Kang, Jinpyo Jeon, Jun-Young Lee, Jun-Hyeong Jeon and Jiwoo Hong
Surface-Wetting Characteristics of DLP-Based 3D Printing Outcomes under Various Printing Conditions for Microfluidic Device Fabrication
Reprinted from: *Micromachines* 2023, 15, 61, https://doi.org/10.3390/mi15010061 91

Victor Lisitsyn, Aida Tulegenova, Mikhail Golkovski, Elena Polisadova, Liudmila Lisitsyna and Dossymkhan Mussakhanov et al.
Radiation Synthesis of High-Temperature Wide-Bandgap Ceramics
Reprinted from: *Micromachines* 2023, 14, 2193, https://doi.org/10.3390/mi14122193 105

Mengyuan Sun, Chufan Chen, Leiyi Wang, Xinling Xie, Yuhang Wang and Min Xu
A Fast Transient Adaptive On-Time Controlled BUCK Converter with Dual Modulation
Reprinted from: *Micromachines* 2023, 14, 1868, https://doi.org/10.3390/mi14101868 134

Junhua Wang, Junfei Xu, Yan Lu, Tancheng Xie, Jianjun Peng and Junliang Chen
Z-Increments Online Supervisory System Based on Machine Vision for Laser Solid Forming
Reprinted from: *Micromachines* 2023, 14, 1558, https://doi.org/10.3390/mi14081558 147

Penghao Zhang, Luyu Wang, Kaiyue Zhu, Qiang Wang, Maolin Pan and Ziqiang Huang et al.
Non-Buffer Epi-AlGaN/GaN on SiC for High-Performance Depletion-Mode MIS-HEMTs Fabrication
Reprinted from: *Micromachines* 2023, 14, 1523, https://doi.org/10.3390/mi14081523 166

Mengyuan Sun, Luyu Wang, Penghao Zhang and Kun Chen
Improving Performance of Al_2O_3/AlN/GaN MIS HEMTs via In Situ N_2 Plasma Annealing
Reprinted from: *Micromachines* 2023, 14, 1100, https://doi.org/10.3390/mi14061100 176

Ke Sun, Gai Wu, Kang Liang, Bin Sun and Jian Wang
Investigation into Photolithography Process of FPCB with 18 m Line Pitch
Reprinted from: *Micromachines* 2023, 14, 1020, https://doi.org/10.3390/mi14051020 184

Eryong Han, Kuanqiang Zhang, Lijuan Chen, Chenfei Guo, Ying Xiong and Yong Guan et al.
Simulation and Experimental Validation of a Pressurized Filling Method for Neutron Absorption Grating
Reprinted from: *Micromachines* 2023, 14, 1016, https://doi.org/10.3390/mi14051016 197

Andre Childs, Jorge Pereira, Charles M. Didier, Aliyah Baksh, Isaac Johnson and Jorge Manrique Castro et al.
Plotter Cut Stencil Masks for the Deposition of Organic and Inorganic Materials and a New Rapid, Cost Effective Technique for Antimicrobial Evaluations
Reprinted from: *Micromachines* 2022, 14, 14, https://doi.org/10.3390/mi14010014 207

Tatiana L. Simonenko, Nikolay P. Simonenko, Philipp Yu. Gorobtsov, Elizaveta P. Simonenko and Nikolay T. Kuznetsov
Microextrusion Printing of Multilayer Hierarchically Organized Planar Nanostructures Based on NiO, $(CeO_2)_{0.8}(Sm_2O_3)_{0.2}$ and $La_{0.6}Sr_{0.4}Co_{0.2}Fe_{0.8}O_{3-\delta}$
Reprinted from: *Micromachines* 2022, 14, 3, https://doi.org/10.3390/mi14010003 222

Guangen Zhao, Yongchao Xu, Qianting Wang, Jun Liu, Youji Zhan and Bingsan Chen
Polishing Performance and Removal Mechanism of Core-Shell Structured Diamond/SiO_2 Abrasives on Sapphire Wafer
Reprinted from: *Micromachines* 2022, 13, 2160, https://doi.org/10.3390/mi13122160 236

Kun Chen, Jingwen Yang, Tao Liu, Dawei Wang, Min Xu and Chunlei Wu et al.
Source/Drain Trimming Process to Improve Gate-All-Around Nanosheet Transistors Switching Performance and Enable More Stacks of Nanosheets
Reprinted from: *Micromachines* 2022, 13, 1080, https://doi.org/10.3390/mi13071080 250

Yingchun Fu, Guowei Han, Jiebin Gu, Yongmei Zhao, Jin Ning and Zhenyu Wei et al.
A High-Performance MEMS Accelerometer with an Improved TGV Process of Low Cost
Reprinted from: *Micromachines* 2022, 13, 1071, https://doi.org/10.3390/mi13071071 256

Sinong Zha, Dongling Li, Quan Wen, Ying Zhou and Haomiao Zhang
Design and Fabrication of Silicon-Blazed Gratings for Near-Infrared Scanning Grating Micromirror
Reprinted from: *Micromachines* 2022, 13, 1000, https://doi.org/10.3390/mi13071000 265

Cuicui Du, Deren Kong and Chundong Xu
Development of a Fault Detection Instrument for Fiber Bragg Grating Sensing System on Airplane
Reprinted from: *Micromachines* 2022, 13, 882, https://doi.org/10.3390/mi13060882 279

Penghao Zhang, Luyu Wang, Kaiyue Zhu, Yannan Yang, Rong Fan and Maolin Pan et al.
High Selectivity, Low Damage ICP Etching of p-GaN over AlGaN for Normally-off p-GaN HEMTs Application
Reprinted from: *Micromachines* 2022, 13, 589, https://doi.org/10.3390/mi13040589 293

Keqin Zhao, Diming Lou, Yunhua Zhang, Liang Fang and Yuanzhi Tang
Experimental Study on Diesel Engine Emission Characteristics Based on Different Exhaust Pipe Coating Schemes
Reprinted from: *Micromachines* **2021**, *12*, 1155, https://doi.org/10.3390/mi12101155 . . . . **301**

About the Editor

Kun Li

Dr. Kun Li is the Distinguished Professor of "Hongshen Young Scholars" at Chongqing University, PhD supervisor, director of High-performance intelligent Additive Manufacturing (HπAM) Lab, associate chairman of Chongqing Key Laboratory of Metal Additive Manufacturing (3D Printing). Dr. Li obtained Bachelor degree from Jilin University, obtained Philosophy degree from Tsinghua University. In February 2017, he went to the University of Texas at El Paso for postdoctoral research and served as Prof. Lawrence E. Murr (The pioneer in additive manufacturing in the USA) and PhD student co-supervisor. In April 2019, he was appointed as a senior researcher in the Department of Mechanical Engineering and Materials Science, University of Pittsburgh. In August 2020, he was hired as a "Hongshen Young Scholar" distinguished professor and PhD supervisor of Chongqing University, engaged in teaching and research work. Mainly engaged on additive manufacturing, intelligent 3D net forming, high-performance materials and phase change, and material computation. He has published more than 80 papers in famous journals and conferences such as *Additive Manufacturing*, *Journal of Materials Science and Technology*, *Small*, and so on. Served as the reviewer of famous journals in the fields of *Acta Materialia*, *Materials Research Letters*, *International Journal of Plasticity*, *Nature Communications*, and other fields. He has participated in many projects such as National Natural Science Foundation of China and International Cooperation Projects with the USA. His scientific research achievements won the Researcher Award of the University of Texas and the Houston Space Agency Technology Application Award.

Editorial

Editorial for the Special Issue on Advanced Micro- and Nano-Manufacturing Technologies

Kun Li [1,2,3]

1. College of Mechanical and Vehicle Engineering, Chongqing University, Chongqing 400044, China; kun.li@cqu.edu.cn
2. Chongqing Key Laboratory of High-Performance Structural Additive Manufacturing, Chongqing 400044, China
3. State Key Laboratory of Mechanical Transmission for Advanced Equipment, Chongqing University, Chongqing 400044, China

Citation: Li, K. Editorial for the Special Issue on Advanced Micro- and Nano-Manufacturing Technologies. *Micromachines* **2024**, *15*, 1479. https://doi.org/10.3390/mi15121479

Received: 4 December 2024
Accepted: 5 December 2024
Published: 8 December 2024

Copyright: © 2024 by the author. Licensee MDPI, Basel, Switzerland. This article is an open access article distributed under the terms and conditions of the Creative Commons Attribution (CC BY) license (https:// creativecommons.org/licenses/by/ 4.0/).

With the continuous advancement of science and technology, micro- and nano-manufacturing technologies have become frontier fields in modern manufacturing [1,2]. These advanced manufacturing processes enable the creation of materials with special functions at the micro and nano scales and significantly improve manufacturing precision and efficiency [3]. High-energy, nano-, and micrometer-scale manufacturing processes have demonstrated great potential in various applications, particularly in optimizing surface properties of mechanical components and achieving complex performance regulation in the design and fabrication of functional and gradient materials [4–8].

Furthermore, functional materials developed through advanced manufacturing methods have broad application prospects in electronics, biomedical engineering, and aerospace [9–12]. Applying modeling and numerical analysis techniques provides a solid theoretical foundation for optimizing manufacturing processes, allowing engineers to predict and address potential issues during the design stage [13–16]. With the rise of smart manufacturing, the integration of advanced detection, monitoring, and control technologies further ensures the realization of efficient, precise, and adaptive production [17–20].

This Special Issue, Advanced Micro- and Nano-Manufacturing Technologies, brings together 21 original research articles and 1 review paper, covering the following topics: (1) high-energy/nano/micro advanced manufacturing processes; (2) material post-processing technologies; (3) advanced manufacturing design, simulation, and numerical analysis; and (4) advanced detection, monitoring, and intelligent control equipment and methods. These studies deepen our understanding of advanced manufacturing technologies and lay a solid foundation for the manufacturing industry's future development. The following summarizes the key contributions and their scientific significance.

High-energy/nano/micro advanced manufacturing processes: To meet the high-precision manufacturing and assembly requirements of micro-mechanical systems, exploring and applying micro- and nanoscale advanced manufacturing processes is crucial [21]. Optimizing techniques, such as continuous casting (CC) and the plasma rotating electrode process (PREP), are key in minimizing material waste while enhancing material densification and mechanical properties. These advancements are essential for fabricating high-density, high-quality alloys. For instance, Lihao et al. [22] demonstrated the production of spherical silver alloy powders suitable for laser powder bed fusion (LPBF), while Likai et al. [23] used these optimized processes to fabricate dense high-entropy alloys like $Fe_{30}Mn_{50}Co_{10}Cr_{10}$ and $Fe_{50}Mn_{30}Co_{10}Cr_{10}$, showcasing significant improvements in material quality and performance.

In advanced epitaxial structures, Zhang et al. [24] designed an unbuffered AlGaN/GaN epitaxial structure on a SiC substrate, achieving superior crystallinity and surface morphology compared to traditional methods. This innovation improves control over growth stress and contributes to the fabrication of high-performance devices. Their study also emphasized the importance of precise control over printing parameters to optimize material properties in advanced manufacturing.

In digital light processing (DLP), Kang et al. [25] explored the effects of print layer thickness and orientation on surface wettability, contributing valuable insights for developing DLP-based microfluidic devices. Similarly, Sun et al. [26] developed a model to predict photoresist profile evolution in photolithography, considering factors such as incident light intensity and air gaps. This model enhances control over photoresist morphology, which is crucial for achieving precision in microfabrication.

This Special Issue also highlights other breakthroughs. For example, Tatiana L. Simonenko et al. [27] synthesized NiO, $La_{0.6}Sr_{0.4}Co_{0.2}Fe_{0.8}O_3$-$\delta$ (LSCF), and $(CeO_2)_{0.8}(Sm_2O_3)_{0.2}$ (SDC) nanoparticles using hydrothermal and citrate-ethylene glycol methods. These nanoparticles were applied in continuous micro-extrusion printing to create a functional ink structure with a three-layer (NiO-SDC)/SDC/(LSCF-SDC) configuration. Chen et al. [28] introduced a new source/drain (S/D) micro-adjustment process enabling vertical stacking of more than four nanosheet layers, pushing the boundaries of GAA technology beyond the 3 nm CMOS process. Zhang et al. [29] optimized the BCl_3/SF_6 inductively coupled plasma (ICP) etching process, achieving an impressive etching selectivity 41:1 for p-GaN. Lastly, Victor Lisitsyn et al. [30] summarized research on ceramic synthesis using high-energy electron flux, demonstrating that ceramics produced through this method exhibit similar properties to those created by traditional thermal methods, including enhanced luminescent characteristics.

Post-processing Techniques for Materials: Post-processing techniques are crucial in optimizing product performance and enhancing surface quality, particularly in micro- and nano-manufacturing processes. These techniques are essential for achieving the desired material properties and device performance [31]. This Special Issue features two research articles exploring innovative post-processing methods.

First, Sun et al. [32] introduced a novel method for forming a single-crystal AlN interface layer for Al_2O_3/AlN/GaN metal-insulator-semiconductor high electron mobility transistors (MIS-HEMTs). This method combines plasma-enhanced atomic layer deposition (PEALD) with in situ N_2 plasma annealing (NPA), significantly improving the gate reliability of Al_2O_3/AlN/GaN MIS-HEMT devices. The synergy of these two techniques effectively enhances the overall device performance, providing a promising solution for high-reliability MIS-HEMT applications. Second, Zhao et al. [33] developed a new diamond/silica composite abrasive using a simplified sol–gel method and applied it to the semi-fixed polishing of sapphire wafers. Unlike traditional polishing slurries, this composite slurry contains only deionized water and no additional chemical additives, making it environmentally friendly. In addition to its environmental benefits, this composite slurry significantly improves polishing performance, with experimental results showing a 27.2% reduction in surface roughness and an 8.8% increase in material removal rate compared to pure diamond polishing.

Advanced Manufacturing Design, Simulation, and Numerical Analysis: The widespread use of computer simulation and numerical analysis in advanced manufacturing has made these technologies essential for reducing experimental costs and enabling in-depth research under complex conditions. This Special Issue presents three studies that showcase the contributions of simulation and numerical analysis to advancing manufacturing processes.

First, Zhaohuibo et al. [34] addressed challenges in SiCp/Al machining by designing a vibration-assisted grinding device (VLD). Through simulation and experimental testing, the team determined the device's optimal working frequency and amplitude, validating its performance and demonstrating its potential to enhance grinding efficiency. Zhaobo et al. [35] used molecular dynamics (MD) simulations to study surface generation and subsurface damage. By analyzing surface morphology, mechanical response, and amorphization of different crystal orientations, they provided insights into how crystal orientation affects the severity of deformation and amorphization, enriching our understanding of material behavior under stress. Lastly, Han et al. [36] applied a particle filling method to manufacture neutron absorption gratings, introducing a pressurized filling technique to increase filling rates. Simulation studies revealed the effects of pressure, groove width, and Young's modulus on filling rates. Based on these results, the team optimized the process, significantly improving manufacturing efficiency.

Advanced Detection, Monitoring, Intelligent Control Equipment, and Methods: In micro- and nano-manufacturing, advanced detection, monitoring, and intelligent control equipment are essential for effectively controlling production processes, detecting material properties, and ensuring product quality [37]. This Special Issue includes four research articles that explore innovations in these advanced technologies and methods.

Chen et al. [38] proposed a novel low-dropout voltage regulator (CL-LDO) for digital simulation hybrid circuits. Designed with a 180-nanometer process, this regulator eliminates the need for on-chip capacitors while maintaining steady-state performance, improving transient response during load current variations. Sun et al. [39] introduced a fully integrated adaptive on-time (AOT) control buck converter, capable of switching between pulse-width modulation (PWM) and pulse-skip modulation (PSM) based on load current variations, optimizing efficiency under light-load conditions. Fu et al. [40] proposed a method for manufacturing high-performance MEMS accelerometers using the TGV process. This approach reduces manufacturing costs while ensuring low-noise characteristics. Zha et al. [41] designed and fabricated a high diffraction efficiency silicon-glass grating with a working wavelength range of 800–2500 nm, suitable for near-infrared spectrometers.

These studies offer new design concepts and technical approaches for sensors and electronic devices in micro-nano manufacturing. Du et al. [42] also developed a fault detection device and method for fiber Bragg grating (FBG) sensor systems, enabling fast detection and localization of faults in aircraft FBG systems. Wang et al. [43] proposed a Z-axis incremental closed-loop control system based on machine vision in laser additive manufacturing, achieving real-time monitoring of precise coating heights.

Furthermore, this Special Issue also includes other innovative research. Andre Childs et al. [44] demonstrated how a traditional plotter-cutting machine can be adapted as an efficient desktop tool for rapidly producing steel mesh masks with a range below 250 μm and explored its application in healthcare. Zhao et al. [45] conducted engine bench tests to investigate the thermal insulation performance of exhaust pipes coated with basalt and fiberglass materials in different braiding forms (sleeve, wound, and felt). They also analyzed the impact of these materials on the emission characteristics of diesel engines.

Finally, the guest editors would like to express their sincere gratitude to all the authors for their valuable contributions to this Special Issue, which have greatly advanced the dissemination and development of cutting-edge micro- and nano-manufacturing technologies. We also thank the dedicated reviewers whose time and effort in reviewing the manuscripts have enhanced the quality of the submissions. We sincerely invite all researchers and professionals to continue submitting their work to this Special Issue. Your insightful contributions will further enrich the ongoing discussions and progress in this field.

Conflicts of Interest: The author declares no conflicts of interest.

References

1. Huang, Z.; Shao, G.; Li, L. Micro/nano functional devices fabricated by additive manufacturing. *Prog. Mater. Sci.* **2023**, *131*, 101020. [CrossRef]
2. Gao, S.; Huang, H. Recent advances in micro- and nano-machining technologies. *Front. Mech. Eng.* **2017**, *12*, 18–32. [CrossRef]
3. Chen, Z.; Lin, Y.-T.; Salehi, H.; Che, Z.; Zhu, Y.; Ding, J.; Sheng, B.; Zhu, R.; Jiao, P. Advanced Fabrication of Mechanical Metamaterials Based on Micro/Nanoscale Technology. *Adv. Eng. Mater.* **2023**, *25*, 2300750. [CrossRef]
4. Ma, Z.; Wang, W.; Xiong, Y.; Long, Y.; Shao, Q.; Wu, L.; Wang, J.; Tian, P.; Khan, A.U.; Yang, W. Carbon Micro/Nano Machining toward Miniaturized Device: Structural Engineering, Large-Scale Fabrication, and Performance Optimization. *Small* **2024**, 2400179. [CrossRef] [PubMed]
5. Shi, H.; Zhou, P.; Li, J.; Liu, C.; Wang, L. Functional gradient metallic biomaterials: Techniques, current scenery, and future prospects in the biomedical field. *Front. Bioeng. Biotechnol.* **2021**, *8*, 616845. [CrossRef]

6. Duan, Y.; You, G.; Sun, K.; Zhu, Z.; Liao, X.; Lv, L.; Tang, H.; Xu, B.; He, L. Advances in wearable textile-based micro energy storage devices: Structuring, application and perspective. *Nanoscale Adv.* **2021**, *3*, 6271–6293. [CrossRef]
7. Zhang, S.; Ke, X.; Jiang, Q.; Chai, Z.; Wu, Z.; Ding, H. Fabrication and Functionality Integration Technologies for Small-Scale Soft Robots. *Adv. Mater.* **2022**, *34*, 2200671. [CrossRef] [PubMed]
8. Guan, Z.; Wang, L.; Bae, J. Advances in 4D printing of liquid crystalline elastomers: Materials, techniques, and applications. *Mater. Horiz.* **2022**, *9*, 1825–1849. [CrossRef]
9. Yang, Y.; Song, X.; Li, X.; Chen, Z.; Zhou, C.; Zhou, Q.; Chen, Y. Recent Progress in Biomimetic Additive Manufacturing Technology: From Materials to Functional Structures. *Adv. Mater.* **2018**, *30*, 1706539. [CrossRef]
10. Li, Y.; Feng, Z.; Hao, L.; Huang, L.; Xin, C.; Wang, Y.; Bilotti, E.; Essa, K.; Zhang, H.; Li, Z.; et al. A Review on Functionally Graded Materials and Structures via Additive Manufacturing: From Multi-Scale Design to Versatile Functional Properties. *Adv. Mater. Technol.* **2020**, *5*, 1900981. [CrossRef]
11. Zhu, Y.; Tang, T.; Zhao, S.; Joralmon, D.; Poit, Z.; Ahire, B.; Keshav, S.; Raje, A.R.; Blair, J.; Zhang, Z.; et al. Recent advancements and applications in 3D printing of functional optics. *Addit. Manuf.* **2022**, *52*, 102682. [CrossRef]
12. Judy, J.W. Microelectromechanical systems (MEMS): Fabrication, design and applications. *Smart Mater. Struct.* **2001**, *10*, 1115. [CrossRef]
13. Jayal, A.D.; Badurdeen, F.; Dillon, O.W.; Jawahir, I.S. Sustainable manufacturing: Modeling and optimization challenges at the product, process and system levels. *CIRP J. Manuf. Sci. Technol.* **2010**, *2*, 144–152. [CrossRef]
14. Hashemi, S.M.; Parvizi, S.; Baghbanijavid, H.; Tan, A.T.L.; Nematollahi, M.; Ramazani, A.; Fang, N.X.; Elahinia, M. Computational modelling of process–structure–property–performance relationships in metal additive manufacturing: A review. *Int. Mater. Rev.* **2022**, *67*, 1–46. [CrossRef]
15. Alabi, T.M.; Aghimien, E.I.; Agbajor, F.D.; Yang, Z.; Lu, L.; Adeoye, A.R.; Gopaluni, B. A review on the integrated optimization techniques and machine learning approaches for modeling, prediction, and decision making on integrated energy systems. *Renew. Energy* **2022**, *194*, 822–849. [CrossRef]
16. Li, K.; Zhan, J.; Wang, Y.; Qin, Y.; Gong, N.; Zhang, D.Z.; Tan, S.; Murr, L.E.; Liu, Z. Application of data-driven methods for laser powder bed fusion of Ni-based superalloys: A review. *J. Manuf. Process.* **2025**, *133*, 285–321. [CrossRef]
17. Ding, H.; Gao, R.X.; Isaksson, A.J.; Landers, R.G.; Parisini, T.; Yuan, Y. State of AI-Based Monitoring in Smart Manufacturing and Introduction to Focused Section. *IEEE/ASME Trans. Mechatron.* **2020**, *25*, 2143–2154. [CrossRef]
18. Lee, J.; Ni, J.; Singh, J.; Jiang, B.; Azamfar, M.; Feng, J. Intelligent Maintenance Systems and Predictive Manufacturing. *J. Manuf. Sci. Eng.* **2020**, *142*, 110805. [CrossRef]
19. Wan, J.; Li, X.; Dai, H.N.; Kusiak, A.; Martínez-García, M.; Li, D. Artificial-Intelligence-Driven Customized Manufacturing Factory: Key Technologies, Applications, and Challenges. *Proc. IEEE* **2021**, *109*, 377–398. [CrossRef]
20. Lu, Y.; Xu, X.; Wang, L. Smart manufacturing process and system automation—A critical review of the standards and envisioned scenarios. *J. Manuf. Syst.* **2020**, *56*, 312–325. [CrossRef]
21. Boopathy, G.; Sathish, S.; Kumar, M.H. Introduction to precision manufacturing for micro-and nanofabrication. In *Microfabrication and Nanofabrication: Precision Manufacturing*; Walter de Gruyter GmbH: Berlin, Germany, 2024; Volume 11, p. 1. [CrossRef]
22. Li, H.; Zhang, S.; Chen, Q.; Du, Z.; Chen, X.; Chen, X.; Zhou, S.; Mei, S.; Ke, L.; Sun, Q.; et al. High-Quality Spherical Silver Alloy Powder for Laser Powder Bed Fusion Using Plasma Rotating Electrode Process. *Micromachines* **2024**, *15*, 396. [CrossRef] [PubMed]
23. Li, K.; Trofimov, V.; Han, C.; Hu, G.; Dong, Z.; Zou, Y.; Wang, Z.; Yan, F.; Fu, Z.; Yang, Y. The Printability, Microstructure, and Mechanical Properties of $Fe_{80-x}Mn_xCo_{10}Cr_{10}$ High-Entropy Alloys Fabricated by Laser Powder Bed Fusion Additive Manufacturing. *Micromachines* **2024**, *15*, 123. [CrossRef] [PubMed]
24. Zhang, P.; Wang, L.; Zhu, K.; Wang, Q.; Pan, M.; Huang, Z.; Yang, Y.; Xie, X.; Huang, H.; Hu, X.; et al. Non-Buffer Epi-AlGaN/GaN on SiC for High-Performance Depletion-Mode MIS-HEMTs Fabrication. *Micromachines* **2023**, *14*, 1523. [CrossRef] [PubMed]
25. Kang, J.-W.; Jeon, J.; Lee, J.-Y.; Jeon, J.-H.; Hong, J. Surface-Wetting Characteristics of DLP-Based 3D Printing Outcomes under Various Printing Conditions for Microfluidic Device Fabrication. *Micromachines* **2024**, *15*, 61. [CrossRef]
26. Sun, K.; Wu, G.; Liang, K.; Sun, B.; Wang, J. Investigation into Photolithography Process of FPCB with 18 μm Line Pitch. *Micromachines* **2023**, *14*, 1020. [CrossRef] [PubMed]
27. Simonenko, T.L.; Simonenko, N.P.; Gorobtsov, P.Y.; Simonenko, E.P.; Kuznetsov, N.T. Microextrusion Printing of Multilayer Hierarchically Organized Planar Nanostructures Based on NiO, $(CeO_2)_{0.8}(Sm_2O_3)_{0.2}$ and $La_{0.6}Sr_{0.4}Co_{0.2}Fe_{0.8}O_{3-\delta}$. *Micromachines* **2023**, *14*, 3. [CrossRef]
28. Chen, K.; Yang, J.; Liu, T.; Wang, D.; Xu, M.; Wu, C.; Wang, C.; Xu, S.; Zhang, D.W.; Liu, W. Source/Drain Trimming Process to Improve Gate-All-Around Nanosheet Transistors Switching Performance and Enable More Stacks of Nanosheets. *Micromachines* **2022**, *13*, 1080. [CrossRef]
29. Zhang, P.; Wang, L.; Zhu, K.; Yang, Y.; Fan, R.; Pan, M.; Xu, S.; Xu, M.; Wang, C.; Wu, C.; et al. High Selectivity, Low Damage ICP Etching of p-GaN over AlGaN for Normally-off p-GaN HEMTs Application. *Micromachines* **2022**, *13*, 589. [CrossRef]
30. Lisitsyn, V.; Tulegenova, A.; Golkovski, M.; Polisadova, E.; Lisitsyna, L.; Mussakhanov, D.; Alpyssova, G. Radiation Synthesis of High-Temperature Wide-Bandgap Ceramics. *Micromachines* **2023**, *14*, 2193. [CrossRef]
31. Jamaludin, A.S.; Mhd Razali, M.N.; Nor Hamran, N.N.; Mohd Zawawi, M.Z.; Md Ali, M.A. Current and Future Challenges of Hybrid Electrochemical-Mechanical Machining Process for Micro- and Nano-Manufacturing. In *Intelligent Manufacturing and Mechatronics*; Springer: Singapore, 2024; pp. 81–89. [CrossRef]

32. Sun, M.; Wang, L.; Zhang, P.; Chen, K. Improving Performance of Al_2O_3/AlN/GaN MIS HEMTs via In Situ N_2 Plasma Annealing. *Micromachines* **2023**, *14*, 1100. [CrossRef]
33. Zhao, G.; Xu, Y.; Wang, Q.; Liu, J.; Zhan, Y.; Chen, B. Polishing Performance and Removal Mechanism of Core-Shell Structured Diamond/SiO_2 Abrasives on Sapphire Wafer. *Micromachines* **2022**, *13*, 2160. [CrossRef] [PubMed]
34. Zhao, H.; Gu, Y.; Xi, Y.; Fu, X.; Gao, Y.; Wang, J.; Xie, L.; Liang, G. Simulation and Experimental Study of Non-Resonant Vibration-Assisted Lapping of SiCp/Al. *Micromachines* **2024**, *15*, 113. [CrossRef] [PubMed]
35. Zhao, B.; Gao, X.; Pan, J.; Liu, H.; Zhao, P. Investigation of Gallium Arsenide Deformation Anisotropy during Nanopolishing via Molecular Dynamics Simulation. *Micromachines* **2024**, *15*, 110. [CrossRef]
36. Han, E.; Zhang, K.; Chen, L.; Guo, C.; Xiong, Y.; Guan, Y.; Tian, Y.; Liu, G. Simulation and Experimental Validation of a Pressurized Filling Method for Neutron Absorption Grating. *Micromachines* **2023**, *14*, 1016. [CrossRef]
37. Chen, Y.; Chen, D.; Liang, S.; Dai, Y.; Bai, X.; Song, B.; Zhang, D.; Chen, H.; Feng, L. Recent Advances in Field-Controlled Micro–Nano Manipulations and Micro–Nano Robots. *Adv. Intell. Syst.* **2022**, *4*, 2100116. [CrossRef]
38. Chen, C.; Sun, M.; Wang, L.; Huang, T.; Xu, M. A Fast Transient Response Capacitor-Less LDO with Transient Enhancement Technology. *Micromachines* **2024**, *15*, 299. [CrossRef] [PubMed]
39. Sun, M.; Chen, C.; Wang, L.; Xie, X.; Wang, Y.; Xu, M. A Fast Transient Adaptive On-Time Controlled BUCK Converter with Dual Modulation. *Micromachines* **2023**, *14*, 1868. [CrossRef]
40. Fu, Y.; Han, G.; Gu, J.; Zhao, Y.; Ning, J.; Wei, Z.; Yang, F.; Si, C. A High-Performance MEMS Accelerometer with an Improved TGV Process of Low Cost. *Micromachines* **2022**, *13*, 1071. [CrossRef]
41. Zha, S.; Li, D.; Wen, Q.; Zhou, Y.; Zhang, H. Design and Fabrication of Silicon-Blazed Gratings for Near-Infrared Scanning Grating Micromirror. *Micromachines* **2022**, *13*, 1000. [CrossRef]
42. Du, C.; Kong, D.; Xu, C. Development of a Fault Detection Instrument for Fiber Bragg Grating Sensing System on Airplane. *Micromachines* **2022**, *13*, 882. [CrossRef]
43. Wang, J.; Xu, J.; Lu, Y.; Xie, T.; Peng, J.; Chen, J. Z-Increments Online Supervisory System Based on Machine Vision for Laser Solid Forming. *Micromachines* **2023**, *14*, 1558. [CrossRef] [PubMed]
44. Childs, A.; Pereira, J.; Didier, C.M.; Baksh, A.; Johnson, I.; Castro, J.M.; Davidson, E.; Santra, S.; Rajaraman, S. Plotter Cut Stencil Masks for the Deposition of Organic and Inorganic Materials and a New Rapid, Cost Effective Technique for Antimicrobial Evaluations. *Micromachines* **2023**, *14*, 14. [CrossRef] [PubMed]
45. Zhao, K.; Lou, D.; Zhang, Y.; Fang, L.; Tang, Y. Experimental Study on Diesel Engine Emission Characteristics Based on Different Exhaust Pipe Coating Schemes. *Micromachines* **2021**, *12*, 1155. [CrossRef] [PubMed]

Disclaimer/Publisher's Note: The statements, opinions and data contained in all publications are solely those of the individual author(s) and contributor(s) and not of MDPI and/or the editor(s). MDPI and/or the editor(s) disclaim responsibility for any injury to people or property resulting from any ideas, methods, instructions or products referred to in the content.

Article

High-Quality Spherical Silver Alloy Powder for Laser Powder Bed Fusion Using Plasma Rotating Electrode Process

Hao Li [1,2], Shenghuan Zhang [3], Qiaoyu Chen [1,2], Zhaoyang Du [1,2], Xingyu Chen [1,2], Xiaodan Chen [1,2], Shiyi Zhou [1,2], Shuwen Mei [4], Linda Ke [5], Qinglei Sun [1,2], Zuowei Yin [1,2], Jie Yin [1,2] and Zheng Li [1,2,*]

1. Gemmological Institute, China University of Geosciences, Wuhan 430074, China; haoli@cug.edu.cn (H.L.); 20121001484@cug.edu.cn (Q.C.); zhaoyang_du@cug.edu.cn (Z.D.); 945365445@cug.edu.cn (X.C.); 17719396590@cug.edu.cn (X.C.); zhoushiyi@cug.edu.cn (S.Z.); sunqinglei@cug.edu.cn (Q.S.); yinzuowei1025@163.com (Z.Y.); yinjie@cug.edu.cn (J.Y.)
2. Hubei Engineering Research Centre of Jewellery, Wuhan 430074, China
3. Key Laboratory of Superlight Materials and Surface Technology, Ministry of Education, College of Materials Science and Chemical Engineering, Harbin Engineering University, Harbin 150001, China; zhangshenghuan15@163.com
4. Nantong Jinyuan Intelligence Manufacturing Technology Co., Ltd., Nantong 226010, China; meishuwen@jyznjs.com
5. Shanghai Engineering Technology Research Centre of Near-Net-Shape Forming for Metallic Materials, Shanghai Spaceflight Precision Machinery Institute, Shanghai 201860, China; kelinda_casc@163.com
* Correspondence: lizheng@cug.edu.cn

Citation: Li, H.; Zhang, S.; Chen, Q.; Du, Z.; Chen, X.; Chen, X.; Zhou, S.; Mei, S.; Ke, L.; Sun, Q.; et al. High-Quality Spherical Silver Alloy Powder for Laser Powder Bed Fusion Using Plasma Rotating Electrode Process. *Micromachines* **2024**, *15*, 396. https://doi.org/10.3390/mi15030396

Academic Editor: Antonio Ancona

Received: 21 February 2024
Revised: 9 March 2024
Accepted: 11 March 2024
Published: 14 March 2024

Copyright: © 2024 by the authors. Licensee MDPI, Basel, Switzerland. This article is an open access article distributed under the terms and conditions of the Creative Commons Attribution (CC BY) license (https:// creativecommons.org/licenses/by/ 4.0/).

Abstract: The plasma rotating electrode process (PREP) is an ideal method for the preparation of metal powders such as nickel-based, titanium-based, and iron-based alloys due to its low material loss and good degree of sphericity. However, the preparation of silver alloy powder by PREP remains challenging. The low hardness of the mould casting silver alloy leads to the bending of the electrode rod when subjected to high-speed rotation during PREP. The mould casting silver electrode rod can only be used in low-speed rotation, which has a negative effect on particle refinement. This study employed continuous casting (CC) to improve the surface hardness of S800 Ag (30.30% higher than mould casting), which enables a high rotation speed of up to 37,000 revolutions per minute, and silver alloy powder with an average sphericity of 0.98 (5.56% higher than gas atomisation) and a sphericity ratio of 97.67% (36.28% higher than gas atomisation) has been successfully prepared. The dense S800 Ag was successfully fabricated by laser powder bed fusion (LPBF), which proved the feasibility of preparing high-quality powder by the "CC + PREP" method. The samples fabricated by LPBF have a Vickers hardness of up to 271.20 HV (3.66 times that of mould casting), leading to a notable enhancement in the strength of S800 Ag. In comparison to GA, the S800 Ag powder prepared by "CC + PREP" exhibits greater sphericity, a higher sphericity ratio and less satellite powder, which lays the foundation for dense LPBF S800 Ag fabrication.

Keywords: Ag alloy; plasma rotating electrode process; continuous casting; laser powder bed fusion; hardness

1. Introduction

Silver (Ag) and silver alloys have high thermal conductivity, electrical conductivity and antibacterial properties with broad application prospects in various fields such as aerospace, medical, jewellery and electric devices. Silver alloy with 80% silver content (S800 Ag) has high toughness and oxidation resistance whilst retaining the ductility of pure silver and is widely used in the manufacture of large silverware and silver jewellery [1–6]. However, the limitations of the traditional casting process in terms of the freedom to shape the part limits the designer's creative freedom and the range of applications for the material. Laser powder bed fusion (LPBF) is a metal additive manufacturing technology that melts discrete powders materials and deposits them layer by layer [7–13]. The layer-by-layer forming feature of the powder bed enables the manufacture of parts with complex shapes [14], expanding the freedom of part design [15–19]. The utilisation of silver alloy materials, which possess excellent properties, combined with the shaping benefits of LPBF, will expand the range of applications for silver alloys in smart electronic products, wearable devices and medical devices.

In recent years, studies on the additive manufacturing of silver and silver alloys have mainly focused on the LPBF forming process [20–22], microstructure, performance [23–26] and microscopic defects [27], but few studies have been carried out on powder's preparation. Gas atomisation (GA), the plasma rotating electrode process (PREP) and plasma atomisation (PA) are the three main methods for the preparation of LPBF metal powders [28]. In particular, the GA and PREP methods have been widely used in the industrial production of various alloy powders such as nickel-based, titanium-based and iron-based alloys [29]. The use of wire as a raw material in PA results in lesser productivity and higher costs compared to the previous two methods [30]. Currently, the silver-based powders used in LPBF are mainly prepared by GA. The powders prepared by GA have about 40% non-spherical particles and satellite particles [31,32], which affects the flowability, adhesion and filling properties of the powders. These powders particles are considered to be a major obstacle to the formation of a uniform and dense powders layer in the process of powders recoating [29]. Gao et al. compared non-spherical and spherical particles at a layer thickness of 80 μm by simulation and found that non-spherical particles reduce the packing density of the powder and increase the surface roughness of the powder bed, and the elongated gaps formed among the joined particles lead to severe breakups of the liquid metal, contributing to the destabilisation of the melt pool [32]. Chu et al. conducted a comparison of the forming effect of LPBF with and without satellite particles. They found that the porosity of the formed sample with satellite particles (0.2~1.7%) was consistently larger than that of the formed sample without satellite particles (0.1~0.2%). Satellite particles affect the uniformity of powder deposition, resulting in the formation of unfused or unfilled defects within the parts [33].

The plasma rotating electrode process (PREP) is a suitable method for producing metal powders used in additive manufacturing. PREP powders have the advantages of good sphericity, fewer satellite powders and narrow particle size distribution [34,35], which is superior to GA powders. However, there is no relevant research on the preparation of PREP powders for silver-based materials.

The conventional silver alloy used in mould casting is constrained by its material and process, resulting in low hardness, poor density and the inadequate uniformity of the crystalline structure. Additionally, the casting electrode rod is prone to defects such as air holes, looseness and shrinkage holes, making it unable to withstand the high-speed rotation during the PREP process. The mould casting silver electrode rod can only be used in low-speed rotation, which negatively affects particle refinement.

This research addresses the issues of low hardness and numerous defects in the traditional mould-casting of silver alloy, which render it unsuitable for PREP powder production. This study provides a new way to enhance the hardness of the silver alloy electrode rod through continuous casting, which allows it to be used in PREP with high-speed rotation. PREP silver alloy powders with high sphericity and a high sphericity ratio have been successfully prepared. Finally, dense silver alloy samples have been prepared using LPBF, which verified the high quality of the PREP powders.

2. Materials and Methods

2.1. Materials

Pure silver ingots and copper alloy particles (Noble metal of Huanggang Co., Ltd., Huanggang, China) were used as raw materials; silver alloy rods of $\varphi 30$ mm were prepared by continuous casting for powder preparation.

2.2. Preparation of Silver Alloy Rod

Continuous casting is a technology in which melted metal is continuously poured into a water-cooled crystalliser and the solidified casting is continuously pulled out from the other end of the crystalliser [36,37], as shown in Figure 1. The quickly cooled metal has a uniform structure with high density. It has better mechanical properties and higher hardness, which provides support for stable high-speed rotation during subsequent PREP.

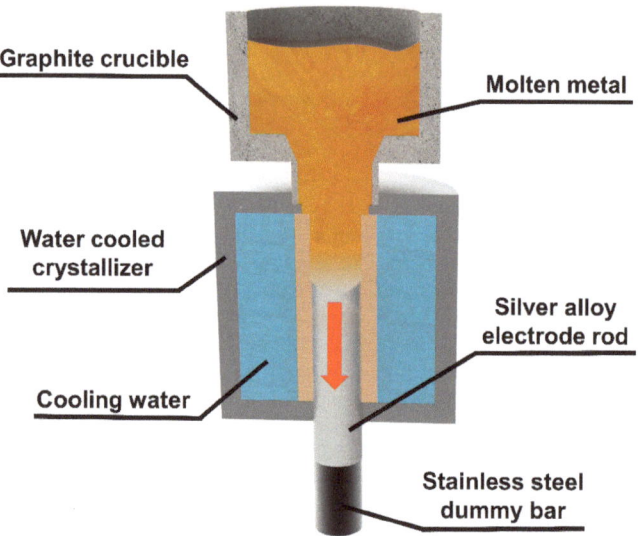

Figure 1. Schematic diagram of the principle of continuous casting.

In this study, φ30 mm rods were prepared by continuous casting, and the casting was carried out under nitrogen environment. The ceramic crucible of the continuous casting machine was preheated to 1000 °C, and pure silver ingots and copper alloy particles were weighed into the graphite crucible according to the ratio (Ag:Cu alloy = 80:20) until complete melting. Then, the cooling water was turned on and the pressure was set to 0.15 MPa. The solidified silver alloy was pulled out by stainless steel dummy bar, and pulling speed was set to 30 mm/min. A silver alloy electrode rod was formed.

2.3. PREP Preparation of Silver Alloy Powders

Sailong Additive Technology Co., Ltd., (Xi'an, China) of SLPA-D desktop PREP system was selected to prepare the S800 Ag powder (Figure 2). In this system, the electrode rod is inserted into the atomisation chamber through the mechanical shaft, and the electrode is driven to rotate at high speed; the electrode rod work as the anode in the powder preparation process, and the plasma gun in the atomisation chamber forms a conduction circuit, generating high-temperature plasma torch. The end face of the rods rotating at high speed is melted, crushed and condensed by the high-temperature plasma torch to form powders [38]. The parameters of PREP silver alloy powder preparation obtained through research and exploration are shown in Table 1. The powder preparation process was carried out in argon gas. The electrode rotation speed was 25,000–37,000 rpm, the DC current was 500–700 A and the feed rate was 3.5–4.5 mm/s.

Table 1. Parameters of silver alloy powders prepared by PREP used in this paper.

Plasma Gas	Rotational Speed (rpm)	DC Current (A)	Feeding Rate (mm/s)
Ar	25,000–37,000	500–700	3.5–4.5

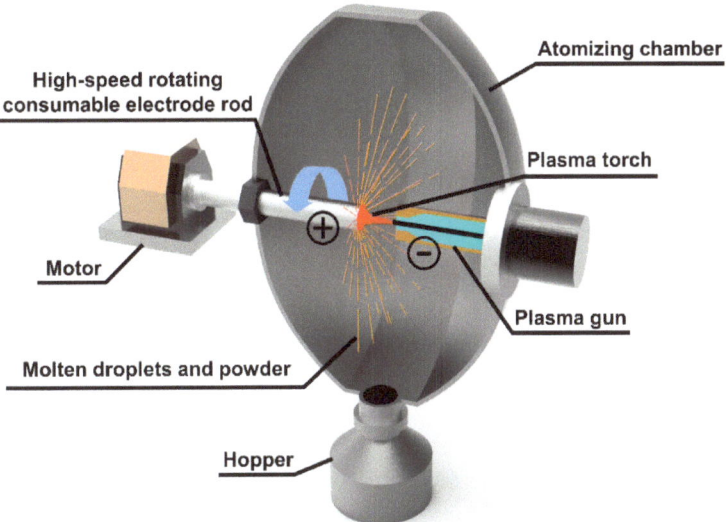

Figure 2. PREP principal diagram.

2.4. Preparation of Samples for Laser Powder Bed Fusion

The SISMA MYSINT100 platform was employed to LPBF fabrication, which was equipped with a Nd: YAG fibre laser with a wavelength of 1064 nm, a laser power of up to 200 W and a spot diameter of 30 μm. All samples were constructed in an argon atmosphere (with a residual oxygen content of 0.5 vol%).

The process parameters and scanning strategy were designed by the Materialise AutoFab mysint 2.0 (b424336) software. Two sets of samples were used for forming, as shown in Figure 3. The single-track sample (Figure 3a) was constructed using a single-track scanning strategy to profile the outer edge of the rectangular body, which was used to simulate the actual melt pool width in order to calculate the hatch distance of bulk sample (Figure 3b). The bulk sample was scanned using a checkerboard scanning strategy (layer thickness of 40 μm), with each layer divided into 4 × 4 mm square bulks, each layer (n + 1) rotated by 45° with respect to the previous layer (n) and each bulk translated by 4 mm along the positive direction of the X-axis and the positive direction of the Y-axis [39–41], as shown in Figure 3b.

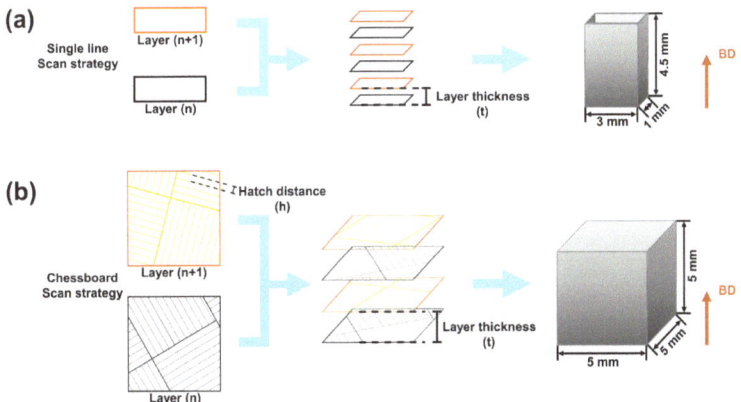

Figure 3. Sample manufacturing strategy: (**a**) single-track scanning strategy; (**b**) checkerboard scanning strategy.

2.5. Measurement of Microstructure and Physical Property

The morphology of the silver alloy powders (acceleration voltage 20 kV) and the grain structure of the silver alloy rod (acceleration voltage 15 kV) were determined by scanning electron microscope (TESCAN MIRA LMS, TESCAN Ltd., Brno, Czech Republic). The particle size distribution statistics of the powders were analysed by Mastersizer 3000 (Malvern Panalytical Ltd., Malvern, UK) laser diffraction particle size analyser. The laser absorptivity of the powders was analysed by JASCO MSV 520 UV-Vis/NIR Micro Spectrophotometers (JASCO International Co., Ltd., Tokyo, Japan).

Image-Pro Plus was used to calculate the sphericity and sphericity ratio of the powders. The sphericity was calculated according to the roundness of the powder particles in the SEM image (Formula (1)). S is the area of the powder particles, and C is the circumference of the powder particles; we measured the long axis and short axis of the powder particles. Particles with a ratio of the long and short axes ≤ 1.2 were considered spherical. Sphericity ratio was calculated by dividing the number of spherical particles by the total number of samples [42].

$$\text{Sphericity} = \frac{4\pi \times S}{C^2} \quad (1)$$

Silver alloy rods and LPBF samples were polished and etched (50 mL ammonia water + 100 mL H_2O_2 (3 vol%) + 50 mL distilled water) for microstructure analysis. The surface of the sample was observed with an optical metallographic microscope (Nikon ECLIPSE MA100N, Nikon Corp., Tokyo, Japan) to characterise the sample defects, microstructure and morphology. The density of LPBF samples were measured by Archimedes' method. The hardness of silver alloy rods and LPBF samples (constant load 1 N, dwell time 10 s) were obtained using Wilson VH1102 Vickers hardness tester (Buehler Ltd., Lake Bluff, IL, USA). The NETZSCH LFA 467 (Netzsch, Bavaria, Germany) laser thermal conductivity meter was used to measure the thermal conductivity of LPBF samples.

3. Results and Discussion

3.1. Continuous Casting of Silver Alloy Rods

3.1.1. Microstructure of Silver Alloy Rods

Figure 4a displays the optical micro-metallographic organisation of the cross-section of the S800 Ag rod. The image reveals a fine dendritic structure, consisting of three phases: the primary phase ($\alpha - Ag$), the degenerated eutectic copper ($\beta - Cu$) and the eutectic phase ($\alpha + \beta$) [43]. The dendrites in the light-coloured area are the primary phase ($\alpha - Ag$), while the darker region between the primary phases contains the eutectic phase ($\alpha + \beta$) and the degenerated eutectic copper ($\beta - Cu$).

Cu-rich nanosized particles and continuous lamellar structures were observed inside the $\alpha - Ag$ phase, and EDS analysis showed that the Cu content of the $\alpha - Ag$ phase was 9.1 At% (Figure 4c). During the process of cooling and solidification, the primary phase ($\alpha - Ag$) grows first in the form of dendrites and Cu is enriched at the edge of head area of the dendrites, resulting in the formation of the eutectic and $\beta - Cu$ phases. In Figure 4b, the yellow dotted line represents the boundary of the $\alpha - Ag$ phase. At the edge of the $\alpha - Ag$ phase, there are noticeable large black phases formed by aggregates of the $\beta - Cu$ solid solution with a copper content of 87.4 At% (Figure 4d). The eutectic phase ($\alpha + \beta$) is present in a stratified eutectic structure located between the dendrites of the $\alpha - Ag$ phase (Figure 4d).

3.1.2. Vickers Hardness of Silver Alloy Rod

Figure 5 shows the micro-metallographic structure of the S800 Ag electrode rod cross-section at various positions ranging from the centre to the edges. The grain size of the centre region (Figure 5a) is bigger than that of edge region (Figure 5b) due to the water cooling during continuous casting process. The water-cooled crystalliser plays a crucial role in rapidly cooling the outer edge, thereby contributing to grain refinement, which significantly enhances the hardness of silver alloys.

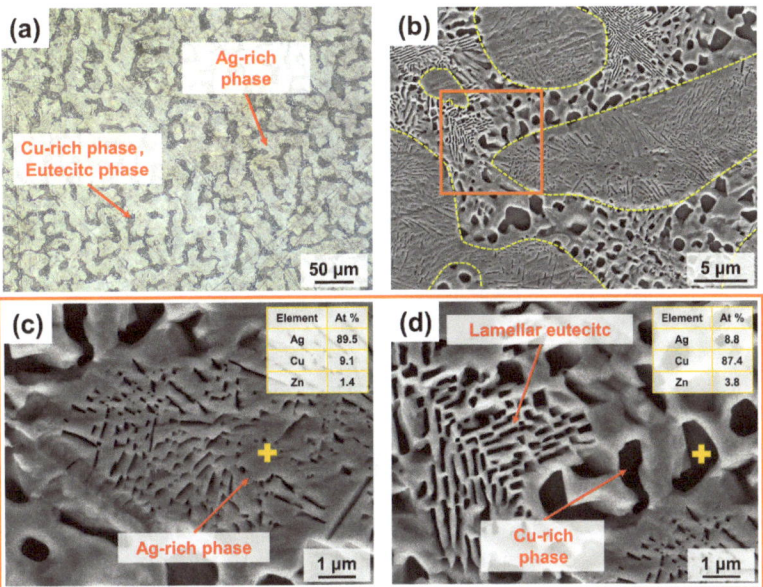

Figure 4. Microstructure of S800 Ag rod cross section; (**a**) Optical microstructure; (**b**) SEM microstructure, α − Ag phase was marked by yellow dashed box; Area of EDS analysis was marked by orange box (**c**) EDS point analysis of α − Ag phase; (**d**) EDS point analysis of eutectic phase and β − Cu phase. The EDS point was marked by yellow +.

The hardness of the S800 Ag electrode rod was measured at different locations of the cross-section from the centre to the edge in Figure 5c. The hardness of the electrode rod increases as the distance from the centre of the circle increases. The average hardness at the centre is 91.82 HV, while at the outer edges, it is 96.42 HV, with a maximum of 98.90 HV. These findings align with the microstructure above, which shows that the surface hardness of the electrode rod is enhanced by fine grain reinforcement [37] using a water-cooled crystalliser. The traditional mould casting of S800 Ag results in a hardness of 74 HV [44]. The hardness of S800 Ag in this study is 24.08% higher in the centre region and 30.30% higher in the outside edge region compared to mould casting. It was proven that continuous casting provides an effective way to improve the hardness of silver alloys. The improvement in surface hardness is beneficial in preventing the rod from deflecting and bending during high-speed rotation. This ensures stable high-speed rotation during the following PREP powder preparation process.

3.2. Physical Properties of Silver Alloy Powders

Figure 6 shows the SEM morphology and particle size distribution of the S800 Ag powders prepared by PREP. The corresponding data for the powders can be found in Table 2. The particle size of the powders is mostly in the range of 15–60 μm, which is in line with the requirements of LPBF printing. The primary constituents of S800 Ag powders consist of 78.8 wt.% Ag, 18.5 wt.% Cu and 2.7 wt.% Zn, and the powder laser absorptivity (wavelength: 1064 nm) is 64.60%.

Figure 7 shows the structure of silver alloy powders (S800 Ag) prepared using PREP and silver alloy powders (S925 Ag) prepared using gas atomisation (GA), provided by Legor Group. The majority of powders prepared by PREP show a high degree of sphericity, and satellite powders are rarely observed. Fine dendrites can be observed on the surface of the powders (Figure 7b), which is due to the rapid cooling of droplets by centrifugal rate and thermal exchange with the atmosphere during the PREP. The molten metal liquid film breaks up due to centrifugal force, resulting in the formation of small droplets. These little liquid particles have thermal exchange with the protective atmosphere, swiftly cooling to form fine dendrites. Compared to PREP powders, GA powders are less spherical (Figure 7d) and have more satellite particles, most of which have irregular small particles adhering to the surface of the powders (Figure 7e), seriously affecting the sphericity of the powders.

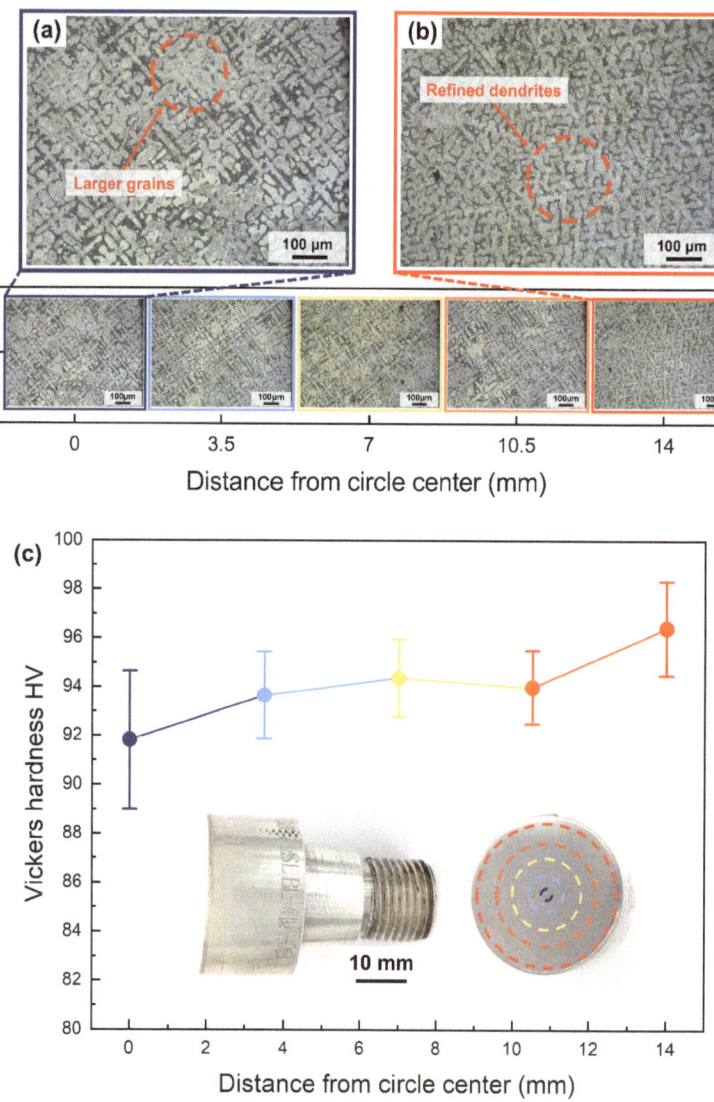

Figure 5. Vickers hardness and microstructure of continuously cast S800 Ag rods at different distances from the centre of the circle, (**a**) microstructure of the central region of the rod, (**b**) microstructure of the edge region of the rod. (**c**) Vickers hardness of continuously cast electrode rod at different distances from the centre of the circle, the test points are marked by dashed circle on the cross-section of the continuously cast rod.

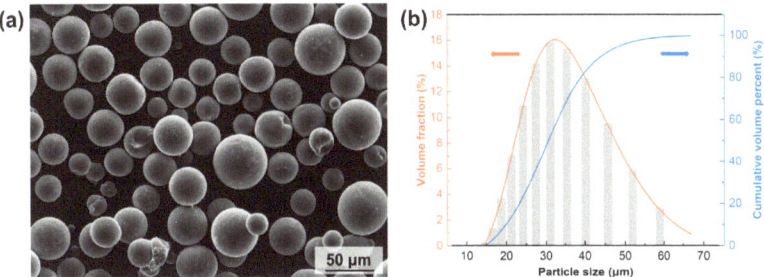

Figure 6. SEM image and particle size distribution of silver alloy powders: (**a**) SEM image of S800 Ag powders, (**b**) particle size distribution of S800 Ag powders.

Table 2. Data of S800 Ag powders prepared by PREP.

	Particle Size Distributions	Apparent Density	Tap Density
S800 Ag	D_{10} = 23.08 μm D_{50} = 34.42 μm D_{90} = 51.04 μm	5.92	6.20

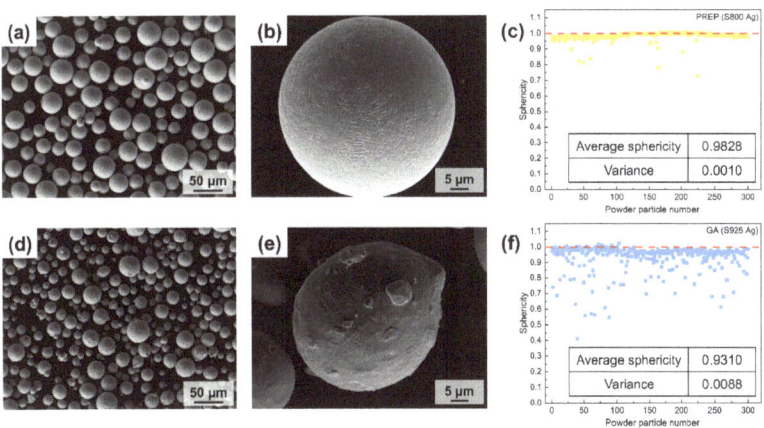

Figure 7. SEM morphology and powders sphericity of PREP silver alloy powders and GA silver alloy powders; (**a**,**b**) PREP S800 Ag powders morphology; (**c**) PREP S800 Ag powders sphericity, red dotted line is the sphericity value corresponding to perfect spherical; (**d**,**e**) GA S925 Ag powders morphology; (**f**) GA S925 Ag powders sphericity, red dotted line is the sphericity value corresponding to perfect spherical.

The sphericity of the PREP and GA silver alloy powders was calculated by computationally analysing the roundness of the particles (Figure 7c,f). A sphericity value of 1 indicates that the particles are perfectly spherical, and the closer the sphericity value is to 1.00, the better the sphericity of the powders. The majority of the PREP silver alloy powders have a sphericity near to 1 and only a small number of powders particles have a sphericity lower than 0.90. The average sphericity is 0.98, which indicates that the preparation of PREP powders has good sphericity. The few particles with sphericity lower than 0.9 are mostly satellite powders and long stripes of powders. Figure 7f shows the sphericity of the GA silver alloy powders. The figure reveals that there are numerous powders with sphericity values below 0.90. The average sphericity of these powders is 0.93, which is lower than that of PREP. This finding aligns with the observations obtained from the SEM images, which indicate that GA preparation results in more irregular and satellite powders. Additionally, the variance calculation of the sphericity values further confirms that the distribution of sphericity in GA powders is more dispersed and heterogeneous compared to PREP powders.

The sphericity ratio calculation findings for silver alloy powders generated using PREP and GA are displayed in Figure 8a. The sphericity ratio refers to the proportion of spherical particles (the ratio of the long and short axes of the particle image ≤ 1.2) in the whole sample. Only 71.67% of GA silver alloy powders satisfied the spherical requirement. In contrast, the silver alloy powders prepared by PREP in this study had a sphericity ratio of 97.67%, which resulted in superior apparent density and tap density compared to the GA powders (Figure 8b). The presence of non-spherical particles and satellite particles in the GA powders resulted in numerous gaps in the powder packing, leading to decreased apparent density and tap density. In conclusion, the silver alloy powders prepared by PREP were significantly better than those prepared by GA in terms of the sphericity, sphericity ratio, apparent density and tap density.

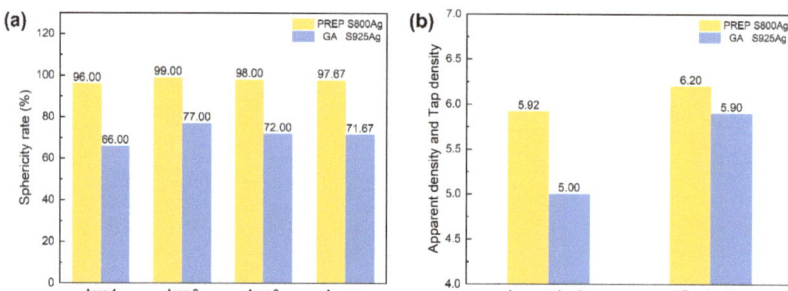

Figure 8. (**a**) sphericity ratio of PREP S800 Ag and GA S925 Ag; (**b**) apparent density and tap density of PREP S800 Ag and GA S925 Ag.

3.3. Optimisation of LPBF Process

3.3.1. Single-Track Samples

The cross-section of the S800 Ag single-track sample is shown in Figure 9. The single line width is employed to simulate the width of the melt pool, which is used to control the forming parameters of the bulk. The processing window has been divided into three regions based on the breadth and continuity of the melt pool (see the domain Figure 9b), excessive melting (red dotted rectangle), good melting (green dotted rectangle) and weak sintering (blue dotted rectangle). Typically, the width of the single line narrows as the laser power decreases and the scanning speed increases. Figure 9a clearly illustrates that the majority of the area displays a consistent and uninterrupted path, which provides the basis for subsequent bulk sample forming.

3.3.2. Bulk Samples

The forming parameters of bulk samples depend on the process window of good melting in single-track samples. The hatch distance is set according to the melt pool overlap rate of 60%. The hatch distance is calculated using the following Formula (2). Hr represents the melt pool overlap rate, and w represents the width of the melt pool.

$$h = (1 - Hr) \times w \qquad (2)$$

As shown in Figure 10, dense LPBF sample have been successfully fabricated using the PREP S800 Ag powder. The scanning trajectory can be observed in the XY plane of S800 Ag by etching, which is consistent with the set scanning strategy and hatch distance. The overlapping melt pool in the Z plane (build direction) exhibits a "fish scale" characteristic. During LPBF, the high energy in the central region of the laser beam leads to the maximum depth of the melt pool, with the energy decreasing from the central region to the edges, thereby forming an arcuate "fish scale" cross-section [45]. As a new layer of powder is deposited on the top of the part, the laser melts the powder, causing the previously layer to be remelted. Variations in melt pool depth and shape lead to irregular wavy-shaped tracks.

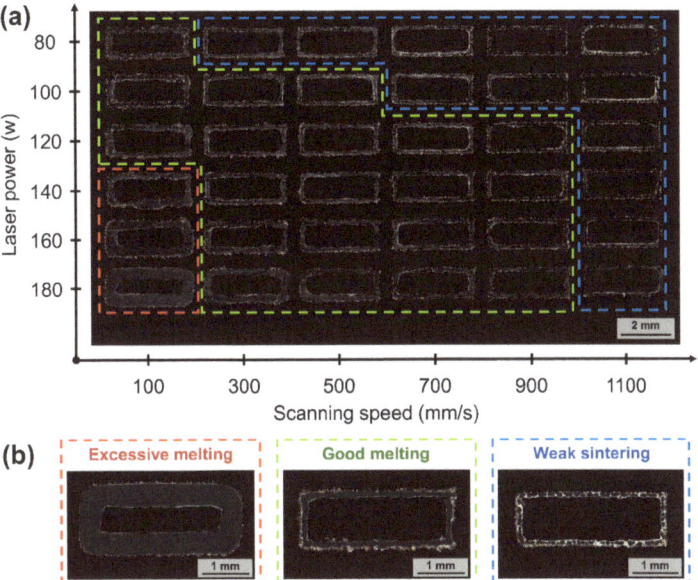

Figure 9. Process parameters of S800 Ag single-track morphology and processing window schematics, (**a**) cross-sectional morphology of single-track samples at different process parameters; (**b**) three processing windows of single-track schematics.

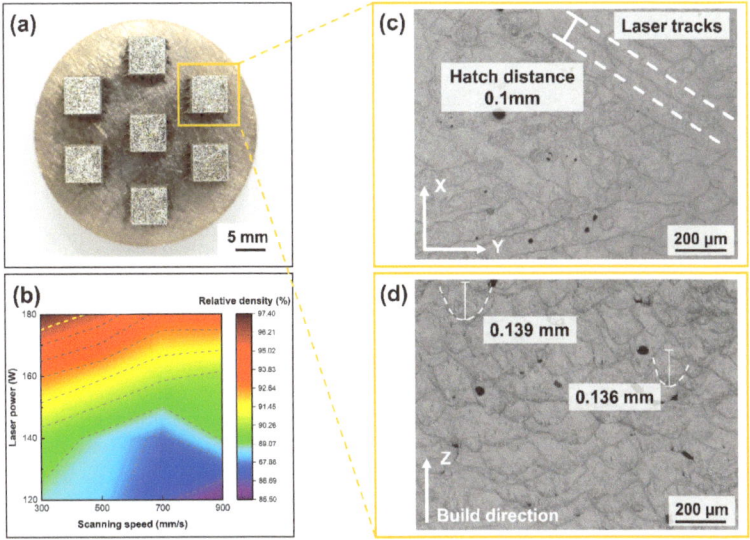

Figure 10. S800 Ag bulk sample, melt channel and melt pool images, (**a**) bulk sample; (**b**) relative density depends on laser power and scanning speed (**c**) melt pool morphology in XY plane; (**d**) melt pool morphology in Z plane.

The relative density of S800 Ag depends on the laser power and scanning speed shown in Figure 10b, which shows that the densities of the components increase with increasing laser power and decreasing scanning speed. The highest relative density (97.36%) was obtained using a laser power of 180 W and a scanning speed of 300 mm/s. It was proven that the S800 Ag powder prepared in this study for LPBF is completely feasible and can be used to fabricate LPBF parts with high density.

3.4. Microstructure and Defects

Optical microscope images and the microstructure of the LPBF S800 Ag bulk samples are shown in Figure 11. Several internal defects can be found, which led to a decrease of relative density of the LPBF S800 Ag. In this study, the type of defect is indicated by roundness (Formula (1)). A roundness value of 0.8 or higher is classified as a keyhole pore defect, between 0.6 and 0.8 is classified as a lack of fusion defect and below 0.6 is classified as a depression wall collapse defect.

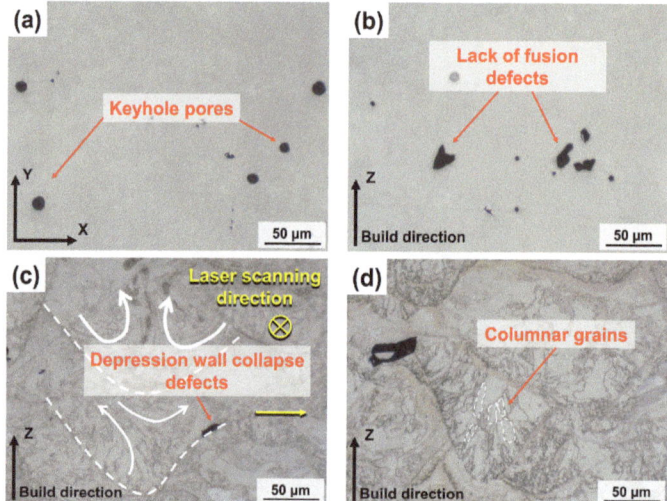

Figure 11. Internal defects and microstructure of S800 Ag LPBF samples; (**a**) keyhole pore defects; (**b**) lack of fusion defects; (**c**) depression wall collapse defects; (**d**) columnar grains growing in the direction of the thermal gradient.

Keyhole pore defects can be found in Figure 11a. The keyhole pores are caused by the destabilisation of the melt pool due to the laser energy set up not being perfectly matched to the metal powder [46]. The formation of keyhole in the melt pool is a result of the metal vapour recoil pressure. As the metal vapour recoil pressure decreases and the surface tension of the metal liquid increases, the keyholes collapse. This collapse leads to the generation of pores at the bottom of the melt pool. However, these pores are unable to escape due to the rapid solidification of the melt pool, resulting in the formation of internal pores [47,48]. Lack of fusion defects can be observed in Figure 11b, which are caused by an insufficient laser energy input. The scanning tracks formed by low laser energy cannot form a good combination with the building layers, resulting in the formation of lack of fusion defects [49].

The formation of depression wall collapse defects at the edge of the melt pool is observed in Figure 11c. The molten metal first flows towards the outer edge of the melt pool under the action of the Marangoni flow and then returns to the centre of the melt pool. This movement causes the collapse of the melt pool, leading to the formation of pores at the front of the melt pool depression wall [50]. Many elongated columnar grains were observed in the Z-plane (Figure 11d), and the grains grew towards the centre of the melt pool along the build direction, which was due to the preferential growth of columnar grains along the direction of the thermal gradient during the gradual cooling [51].

3.5. Thermal Conductivity and Vickers Hardness

The Vickers hardness of the same material is mainly affected by the macroscopic internal defects and microscopic grain size of the sample. Figure 12a shows the relationship between process parameters and Vickers hardness. The LPBF S800 Ag Vickers hardness decreases with decreasing laser power and increasing scanning speed. According to Figure 12b, there is a clear positive relationship between Vickers hardness and relative density. The sample with the highest density (97.36%) exhibited the highest Vickers hardness (271.20 HV). This suggests that the primary factor influencing hardness is the presence of macroscopic internal defects. Specifically, the internal pores are susceptible to collapsing when subjected to external forces, leading to a decrease in Vickers hardness. The disparity in Vickers hardness among samples with similar relative density can be attributed to variations in the size and morphology of the internal grains.

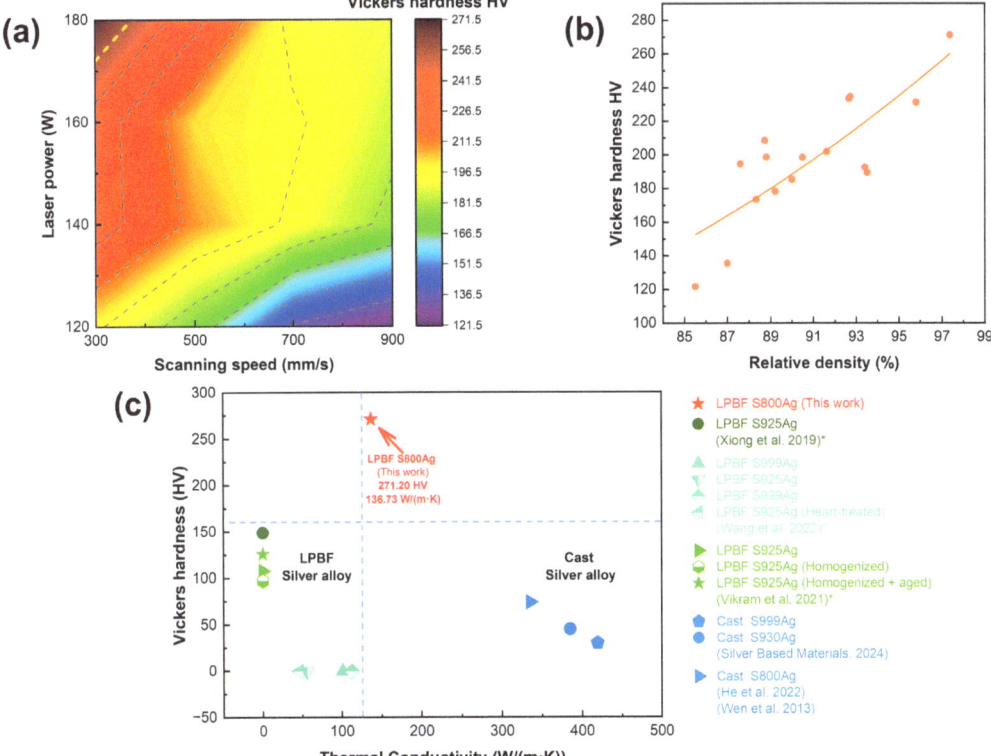

Figure 12. (a) Vickers hardness depend on laser power and scanning speed, (b) Vickers hardness depends on relative density; (c) Hardness and thermal conductivity of silver alloys in this and previous studies, *: Xiong et al. and Vikram et al. did not conduct research on thermal conductivity; Wang et al. did not conduct research on Vickers hardness [20,21,25,44,52,53].

Figure 12c shows a comparison of the Vickers hardness and thermal conductivity of LPBF S800 Ag in this study with other silver alloys investigated in previous studies. The LPBF S800 Ag samples fabricated in this work exhibits superior Vickers hardness and thermal conductivity properties compare to the S925 Ag and S925 Ag LPBF samples in previous studies. However, the thermal conductivity remains inferior to that of conventionally cast parts. The thermal conductivity of traditionally cast S800 Ag is 334.9 W/(m·K) [52], whereas the LPBF printed S800 Ag exhibits a thermal conductivity of 136.73 W/(m·K). The primary mode of heat conduction in metal is through the movement of electrons. The presence of numerous small, closed circular holes and irregular pores on both the interior and surface of the LPBF samples lead to a lower relative density (97.36%) compared to conventionally casted samples. These features hinder the movement of electrons, thereby reducing the overall thermal conductivity. The rapid melting and cooling rate during LPBF induced the grain refinement and increased residual stresses [54,55]. As a result, the Vickers hardness of LPBF S800 Ag, which is 3.66 times higher than that of mould casting, significantly enhances material strength and abrasion resistance, the application range of silver alloys will be further expanded.

4. Conclusions

The hardness of the S800 Ag electrode rod for PREP was improved using continuous casting. The employment of continuous casting technology effectively solved the problems of inadequate hardness and defects frequently observed in traditional silver mould casting, as well as the issue of bending during high-speed rotation. S800 Ag powders with good sphericity and high sphericity ratio were successfully prepared by "CC + PREP". Furthermore, dense S800 Ag was successfully fabricated by LPBF.

1. The hardness of the silver alloy electrode rod was effectively enhanced through the continuous casting process. The fine dendrites were induced due to the outer edge of the rods undergoing rapid cooling using the water-cooled crystalliser. The hardness was enhanced by the reinforcement of these fine crystals, resulting in a 30.30% increase in hardness compared to S800 Ag prepared by traditional mould casting.
2. The S800 Ag electrode rod with enhanced hardness enabled stable rotation with the speed up to 25,000–37,000 rpm. The silver alloy powders prepared by "CC + PREP", which has a small particle size (15–60 μm), satisfy the requirements for LPBF fabrication. Furthermore, S800 Ag powder showed a 5.56% increase in average sphericity (0.98) and a 36.28% increase in sphericity ratio (97.67%) compared to the GA silver alloy powders.
3. The LPBF process rapidly cooled the material, resulting in a crystalline reinforcement that provided S800 Ag with a Vickers hardness (271.20 HV) 3.66 times higher than mould casting. The strength was significantly improved, hence facilitating the development of silver alloy components with high strength and complex structures.

The present study proposes the "CC + PREP" approach as a solution to the issue of low speed of rotation during PREP due to the low hardness of silver alloy electric rods using mould casting. Further research will be conducted to explore in typical silver alloys such as S925 Ag and S999 Ag. The optimisation of the LPBF parameters will be applied using the high power of multi lasers to improve the density for the application of large silverware and electric device.

Author Contributions: Conceptualisation, Z.L. and H.L.; methodology, Z.L., S.Z. (Shenghuan Zhang) and J.Y.; investigation, S.Z. (Shenghuan Zhang), Z.D., Q.C., X.C. (Xingyu Chen) and X.C. (Xiaodan Chen); resources, Z.L., J.Y. and Z.Y.; writing—original draft preparation, H.L.; writing—review and editing, Z.L., J.Y., H.L., S.Z. (Shenghuan Zhang), Q.C., L.K., Z.D., Q.S., S.Z. (Shiyi Zhou), S.M., X.C. (Xingyu Chen) and X.C. (Xiaodan Chen); supervision, Z.L.; project administration, Z.L. and Z.Y.; funding acquisition, Z.L., S.Z. (Shenghuan Zhang) and J.Y. All authors have read and agreed to the published version of the manuscript.

Funding: This research was funded by the Knowledge Innovation Program of Wuhan-Basic Research (No. 2022020801010196), The National Natural Science Foundation of China (61805095, 52302032), Natural Science Foundation Grant of Hubei Province (2023AFB007), Foundation of Hubei Jewelry Engineering Technology Research Center (CIGTXM-03-202305, CIGTXM-03-202307).

Data Availability Statement: Data are contained within the article.

Conflicts of Interest: Author Shuwen Mei was employed by the company Nantong Jinyuan Intelligence Manufacturing Technology Co., Ltd. The remaining authors declare that the research was conducted in the absence of any commercial or financial relationships that could be construed as a potential conflict of interest.

References

1. Bose, S.; Vahabzadeh, S.; Bandyopadhyay, A. Bone tissue engineering using 3D printing. *Mater. Today* **2013**, *16*, 496–504. [CrossRef]
2. Praiphruk, S.; Lothongkum, G.; Nisaratanaporn, E.; Lohwongwatana, B. Investigation of supersaturated silver alloys for high hardness jewelry application. *J. Met. Mater. Miner.* **2013**, *23*, 67–73.
3. Li, W.; Hu, D.; Li, L.; Li, C.-F.; Jiu, J.; Chen, C.; Ishina, T.; Sugahara, T.; Suganuma, K. Printable and Flexible Copper–Silver Alloy Electrodes with High Conductivity and Ultrahigh Oxidation Resistance. *ACS Appl. Mater. Interfaces* **2017**, *9*, 24711–24721. [CrossRef] [PubMed]
4. Yu, Q.; Meng, K.; Guo, J. Research on innovative application of silver material in modern jewelry design. *MATEC Web Conf.* **2018**, *176*, 02013. [CrossRef]
5. Mijnendonckx, K.; Leys, N.; Mahillon, J.; Silver, S.; Van Houdt, R. Antimicrobial silver: Uses, toxicity and potential for resistance. *Biometals* **2013**, *26*, 609–621. [CrossRef]
6. Reti, A.; Mridha, S. Silver: Alloying, Properties, and Applications. In *Reference Module in Materials Science and Materials Engineering*; Elsevier: Amsterdam, The Netherlands, 2016.
7. Zhu, S.; Du, W.; Wang, X.; Han, G.; Ren, Z.; Zhou, K. Advanced Additive Remanufacturing Technology. *Chin. J. Mech. Eng. Addit. Manuf. Front.* **2023**, *2*, 100066. [CrossRef]
8. Wang, Z.; Zhou, M.; Xiao, H.; Yuan, S. Development and Evaluation of Multiscale Fiber-reinforced Composite Powders for Powder-bed Fusion Process. *Chin. J. Mech. Eng. Addit. Manuf. Front.* **2023**, *2*, 100079. [CrossRef]
9. Li, K.; Chen, W.; Yin, B.; Ji, C.; Bai, S.; Liao, R.; Yang, T.; Wen, P.; Jiang, B.; Pan, F. A comparative study on WE43 magnesium alloy fabricated by laser powder bed fusion coupled with deep cryogenic treatment: Evolution in microstructure and mechanical properties. *Addit. Manuf.* **2023**, *77*, 103814. [CrossRef]
10. Zenou, M.; Grainger, L. 3–Additive manufacturing of metallic materials. In *Additive Manufacturing*; Butterworth-Heinemann: Oxford, UK, 2018; pp. 53–103. [CrossRef]
11. Li, K.; Ma, R.; Qin, Y.; Gong, N.; Wu, J.; Wen, P.; Tan, S.; Zhang, D.Z.; Murr, L.E.; Luo, J. A review of the multi-dimensional application of machine learning to improve the integrated intelligence of laser powder bed fusion. *J. Mater. Process. Technol.* **2023**, *318*, 118032. [CrossRef]
12. Li, K.; Ma, R.; Zhang, M.; Chen, W.; Li, X.; Zhang, D.Z.; Tang, Q.; Murr, L.E.; Li, J.; Cao, H. Hybrid post-processing effects of magnetic abrasive finishing and heat treatment on surface integrity and mechanical properties of additively manufactured Inconel 718 superalloys. *J. Mater. Sci. Technol.* **2022**, *128*, 10–21. [CrossRef]
13. Huang, T.; Tong, C.; Pan, J.; Cheng, Z.; Yu, B.; Yin, J.; Yin, Z.; Chen, S.; Yu, H.; Yan, K. Spatial-multiplexing of nonlinear states in a few-mode-fiber-based Kerr resonator. *Opt. Commun.* **2024**, *555*, 130238. [CrossRef]
14. Li, K.; Chen, W.; Gong, N.; Pu, H.; Luo, J.; Zhang, D.Z.; Murr, L.E. A critical review on wire-arc directed energy deposition of high-performance steels. *J. Mater. Res. Technol.* **2023**, *24*, 9369–9412. [CrossRef]
15. Li, K.; Ji, C.; Bai, S.; Jiang, B.; Pan, F. Selective laser melting of magnesium alloys: Necessity, formability, performance, optimization and applications. *J. Mater. Sci. Technol.* **2023**, *154*, 65–93. [CrossRef]
16. Li, Z.; Li, H.; Yin, J.; Li, Y.; Nie, Z.; Li, X.; You, D.; Guan, K.; Duan, W.; Cao, L.; et al. A Review of Spatter in Laser Powder Bed Fusion Additive Manufacturing: In Situ Detection, Generation, Effects, and Countermeasures. *Micromachines* **2022**, *13*, 1366. [CrossRef] [PubMed]
17. Wei, S.; Hutchinson, C.; Ramamurty, U. Mesostructure engineering in additive manufacturing of alloys. *Scr. Mater.* **2023**, *230*, 115429. [CrossRef]
18. Zhan, J.; Wu, J.; Ma, R.; Li, J.; Huang, T.; Lin, J.; Murr, L.E. Effect of microstructure on the superelasticity of high-relative-density Ni-rich NiTi alloys fabricated by laser powder bed fusion. *J. Mater. Process. Technol.* **2023**, *317*, 117988. [CrossRef]
19. Huang, T.; Zheng, H.; Xu, G.; Pan, J.; Xiao, F.; Sun, W.; Yan, K.; Chen, S.; Huang, B.; Huang, Y. Coexistence of nonlinear states with different polarizations in a Kerr resonator. *Phys. Rev. A* **2024**, *109*, 013503. [CrossRef]
20. Xiong, W.; Hao, L.; Li, Y.; Tang, D.; Cui, Q.; Feng, Z.; Yan, C. Effect of selective laser melting parameters on morphology, microstructure, densification and mechanical properties of supersaturated silver alloy. *Mater. Des.* **2019**, *170*, 107697. [CrossRef]
21. Wang, D.; Wei, Y.; Wei, X.; Khanlari, K.; Wang, Z.; Feng, Y.; Yang, X. Selective Laser Melting of Pure Ag and 925Ag Alloy and Their Thermal Conductivity. *Crystals* **2022**, *12*, 480. [CrossRef]
22. Robinson, J.; Stanford, M.; Arjunan, A. Stable formation of powder bed laser fused 99.9% silver. *Mater. Today Commun.* **2020**, *24*, 101195. [CrossRef]
23. Xiong, W.; Hao, L.; Peijs, T.; Yan, C.; Cheng, K.; Gong, P.; Cui, Q.; Tang, D.; Al Islam, S.; Li, Y. Simultaneous strength and ductility enhancements of high thermal conductive Ag7.5Cu alloy by selective laser melting. *Sci. Rep.* **2022**, *12*, 4250. [CrossRef]

24. Robinson, J.; Stanford, M.; Arjunan, A. Correlation between selective laser melting parameters, pore defects and tensile properties of 99.9% silver. *Mater. Today Commun.* **2020**, *25*, 101550. [CrossRef]
25. Vikram, R.J.; Kollo, L.; Prashanth, K.G.; Suwas, S. Investigating the Structure, Microstructure, and Texture in Selective Laser-Melted Sterling Silver 925. *Metall. Mater. Trans. A* **2021**, *52*, 5329–5341. [CrossRef]
26. Wang, Z.; Xie, M.; Li, Y.; Zhang, W.; Yang, C.; Kollo, L.; Eckert, J.; Prashanth, K.G. Premature failure of an additively manufactured material. *NPG Asia Mater.* **2020**, *12*, 30. [CrossRef]
27. Arjunan, A.; Robinson, J.; Al Ani, E.; Heaselgrave, W.; Baroutaji, A.; Wang, C. Mechanical performance of additively manufactured pure silver antibacterial bone scaffolds. *J. Mech. Behav. Biomed. Mater.* **2020**, *112*, 104090. [CrossRef]
28. Yao, N.; Peng, X. The preparation method of metal powder for 3D printing. *Sichuan Nonferrous Met.* **2013**, *12*, 48–51. (In Chinese)
29. Ruan, G.; Liu, C.; Qu, H.; Guo, C.; Li, G.; Li, X.; Zhu, Q. A comparative study on laser powder bed fusion of IN718 powders produced by gas atomization and plasma rotating electrode process. *Mater. Sci. Eng. A* **2022**, *850*, 143589. [CrossRef]
30. Entezarian, M.; Allaire, F.; Tsantrizos, P.; Drew, R. Plasma atomization: A new process for the production of fine, spherical powders. *JOM* **1996**, *48*, 53–55. [CrossRef]
31. Li, R.; Shi, Y.; Wang, Z.; Wang, L.; Liu, J.; Jiang, W. Densification behavior of gas and water atomized 316L stainless steel powder during selective laser melting. *Appl. Surf. Sci.* **2010**, *256*, 4350–4356. [CrossRef]
32. Gao, X.; Abreu Faria, G.; Zhang, W.; Wheeler, K.R. Numerical analysis of non-spherical particle effect on molten pool dynamics in laser-powder bed fusion additive manufacturing. *Comput. Mater. Sci.* **2020**, *179*, 109648. [CrossRef]
33. Chu, F.; Zhang, K.; Shen, H.; Liu, M.; Huang, W.; Zhang, X.; Liang, E.; Zhou, Z.; Lei, L.; Hou, J.; et al. Influence of satellite and agglomeration of powder on the processability of AlSi10Mg powder in Laser Powder Bed Fusion. *J. Mater. Res. Technol.* **2021**, *11*, 2059–2073. [CrossRef]
34. Chen, G.; Zhao, S.Y.; Tan, P.; Wang, J.; Xiang, C.S.; Tang, H.P. A comparative study of Ti-6Al-4V powders for additive manufacturing by gas atomization, plasma rotating electrode process and plasma atomization. *Powder Technol.* **2018**, *333*, 38–46. [CrossRef]
35. Tang, J.; Nie, Y.; Lei, Q.; Li, Y. Characteristics and atomization behavior of Ti-6Al-4V powder produced by plasma rotating electrode process. *Adv. Powder Technol.* **2019**, *30*, 2330–2337. [CrossRef]
36. Louhenkilpi, S. Continuous casting of steel. In *Treatise on Process Metallurgy*; Elsevier: Amsterdam, The Netherlands, 2014; pp. 373–434.
37. Zhu, X.; Xiao, Z.; An, J.; Jiang, H.; Jiang, Y.; Li, Z. Microstructure and properties of Cu-Ag alloy prepared by continuously directional solidification. *J. Alloy Compd.* **2021**, *883*, 160769. [CrossRef]
38. Luo, X.; Yang, C.; Fu, Z.Q.; Liu, L.H.; Lu, H.Z.; Ma, H.W.; Wang, Z.; Li, D.D.; Zhang, L.C.; Li, Y.Y. Achieving ultrahigh-strength in beta-type titanium alloy by controlling the melt pool mode in selective laser melting. *Mater. Sci. Eng. A* **2021**, *823*, 141731. [CrossRef]
39. Liverani, E.; Gamberoni, A.; Balducci, E.; Ascari, A.; Ceschini, L.; Fortunato, A. Selective Laser Melting of AISI316L: Process Optimization and Mechanical Property Evaluation. In Proceedings of the ASME 2016 11th International Manufacturing Science and Engineering Conference, Blacksburg, VA, USA, 27 June–1 July 2016; Volume 1. [CrossRef]
40. Liverani, E.; Balbo, A.; Monticelli, C.; Leardini, A.; Belvedere, C.; Fortunato, A. Corrosion Resistance and Mechanical Characterization of Ankle Prostheses Fabricated via Selective Laser Melting. *Procedia CIRP* **2017**, *65*, 25–31. [CrossRef]
41. Liverani, E.; Toschi, S.; Ceschini, L.; Fortunato, A. Effect of selective laser melting (SLM) process parameters on microstructure and mechanical properties of 316L austenitic stainless steel. *J. Mater. Process. Technol.* **2017**, *249*, 255–263. [CrossRef]
42. GB/T 39251-2020; Additive Manufacturing-Methods to Characterize Performance of Metal Powders. State Administration for Market Regulation and Standardization Administration of the People's Republic of China: Beijing, China, 2020.
43. Chanmuang, C.; Kongmuang, W.; Pearce, J.; Chairuangsri, T. Influence of casting techniques on hardness, tarnish behavior and microstructure of Ag-Cu-Zn-Si sterling silver jewelry alloys. *J. Met. Mater. Miner.* **2012**, *22*, 19–26.
44. He, X.; Fu, H.; Zhang, H.; Fang, J.; Xie, M.; Xie, J. Machine Learning Aided Rapid Discovery of High Performance Silver Alloy Electrical Contact Materials. *Acta Metall. Sin.* **2022**, *58*, 816–826. [CrossRef]
45. Yao, L.; Huang, S.; Ramamurty, U.; Xiao, Z. On the formation of "Fish-scale" morphology with curved grain interfacial microstructures during selective laser melting of dissimilar alloys. *Acta Mater.* **2021**, *220*, 117331. [CrossRef]
46. Dwivedi, A.; Khurana, M.K.; Bala, Y.G. Heat-treated Nickel Alloys Produced Using Laser Powder Bed Fusion-based Additive Manufacturing Methods: A Review. *Chin. J. Mech. Eng. Addit. Manuf. Front.* **2023**, *2*, 100087. [CrossRef]
47. Zhao, C.; Shi, B.; Chen, S.; Du, D.; Sun, T.; Simonds, B.J.; Fezzaa, K.; Rollett, A.D. Laser melting modes in metal powder bed fusion additive manufacturing. *Rev. Mod. Phys.* **2022**, *94*, 045002. [CrossRef]
48. Snow, Z.; Nassar, A.R.; Reutzel, E.W. Invited Review Article: Review of the formation and impact of flaws in powder bed fusion additive manufacturing. *Addit. Manuf.* **2020**, *36*, 101457. [CrossRef]
49. Qi, Y.; Zhang, H.; Nie, X.; Hu, Z.; Zhu, H.; Zeng, X. A high strength Al–Li alloy produced by laser powder bed fusion: Densification, microstructure, and mechanical properties. *Addit. Manuf.* **2020**, *35*, 101346. [CrossRef]
50. Wang, J.; Zhu, R.; Liu, Y.; Zhang, L. Understanding melt pool characteristics in laser powder bed fusion: An overview of single- and multi-track melt pools for process optimization. *Adv. Powder Mater.* **2023**, *2*, 100137. [CrossRef]
51. Kobryn, P.A.; Semiatin, S.L. Microstructure and texture evolution during solidification processing of Ti–6Al-4V. *J. Mater. Process. Technol.* **2003**, *135*, 330–339. [CrossRef]

52. Wen, B.; Wang, B.; Lu, X. *Handbook of Metal Materials*, 2nd ed.; Publishing House of Electronics Industry: Beijing, China, 2013; pp. 759–761.
53. Silver Based Materials. Available online: https://www.electrical-contacts-wiki.com/index.php/Silver_Based_Materials (accessed on 2 February 2024).
54. Bermingham, M.; StJohn, D.; Krynen, J.; Tedman-Jones, S.; Dargusch, M. Promoting the columnar to equiaxed transition and grain refinement of titanium alloys during additive manufacturing. *Acta Mater.* **2019**, *168*, 261–274. [CrossRef]
55. Rashid, A.; Gopaluni, A. A Review of Residual Stress and Deformation Modeling for Metal Additive Manufacturing Processes. *Chin. J. Mech. Eng. Addit. Manuf. Front.* **2023**, *2*, 100102. [CrossRef]

Disclaimer/Publisher's Note: The statements, opinions and data contained in all publications are solely those of the individual author(s) and contributor(s) and not of MDPI and/or the editor(s). MDPI and/or the editor(s) disclaim responsibility for any injury to people or property resulting from any ideas, methods, instructions or products referred to in the content.

Article

A Fast Transient Response Capacitor-Less LDO with Transient Enhancement Technology

Chufan Chen [1], Mengyuan Sun [1], Leiyi Wang [1,2], Teng Huang [1] and Min Xu [1,*]

[1] State Key Laboratory of ASIC and System, School of Microelectronics, Fudan University, Shanghai 200433, China; 21212020002@m.fudan.edu.cn (C.C.); 20212020003@fudan.edu.cn (M.S.); lywang2018@stu.suda.edu.cn (L.W.); 21212020005@m.fudan.edu.cn (T.H.)
[2] School of Electronic and Information Engineering, Soochow University, Suzhou 215006, China
* Correspondence: xu_min@fudan.edu.cn

Abstract: This paper proposes a fast transient load response capacitor-less low-dropout regulator (CL-LDO) for digital analog hybrid circuits in the 180 nm process, capable of converting input voltages from 1.2 V to 1.8 V into an output voltage of 1 V. The design incorporates a rail-to-rail input and push–pull output (RIPO) amplifier to enhance the gain while satisfying the requirement for low power consumption. A super source follower buffer (SSFB) with internal stability is introduced to ensure loop stability. The proposed structure ensures the steady-state performance of the LDO without an on-chip capacitor. The auxiliary circuit, or transient enhancement circuit, does not compromise the steady-state stability and effectively enhances the transient performance during sudden load current steps. The proposed LDO consumes a quiescent current of 47 µA and achieves 25 µV/mA load regulation with a load current ranging from 0 to 20 mA. The simulation results demonstrate that a settling time of 0.2 µs is achieved for load steps ranging from 0 mA to 20 mA, while a settling time of 0.5 µs is attained for load steps ranging from 20 mA to 0 mA, with an edge time of 0.1 µs.

Keywords: low dropout regulator; capacitor-less; transient enhancement circuit; super source follower

Citation: Chen, C.; Sun, M.; Wang, L.; Huang, T.; Xu, M. A Fast Transient Response Capacitor-Less LDO with Transient Enhancement Technology. *Micromachines* **2024**, *15*, 299. https://doi.org/10.3390/mi15030299

Academic Editor: Kun Li

Received: 26 December 2023
Revised: 17 February 2024
Accepted: 19 February 2024
Published: 22 February 2024

Copyright: © 2024 by the authors. Licensee MDPI, Basel, Switzerland. This article is an open access article distributed under the terms and conditions of the Creative Commons Attribution (CC BY) license (https:// creativecommons.org/licenses/by/ 4.0/).

1. Introduction

Power management ICs (PMICs) are essential components in electronic devices that require efficient power management [1]. Low-dropout regulators (LDOs) are preferred over other voltage regulators due to their low noise, low ripple, low quiescent current and high power supply rejection ratio (PSRR) [2–5]. To satisfy the different voltage regulation requirements of various modules in a single system-on-chip (SOC), multiple voltage regulators can be integrated on the same chip. This approach can reduce the power dissipation and improve the overall efficiency of the system. In SoC designs, LDOs are commonly used to supply power to analog or mixed-signal modules, which are particularly sensitive to noise and voltage fluctuations [6–8]. However, the traditional LDO architecture relies on a large capacitor, which occupies a large area and reduces circuit integration. Removal of this large off-chip capacitor will inevitably degrade the performance requirements of the LDO, especially stability, transient response and PSRR. Therefore, in recent years, capacitor-less LDOs (CL-LDOs) have been widely studied and reported [9–11].

A series of techniques have been proposed to improve the transient response of CL-LDO [12–17]. Using a flipped voltage follower (FVF) to separate the dominant poles in conventional LDO is one of the most popular methods. The FVF and overshoot detection circuit used in [12] reduce the overshoot/undershoot voltages of LDO and achieve fast settling times during load steps. The push–pull amplifier is a kind of architecture that can be used to provide a fast response to load and line transients [13,15]. The LDO with Class-AB OTA in [13] provides not only a fast response to load and line transients, but also handles a wide range of load capacitors, while the push–pull output stage-based LDO in [15] can achieve a 2.7 µs settling time with the load current switching from 100 pA to 100 mA. A dynamic biasing technique is widely used, whereby the bias current of the LDO is adjusted based on the load current. This can improve the efficiency of the LDO and reduce the power dissipation. To enhance both the transient and stability, Li et al. proposed a CL-LDO based on dual-active feedback frequency compensation that ultimately guarantees stable operation in a load range of 0 to 100 mA [14]. In [17], the authors use modified Miller compensation with the insertion of a sensor amplifier stage to inject more transient current in the biasing circuit. This method feeds the regulator to rapidly charge the power PMOS gate capacitance and improves the fast transient response.

This work proposes a novel CL-LDO circuit with a fast transient response. Section 2 demonstrates the complete architecture with a rail-to-rail input, push–pull output (RIPO) two-stage amplifier and a super source follower buffer (SSFB) and analyzes the stability and transient response. Section 3 presents a design example, validated through simulation results, and compares this work with others. Section 4 summarizes the conclusion.

2. Proposed CL-LDO Architecture with RIPO and SSFB

2.1. Conventional Topology of LDO

The traditional LDO topology with an off-chip capacitor depicted in Figure 1 exhibits three poles and a left-half-plane zero without the need for an auxiliary circuit. The stability of this system is ensured by the presence of a left-half-plane zero, which is generated by the output capacitor with a capacitance in the microfarad range and its equivalent resistance. Additionally, Equation (1) establishes that the dominant pole is positioned at the output node. The remaining two poles are positioned at the output of the amplifier and feedback resistance. The removal of this bully output capacitor poses a greater challenge to the stability. To enhance the stability, compensation capacitors are added in the auxiliary circuit.

$$P_1 = \frac{1}{[(R_1 + R_2) || R_L || r_{op}](C_L + C_O)} \tag{1}$$

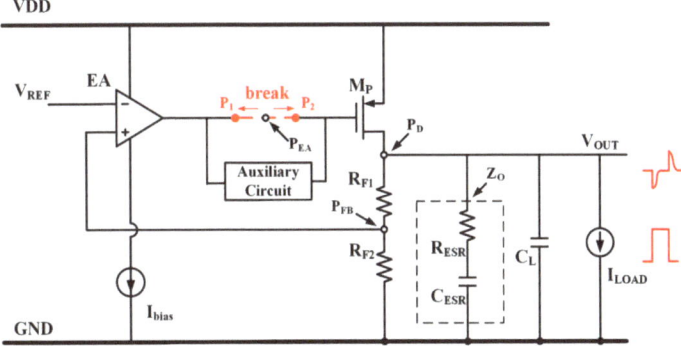

Figure 1. Conventional architecture of LDO.

The spike output voltage in the LDO when there are sudden changes in the load current from other cells is also depicted in Figure 1. The unity gain frequency (UGF) is one of the most essential factors of transient performance. To achieve a large UGF, prior studies have proposed an architecture incorporating a buffer as the auxiliary circuit to decouple the high impedance from the EA's output and the large capacitance from the M_P's input. Additionally, the response time T_R [12] of the LDO can be approximated as Equation (2):

$$T_R \approx \frac{1}{BW} + C_{par}\frac{\Delta V_G}{I_G} \tag{2}$$

where BW denotes the loop bandwidth, C_{par} represents the parasitic cap, ΔV_G refers to the required voltage change and I_G represents the slewing current at M_P's gate. The buffer with a low output resistance can offer a high slewing current to rapidly respond to the load step.

2.2. Proposed RIPO and SSFB

A rail-to-rail input of the RIPO amplifier is shown in Figure 2a. The PMOS input pair M_1–M_2 is utilized to achieve the negative supply rail, while the NMOS input pair M_3–M_4 is employed to reach the positive supply rail. The transistors M_5 to M_8 are used as level shifters for the PMOS input pair, thereby expanding the negative input range to ensure that the PMOS input pair operates in the saturation region. The tail currents of two complementary input pairs are supported by M_9 and M_{10}. The positive supply rail extends from $Vcm+$ to VDD, while the negative supply rail spans from GND to $Vcm-$. The expressions for $Vcm+$ and $Vcm-$ are represented by Equation (3) and Equation (4), respectively:

$$Vcm+ = V_{dsatn} + V_{gs4} \tag{3}$$

$$Vcm- = VDD - V_{dsatp} - V_{gs1} + V_{gs5} \approx VDD - V_{dsatp} \tag{4}$$

where V_{dsatn} and V_{dsatp} are the minimum drain–source voltages that ensure M_9 and M_{10} operate as current sources. When $VDD > V_{gs4} + V_{dsatn} + V_{dsatp} = V_{gs} + 2V_{dsat}$, the input range is obviously from 0 to VDD. The complete RIPO amplifier circuit is depicted in Figure 2b. It is based on the compact cascode amplifier (EA_1) with rail-to-rail input, where the gate voltage of EA_2 is sourced from the output of EA_1. When the voltage of INP, the negative input of EA_1, increases, the voltages of b and d decrease simultaneously. In this case, both gate voltages of M_{24} and M_{25} are reversed from INP and in phase at INN. So, the output stage, EA_2, functions as a push–pull amplifier in the RIPO circuit. The stability analysis of this RIPO amplifier configuration with CC_1 is addressed in Section 2.3.

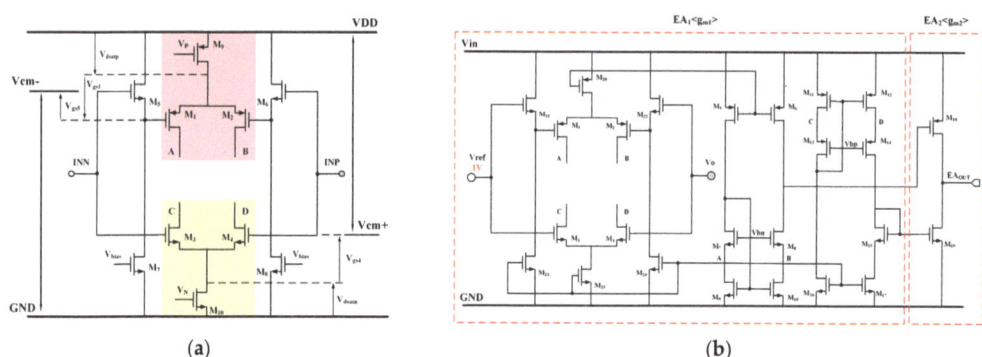

Figure 2. Structures of (**a**) input stage of RIPO amplifier, (**b**) complete schematic of RIPO amplifier.

The proposed SSFB added as an auxiliary circuit is shown in Figure 3. The core components of the SSFB are M_{26}–M_{28}, where the M_{26} is used as the source follower, while M_{27} and M_{28} serve to enhance the following capability. The primary signal transmission pathway involves the passage of signals from the gate of M_{26} through resistor Rz to reach the output terminal. The node at the gate of M_{27} is a high impedance node. Generally, the stability of a buffer solely based on this main signal path is not taken into consideration. However, it should be noted that the proposed buffer also incorporates an inner loop, which may cause stability problems. To deal with the stability issue, compensation is achieved by incorporating capacitor C_B and resistors R_B R_Z. A detailed analysis is provided in Section 2.3. The transistors M_{31} and M_{36} are used as current sources to supply the static operating currents I_{bp} and I_{bn}, respectively. The ratio of I_{bp} to I_{bn} is set at 1:4, with I_{bp} biased at 0.25 µA and I_{bn} biased at 1 µA.

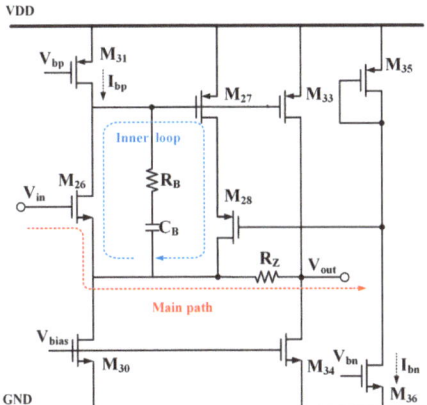

Figure 3. Schematic of super source follower buffer.

2.3. Stability Analysis

The simplified structure of the whole CL-LDO with RIPO and SSFB is illustrated in Figure 4. The main feedback loop of the LDO consistently employs linear feedback, and this paper uses the unit negative feedback. Considering that the transfer function of the buffer is close to unity except at high frequencies, and considering its large input impedance, we temporarily substitute it with $A_{vbuf} \approx 1$ when analyzing the frequency response of CL-LDO. This is discussed separately later. The pole inside EA, which is at an extremely high frequency due to the small parasitic capacitance C_{O1} and is composed of the output of E_{A1} and input of E_{A2}, is disregarded in the frequency response analysis of the proposed CL-LDO. R_{O2} and C_{O2} stand for the output resistance of EA and the input parasitic capacitor of the buffer, respectively. C_{O3} comprises the gate–source capacitor (C_{gs}) of the power PMOS M_P and the output parasitic capacitor of the buffer. Considering that the parasitic capacitance is significantly smaller than C_{gs} by several orders of magnitude, it can be approximated that C_{O3} is approximately equal to C_{gs}. R_{O3} is equal to the output resistance of the super source buffer, which is extremely small. C_{gP} consists of the Miller compensation capacitor, C_P, and the gate–drain capacitor (C_{gP}) of M_P. The resistance R_O denotes the equivalent output resistance, which is influenced by the load current, while C_L represents the load capacitor. The $A_v(s)$ is given by Equations (5)–(9).

$$A_v(s) = \frac{V_{out}(s)}{V_{in}(s)} \approx \frac{A_{dc}\left(1 - \frac{sC_{gP}}{g_{mP}}\right)}{\left(1 + \frac{s}{p_1}\right)\left(1 + \frac{s}{p_2}\right)\left(1 + \frac{s}{p_3}\right)} \tag{5}$$

$$A_{dc} = g_{m_1}R_{O1} \times g_{m_2}R_{O2} \times A_{vbuf} \times g_{mP}R_O \tag{6}$$

$$P_1 \approx \frac{1}{R_{O3}(C_{O3} + (1 + g_{mP}R_O)C_{gP})} \tag{7}$$

$$P_2 \approx \frac{(1 + g_{mP}R_O)C_P + C_{O3}}{R_O(C_LC_P + C_PC_{O3} + C_LC_{O3})} \tag{8}$$

$$P_3 = \frac{1}{R_{O2}C_{O2}} \tag{9}$$

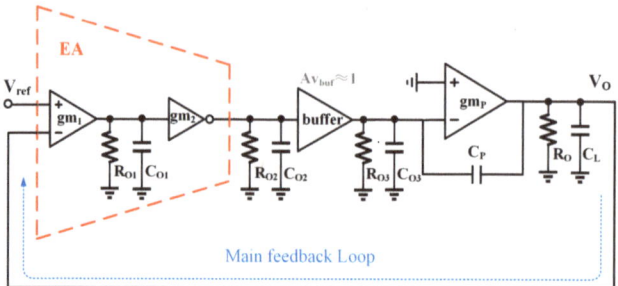

Figure 4. Small-signal modeling of proposed CL-LDO.

P_1 is the dominant pole, while P_2 and P_3 are the non-dominant poles. When the load current I_{load} increases, the output resistance decreases due to its inverse proportionality with the load current. Since g_{mp} is proportional to $\sqrt{I_{load}}$ and R_O is proportional to $1/I_{load}$, P_1 and P_2 are proportional to $\sqrt{I_{load}}$. To guarantee system stability, the phase margin should be above 60°; so, P_2 and P_3 should be placed above the double unity gain frequency under all conditions. The approximate output resistance of EA $R_{O2} = r_{O24} || r_{O25}$ is several megaohms, while the equivalent capacitor at the input of M_P is approximately tens of pF. Without the proposed buffer, the non-dominant pole is generated by the resistance R_{O2} and capacitors C_{gs} and C_{gd}, which are near to the dominant pole, thereby leading to stability issues. However, in this paper, the buffer incorporating a low output resistance separates this low-frequency pole into two high-frequency poles. The frequency response when I_{load} changes is shown in Figure 5. The circuit could keep steady when I_{load} rises to 20 mA without an output capacitor.

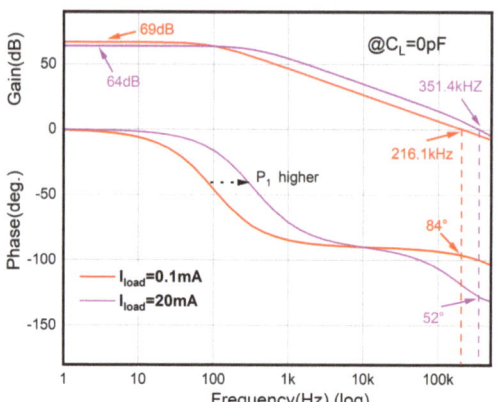

Figure 5. Frequency response at different I_{Load} values.

The stability of the entire proposed loop must be ensured under all conditions, along with the buffer stability, which has been approximately replaced by A_{vbuf}. To analyze the loop stability of the buffer, the block diagram in Figure 6 is proposed. The resistor R_Z is added to generate a zero with the parasitic capacitor C_{gp}. The presence of this zero ensures the stable operation of the inner loop, even when C_{gp} is large and, in turn, generates another pole for the main feedback loop. The capacitor C_B and resistor R_B are also added to compensate. Although the Miller gain applied to C_B is relatively small, it should be noted that one end of C_B is connected to the drain of M_{24}. Consequently, to facilitate a simplified analysis within the block diagram, both C_B and R_B are connected in series and grounded. The gain of the inner loop is approximately given by Equation (10).

$$A_{vloop} \approx \frac{g_{m27}r_{o31}(1+sR_BC_B)(1+sR_ZC_{gp})}{(1+sr_{o31}C_B)(1+sR_ZC_{gp})} \quad (10)$$

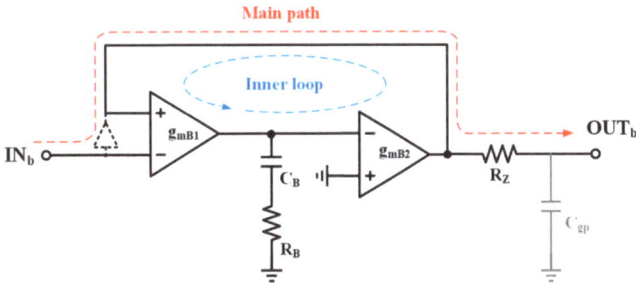

Figure 6. Block diagram of super source follower buffer.

2.4. Transient Response Analysis

The proposed CL-LDO is expected to exhibit an enhanced transient response and reduced undershoot and overshoot spikes. To enhance the transient response, the dynamic charging transistors are added to deal with large transient steps. As depicted in Figure 7, the gate of PMOS M_{DCp} and NMOS M_{DCn} is directly regulated by the voltages V_d of the d node and V_b of the b node in the folded cascode amplifier. The ratio of the size of M_{DCTp} to M_{DCTn} is set as 1:2. At steady state, the M_{DCn} and M_{DCp} transistors are biased in the cutoff region by V_b and V_d because the overdrive voltages of both M_{16} and M_{18} are lower than the threshold voltages of M_{DCn} and M_{DCp}. In this case, this transient enhancement circuit will not influence the stability even with some offset at the input pairs. However, when a large transient step occurs, the output voltage will increase or decrease instantly. Since V_b and V_d are naturally sensitive to the transient response, large current I_{charge} and $I_{discharge}$ can be generated to charge or discharge the large gate parasitic capacitor of M_P without additional sensing circuits. The deviation of the output voltage V_{dev} that causes V_b and V_d to bias dynamic charging transistors in open mode is given in Equations (11) and (12).

$$V_{devp} = \frac{\left|V_{th,MDCTp}\right| - V_{ov12}}{g_{mn}R_d} \quad (11)$$

$$V_{devn} = \frac{V_{th,MDCTn} - V_{ov10}}{g_{mp}R_b} \quad (12)$$

where $V_{th,MDCTp}$ and $V_{th,MDCTn}$ are the threshold of M_{DCp} and M_{DCn}, and where V_{ov12} and V_{ov10} are the overdrive voltages of M_{12} and M_{10}. Here, g_{mp} and g_{mn} stand for the trans-conductance of the input pair consisting of $M_{1,2}$ and $M_{3,4}$, while R_d and R_b denote the equivalent resistance at nodes d and b, respectively. When considering the size of M_{DCp} and M_{DCn}, due to the presence of a small parasitic capacitance, the minimum length is used to ensure a rapid response time. According to the equation, it is evident that $V_{devp,n}$ is controlled by the threshold voltage of $M_{DCp,n}$. However, if the transistor's size and voltages of the b and d nodes are appropriately designed, $V_{devp,n}$ will be constrained by the gain of this transient enhancement circuit and will remain unaffected by $V_{th,MDCp,n}$. Additionally, a smaller $V_{devp,n}$ leads to a reduced ΔV_{OUT}. To effectively regulate the overshoot and undershoot voltage at one-tenth of V_{OUT}, it is recommended that $V_{devp,n}$ be set to approximately 100 mV.

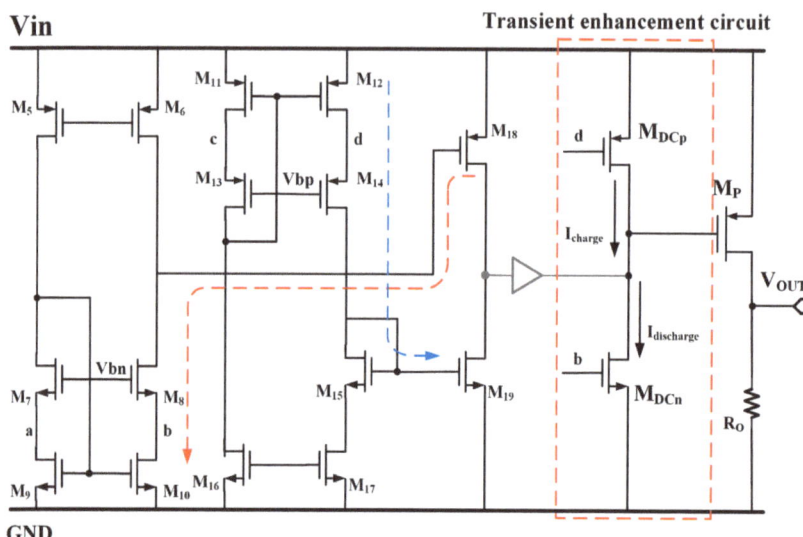

Figure 7. Schematic of transient enhancement circuit.

To control the transient response limitation caused by the finite bandwidth of the main linear regulation loop, a simple operational trans-conductance amplifier (OTA) with a constant small current is incorporated to regulate the high impedance node of the proposed SSFB. By employing this simple OTA for control, the unity gain frequency can be pushed to a higher point and the bandwidth of the main loop can be expanded. The loop frequency response of the whole circuit with added OTA is shown in Section 3 to demonstrate the stability.

3. Simulation Results and Discussion

The proposed CL-LDO is simulated using a TSMC 0.18 µm standard CMOS process. With a supply voltage range of 1.2 V to 1.8 V and a bias current of 2 µA, this CL-LDO is designed to maintain output voltage regulation at 1 V. We will talk about the precise simulation findings for the stability, load regulation, line regulation, and power supply rejection under various conditions.

3.1. Loop Frequency Response

The loop frequency response under different load capacitor and load current combinations is shown in Figure 8. Figure 8a shows the Bode diagram without load capacitor, while Figure 8b shows the load capacitor at 100 pF. Both (a) and (b) show the current load range from 20 mA to 0 mA. As previously analyzed, the bandwidth is pushed from several hundred kilohertz to 1.6 megahertz. On the contrary, the dc gain decreases by approximately 30 dB, which demonstrates the trade-off between gain and speed. It is evident that the load condition has little influence on stability since the node at output is set as the non-dominant pole. The minimum phase margin is 58.12° when the load current is 0 mA and the load capacitor is 100 pF. Meanwhile, the maximum phase margin is 73.33° when the load current is 20 mA and without a load capacitor.

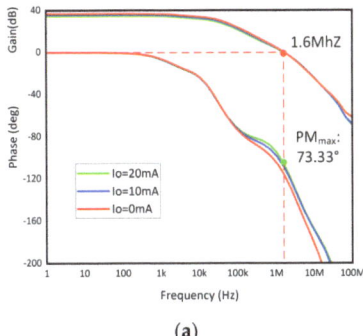 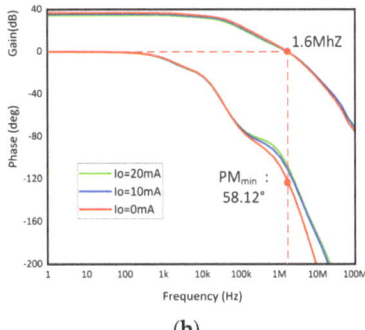

Figure 8. Simulation results of loop frequency response under different I_O and (**a**) C_L = 0 pF; (**b**) C_L = 100 pF.

3.2. Load Transient Response and Load Regulation

The load transient response and load regulation of the proposed CL-LDO are depicted in Figure 9. The load current ranges from 0 A to 20 mA, while the rise and fall times of I_{Load} for emulating the load transient response are set at 100 ns. The simulation results of CL-LDO with and without the transient enhancement circuit are compared and illustrated in Figure 9a. The response time of CL-LDO during load current rise and fall is significantly improved, with a reduction to 0.2 µs and 0.5 µs, respectively, surpassing the performance of the circuit without a transient enhancement circuit. The undershoot voltage drops from 566.5 mV to 238.6 mV, and the overshoot voltage drops from 437.6 mV to 156 mV. Figure 9b shows the load regulation when V_{IN} = 1.8 V and C_L = 0 pF. The V_{OUT} suffers from a 530 µV variation when I_{Load} changes from 0 to 20 mA, resulting in a load regulation of 26.5 µV/mA.

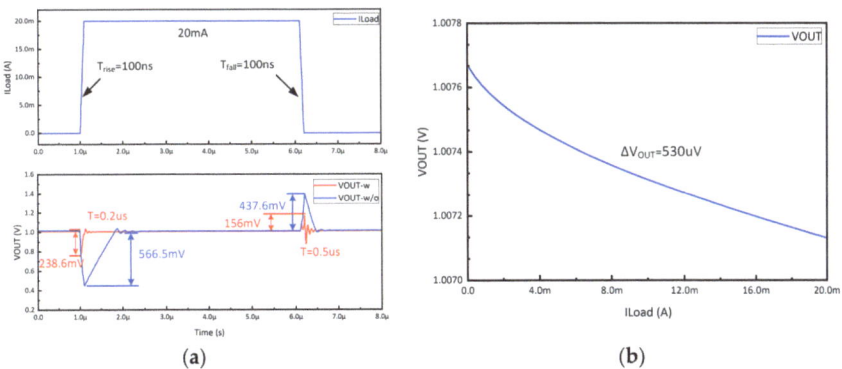

Figure 9. (**a**) Simulated load transient response of the proposed CL-LDO with I_{Load} step between 0 A and 20 mA. (**b**) Load regulation with C_L = 0 pF and V_{IN} = 1.8 V.

3.3. Line Transient Response and Line Regulation

Figure 10a illustrates the line transient response when the V_{IN} step is between 1.2 V and 1.8 V at an edge time of 10 µs of the proposed CL-LDO. The line transient response is simulated at C_L = 100 pF and I_O = 0 mA. The output voltage exhibits an overshoot of 6.3 mV when the V_{IN} steps up. Conversely, it experiences an undershoot of 7.7 mV when the line regulation, which quantifies the deviation in output voltage, is simulated under identical conditions. The voltage output, as depicted in Figure 10b, exhibits a variation of 0.9 mV, resulting in a line regulation of 1.5 mV/V.

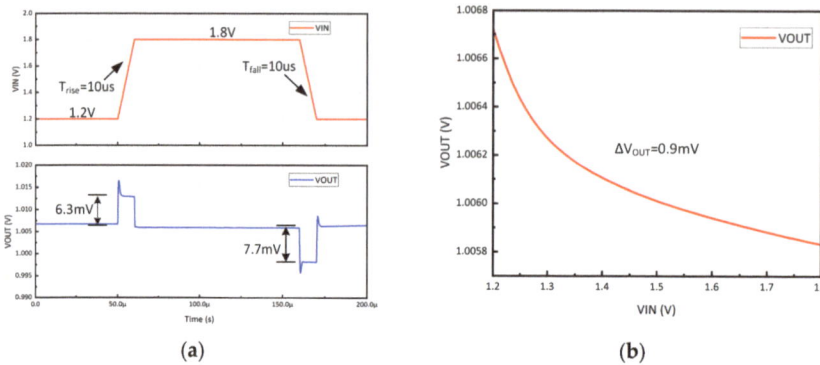

Figure 10. (a) Simulated line transient response of the proposed CL-LDO with VDD step between 1.2 and 1.8 V. (b) Line regulation with C_L = 100 pF and I_O = 0 mA.

3.4. Power-Supply Rejection

The *PSR* of an LDO is given in [18], as shown in Equation (13).

$$PSR = \frac{v_{out}(s)}{v_{in}(s)} = \frac{\frac{R_L}{R_L+r_{ds}}}{\left(1 + \frac{s}{\omega_0}\right)(1 + LG(s))} \quad (13)$$

where $LG(s)$ stands for the loop gain, ω_0 is the pole at the output of the LDO, and R_L and r_{ds} denote the load resistance and the output impedance of M_P, respectively. At low frequency, *PSR* is obviously equal to $1/(1 + LG(s))$. If the ω_0 is the non-dominant pole, the loop gain exhibits a roll-off at -20 dB/decade, resulting in a corresponding decline in the *PSR* at the same rate from the dominant pole. This degradation will persist until the *PSR* remains constant when $LG(s)$ is significantly smaller than 1. At a higher frequency, the *PSR* is primarily influenced by the load capacitor and the M_P's parasitic capacitors, resulting in a reduction in the equivalent resistance. The simulated PSR performance of the proposed CL-LDO at a 20 mA load current and 0 pF load capacitor is shown in Figure 11. The PSR of the proposed CL-LDO is -43 dB at 100 Hz and -9 dB at 1 MHz. The attenuation trend of PSR degrades by -20 dB/decade after the dominant pole, which corresponds to the analysis of Equation (13) and the stability analysis.

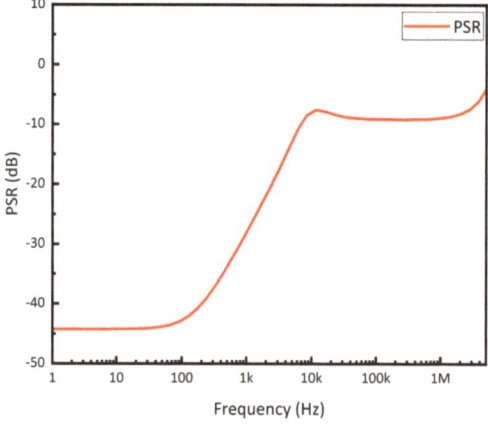

Figure 11. Simulated PSR performance of the proposed CL-LDO.

3.5. Performance Comparison

The figure of merit (*FOM*) in [19,20] is adopted to evaluate the different current efficiencies of the CL-LDOs. The smaller FOM indicates superior performance in terms of the current efficiency and load transient response. The parasitic capacitance of the power transistor is influenced by the minimum channel length (*L*) in different processes. A process with a shorter minimum *L* may result in a smaller *FOM* due to the reduced parasitic capacitance of the transistor. To ensure a fair comparison, the FOM_1 equation, originally proposed in [6] with consideration of the minimum *L*, is used to compare the transient response.

$$FOM_1 = \frac{T_{edge} \cdot \Delta V_{OUT} \cdot \left(I_Q + I_{Load(min)}\right)}{\Delta I_{Load} \cdot L^2} \tag{14}$$

The performance comparison with previously reported CL-LDOs is summarized in Table 1. In this table, the representative study findings from recent years are compared with this design to demonstrate the improved performance. The I_Q row shows that the power consumption of this design is only slightly higher than that proposed in 2020, and significantly lower than other architectures. The Load Reg and T_{settle} rows show that this design can achieve good voltage regulation and a fast response performance when the load current changes. These comparison results demonstrate that even under the relatively backward 180 nm process, the architecture can still have lower power consumption, smaller load regulation and a faster response speed. As a result, this demonstrates a lower FOM_1, indicating a higher performance benefit.

Table 1. Main performance summary and comparison.

Reference	[21]	[22]	[14]	[23]	This Work
Year	2017	2018	2020	2022	**2023**
Process	40 nm	130 nm	65 nm	350 nm	**180 nm**
V_{IN} [V]	1.1	1–1.4	0.95–1.2	2.7–3.3	**1.2–1.8**
V_{OUT} [V]	1	0.8	0.8	2.5	**1**
$I_{Load,max}$ [mA]	200	40	100	100	**20**
$I_{Load,min}$ [mA]	0	9	0	0.1	**0**
C_L [pF]	0–100	0–50	0–100	0–100	**0–100**
I_Q [μA]	275	200	14	66	**47**
ΔV_{OUT} [V]	0.12	0.036	0.23	0.255	**0.15**
Line Reg [mV/V]	0.75	0.857	12	0.8	**1.5**
Load Reg [μV/mA]	19	248	90	60	**25**
T_{edge} [ns]	100	100	220	400	**100**
T_{settle} [μs]	0.8	0.04	3.2	1.2	**0.5**
FOM_1 [ns·V/μm²]	10.65	0.62	1.67	0.63	**1.09**

4. Conclusions

This paper proposes a new capacitor-less LDO structure for digital analog hybrid circuits. The proposed capacitor-less LDO utilizes RIPO and SSFB to satisfy the design challenge of stability typically associated with the absence of on-chip capacitors. This proposed structure is stable at a load current range of 0 mA to 20 mA, with a maximum allowable CL of 100 pF. With the transient enhancement circuit, this structure achieves a good transient response while ensuring stability. The settling time is about 0.22 μs when the load current steps from 0 mA to 20 mA within 100 ns.

Author Contributions: Conceptualization, C.C. and M.S.; Methodology, C.C., M.S., L.W. and T.H.; Software, C.C. and T.H.; Validation, C.C.; Formal analysis, C.C., M.S., L.W. and T.H.; Investigation, C.C.; Resources, M.X.; Data curation, C.C.; Writing—original draft, C.C.; Writing—review & editing, C.C., M.S., T.H. and M.X.; Supervision, M.X.; Project administration, M.X. All authors have read and agreed to the published version of the manuscript.

Funding: This research received no external funding.

Data Availability Statement: Data are contained within the article.

Conflicts of Interest: The authors declare no conflict of interest.

References

1. Ho, E.N.Y.; Mok, P.K.T. A Capacitor-Less CMOS Active Feedback Low-Dropout Regulator with Slew-Rate Enhancement for Portable On-Chip Application. *IEEE Trans. Circuits Syst. II Express Briefs* **2010**, *57*, 80–84. [CrossRef]
2. Lu, Y.; Wang, Y.; Pan, Q.; Ki, W.-H.; Yue, C.P. A Fully-Integrated Low-Dropout Regulator with Full-Spectrum Power Supply Rejection. *IEEE Trans. Circuits Syst. I Regul. Pap.* **2015**, *62*, 707–716. [CrossRef]
3. Patel, A.P.; Rincon-Mora, G.A. High Power-Supply-Rejection (PSR) Current-Mode Low-Dropout (LDO) Regulator. *IEEE Trans. Circuits Syst. II Express Briefs* **2010**, *57*, 868–873. [CrossRef]
4. Lavalle-Aviles, F.; Torres, J.; Sanchez-Sinencio, E. A High Power Supply Rejection and Fast Settling Time Capacitor-Less LDO. *IEEE Trans. Power Electron.* **2019**, *34*, 474–484. [CrossRef]
5. Yadav, B.B.; Mounika, K.; De, K.; Abbas, Z. Low Quiescent Current, Capacitor-Less LDO with Adaptively Biased Power Transistors and Load Aware Feedback Resistance. In Proceedings of the 2020 IEEE International Symposium on Circuits and Systems (ISCAS), Seville, Spain, 12–14 October 2020; pp. 1–5.
6. Hong, S.W.; Cho, G.H. High-Gain Wide-Bandwidth Capacitor-Less Low-Dropout Regulator (LDO) for Mobile Applications Utilizing Frequency Response of Multiple Feedback Loops. *IEEE Trans. Circuits Syst. I Regul. Pap.* **2016**, *63*, 46–57. [CrossRef]
7. Kim, S.J.; Chang, S.B.; Seok, M. A High PSRR, Low Ripple, Temperature-Compensated, 10-μA-Class Digital LDO Based on Current-Source Power-FETs for a Sub-mW SoC. *IEEE Solid-State Circuits Lett.* **2021**, *4*, 88–91. [CrossRef]
8. Lu, Y.; Ki, W.-H.; Yue, C.P. 17.11 A 0.65ns-response-time 3.01ps FOM fully-integrated low-dropout regulator with full-spectrum power-supply-rejection for wideband communication systems. In Proceedings of the 2014 IEEE International Solid- State Circuits Conference (ISSCC), San Francisco, CA, USA, 9–13 February 2014; pp. 306–307.
9. Torres, J.; El-Nozahi, M.; Amer, A.; Gopalraju, S.; Abdullah, R.; Entesari, K.; Sanchez-Sinencio, E. Low Drop-Out Voltage Regulators: Capacitor-less Architecture Comparison. *IEEE Circuits Syst. Mag.* **2014**, *14*, 6–26. [CrossRef]
10. Khan, M.; Chowdhury, M.H. Capacitor-less Low-Dropout Regulator (LDO) with Improved PSRR and Enhanced Slew-Rate. In Proceedings of the 2018 IEEE International Symposium on Circuits and Systems (ISCAS), Florence, Italy, 27–30 May 2018; pp. 1–5.
11. Leo, C.J.; Raja, M.K.; Minkyu, J. An ultra low-power capacitor-less LDO with high PSR. In Proceedings of the 2013 IEEE MTT-S International Microwave Workshop Series on RF and Wireless Technologies for Biomedical and Healthcare Applications (IMWS-BIO), Singapore, 9–11 December 2013; pp. 1–3.
12. Liu, N.; Chen, D. A Transient-Enhanced Output-Capacitorless LDO with Fast Local Loop and Overshoot Detection. *IEEE Trans. Circuits Syst. I Regul. Pap.* **2020**, *67*, 3422–3432. [CrossRef]
13. Raducan, C.; Grajdeanu, A.-T.; Plesa, C.-S.; Neag, M.; Negoita, A.; Topa, M.D. LDO With Improved Common Gate Class-AB OTA Handles any Load Capacitors and Provides Fast Response to Load Transients. *IEEE Trans. Circuits Syst. I Regul. Pap.* **2020**, *67*, 3740–3752. [CrossRef]
14. Li, G.; Qian, H.; Guo, J.; Mo, B.; Lu, Y.; Chen, D. Dual Active-Feedback Frequency Compensation for Output-Capacitorless LDO with Transient and Stability Enhancement in 65-nm CMOS. *IEEE Trans. Power Electron.* **2020**, *35*, 415–429. [CrossRef]
15. Qu, X.; Zhou, Z.-K.; Zhang, B.; Li, Z.-J. An Ultralow-Power Fast-Transient Capacitor-Free Low-Dropout Regulator with Assistant Push–Pull Output Stage. *IEEE Trans. Circuits Syst. II Express Briefs* **2013**, *60*, 96–100. [CrossRef]
16. Ho, M.; Leung, K.N.; Mak, K.-L. A Low-Power Fast-Transient 90-nm Low-Dropout Regulator with Multiple Small-Gain Stages. *IEEE J. Solid-State Circuits* **2010**, *45*, 2466–2475. [CrossRef]
17. Ameziane, H.; Hassan, Q.; Kamal, Z.; Mohcine, Z. An enhancement transient response of capless LDO with improved dynamic biasing control for SoC applications. In Proceedings of the 2015 27th International Conference on Microelectronics (ICM), Casablanca, Morocco, 20–23 December 2015; pp. 122–125.
18. Yang, B.; Drost, B.; Rao, S.; Hanumolu, P.K. A high-PSR LDO using a feedforward supply-noise cancellation technique. In Proceedings of the 2011 IEEE Custom Integrated Circuits Conference—CICC, San Jose, CA, USA, 19–21 September 2011; pp. 1–4.
19. Guo, J.; Leung, K.N. A 6-μW Chip-Area-Efficient Output-Capacitorless LDO in 90-nm CMOS Technology. *IEEE J. Solid-State Circuits* **2010**, *45*, 1896–1905. [CrossRef]
20. Cai, G.; Lu, Y.; Zhan, C.; Martins, R.P. A Fully Integrated FVF LDO with Enhanced Full-Spectrum Power Supply Rejection. *IEEE Trans. Power Electron.* **2021**, *36*, 4326–4337. [CrossRef]
21. Lim, Y.; Lee, J.; Lee, Y.; Song, S.-S.; Kim, H.-T.; Lee, O.; Choi, J. An External Capacitor-Less Ultralow-Dropout Regulator Using a Loop-Gain Stabilizing Technique for High Power-Supply Rejection Over a Wide Range of Load Current. *IEEE Trans. Very Large Scale Integr. (VLSI) Syst.* **2017**, *25*, 3006–3018. [CrossRef]

22. Desai, C.; Mandal, D.; Bakkaloglu, B.; Kiaei, S. A 1.66 mV FOM Output Cap-Less LDO with Current-Reused Dynamic Biasing and 20 ns Settling Time. *IEEE Solid-State Circuits Lett.* **2018**, *1*, 50–53. [CrossRef]
23. Ming, X.; Kuang, J.-J.; Gong, X.-C.; Lin, Z.; Xiong, J.; Qin, Y.; Wang, Z.; Zhang, B. A Fast-Transient Capacitorless LDO with Dual Paths Active-Frequency Compensation Scheme. *IEEE Trans. Power Electron.* **2022**, *37*, 10332–10347. [CrossRef]

Disclaimer/Publisher's Note: The statements, opinions and data contained in all publications are solely those of the individual author(s) and contributor(s) and not of MDPI and/or the editor(s). MDPI and/or the editor(s) disclaim responsibility for any injury to people or property resulting from any ideas, methods, instructions or products referred to in the content.

Article

The Printability, Microstructure, and Mechanical Properties of $Fe_{80-x}Mn_xCo_{10}Cr_{10}$ High-Entropy Alloys Fabricated by Laser Powder Bed Fusion Additive Manufacturing

Kai Li, Vyacheslav Trofimov [ID], Changjun Han *[ID], Gaoling Hu, Zhi Dong, Yujin Zou, Zaichi Wang, Fubao Yan, Zhiqiang Fu * and Yongqiang Yang

School of Mechanical and Automotive Engineering, South China University of Technology, Guangzhou 510641, China; 202120100662@mail.scut.edu.cn (K.L.); trofimov@scut.edu.cn (V.T.); menhgling@mail.scut.edu.cn (G.H.); medongzhi@mail.scut.edu.cn (Z.D.); 202220100495@mail.scut.edu.cn (Y.Z.); 202220100483@mail.scut.edu.cn (Z.W.); mefbyan@mail.scut.edu.cn (F.Y.); meyqyang@scut.edu.cn (Y.Y.)
* Correspondence: cjhan@scut.edu.cn (C.H.); zhiqiangfu2019@scut.edu.cn (Z.F.)

Citation: Li, K.; Trofimov, V.; Han, C.; Hu, G.; Dong, Z.; Zou, Y.; Wang, Z.; Yan, F.; Fu, Z.; Yang, Y. The Printability, Microstructure, and Mechanical Properties of $Fe_{80-x}Mn_xCo_{10}Cr_{10}$ High-Entropy Alloys Fabricated by Laser Powder Bed Fusion Additive Manufacturing. Micromachines 2024, 15, 123. https:// doi.org/10.3390/mi15010123

Academic Editor: Kun Li

Received: 17 December 2023
Revised: 5 January 2024
Accepted: 8 January 2024
Published: 11 January 2024

Copyright: © 2024 by the authors. Licensee MDPI, Basel, Switzerland. This article is an open access article distributed under the terms and conditions of the Creative Commons Attribution (CC BY) license (https:// creativecommons.org/licenses/by/ 4.0/).

Abstract: This work investigated the effect of Fe/Mn ratio on the microstructure and mechanical properties of non-equimolar $Fe_{80-x}Mn_xCo_{10}Cr_{10}$ (x = 30% and 50%) high-entropy alloys (HEAs) fabricated by laser powder bed fusion (LPBF) additive manufacturing. Process optimization was conducted to achieve fully dense $Fe_{30}Mn_{50}Co_{10}Cr_{10}$ and $Fe_{50}Mn_{30}Co_{10}Cr_{10}$ HEAs using a volumetric energy density of 105.82 J·mm^{-3}. The LPBF-printed $Fe_{30}Mn_{50}Co_{10}Cr_{10}$ HEA exhibited a single face-centered cubic (FCC) phase, while the $Fe_{50}Mn_{30}Co_{10}Cr_{10}$ HEA featured a hexagonal close-packed (HCP) phase within the FCC matrix. Notably, the fraction of HCP phase in the $Fe_{50}Mn_{30}Co_{10}Cr_{10}$ HEAs increased from 0.94 to 28.10%, with the deformation strain ranging from 0 to 20%. The single-phase $Fe_{30}Mn_{50}Co_{10}Cr_{10}$ HEA demonstrated a remarkable combination of high yield strength (580.65 MPa) and elongation (32.5%), which surpassed those achieved in the FeMnCoCr HEA system. Comparatively, the dual-phase $Fe_{50}Mn_{30}Co_{10}Cr_{10}$ HEA exhibited inferior yield strength (487.60 MPa) and elongation (22.3%). However, it displayed superior ultimate tensile strength (744.90 MPa) compared to that in the $Fe_{30}Mn_{50}Co_{10}Cr_{10}$ HEA (687.70 MPa). The presence of FCC/HCP interfaces obtained in the $Fe_{50}Mn_{30}Co_{10}Cr_{10}$ HEA resulted in stress concentration and crack expansion, thereby leading to reduced ductility but enhanced resistance against grain slip deformation. Consequently, these interfaces facilitated an earlier attainment of yield limit point and contributed to increased ultimate tensile strength in the $Fe_{50}Mn_{30}Co_{10}Cr_{10}$ HEA. These findings provide valuable insights into the microstructure evolution and mechanical behavior of LPBF-printed metastable FeMnCoCr HEAs.

Keywords: additive manufacturing; high-entropy alloys; FeMnCoCr; laser powder bed fusion; transformation-induced plasticity

1. Introduction

Laser powder bed fusion (LPBF) is a disruptive additive manufacturing (AM) technology with the potential to revolutionize the fabrication of geometrically complex products from metal powders, overcoming the limitations of conventional processing methods [1,2]. Its ultra-high cooling rate of 10^3–10^8 K/s during non-equilibrium solidification effectively inhibits grain growth and the segregation of alloying elements within metal materials, which are challenging to achieve through conventional processing methods [3]. LPBF enables the production of metallic materials with microstructures characterized by highly heterogeneous grain geometries, sub-grain dislocation structures, and chemical segregation [4,5], thus resulting in exceptional mechanical properties.

High-entropy alloys (HEAs) have recently emerged as a novel paradigm for designing multi-principle-element alloys and attract significant attention from both academia and industry due to their distinctive microstructures and exceptional structural-functional properties [6]. According to phase types, HEAs can be generally categorized into single face-centered cubic (FCC) phase, such as the CoCrFeMnNi system (also known as Cantor alloys [7]); body-centered cubic (BCC) phase, such as the refractory NbMoTaW system; dual-phase such as FCC + BCC; or hexagonal close-packed (HCP) phase based on the structure of the phases [8]. Through optimizing the combination of constituent elements and modulating their proportions, HEAs can exhibit excellent mechanical properties. However, most of the reported studies have focused on conventional processes used for processing HEAs such as FeMnCoCrNi [9–14] and FeMnCoCr [15–18], resulting in inadequate yield strengths that limit practical engineering applications including jet aircraft turbine blades, nuclear fusion reactors, etc. Fortunately, LPBF offers rapid melting and solidification rates that promote the formation of refined grains [19], segregated phases [20], and high-density dislocation [21], which play a crucial role in strengthening the material by regulating dislocation motion during deformation.

The primary strengthening mechanisms of single-phase HEAs are solid-solution strengthening and dislocation strengthening, which enhance strength at the expense of plasticity. However, with the development of HEAs, design concepts based on equimolar ratios and single-phase microstructures have become limited. Therefore, the concept of "metastability-engineering" has been introduced in certain non-equimolar HEAs. Li et al. further expanded this concept by adjusting the Mn content in metastable FeMnCoCr HEAs to modulate the stacking fault energy of the matrix phase, thereby increasing the driving force (ΔG) for austenite-to-martensite ($\gamma \rightarrow \varepsilon$) transformation within the matrix [16]. By promoting a dual-phase microstructure through composition design in FeMnCoCr HEAs, various mechanisms such as transformation-induced plasticity (TRIP), solid-solution lattice distortion, and dislocations can be utilized to strengthen their mechanical properties and overcome trade-offs between strength and ductility [22,23].

Engineering the martensitic transformation or TRIP effect has been validated as a viable and effective method of achieving superior strength–ductility synergy. Antolovich et al. reported that austenite-to-martensite at the crack of TRIP steels could assimilate additional energy [24]. Agrawal et al. demonstrated that the martensitic transformation delayed the crack initiation and prolonged crack extension life in the high cycle fatigue response of $Fe_{40}Mn_{20}Co_{20}Cr_{15}Si_5$ HEA fabricated via LPBF [25]. Wang et al. discovered that the generation of compressive stress during martensitic transformation inhibited crack initiation [26]. However, some studies have shown that the transformation effect may degrade the ductility of alloys [27–30]. Unfortunately, the optimal combination of ductility and strength in HEAs remains poorly understood, due to potential deformation-induced transformation probably compromising their mechanical properties. Additionally, it has been observed that such transformation not only enhance plasticity but also induce local brittleness and nucleation damages [29]. Consequently, significant efforts were made to fabricate high-performance FeMnCoCr HEAs via LPBF while further exploring the mechanisms of martensitic transformation.

The FeMnCoCr high-entropy alloy system exhibits excellent mechanical properties overcoming the strength–ductility trade-off. Moreover, the Mn content plays an important role in the phase composition and phase stability of the alloy system. Specifically, the Mn content of 40 at% represents a critical point for the transition in deformation mechanisms, shifting from dislocation-dominated plasticity to twinning-induced plasticity and transformation-induced plasticity [16]. Therefore, $Fe_{30}Mn_{50}Co_{10}Cr_{10}$ (with a Mn content of 50 at%) and $Fe_{50}Mn_{30}Co_{10}Cr_{10}$ (with a Mn content of 30 at%) were selected for further investigation. However, at present, while a large amount of progress has been made towards LPBF-printed single-phase system HEAs, few studies have been carried out on single- and dual-phase FeMnCoCr systems. Furthermore, this is the first simultaneous study of the additive manufacturing of stable and metastable non-equimolar FeMnCoCr HEAs by tuning Mn elemental composition. This study investigated the processing optimization, microstructure, and mechanical properties of $Fe_{80-x}Mn_xCo_{10}Cr_{10}$ HEAs fabricated by LPBF. Specifically, the effect of process parameters as well as Fe/Mn contents on the printability, microstructure, and mechanical properties of the LPBF-printed $Fe_{80-x}Mn_xCo_{10}Cr_{10}$ HEAs were analyzed. A comparison of the microstructure and mechanical properties between $Fe_{30}Mn_{50}Co_{10}Cr_{10}$ and $Fe_{50}Mn_{30}Co_{10}Cr_{10}$ HEAs is discussed, while deformation-strengthening damage mechanisms are further elucidated. The underlying mechanisms behind TRIP effect and strengthening behavior during tensile deformation are revealed.

2. Materials and Experiments

2.1. Materials

The pre-alloyed HEA powders used were $Fe_{30}Mn_{50}Co_{10}Cr_{10}$ (at%) and $Fe_{50}Mn_{30}Co_{10}Cr_{10}$ (at%) with a particle size distribution from 5 to 75 μm, which exhibited a stabilized single FCC phase and a metastable dual phase, respectively. The morphologies and element distributions of the powders were detected using a scanning electron microscope (SEM, FEI Quanta 250, Eindhoven, The Netherlands) with energy-dispersive spectroscopy (EDS) analysis. Both powder particles displayed near-spherical shapes (Figure 1a,c), as shown in the inserted images depicting the measured compositions of the powders, while three randomly selected powder particles were subjected to compositional analysis, indicating homogeneous elemental distribution. The average particle size D50 of the powders, measured using a laser particle size analyzer, increased from 28 to 30 μm upon decreasing the Mn content from 30 to 50 at% (Figure 1b,d).

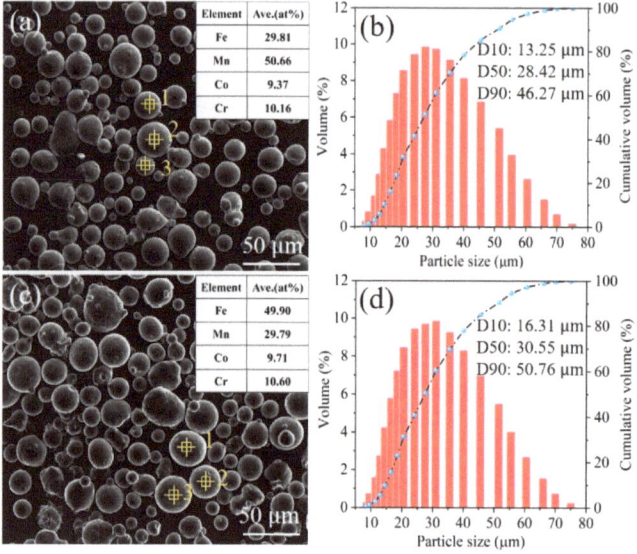

Figure 1. Characteristics of the $Fe_{80-x}Mn_xCo_{10}Cr_{10}$ powders: (**a**,**c**) SEM images showing the morphology of the $Fe_{30}Mn_{50}Co_{10}Cr_{10}$ and $Fe_{50}Mn_{30}Co_{10}Cr_{10}$ powders, respectively, and inserted tables indicate the composition of the powders, (**b**,**d**) particle size distribution of the powders, respectively. Numbers 1, 2, and 3 represent EDS spectral point measurements.

2.2. Sample Fabrication by LPBF

The LPBF process was conducted using a Dimetal-100 machine (Laseradd Technology, Guangzhou, China) with a 1064 nm Nd: YAG fiber laser and a two-axis scanner. A schematic illustration of the HEA samples produced by LPBF is shown in Figure 2a. The two types of $Fe_{80-x}Mn_xCo_{10}Cr_{10}$ samples were printed on 316L stainless steel substrates and subsequently separated from the substrates by electrical discharge machining. The samples with sizes of 6 × 6 × 6 mm³ were printed using various combinations of process parameters as shown in Figure 2b, which included laser power (*P*) ranging from 100 W to

300 W, scanning speed (v) ranging from 500 mm/s to 1300 mm/s, hatch space (h) of 70 μm, a layer thickness (t) of 30 μm, and a scanning strategy involving rotating neighboring layers by 67° that can effectively mitigate residual stresses, thereby minimizing the risk of cracking in the fabricated samples. Volumetric energy densities (VEDs) were evaluated based on these parameters, i.e., VED = $P/(vht)$. The same VEDs were used to print $Fe_{30}Mn_{50}Co_{10}Cr_{10}$ and $Fe_{50}Mn_{30}Co_{10}Cr_{10}$ HEAs for comparing their printability, microstructural evolution, and mechanical properties. The dog bone-shaped samples measuring 42 × 9 × 6 mm³ using different VEDs were fabricated and subsequently cut into tensile samples with dimensions of 42 × 9 × 1.5 mm³.

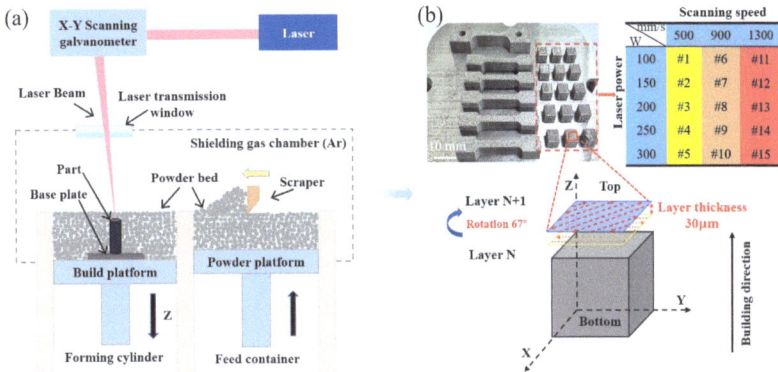

Figure 2. (**a**) Schematic illustration of the LPBF process, and (**b**) printed HEA samples and process parameters used.

2.3. Characterizations

The relative density of the LPBF-printed HEAs depended on the ratio of the experimental density determined using Archimedes' principle to the theoretical density, and each sample was tested three times to ensure that the results were reproducible. For microstructural characterizations, the samples were sequentially polished with 200-, 400-, 800-, 1200-, 2000-, and 3000-grit SiC sandpaper, and then polished using 2 μm sized diamond polishing particles until the surfaces were smooth without any scratches, and the polished sample surfaces were etched with aqua regia solution (HNO_3:HCl = 1:3) for 30 s. The phase identification of the HEAs was performed via X-ray diffraction (XRD, Malvern Panalytical, Shanghai, China). The surface morphologies were characterized using an optical microscope (OM, Zeiss, Munich, Germany) and SEM, and composition characterization was conducted using an SEM equipped with an energy-dispersive spectrometer. The samples were polished by mechanical vibration on an electron-backscattered diffraction apparatus (EBSD, FEI Nova NanoSEM230, Hillsboro, OR, USA) for testing the microstructures of kernel average misorientation (KAM), inverse polar figure (IPF), and phase diagram used for analyzing the results with channel5 software.

2.4. Mechanical Testing

In microhardness testing, the indentation of the sample was performed on a digital microhardness tester (MVS-1000D1, Shanghai, China) with a Vickers indenter, applying a load of 0.2 kg for 15 s and ensuring the uniform distribution of five test points on the samples. The universal testing machine (SUST CMT5504, Zhuhai, China) was used to explore the tensile properties of the HEA samples, with a loading speed of 1 mm/min and a mechanical extensometer with a gauge length of 10 mm. Three samples were tested at room temperature for each printed HEA to verify the reproducibility of the mechanical properties, such as its yield strength (σ_y), ultimate tensile strength (σ_u), and elongation (δ). Intermittent tensile tests at strains of 10% and 20% were performed on the same samples, followed by cutting the fractured center region for an analysis of the deformed microstructure. The fracture morphology of the tensile samples was determined using a scanning electron microscope.

3. Results and Discussion

3.1. The Printability of LPBF-Printed $Fe_{80-x}Mn_xCo_{10}Cr_{10}$ HEAs

The relative density of the $Fe_{30}Mn_{50}Co_{10}Cr_{10}$ HEA samples exhibited an initial increase followed by a decrease as the VED increased from 36.63 to 285.71 J·mm^{-3} (Figure 3a). Notably, the relative density remained above 99% within a VED range of 75~150 J·mm^{-3}. However, a slight decline in the relative density was observed with an increase in the VED, which can be attributed to the formation of pores in the samples under high-VED conditions. The augmentation of the Marangoni effect and elemental evaporation at elevated VEDs leads to increased porosity and a resultant reduction in relative density. Figure 3b–d illustrate the typical surface morphology of the printed samples. In the case of a VED of 52.91 J·mm^{-3}, unmelted powder particles and pores were discernible on the sample surface along with discontinuities in melt track formation. With the VED increasing to 105.82 J·mm^{-3}, a flat surface was achieved with continuous melt track formation, resulting in improved printing quality. However, at higher VEDs, overlapping melt tracks became evident along with pore formation on the sample surface and an increase in splashing phenomena [31].

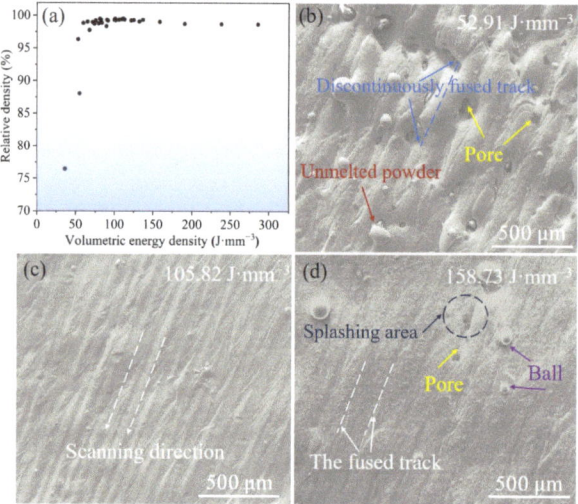

Figure 3. (a) Relative density of as-printed $Fe_{30}Mn_{50}Co_{10}Cr_{10}$ HEA samples with various volumetric energy densities. SEM images exhibiting the representative surface morphology of the as-printed samples under different VEDs: (b) 52.91 J·mm^{-3}, (c) 105.82 J·mm^{-3}, and (d) 158.73 J·mm^{-3}.

Figure 4 presents optical microscope images of the LPBF-printed $Fe_{30}Mn_{50}Co_{10}Cr_{10}$ samples with the combination of laser power and scan speed. The surface morphology of the printed samples exhibits defects such as unmelted powder particles, cracks, and pores. Initially, an increase in VED led to a decrease in the number of defects before eventually increasing, which is consistent with the trend observed for relative density. At high VEDs (above 190 J·mm^{-3}), numerous small circular pores appeared on the sample surface due to the low boiling point of Mn (2324 °C) and the smallest heat of vaporization (220.7 kJ/mol), making it susceptible to vaporization under higher laser energy densities. Consequently, this results in intensified recoil pressure within the melt pool, promoting the formation of pores that are prone to axial fluctuations and radial perturbations governed by energy and pressure balances [32]. Moreover, this phenomenon induces instability and poses a risk of pore collapse, which can lead to gas bubble formation within the melt pool that may become trapped by a solidification front [33]. These small pores act as sources for stress concentration while also serving as locations for crack initiation and propagation that can degrade mechanical properties.

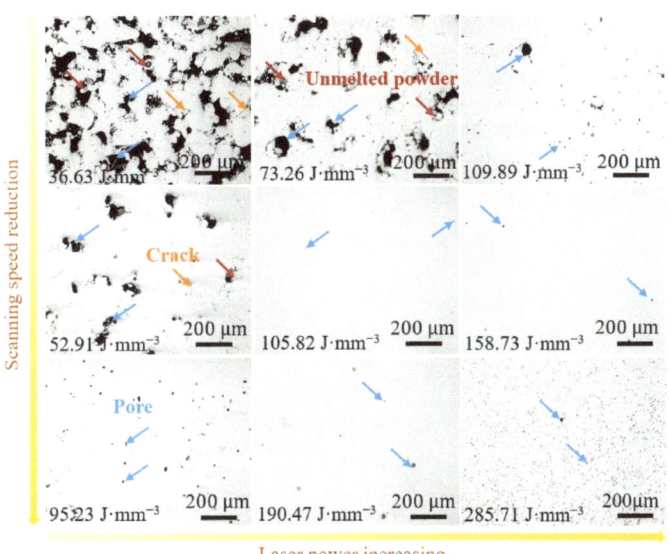

Figure 4. Optical microscopic images of the LPBF-printed $Fe_{30}Mn_{50}Co_{10}Cr_{10}$ HEAs with different VEDs. The red, blue and orange arrows represent unmelt powder, pore and crack, respectively.

The volumetric energy density increased from 36.63 to 333.33 J·mm^{-3} (Figure 5), with an increase in the relative density of the LPBF-printed $Fe_{50}Mn_{30}Co_{10}Cr_{10}$ HEAs. Particularly, relative densities exceeding 99% were achieved at VEDs above 85 J·mm^{-3}. Figure 5 demonstrates that at a VED of 36.63 J·mm^{-3}, numerous unmelted powder particles and pores are observed on the sample surface, while discontinuous melt tracks are formed. However, when the VED reaches 190.47 J·mm^{-3}, a flat surface morphology with visible melt tracks devoid of significant defects is attained.

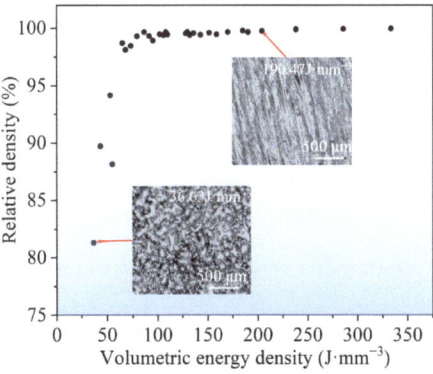

Figure 5. Relative density of as-printed $Fe_{50}Mn_{30}Co_{10}Cr_{10}$ HEA samples, and inserted image presents representative surface morphology of as-printed samples under different VEDs.

The surface morphology of LPBF-printed $Fe_{50}Mn_{30}Co_{10}Cr_{10}$ HEA samples with various laser powers and scanning speeds is depicted in Figure 6. The results reveal that the printed samples primarily exhibited defects including unmelted powder particles and pores. An inverse relationship between defects and VED is observable, which aligns to the trend in the relative density of the HEA samples. In comparison to $Fe_{30}Mn_{50}Co_{10}Cr_{10}$, the $Fe_{50}Mn_{30}Co_{10}Cr_{10}$ samples printed at high VEDs (above 190 J·mm^{-3}) displayed fewer round pores on their surfaces. This can be attributed to the higher content of Fe, which possesses a high melting and boiling point compared to Mn, thereby promoting homogeneity in heat conduction at the two-phase interface during powder melting and the uniform distribution of elements within the melt pool. These factors contribute to reducing temperature gradients and solute solubility differences across different regions of the melt pool, inhibiting Marangoni effect formation [34]. Consequently, there were minimal defects on the surface of the $Fe_{50}Mn_{30}Co_{10}Cr_{10}$ samples under high-VED conditions.

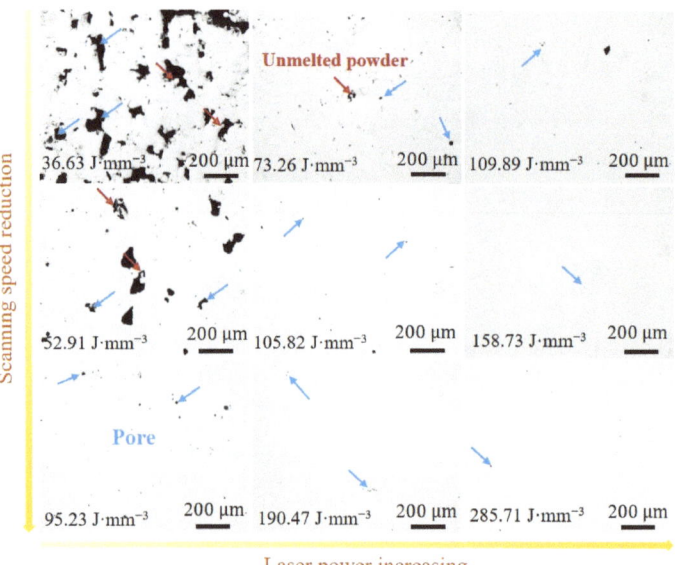

Figure 6. Optical microscopic images of the LPBF-printed $Fe_{50}Mn_{30}Co_{10}Cr_{10}$ HEA samples processed with different VEDs. The red and blue arrows represent unmelt powder and pore, respectively.

3.2. The Microstructure of LPBF-Printed $Fe_{80-x}Mn_xCo_{10}Cr_{10}$ HEAs

The $Fe_{30}Mn_{50}Co_{10}Cr_{10}$ HEA powders and their LPBF-printed samples exhibit a single-phase austenitic crystal structure (Figure 7a). The shift towards higher diffraction angles for peak (111), which changed from 43.15° to 43.25°, indicates a decrease in lattice parameters within the HEA as the VED increases (Figure 7b). This occurrence of a shifted diffraction angle can be attributed to the significant evaporation of Mn, with a lower boiling point compared to other elements in the FeMnCoCr HEA system, leading to a significantly lower elemental concentration at different volumetric energy densities. Additionally, due to its larger atomic diameter compared to Fe, Co, and Cr, the volatilization of Mn results in decreased lattice parameters within this HEA [21]. The diffraction pattern in Figure 7c exclusively demonstrates characteristic peaks corresponding to the austenitic crystal structure of $Fe_{50}Mn_{30}Co_{10}Cr_{10}$ powders, and the samples fabricated with different VEDs reveal the presence of martensite and austenite phase. Figure 7d demonstrates that the diffraction angles for peak (111) did not change as the VED increased except for the powder, and compared to the $Fe_{30}Mn_{50}Co_{10}Cr_{10}$ sample, its diffraction peak angle shows a more stabilized state at different VEDs. In summary, decreasing the Mn content from 50 to 30 at% transformed the lattice structure of the $Fe_{80-x}Mn_xCo_{10}Cr_{10}$ HEA from FCC single-phase to FCC/HCP dual-phase. Gradual heat accumulation during the LPBF process results in a reduced temperature gradient and cooling rate in the melt pool, thereby promoting a thermotropic martensite phase transition. Notably, an increased intensity in the diffraction peak (220) was observed for the LPBF-printed $Fe_{30}Mn_{50}Co_{10}Cr_{10}$ and $Fe_{50}Mn_{30}Co_{10}Cr_{10}$ HEA samples when increasing the VED, indicating a preferential grain orientation within these samples.

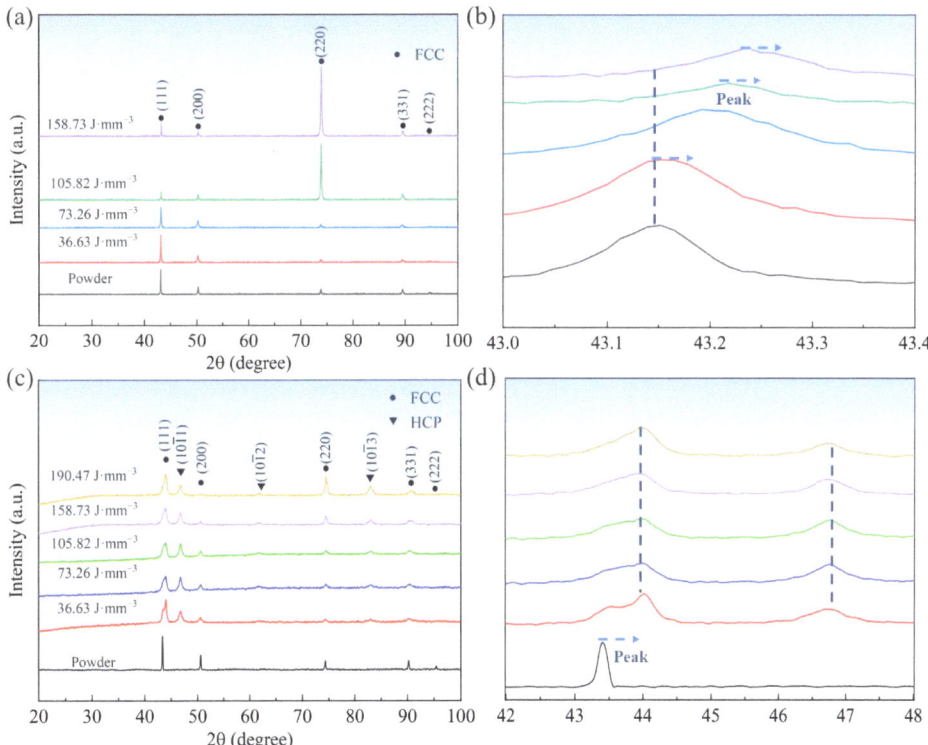

Figure 7. XRD patterns of the $Fe_{80-x}Mn_xCo_{10}Cr_{10}$ powders and their LPBF-printed samples: (**a**) different VEDs of $Fe_{30}Mn_{50}Co_{10}Cr_{10}$, (**b**) enlarged drawing of peak (111) in (**a**), (**c**) different VEDs of $Fe_{50}Mn_{30}Co_{10}Cr_{10}$, (**d**) enlarged drawing of peak (111) in (**c**).

The microstructure of the $Fe_{30}Mn_{50}Co_{10}Cr_{10}$ HEA sample printed using a volumetric energy density of 105.82 J·mm^{-3} reveals the presence of corrosion pits at the melt pool boundary, as depicted in Figure 8a,b. The accelerated corrosion rate observed at this boundary can be attributed to the existence of significant thermal stresses. Within the melt pool, columnar crystal structures grew along the heat flow direction at both the bottom boundary and inner area, while a cellular crystal structure was observed at the top region. Additionally, limited spherical pores were detected within the melt pool, which may have resulted from the volatilization of low-melting-point elements such as Mn during the LPBF process. Elemental mapping analysis depicted in Figure 8c demonstrates the uniform distribution of Fe, Mn, Co, and Cr elements within the $Fe_{30}Mn_{50}Co_{10}Cr_{10}$ HEA corresponding to locations where corrosion pits occurred at the melt pool boundary (highlighted by the yellow box), without any significant elemental segregation.

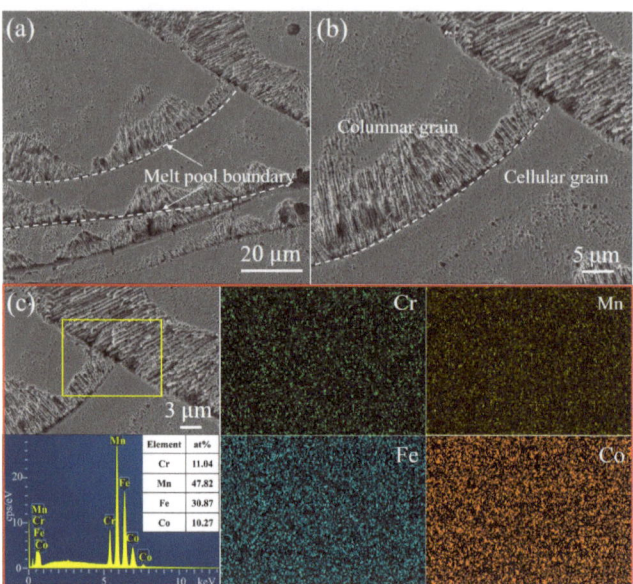

Figure 8. Microstructure and element distribution of the LPBF-printed $Fe_{30}Mn_{50}Co_{10}Cr_{10}$ HEA sample using a VED of 105.82 J·mm^{-3}: (**a**) SEM image, (**b**) high-magnification image of (**a**), and (**c**) EDS mapping of element distributions.

The microstructure of the LPBF-printed $Fe_{50}Mn_{30}Co_{10}Cr_{10}$ HEA using an identical VED to $Fe_{30}Mn_{50}Co_{10}Cr_{10}$ is presented in Figure 9, showing primarily columnar and cellular grains with a distinct melt pool boundary. The morphology of the melt pool exhibited a semi-elliptical shape due to the Gaussian beam profile of the laser. Additionally, a significant heat accumulation led to large temperature gradients between solidified layers, facilitating the growth of columnar grains perpendicular or at an angle to the melt pool boundary. Numerous columnar grains extended across multiple solidified layers, while some gradually transformed into cellular grains as they propagated from perpendicular to the melt pool boundary to the direction of the heat flow. Figure 9c demonstrates a homogeneous distribution of HEA elements throughout the cross-section of the printed sample without any noticeable element segregation, except for minimal volatilization observed for Mn.

To further investigate the microstructure of the LPBF-printed $Fe_{80-x}Mn_xCo_{10}Cr_{10}$ HEAs with different compositions in terms of grain size, crystal orientation, and grain texture, representative samples printed with the same VED (105.82 J·mm^{-3}) were selected for microstructure analysis using EBSD (Figure 10). The $Fe_{30}Mn_{50}Co_{10}Cr_{10}$ HEA sample exhibits a smaller grain size compared to the $Fe_{50}Mn_{30}Co_{10}Cr_{10}$ sample (Figure 10a,g), with average grain sizes of 6.04 μm and 9.18 μm, respectively (Figure 10d,j). According to the Hall–Petch theory [35], increasing grain size is generally detrimental to the mechanical properties of metallic materials. A comparison between the KAM color map and KAM distributions reveals that the $Fe_{30}Mn_{50}Co_{10}Cr_{10}$ sample has a higher KAM value (0.69°) than the $Fe_{50}Mn_{30}Co_{10}Cr_{10}$ sample (0.61°) (Figure 10b,e,h,k), exhibiting a higher value of geometrically necessary dislocations (GNDs) in the $Fe_{30}Mn_{50}Co_{10}Cr_{10}$ HEA. Figure 10c,f,i,l demonstrate that the content of low-angle grain boundaries (LAGBs), ranging from 2° to 15° in orientation angles, is lower in the $Fe_{30}Mn_{50}Co_{10}Cr_{10}$ sample (62.5%) compared to that in the $Fe_{50}Mn_{30}Co_{10}Cr_{10}$ sample (75.3%). The reduction in the LAGBs contributes to improvements in mechanical properties due to their significant advantage as nucleation sites during creep and fatigue processes [36].

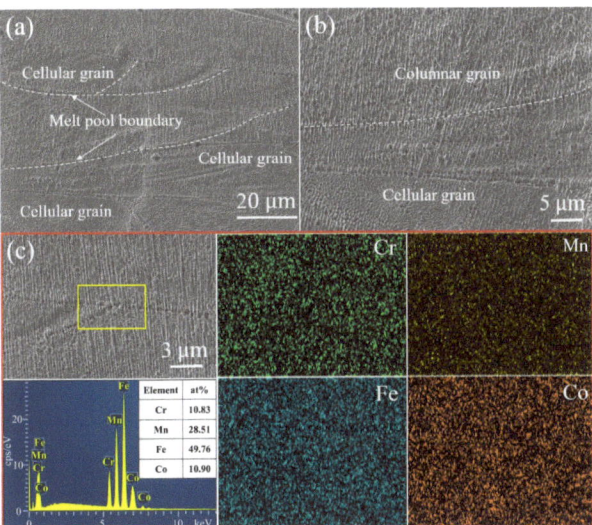

Figure 9. Microstructure and element distribution of the LPBF-printed $Fe_{50}Mn_{30}Co_{10}Cr_{10}$ HEA sample using a VED of 105.82 J·mm^{-3}: (**a**) SEM image, (**b**) high-magnification image of (**a**), and (**c**) EDS mapping of element distribution.

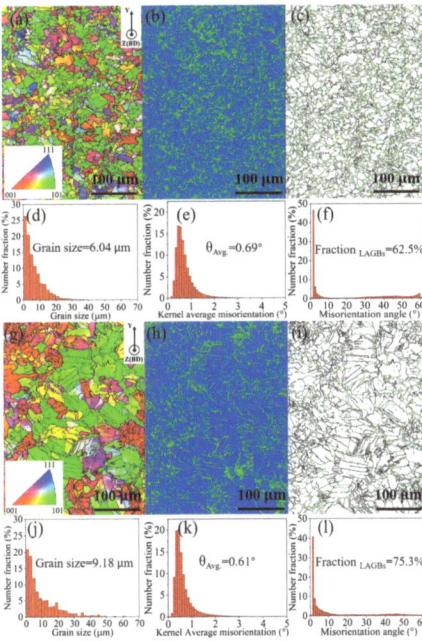

Figure 10. EBSD analyses of the $Fe_{80-x}Mn_xCo_{10}Cr_{10}$ HEA samples: (**a**–**c**) IPF, KAM color map, and misorientation map of the $Fe_{30}Mn_{50}Co_{10}Cr_{10}$ sample, respectively; (**d**–**f**) its grain size distribution, KAM distribution, and misorientation angle distribution, respectively. (**g**–**i**) IPF, KAM color map, and misorientation map of the $Fe_{50}Mn_{30}Co_{10}Cr_{10}$ sample, respectively, and (**j**–**l**) its grain size distribution, KAM distribution, and misorientation angle distribution, respectively.

The Fe$_{50}$Mn$_{30}$Co$_{10}$Cr$_{10}$ sample exhibits a more disordered grain orientation, as depicted in Figure 11a,c, while both HEAs demonstrate the preferential growth of grains with an orientation parallel to <110>. This is evidenced by the prominent peak of intensity that was observed at the central region of the (110) polar figure, which corresponds well with the strong (220) diffraction peaks observed in the XRD pattern (Figure 7). In Figure 11b,d, both HEAs exhibit a strong texture with a <101> orientation along the build direction, thus highlighting the influence of direction of heat flow on the resulting texture characteristics. Typically, austenitic crystal structures tend to grow with a prioritized texture with <001> oriented along the build direction [37]. However, this preferred orientation and microstructure can be altered by adjusting the rotating angle of successive solidified layers.

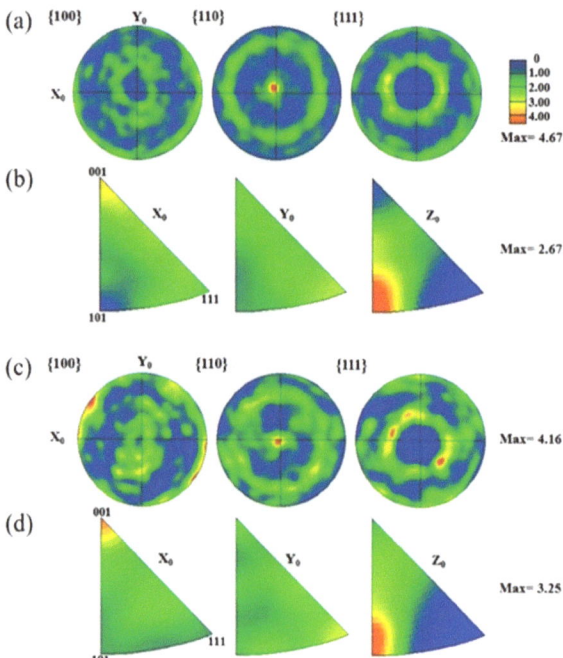

Figure 11. Grain orientation of LPBF-printed Fe$_{80-x}$Mn$_x$Co$_{10}$Cr$_{10}$ HEAs: (**a**,**b**) polar figure and IPF of the Fe$_{30}$Mn$_{50}$Co$_{10}$Cr$_{10}$ alloy along the X-Y plane, respectively, and (**c**,**d**) polar figure and IPF of the Fe$_{50}$Mn$_{30}$Co$_{10}$Cr$_{10}$ alloy along the X-Y plane.

3.3. Mechanical Properties

Figure 12 depicts the microhardness of as-printed Fe$_{80-x}$Mn$_x$Co$_{10}$Cr$_{10}$ HEAs with various VEDs. It is evident that with increasing VED, the microhardness of the Fe$_{50}$Mn$_{30}$Co$_{10}$Cr$_{10}$ samples and the Fe$_{30}$Mn$_{50}$Co$_{10}$Cr$_{10}$ samples tend to initially increase and then decrease. Moreover, the microhardness of Fe$_{50}$Mn$_{30}$Co$_{10}$Cr$_{10}$ remains consistently higher than that of Fe$_{30}$Mn$_{50}$Co$_{10}$Cr$_{10}$. In the case of a VED of 36.63 J·mm^{-3}, the minimum microhardness values for the Fe$_{50}$Mn$_{30}$Co$_{10}$Cr$_{10}$ and Fe$_{30}$Mn$_{50}$Co$_{10}$Cr$_{10}$ samples are 176.66 HV and 116.74 HV, respectively. Furthermore, with the VED increasing to 105.82 J·mm^{-3}, there is an increase in maximum microhardness in the Fe$_{30}$Mn$_{50}$Co$_{10}$Cr$_{10}$ sample, reaching up to 184.98 HV; however, this increase in microhardness is accompanied by a significant presence of microporous defects in samples at higher VEDs, leading to an overall decrease in microhardness. In the case of a VED of 190.47 J·mm^{-3}, the maximum microhardness on the Fe$_{50}$Mn$_{30}$Co$_{10}$Cr$_{10}$ sample reaches 234.96 HV, due to martensitic ε-phase formation resulting in higher microhardness.

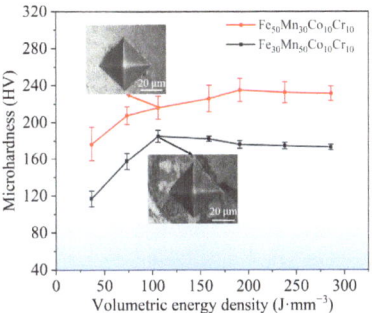

Figure 12. Microhardness of LPBF-printed $Fe_{80-x}Mn_xCo_{10}Cr_{10}$ HEA samples under different VEDs.

Representative stress–strain curves of $Fe_{30}Mn_{50}Co_{10}Cr_{10}$ and $Fe_{50}Mn_{30}Co_{10}Cr_{10}$ HEAs fabricated by LPBF at identical VED are illustrated in Figure 13a. The yield strength and ultimate tensile strength of the $Fe_{50}Mn_{30}Co_{10}Cr_{10}$ samples are 487.60 MPa and 744.90 MPa, respectively. In contrast, the $Fe_{30}Mn_{50}Co_{10}Cr_{10}$ samples exhibit an σ_y value of 580.65 MPa and an σ_u value of 687.70 MPa. In addition, it is noteworthy that the elongation of the $Fe_{50}Mn_{30}Co_{10}Cr_{10}$ sample (22.3%) is lower than that observed for the $Fe_{30}Mn_{50}Co_{10}Cr_{10}$ sample (32.5%), as presented in Table 1. Interestingly, despite possessing a slightly higher σ_y value due to its smaller grain size, grain-refined single-phase FCC-structured $Fe_{30}Mn_{50}Co_{10}Cr_{10}$ samples exhibit significantly lower σ_u values compared to dual-phase $Fe_{50}Mn_{30}Co_{10}Cr_{10}$ samples with various grain sizes. The discrepancy can be attributed to an increase in martensitic phase fraction and higher residual stresses within the austenitic phase matrix during the tensile deformation process [38], causing an earlier attainment of yield limit point for the $Fe_{50}Mn_{30}Co_{10}Cr_{10}$ HEA.

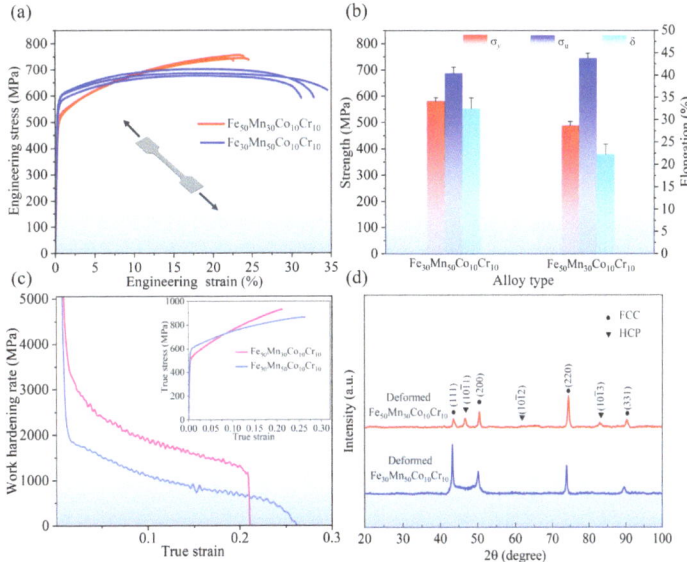

Figure 13. Tensile properties of LPBF-printed $Fe_{30}Mn_{50}Co_{10}Cr_{10}$ and $Fe_{50}Mn_{30}Co_{10}Cr_{10}$ HEAs under an identical VED of 105.82 J·mm^{-3}: (**a**) stress–strain curves, (**b**) histogram of yield strength, ultimate tensile strength, and elongation, (**c**) responses between work hardening rate and true strain, and inserted image presents true stress–strain curves, and (**d**) XRD patterns of the deformed samples printed by VED of the two HEAs.

Table 1. Tensile properties of $Fe_{80-x}Mn_xCo_{10}Cr_{10}$ alloy (at%).

HEA	σ_y/(MPa)	σ_u/(MPa)	δ/(%)
$Fe_{30}Mn_{50}Co_{10}Cr_{10}$	580.65 ± 13.48	687.70 ± 23.25	32.5 ± 2.4
$Fe_{50}Mn_{30}Co_{10}Cr_{10}$	487.60 ± 15.23	744.90 ± 18.47	22.3 ± 2.2

The $Fe_{50}Mn_{30}Co_{10}Cr_{10}$ sample displays a higher work hardening rate compared to the $Fe_{30}Mn_{50}Co_{10}Cr_{10}$ sample (Figure 13c, attributed to the occurrence of a martensitic phase transition in the high dislocation region of austenite. This phase transition impedes dislocation slip and results in a sustained higher work hardening rate. Conversely, the fracture behavior of the $Fe_{50}Mn_{30}Co_{10}Cr_{10}$ sample is characterized by an abrupt decline in work hardening rate, indicating a "hard and brittle" effect during late strain stages. Although the martensitic phase transformation improves the ultimate tensile strength of the $Fe_{50}Mn_{30}Co_{10}Cr_{10}$ sample, it adversely affects its ductility. The deformed $Fe_{50}Mn_{30}Co_{10}Cr_{10}$ HEA samples in Figure 13d exhibit the presence of both martensite and austenite phases, while the deformed $Fe_{30}Mn_{50}Co_{10}Cr_{10}$ HEA samples consist of a sole austenite single phase. In conclusion, this study highlights the excellent mechanical properties achieved through LPBF printing for both $Fe_{30}Mn_{50}Co_{10}Cr_{10}$ and $Fe_{50}Mn_{30}Co_{10}Cr_{10}$ HEAs.

As shown in Figure 14, the LPBF-printed FeMnCoCr HEA systems exhibit an an exceptional combination of ductility and strength compared with previously reported additively manufactured FeMnCoCr HEAs [4,21,39–50] and cast FeMnCoCr-based HEAs [9–18,29,50–55]. This study shows that the elongation is over ~30% and the yield strength is over ~550 MPa. Meanwhile, the yield strength is significantly higher than the cast counterparts, and its ductility is comparable to pure and elementally added alloys.

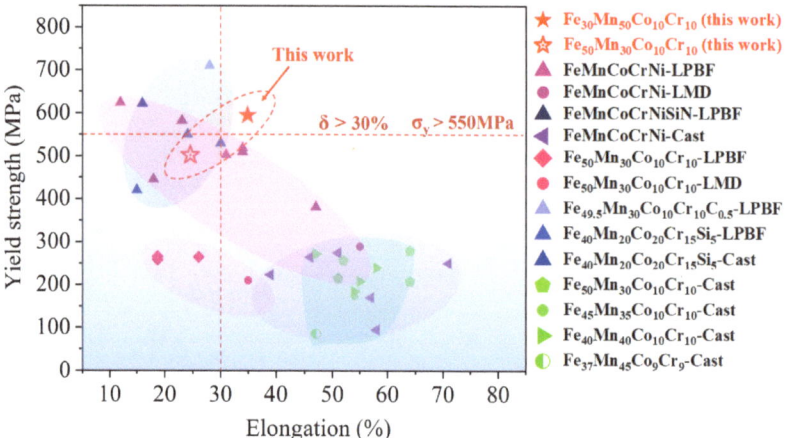

Figure 14. Comparison chart of the mechanical properties of the typical $Fe_{30}Mn_{50}Co_{10}Cr_{10}$, $Fe_{50}Mn_{30}Co_{10}Cr_{10}$ HEAs, and FeMnCoCr-based HEAs fabricated by AM [4,21,39–50] and traditional manufacturing process [9–18,29,50–55].

The fracture surface of the $Fe_{30}Mn_{50}Co_{10}Cr_{10}$ sample exhibits a multitude of equiaxial and shear dimples, indicating typical ductile fracture behavior (Figure 15a,b). Additionally, sporadic precipitated particles are observed at the center of these dimples, contributing to stress concentration and promoting micro-void expansion during tensile deformation, thereby influencing the mechanical properties of the HEAs. Figure 15c,d demonstrate that the $Fe_{50}Mn_{30}Co_{10}Cr_{10}$ sample displays a characteristic ductile fracture with numerous dimples containing pore and crack defects. This can be attributed to the presence of heterogeneous interfacial stresses between austenite and newly formed martensite, which induce the initiation and propagation of microcracks and pores within the dual-phase structure, consequently resulting in reduced ductility.

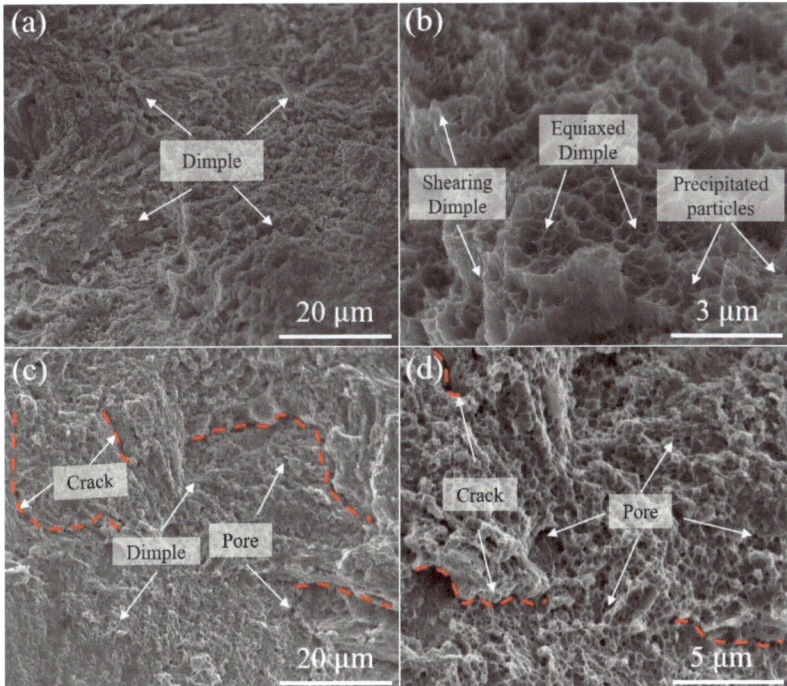

Figure 15. SEM images of fractured morphology of LPBF-fabricated $Fe_{80-x}Mn_xCo_{10}Cr_{10}$ samples after tensile deformation: (**a**) $Fe_{30}Mn_{50}Co_{10}Cr_{10}$ sample, (**b**) high-magnification image of (**a**), (**c**) $Fe_{50}Mn_{30}Co_{10}Cr_{10}$ sample, and (**d**) high-magnification image of (**c**). The red dotted lines represent crack.

In summary, the two HEAs have extraordinary tensile properties, especially the characteristics (brittleness) of the $Fe_{50}Mn_{30}Co_{10}Cr_{10}$ HEA during tensile deformation. Moreover, the $Fe_{30}Mn_{50}Co_{10}Cr_{10}$ HEA has better ductility and does not present transition behavior during deformation (Figure 13d). Therefore, it is valuable to further explore the microstructural evolution of $Fe_{50}Mn_{30}Co_{10}Cr_{10}$ HEAs under different strains. The microstructure deformation in the LPBF-printed $Fe_{50}Mn_{30}Co_{10}Cr_{10}$ sample was investigated using EBSD at various regions of its fracture with different strain levels. The melt pools experienced significant temperature gradients and cooling rates during LPBF. The appearance of a minimal presence of HCP phase in the LPBF-fabricated $Fe_{50}Mn_{30}Co_{10}Cr_{10}$ HEA can be attributed to a non-equilibrium solidification of the melt pools significantly promoting a thermal martensitic transformation in this HEA (Figure 16a). With an increase in local strain from 0 to 20%, the average proportion of the HCP phase obtained from EBSD increases accordingly, ranging from 0.94% to 28.1% (Figure 16b,c). At the initial stage of deformation, regions with or neighboring to the FCC phase present significantly larger KAM values (Figure 16e), while regions with the HCP phase exhibit less noticeable KAM values. The KAM values indicate the density of GNDs to be within a specified area [56]; thus, larger KAM values in the FCC phase regions indicate a higher density of GNDs and pronounced plastic strains within these areas. These results illustrate that it is predominantly the FCC phase that accommodates plastic strain at this stage. Regarding strain partitioning behavior during the intermediate and late stages of deformation, the FCC phase regions still obtain larger KAM values, while most of the HCP phase regions exhibit even larger KAM values in comparison to those observed in the initial stage (Figure 16f). Generally, the HCP phase is known to be more rigid and less susceptible to deformation compared to FCC phase observed in most industrial alloys. Moreover, the GND at the FCC/HCP interface that contributes to the tendency of microporous generation is significantly higher compared with the other regions. This is an explanation for why it is more challenging for GND arrays to form within internal regions of the HCP-phase grains; instead, dislocations and stacking faults occur as slip bands within the HCP phase [57].

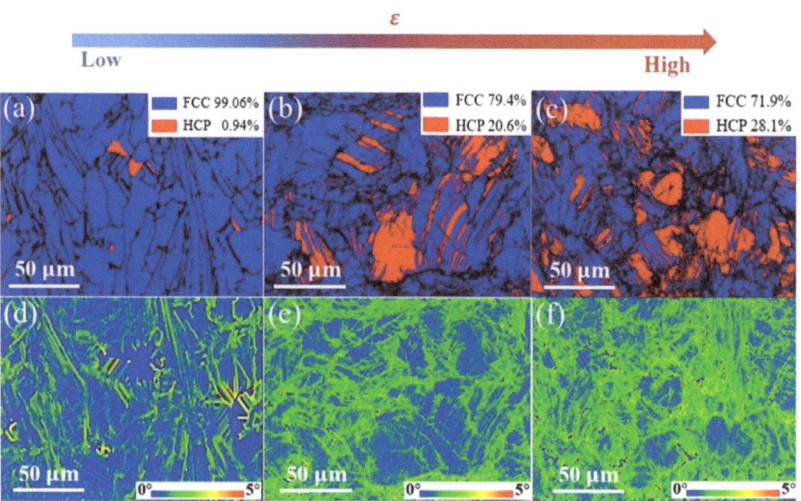

Figure 16. EBSD results illustrating the microstructural evolution in the $Fe_{50}Mn_{30}Co_{10}Cr_{10}$ HEA sample under increasing tensile strains: (**a**–**c**) phase distributions at strains of 0%, 10%, and 20%, respectively, and (**d**–**f**) KAM maps at strains of 0%, 10%, and 20%, respectively.

The phase transformations in the $Fe_{50}Mn_{30}Co_{10}Cr_{10}$ HEA play a dual influence mechanism on its mechanical properties, as illustrated in Figure 17. During early deformation, the phase transformation enhances the strain hardening rate and the mechanical strength of the HEA. However, during late deformation stages, the dual-phase deformation of martensite and austenite causes a local high degree of incompatibility, leading to blocked dislocation movement and increased dislocation density near the interphase, generating stress concentration [29]. This promotes the nucleation of pores and the expansion of cracks. The formation of a martensite phase transition is more brittle, making it susceptible to induced crack expansion and damaging the HEA's ductility. The $Fe_{50}Mn_{30}Co_{10}Cr_{10}$ HEA has lower stacking fault energy than the $Fe_{30}Mn_{50}Co_{10}Cr_{10}$ HEA [16], allowing for the easy bundling of incomplete dislocations to ensure cross slip as the stacking fault energy decreases. The synergistic deformation of these two phases generates a beneficial dynamic strain–stress partitioning effect [58]. During the deformation process, the GNDs accumulated at the interface of the dual-phase region generate heterogeneous deformation-induced stresses [59], which further lead to the plastic instability of the alloy. Therefore, the martensitic transformation plays a dual role in strength and ductility; i.e., one is useful for tensile strength and initial ductile elongation, while the other is harmful for ductility fracture. This contradicts the general view that the martensite phase transition is beneficial for ductility [60].

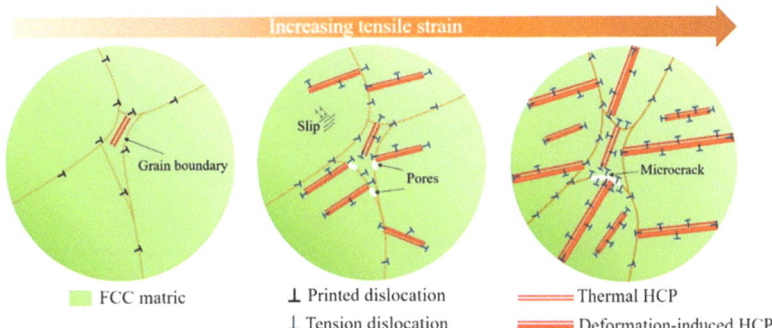

Figure 17. Schematic diagram of microstructural evolution of the $Fe_{50}Mn_{30}Co_{10}Cr_{10}$ HEA at increased strains.

4. Conclusions

In this study, gas-atomized non-equimolar $Fe_{30}Mn_{50}Co_{10}Cr_{10}$ and $Fe_{50}Mn_{30}Co_{10}Cr_{10}$ powders were used for laser powder bed fusion. The microstructural evolution and mechanical properties of non-equimolar $Fe_{80-x}Mn_xCo_{10}Cr_{10}$ (x = 30%, 50%) HEAs fabricated by LPBF were comprehensively characterized. The findings are presented as follows.

The increased volumetric energy density led to a higher relative density for the LPBF-printed $Fe_{80-x}Mn_xCo_{10}Cr_{10}$ sample. At a VED of 75 and 85 J·mm^{-3}, the $Fe_{30}Mn_{50}Co_{10}Cr_{10}$ and $Fe_{50}Mn_{30}Co_{10}Cr_{10}$ samples achieved a relative density of 99%, respectively. Surface morphological observations revealed the formation of circular pores in the $Fe_{30}Mn_{50}Co_{10}Cr_{10}$ sample when Mn element evaporation occurred above a VED of 190 J·mm^{-3}, resulting in a decrease in relative density.

The LPBF-printed and deformed $Fe_{30}Mn_{50}Co_{10}Cr_{10}$ samples exhibited single-phase FCC structures, but the $Fe_{50}Mn_{30}Co_{10}Cr_{10}$ samples exhibited dual-phase FCC-HCP structures under different VEDs. The diffraction peak intensities of both HEAs shifted with the increase in VED. The grain orientation of both HEAs demonstrated preferential growth of grains with an orientation parallel to <110>, which corresponds to the strong (220) diffraction peaks observed in the XRD pattern. The microstructure of both HEAs consisted of columnar and cellular grains. Moreover, with an increase in the strain of the $Fe_{50}Mn_{30}Co_{10}Cr_{10}$ sample from 0 to 20%, there was a corresponding rise in the average fraction of the martensitic phase range from 0.94% to 28.1%.

At a VED of 105.82 and 190.47 J·mm^{-3}, the maximum microhardness of the $Fe_{30}Mn_{50}Co_{10}Cr_{10}$ and $Fe_{50}Mn_{30}Co_{10}Cr_{10}$ samples reached 184.98 HV and 234.96 HV, respectively. The yield strength, ultimate tensile strength, and elongation of the $Fe_{50}Mn_{30}Co_{10}Cr_{10}$ samples were 487.60 MPa, 744.90 MPa, and 22.3%, respectively. In contrast, the $Fe_{30}Mn_{50}Co_{10}Cr_{10}$ samples exhibited an σ_y value of 580.65 MPa, an σ_u value of 687.70 MPa, and an δ value of 32.5%. Despite the fact that the $Fe_{50}Mn_{30}Co_{10}Cr_{10}$ HEA possessed larger grains, lower dislocation density, and fewer LAGBs compared to the $Fe_{30}Mn_{50}Co_{10}Cr_{10}$ HEA in the horizontal plane, the ultimate tensile strength of the former was superior to that of the latter. Such an abnormal phenomenon can be attributed to the HCP phase formed during tensile deformation in the $Fe_{50}Mn_{30}Co_{10}Cr_{10}$ HEA, enhancing its mechanical strength. Moreover, the elongation of the $Fe_{50}Mn_{30}Co_{10}Cr_{10}$ HEA was lower compared with the $Fe_{30}Mn_{50}Co_{10}Cr_{10}$ HEA.

Author Contributions: Conceptualization, methodology, validation, writing—original draft preparation, K.L.; supervision, writing—review and editing, funding acquisition, V.T. and C.H.; data curation, G.H. and Z.D.; visualization, Y.Z. and Z.W.; formal analysis, F.Y.; investigation, writing—review and editing, Z.F.; project administration, Y.Y. All authors have read and agreed to the published version of the manuscript.

Funding: This research was funded by the Fundamental Research Funds for the Central Universities (Grant No. 2023ZYGXZR061), Science and Technology Program of Guangzhou (Grant No.202201010362), Guangdong Basic and Applied Basic Research Foundation (Grant No. 2022A1515010304).

Data Availability Statement: The data presented in this paper are available on request from the corresponding author. The data are not publicly available due to privacy restrictions.

Conflicts of Interest: The authors declare no conflicts of interest.

References

1. Shi, W.; Li, J.; Liu, Y.; Liu, S.; Lin, Y.; Han, Y. Experimental study on mechanism of influence of laser energy density on surface quality of Ti-6Al-4V alloy in selective laser melting. *J. Cent. South Univ.* **2022**, *29*, 3447–3462. [CrossRef]
2. Ren, J.; Zhang, Y.; Zhao, D.; Chen, Y.; Guan, S.; Liu, Y.; Liu, L.; Peng, S.; Kong, F.; Poplawsky, J.D.; et al. Strong yet ductile nanolamellar high-entropy alloys by additive manufacturing. *Nature* **2022**, *608*, 62–68. [CrossRef]
3. DebRoy, T.; Wei, H.L.; Zuback, J.S.; Mukherjee, T.; Elmer, J.W.; Milewski, J.O.; Beese, A.M.; Wilson-Heid, A.D.; De, A.; Zhang, W. Additive manufacturing of metallic components–process, structure and properties. *Prog. Mater. Sci.* **2018**, *92*, 112–224. [CrossRef]
4. Li, R.; Niu, P.; Yuan, T.; Li, Z. Displacive transformation as pathway to prevent micro-cracks induced by thermal stress in additively manufactured strong and ductile high-entropy alloys. *Trans. Nonferrous Met. Soc. China* **2021**, *31*, 1059–1073. [CrossRef]
5. Kartikeya Sarma, I.; Selvaraj, N.; Kumar, A. Parametric investigation and characterization of 17-4 PH stainless steel parts fabricated by selective laser melting. *J. Cent. South Univ.* **2023**, *30*, 855–870. [CrossRef]
6. Han, C.; Fang, Q.; Shi, Y.; Tor, S.B.; Chua, C.K.; Zhou, K. Recent advances on high-entropy alloys for 3D printing. *Adv. Mater.* **2020**, *32*, 1903855. [CrossRef]
7. Yeh, J.W.; Chen, S.K.; Lin, S.J.; Gan, J.Y.; Chin, T.S.; Shun, T.T.; Tsau, C.H.; Chang, S.Y. Nanostructured high-entropy alloys with multiple principal elements: Novel alloy design concepts and outcomes. *Adv. Eng. Mater.* **2004**, *6*, 299–303. [CrossRef]
8. Li, Z.; Zhao, S.; Ritchie, R.O.; Meyers, M.A. Mechanical properties of high-entropy alloys with emphasis on face-centered cubic alloys. *Prog. Mater. Sci.* **2019**, *102*, 296–345. [CrossRef]
9. Otto, F.; Dlouhý, A.; Somsen, C.; Bei, H.; Eggeler, G.; George, E.P. The influences of temperature and microstructure on the tensile properties of a CoCrFeMnNi high-entropy alloy. *Acta Mater.* **2013**, *61*, 5743–5755. [CrossRef]
10. Gali, A.; George, E.P. Tensile properties of high- and medium-entropy alloys. *Intermetallics* **2013**, *39*, 74–78. [CrossRef]
11. Salishchev, G.A.; Tikhonovsky, M.A.; Shaysultanov, D.G.; Stepanov, N.D.; Kuznetsov, A.V.; Kolodiy, I.V.; Tortika, A.S.; Senkov, O.N. Effect of Mn and V on structure and mechanical properties of high-entropy alloys based on CoCrFeNi system. *J. Alloys Compd.* **2014**, *591*, 11–21. [CrossRef]
12. Chen, J.; Yao, Z.; Wang, X.; Lu, Y.; Wang, X.; Liu, Y.; Fan, X. Effect of C content on microstructure and tensile properties of as-cast CoCrFeMnNi high entropy alloy. *Mater. Chem. Phys.* **2018**, *210*, 136–145. [CrossRef]
13. Laplanche, G.; Kostka, A.; Horst, O.M.; Eggeler, G.; George, E.P. Microstructure evolution and critical stress for twinning in the CrMnFeCoNi high-entropy alloy. *Acta Mater.* **2016**, *118*, 152–163. [CrossRef]
14. Yao, M.J.; Pradeep, K.G.; Tasan, C.C.; Raabe, D. A novel, single phase, non-equiatomic FeMnNiCoCr high-entropy alloy with exceptional phase stability and tensile ductility. *Scr. Mater.* **2014**, *72*, 5–8. [CrossRef]
15. Kim, J.; Kim, J. Grain size-dependent phase-specific deformation mechanisms of the $Fe_{50}Mn_{30}Co_{10}Cr_{10}$ high entropy alloy. *Mater. Sci. Eng. A* **2022**, *854*, 143867. [CrossRef]
16. Li, Z.; Pradeep, K.G.; Deng, Y.; Raabe, D.; Tasan, C.C. Metastable high-entropy dual-phase alloys overcome the strength–ductility trade-off. *Nature* **2016**, *534*, 227–230. [CrossRef]
17. He, Z.F.; Jia, N.; Wang, H.W.; Liu, Y.; Li, D.Y.; Shen, Y.F. The effect of strain rate on mechanical properties and microstructure of a metastable FeMnCoCr high entropy alloy. *Mater. Sci. Eng. A* **2020**, *776*, 138982. [CrossRef]
18. Wang, H.; He, Z.; Jia, N. Microstructure and mechanical properties of a FeMnCoCr high-entropy alloy with heterogeneous structure. *Acta Met. Sin.* **2020**, *57*, 632–640.
19. Guan, S.; Wan, D.; Solberg, K.; Berto, F.; Welo, T.; Yue, T.M.; Chan, K.C. Additive manufacturing of fine-grained and dislocation-populated CrMnFeCoNi high entropy alloy by laser engineered net shaping. *Mater. Sci. Eng. A* **2019**, *761*, 138056. [CrossRef]
20. Guo, C.; Wei, S.; Wu, Z.; Wang, P.; Zhang, B.; Ramamurty, U.; Qu, X. Effect of dual phase structure induced by chemical segregation on hot tearing reduction in additive manufacturing. *Mater. Des.* **2023**, *228*, 111847. [CrossRef]
21. Li, R.; Niu, P.; Yuan, T.; Cao, P.; Chen, C.; Zhou, K. Selective laser melting of an equiatomic CoCrFeMnNi high-entropy alloy: Processability, non-equilibrium microstructure and mechanical property. *J. Alloys Compd.* **2018**, *746*, 125–134. [CrossRef]
22. Huang, H.; Wu, Y.; He, J.; Wang, H.; Liu, X.; An, K.; Wu, W.; Lu, Z. Phase-transformation ductilization of brittle high-entropy alloys via metastability engineering. *Adv. Mater.* **2017**, *29*, 1701678. [CrossRef] [PubMed]
23. Wang, X.; De Vecchis, R.R.; Li, C.; Zhang, H.; Hu, X.; Sridar, S.; Wang, Y.; Chen, W.; Xiong, W. Design metastability in high-entropy alloys by tailoring unstable fault energies. *Sci. Adv.* **2022**, *8*, o7333. [CrossRef] [PubMed]
24. Antolovich, S.D.; Singh, B. On the toughness increment associated with the austenite to martensite phase transformation in TRIP steels. *Metall. Mater. Trans. B* **1971**, *2*, 2135–2141. [CrossRef]
25. Agrawal, P.; Haridas, R.S.; Thapliyal, S.; Yadav, S.; Mishra, R.S.; McWilliams, B.A.; Cho, K.C. Metastable high entropy alloys: An excellent defect tolerant material for additive manufacturing. *Mater. Sci. Eng. A* **2021**, *826*, 142005. [CrossRef]

26. Wang, X.; Liu, C.; Sun, B.; Ponge, D.; Jiang, C.; Raabe, D. The dual role of martensitic transformation in fatigue crack growth. *Proc. Natl. Acad. Sci. USA* **2022**, *119*, e2110139119. [CrossRef] [PubMed]
27. Jacques, P.; Furnemont, Q.; Pardoen, T.; Delannay, F. On the role of martensitic transformation on damage and cracking resistance in TRIP-assisted multiphase steels. *Acta Mater.* **2001**, *49*, 139–152. [CrossRef]
28. Lacroix, G.; Pardoen, T.; Jacques, P.J. The fracture toughness of TRIP-assisted multiphase steels. *Acta Mater.* **2008**, *56*, 3900–3913. [CrossRef]
29. Wei, S.; Kim, J.; Tasan, C.C. Boundary micro-cracking in metastable Fe45Mn35Co10Cr10 high-entropy alloys. *Acta Mater.* **2019**, *168*, 76–86. [CrossRef]
30. Hu, C.; Huang, C.P.; Liu, Y.X.; Perlade, A.; Zhu, K.Y.; Huang, M.X. The dual role of TRIP effect on ductility and toughness of a medium Mn steel. *Acta Mater.* **2023**, *245*, 118629. [CrossRef]
31. Li, Z.; Li, H.; Yin, J.; Li, Y.; Nie, Z.; Li, X.; You, Q.; Guan, K.; Duan, W.; Cao, L.; et al. A review of spatter in laser powder bed fusion additive manufacturing: In situ detection, generation, effects, and countermeasures. *Micromachines* **2022**, *13*, 1366. [CrossRef] [PubMed]
32. Guo, L.; Liu, H.; Wang, H.; Wei, Q.; Xiao, Y.; Tang, Z.; Wu, Y.; Wang, H. Identifying the keyhole stability and pore formation mechanisms in laser powder bed fusion additive manufacturing. *J. Mater. Process. Technol.* **2023**, *321*, 118153. [CrossRef]
33. Wang, L.; Zhang, Y.; Chia, H.Y.; Yan, W. Mechanism of keyhole pore formation in metal additive manufacturing. *NPJ Comput. Mater.* **2022**, *8*, 22. [CrossRef]
34. Michi, R.A.; Plotkowski, A.; Shyam, A.; Dehoff, R.R.; Babu, S.S. Towards high-temperature applications of aluminium alloys enabled by additive manufacturing. *Int. Mater. Rev.* **2022**, *67*, 298–345. [CrossRef]
35. Hall, E.O. The deformation and ageing of mild steel: III discussion of results. *Proc. Phys. Society. Sect. B* **1951**, *64*, 747. [CrossRef]
36. Tan, L.; Sridharan, K.; Allen, T.R. Effect of thermomechanical processing on grain boundary character distribution of a Ni-based superalloy. *J. Nucl. Mater.* **2007**, *371*, 171–175. [CrossRef]
37. Zhou, K.; Han, C. Laser Powder Bed Fusion. *Met. Powder-Based Addit. Manuf.* **2023**, 75–159. [CrossRef]
38. Calcagnotto, M.; Ponge, D.; Demir, E.; Raabe, D. Orientation gradients and geometrically necessary dislocations in ultrafine grained dual-phase steels studied by 2D and 3D EBSD. *Mater. Sci. Eng. A* **2010**, *527*, 2738–2746. [CrossRef]
39. Zhu, Z.G.; Nguyen, Q.B.; Ng, F.L.; An, X.H.; Liao, X.Z.; Liaw, P.K.; Nai, S.; Wei, J. Hierarchical microstructure and strengthening mechanisms of a CoCrFeNiMn high entropy alloy additively manufactured by selective laser melting. *Scr. Mater.* **2018**, *154*, 20–24. [CrossRef]
40. Wang, B.; Sun, M.; Li, B.; Zhang, L.; Lu, B. Anisotropic response of CoCrFeMnNi high-entropy alloy fabricated by selective laser melting. *Materials* **2020**, *13*, 5687. [CrossRef]
41. Chen, P.; Yang, C.; Li, S.; Attallah, M.M.; Yan, M. In-situ alloyed, oxide-dispersion-strengthened CoCrFeMnNi high entropy alloy fabricated via laser powder bed fusion. *Mater. Des.* **2020**, *194*, 108966. [CrossRef]
42. Haase, C.; Tang, F.; Wilms, M.B.; Weisheit, A.; Hallstedt, B. Combining thermodynamic modeling and 3D printing of elemental powder blends for high-throughput investigation of high-entropy alloys–Towards rapid alloy screening and design. *Mater. Sci. Eng. A* **2017**, *688*, 180–189. [CrossRef]
43. Xiang, S.; Luan, H.; Wu, J.; Yao, K.; Li, J.; Liu, X.; Tian, Y.; Mao, W.; Bai, H.; Le, G.; et al. Microstructures and mechanical properties of CrMnFeCoNi high entropy alloys fabricated using laser metal deposition technique. *J. Alloys Compd.* **2019**, *773*, 387–392. [CrossRef]
44. Guo, L.; Gu, J.; Gan, B.; Ni, S.; Bi, Z.; Wang, Z.; Song, M. Effects of elemental segregation and scanning strategy on the mechanical properties and hot cracking of a selective laser melted FeCoCrNiMn-(N, Si) high entropy alloy. *J. Alloys Compd.* **2021**, *865*, 158892. [CrossRef]
45. Hou, Y.; Liu, T.; He, D.; Li, Z.; Chen, L.; Su, H.; Fu, P.; Dai, P.; Huang, W. Sustaining strength-ductility synergy of SLM Fe50Mn30Co10Cr10 metastable high-entropy alloy by Si addition. *Intermetallics* **2022**, *145*, 107565. [CrossRef]
46. Hou, Y.; Li, Z.; Chen, L.; Xiang, Z.; Dai, P.; Chen, J. SLM Fe50Mn30Co10Cr10 metastable high entropy alloy with Al-Ti addition: Synergizing strength and ductility. *J. Alloys Compd.* **2023**, *941*, 168830. [CrossRef]
47. Niu, P.; Li, R.; Fan, Z.; Yuan, T.; Zhang, Z. Additive manufacturing of TRIP-assisted dual-phases Fe50Mn30Co10Cr10 high-entropy alloy: Microstructure evolution, mechanical properties and deformation mechanisms. *Mater. Sci. Eng. A* **2021**, *814*, 141264. [CrossRef]
48. Zhu, Z.G.; An, X.H.; Lu, W.J.; Li, Z.M.; Ng, F.L.; Liao, X.Z.; Ramamurty, U.; Nai, S.M.L.; Wei, J. Selective laser melting enabling the hierarchically heterogeneous microstructure and excellent mechanical properties in an interstitial solute strengthened high entropy alloy. *Mater. Res. Lett.* **2019**, *7*, 453–459. [CrossRef]
49. Thapliyal, S.; Agrawal, P.; Agrawal, P.; Nene, S.S.; Mishra, R.S.; McWilliams, B.A.; Cho, K.C. Segregation engineering of grain boundaries of a metastable Fe-Mn-Co-Cr-Si high entropy alloy with laser-powder bed fusion additive manufacturing. *Acta Mater.* **2021**, *219*, 117271. [CrossRef]
50. Agrawal, P.; Thapliyal, S.; Nene, S.S.; Mishra, R.S.; McWilliams, B.A.; Cho, K.C. Excellent strength-ductility synergy in metastable high entropy alloy by laser powder bed additive manufacturing. *Addit. Manuf.* **2020**, *32*, 101098. [CrossRef]
51. Kim, J.; Kim, J.H.; Park, H.; Kim, J.; Yang, G.; Kim, R.; Song, T.; Suh, D.; Kim, J. Temperature-dependent universal dislocation structures and transition of plasticity enhancing mechanisms of the Fe40Mn40Co10Cr10 high entropy alloy. *Int. J. Plast.* **2022**, *148*, 103148. [CrossRef]

52. He, Z.F.; Jia, N.; Ma, D.; Yan, H.; Li, Z.M.; Raabe, D. Joint contribution of transformation and twinning to the high strength-ductility combination of a FeMnCoCr high entropy alloy at cryogenic temperatures. *Mater. Sci. Eng. A* **2019**, *759*, 437–447. [CrossRef]
53. You, D.; Yang, G.; Choa, Y.; Kim, J. Crack-resistant σ/FCC interfaces in the $Fe_{40}Mn_{40}Co_{10}Cr_{10}$ high entropy alloy with the dispersed σ-phase. *Mater. Sci. Eng. A* **2022**, *831*, 142039. [CrossRef]
54. Deng, Y.; Tasan, C.C.; Pradeep, K.G.; Springer, H.; Kostka, A.; Raabe, D. Design of a twinning-induced plasticity high entropy alloy. *Acta Mater.* **2015**, *94*, 124–133. [CrossRef]
55. Tasan, C.C.; Deng, Y.; Pradeep, K.G.; Yao, M.J.; Springer, H.; Raabe, D. Composition dependence of phase stability, deformation mechanisms, and mechanical properties of the CoCrFeMnNi high-entropy alloy system. *Jom* **2014**, *66*, 1993–2001. [CrossRef]
56. Gao, H.; Huang, Y.; Nix, W.D.; Hutchinson, J.W. Mechanism-based strain gradient plasticity—I. Theory. *J. Mech. Phys. Solids* **1999**, *47*, 1239–1263. [CrossRef]
57. Fang, Q.; Chen, Y.; Li, J.; Jiang, C.; Liu, B.; Liu, Y.; Liaw, P.K. Probing the phase transformation and dislocation evolution in dual-phase high-entropy alloys. *Int. J. Plast.* **2019**, *114*, 161–173. [CrossRef]
58. Yang, K.; Li, Y.; Hong, Z.; Du, S.; Ma, T.; Liu, S.; Jin, X. The dominating role of austenite stability and martensite transformation mechanism on the toughness and ductile-to-brittle-transition temperature of a quenched and partitioned steel. *Mater. Sci. Eng. A* **2021**, *820*, 141517. [CrossRef]
59. Li, A.X.; Liu, X.S.; Li, R.; Yu, S.B.; Jiang, M.H.; Zhang, J.S.; Che, C.N.; Huang, D.; Yu, P.F.; Li, G. Double heterogeneous structures induced excellent strength-ductility synergy in Ni40Co30Cr20Al5Ti5 medium-entropy alloy. *J. Mater. Sci. Technol.* **2023**, *181*, 176–188. [CrossRef]
60. Tian, C.; Ouyang, D.; Wang, P.; Zhang, L.; Cai, C.; Zhou, K.; Shi, Y. Strength-ductility synergy of an additively manufactured metastable high-entropy alloy achieved by transformation-induced plasticity strengthening. *Int. J. Plast.* **2024**, *172*, 103823. [CrossRef]

Disclaimer/Publisher's Note: The statements, opinions and data contained in all publications are solely those of the individual author(s) and contributor(s) and not of MDPI and/or the editor(s). MDPI and/or the editor(s) disclaim responsibility for any injury to people or property resulting from any ideas, methods, instructions or products referred to in the content.

Article

Simulation and Experimental Study of Non-Resonant Vibration-Assisted Lapping of SiCp/Al

Huibo Zhao [1,2], Yan Gu [1,2,*], Yuan Xi [1,2], Xingbao Fu [1,2], Yinghuan Gao [1,2], Jiali Wang [1,2], Lue Xie [1,2] and Guangyu Liang [1,2]

1. Jilin Provincial Key Laboratory of Micro-Nano and Ultra-Precision Manufacturing, School of Mechatronic Engineering, Changchun University of Technology, Yan'an Ave 2055, Changchun 130012, China
2. Jilin Provincial Key Laboratory of International Science and Technology Cooperation for High Performance Manufacturing and Testing, School of Mechatronic Engineering, Changchun University of Technology, Yan'an Ave 2055, Changchun 130012, China
* Correspondence: guyan@ccut.edu.cn; Tel.: +86-431-8571-6288

Abstract: SiCp/Al is a difficult-to-machine material that makes it easy to produce surface defects during machining, and researchers focus on reducing the surface defects. Vibration-assisted machining technology is considered an effective method to reduce surface defects by changing the trajectory and contact mode of the abrasive. Aiming at the problem of SiCp/Al processing technology, a vibration-assisted lapping device (VLD) is designed, and elliptical motion is synthesized by a set of parallel symmetrical displacement output mechanisms. The working parameters of the device were tested by simulation and experiment, and the lapping performance was verified. Then, the effects of removal characteristics and process parameters on surface roughness and lapping force were analyzed by simulation and experiment. Simulation and experimental results show that frequency and amplitude that are too low or too high are not conducive to the advantages of NVL. The best surface quality was 54 nm, obtained at A = 8 μm and f = 850 Hz.

Keywords: non-resonant vibration-assisted lapping; SiCp/Al; vibration-assisted lapping device; surface integrity

Citation: Zhao, H.; Gu, Y.; Xi, Y.; Fu, X.; Gao, Y.; Wang, J.; Xie, L.; Liang, G. Simulation and Experimental Study of Non-Resonant Vibration-Assisted Lapping of SiCp/Al. *Micromachines* **2024**, *15*, 113. https://doi.org/10.3390/mi15010113

Academic Editor: Kun Li

Received: 3 December 2023
Revised: 3 January 2024
Accepted: 5 January 2024
Published: 9 January 2024

Copyright: © 2024 by the authors. Licensee MDPI, Basel, Switzerland. This article is an open access article distributed under the terms and conditions of the Creative Commons Attribution (CC BY) license (https://creativecommons.org/licenses/by/4.0/).

1. Introduction

SiCp/Al has broad application prospects and excellent material properties, and its special strengthening mechanism makes it easy to form surface defects during processing [1,2]. The high hardness of SiC particles in SiCp/Al will increase tool wear and reduce processing efficiency and processing economy [3,4]. At the same time, particle breakage, particle fracture, and interfacial debonding will occur in the process of SiC particles; these phenomena constitute the main form of its surface defects.

The abrasive in the conventional lapping (CL) method is in continuous contact with SiCp/Al, and the strength difference between the aluminum matrix and SiC particles leads to surface damage of SiC particles during processing, so there are large heating and pits in the processing process [5,6]. Many studies have pointed out that ultrasonic vibration-assisted machining (UVM) technology and non-resonant vibration-assisted machining (NVM) technology can reduce force and surface damage by changing the motion path of the abrasive or tool [7–9]. Vibration-assisted machining technology based on ultrasonic vibration devices improves surface quality and surface integrity by making abrasive or tools hit the workpiece surface at high frequency [10]. This technique has received much attention [11–13]. Gu et al. predicted the surface roughness of SiCp/Al by ultrasonic vibration-assisted lapping (UVL). The experimental results verified the effectiveness of UVL in improving surface roughness [14]. Zheng et al. compared the material removal, friction coefficient, and scratch force behavior between the CL method and the UVL method through scratch experiments. The experimental results show that the UVL reduces the scratch force and friction coefficient [15]. Li et al. carried out an ultrasonic vibration-assisted scratch experiment and analyzed the primary forms of sub-surface damage. The experimental results showed that the UVM method had an inhibiting effect on sub-surface damage. According to the above results, NVM can help improve surface quality and reduce surface defects [16]. Wu et al. designed a rotary ultrasonic lapping spindle and carried out SiCp/Al machining experiments. Observing and analyzing the surface morphology of SiCp/Al verified that the axial UVM method can also reduce the lapping force and surface roughness [17].

Gu et al. established a surface roughness prediction model of UVL with a maximum error of 3.64% and obtained the optimal process parameters through the roughness prediction model [13]. Therefore, appropriate process parameters are essential in optimizing surface quality [18]. Although UVL has a series of advantages compared to CL, it also has some unavoidable problems. The ultrasonic vibration device usually works near the resonance point, and its high-frequency vibration will produce much heat, affecting the processing effect of SiCp/Al composite materials [19,20]. In addition, resonant ultrasonic vibration suffers from many limitations in vibration frequency and amplitude adjustment, which makes it difficult to achieve a wide range of free adjustments, and the selection of processing parameters will be limited [21,22].

Researchers have tried to apply the non-resonant vibration device to vibration-assisted cutting (VC) [23,24], vibration-assisted milling (VM) [25], and vibration-assisted polishing (VP) [26,27] and achieved good results. Zhou et al. designed an elliptical VC device and studied the removal mechanism, cutting force, surface roughness, and microstructure in elliptical VC. The orthogonal cutting experiments showed that the periodically changing cutting trajectory reduced the average cutting force and improved the surface quality [24]. Zheng et al. showed that applying vibration on a micro-milling cutter can help improve tool life, and this method can improve surface roughness. The above research results demonstrate the positive effect of non-resonant vibration-assisted machining (NVM) on improving surface quality and reducing surface defects [25]. Gu et al. proposed a non-resonant VP method to investigate the effect of plane vibration on machining processes. Research showed that introducing plane vibration changes the abrasive trajectory and force, thereby improving the abrasive removal capability [26,27]. Wang et al. designed a three-dimensional VP device for structured surface polishing. Combined with the workpiece surface and two-dimensional vibration parameters, the three-dimensional space vibration trajectory and polishing trajectory model were established. The influence of phase and frequency on the polishing trajectory was analyzed by simulation. The experimental results proved the feasibility and effectiveness of three-dimensional VP technology for structured surface polishing [28].

In the NVM method, by adjusting the amplitude and frequency, the tool moves in a specific trajectory, making the tool or the workpiece periodically separated [29], and the contact mode of abrasive is changed from continuous contact to periodic contact so that the processing force is reduced, and the chips can be discharged quickly. Although the NVM method has been introduced into many fields, the processing mechanism and technology of non-resonant vibration-assisted lapping (NVL) SiCp/Al are not fully studied. The key to obtaining better surface quality and surface integrity in the processing of SiCp/Al is to avoid large area fracture and pulling off of SiC particles, so amplitude and frequency processing parameters are essential.

This paper mainly analyzes the removal characteristics and processing technology of NVL SiCp/Al. Firstly, a two-dimensional VLD was designed, and its performance was simulated and tested. Then, a single abrasive NVL SiCp/Al finite element model was established. Based on the relationship between the tool path and the relative position of silicon carbide, the removal characteristics of SiCp/Al during NVL were discussed. Finally, the effects of spindle speed n, feed speed v_f, lapping depth a_p, amplitude A, and frequency f on lapping force and surface quality were studied by simulation and experiment methods.

2. Principle of Processing

The characteristic of NVL SiCp/Al is that the lapping tool moves in the space in the form of elliptical vibration, and the formation of the lapping tool trajectory depends on the output signal of the VLD. Figure 1 shows the motion path of abrasive lapping in the NVL and CL methods. In the traditional lapping method, the abrasive keeps continuous contact with the machined workpiece under pressure, and the higher contact pressure can easily cause surface damage to the workpiece, which is not conducive to improving the Surface quality [30]. In the vibration-assisted lapping method, the motion path of the abrasive changes due to the application of vibration. Abrasive passes through points A, B, C, D, and E. The abrasive is pressed into the material at point A, reaches the lowest point of the trajectory at point B, and separates from the workpiece at point C and point D is the highest point of the processing trajectory, and one cycle of processing is completed when point E is reached.

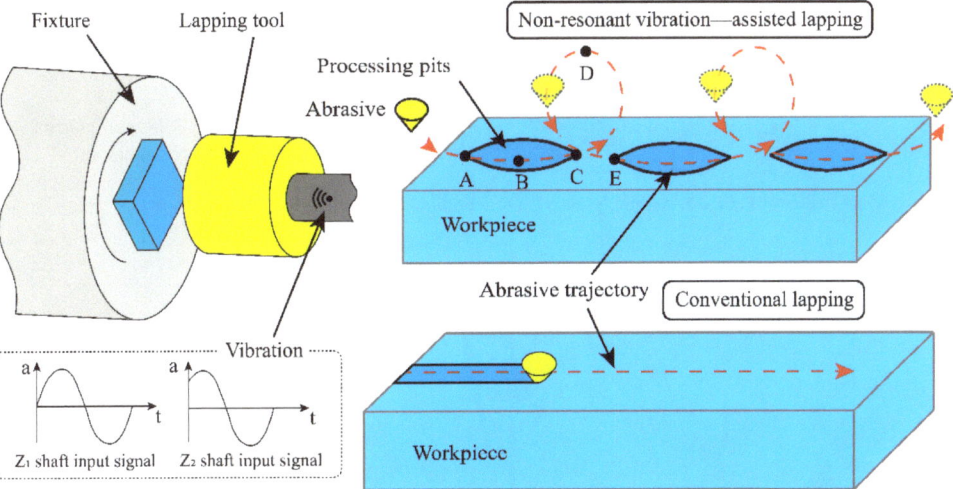

Figure 1. Schematic diagram of non-resonant vibration-assisted lapping.

3. Design and Analysis of VLD

The structural model of VLD is shown in Figure 2. The VLD comprises Z-hinge groups, a straight-axis right circular flexure hinge, a piezoelectric actuator (PZT), a tightening bolt, and a lapping tool. The device is symmetric with respect to the XOZ and the XOY plane, and the two PZTs are placed in parallel. Through the transmission of the Z-shape flexible hinge group and straight-axis right circular flexure hinge, the output displacement will be coupled at the transfer beam. Finally, the movement trajectory will be output at the end of the lapping tool. A fixture that holds the displacement sensor is fixed on the displacement output mechanism, and the VLD is set on the multicomponent dynamometer. Such a structure not only ensures the stability of the device but also ensures that the VLD has no output displacement in the longitudinal direction.

3.1. Vibration Trajectory Analysis

By feeding electrical signals with phase differences to the two piezoelectric actuator sets, the piezoelectric actuator's output displacement can synthesize an elliptical motion. In the kinematic analysis, the lapping tool and transfer beam are assumed to be rigid T-bars, which do not deform during processing. As depicted in Figure 3, define the coordinate system $O - xz$ for reference. Within this coordinate system, $P(x_0, z_0)$ signifies the equivalent coordinate of the lapping head in the absence of input displacement. Furthermore, $P(x_{CT}, z_{CT})$ represents the coordinate associated with the lapping tool when $PEA - 1$ and

$PEA-2$ receive Z_1 and Z_2 displacements, respectively. Additionally, L_x denotes the half-length of the transfer beam. x_{CT}, z_{CT} and γ is represented by (1).

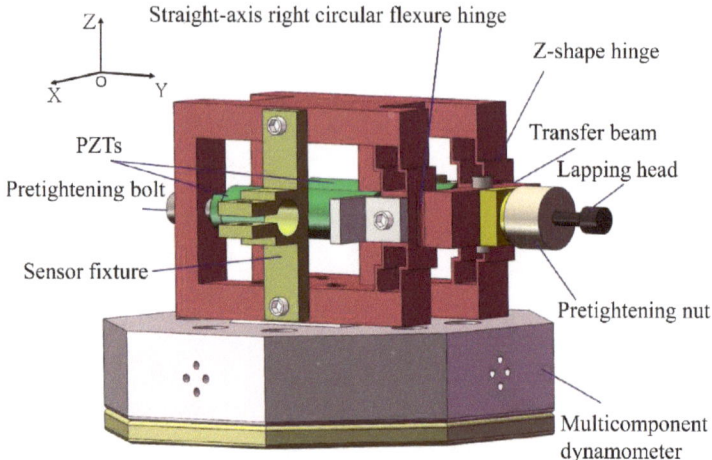

Figure 2. Assembly diagram of vibration-assisted lapping device.

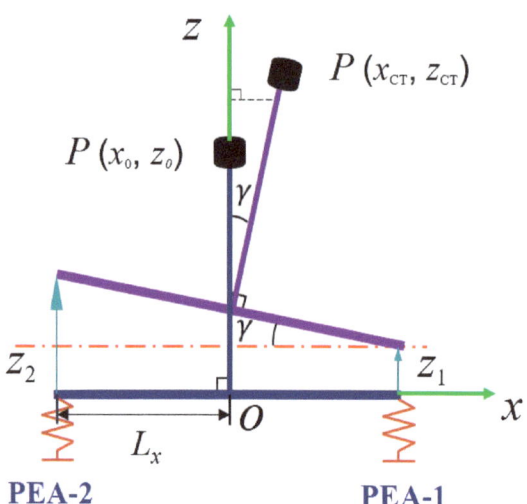

Figure 3. The schematic diagram of device movement.

$$\begin{cases} x_{CT} = z_0 sin\gamma \\ \gamma = sin^{-1}(\frac{z_2-z_1}{2L_x}) \\ z_{CT} = L_x sin\gamma + z_1 + z_0 \cos\gamma \end{cases} \quad (1)$$

Substituting the second Equation in the first and third Equations:

$$\begin{cases} x_{CT} = z_0 \frac{z_2-z_1}{2L_x} \\ z_{CT} = \frac{z_1+z_2}{2} + z_0 \cos\left(sin^{-1}(\frac{z_2-z_1}{2L_x})\right) \end{cases} \quad (2)$$

The movement of the lapping tool is driven by two PZTs, which are excited by two sinusoidal signals with phase differences. The displacement of the piezoelectric actuators after being excited by the signal can be expressed as:

$$\begin{cases} z_1 = A\sin(2\pi ft) \\ z_2 = B\sin(2\pi ft + \psi) \end{cases} \quad (3)$$

where f is the frequency, ψ is the phase difference, t is the time, and A and B are the amplitudes of the two axes, respectively.

In order to easily express the trajectory of the lapping head, let $A = B = U$, $\Psi = 90°$. Since the value of the input displacement is much smaller than the size and length of the transfer beam, substitute Equation (3) into (2),

$$\begin{cases} x_{CT} = z_0 \frac{U\sin(2\pi ft+\psi) - U\sin(2\pi ft)}{2L_x} \\ z_{CT} = \frac{U\sin(2\pi ft+\psi) + U\sin(2\pi ft)}{2} + z_0 \cos\left(\sin^{-1}\left(\frac{U\sin(2\pi ft+\psi) - U\sin(2\pi ft)}{2L_x}\right)\right) \end{cases} \quad (4)$$

According to the equivalent infinitesimal rule, Equation (4) can be expressed:

$$\begin{cases} \frac{x_{CT}}{z_0 U/2L_x} = \sin(2\pi ft) - \cos(2\pi ft) \\ \frac{z_{CT}-z_0}{U/2} = \sin(2\pi ft) + \cos(2\pi ft) \end{cases} \quad (5)$$

The lapping head makes reciprocating motion in space, and its movement trajectory is elliptical. After calculation, the movement trajectory of the lapping head can be obtained as follows:

$$\frac{x_{CT}^2}{(z_0 U/\sqrt{2}L_x)^2} + \frac{(z_{CT}-z_0)^2}{(U/\sqrt{2})^2} = 1 \quad (6)$$

According to the equation, it can be seen that the movement trajectory of the lapping head is an ellipse centered on $(0, z_0)$.

3.2. The Simulation Analysis of VLD

VLD performance was analyzed using ABAQUS 2021 simulation software, and in order to avoid the damage of VLD during processing, the mode analysis and stress analysis were carried out. The material selected for the VLD was 7075Al, and the material parameters are shown in Table 1.

Table 1. The mechanical properties of the 7075Al.

Elastic Modulus E (GPa)	Density ρ (kg·m^{-3})	Poisson's Ratio μ	Yield Strength σ (MPa)
71.7	2810	0.33	503

During the simulation, the bottom surface of the device was completely fixed. The first two modes and the maximum stress are shown in Figure 4. The first two natural frequencies are 1655 Hz and 1705 Hz, respectively, as shown in Figure 4a,b. The operating frequency needs to be set smaller than the natural frequency. When a displacement load of 25 μm was applied on both sides of the device, the maximum stresses on both sides of the device were 69.3 MPa and 69.5 MPa, respectively, lower than the allowable stress of the material, as shown in Figure 4c,d. Therefore, the device could meet the normal working requirements.

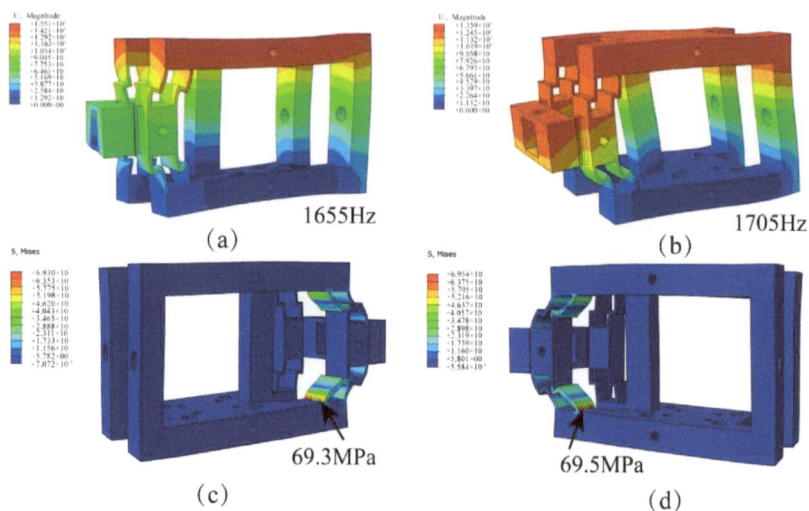

Figure 4. Performance test of vibration-assisted lapping device: (**a**) The first-order nature frequency (1655 Hz). (**b**) The second-order nature frequency (1705 Hz). (**c**) Stress distribution diagram on the right side of the device. (**d**) Stress diagram on the left side of the device.

4. Simulation of NVL for SiCp/Al Composites

The material structure of SiCp/Al has heterogeneous characteristics, and the difference in elastic modulus and hardness between aluminum matrix and SiC particles leads to surface defects during machining. Therefore, an NVL SiCp/Al composite material model is established in this section. The removal mechanism of SiCp/Al composite during diamond particle lapping, the form of surface defects, and the effect of parameters (vibration amplitude A, vibration frequency f, lapping speed v_c, lapping depth a_p) on the surface morphology were analyzed by finite element simulation method.

4.1. Finite Element Model Setting of NVL for SiCp/Al Composites

The formation process of surface defects in the NVL process and the influence of process parameters on surface defects were analyzed by simulation. A SiCp/Al composite material model with SiCp/Al material, diamond abrasive particles, and a two-phase interface was established, as shown in Figure 5. The two-phase interface was used to reflect the stick-slip effect between SiC particles and the Al matrix. A cohesive element with a thickness of 0 was embedded between SiC particles and the Al matrix to simulate the two-phase interface. The volume fraction and averaged particle size of SiC are 20% and 10 μm, respectively. The abrasive material was diamond, and the effect of lapping heat and abrasive wear on the processing was not considered. The diamond abrasive was simplified into a triangle with a peak angle of 120° and set as a rigid body. A velocity load was applied to the abrasive, and the bottom and left side of the workpiece are fixed constraints.

The plastic deformation phenomenon occurring during processing can be represented by the Johnson–Cook constitutive model [31]. The material properties of Aluminum and SiC are shown in Table 2.

4.2. Simulation Study on SiCp/Al Removal Characteristics of NVL

Due to the special structure of SiCp/Al, the shear position of SiC particles will directly affect the generation of surface defects. As shown in Figure 6, the shear position of the SiC particle can be divided into three cases: the abrasive is located above the SiC particle (Position A), the abrasive is located in the middle of the SiC particle (Position C), and the abrasive is located below the SiC particles (Position B).

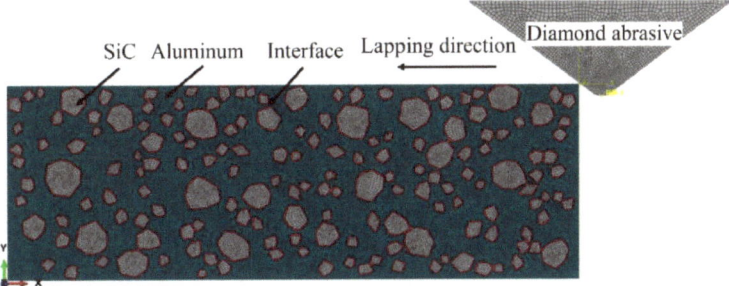

Figure 5. The assembly drawing of abrasive and workpiece.

Table 2. The physical and mechanical properties of 6005 aluminum and SiC particles.

Materials	Sic Particles	6005 Al
Young's modulus (GPa)	183	70
Poisson's ratio	0.2	0.3
Density (kg/m^3)	3163	2700
Thermal conductivity (W/mk)	81	190
Coefficient of thermal expansion (K^{-1})	4.9×10^{-6}	2.3×10^{-5}
specific heat capacity (J/gk)	0.427	0.91

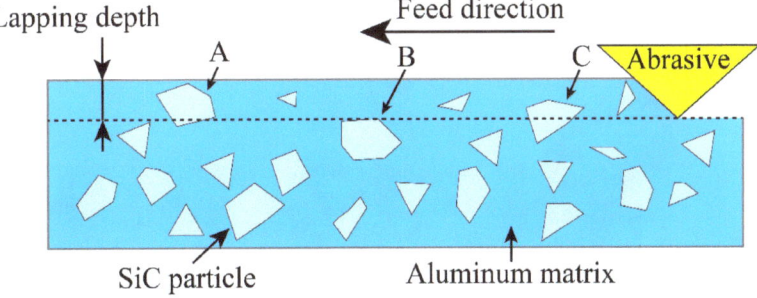

Figure 6. Three positions of abrasive particles and SiC particles.

As shown in Figure 7(a1–a4), when the abrasive was located above the SiC particle (Position A), slight stress concentration occurred in the SiC particle, but no particle breakage occurred. As the abrasive continued to move, the aluminum matrix became stress concentration and plastic deformation due to the impact and extrusion of the abrasive. The increasing stress leads to the formation of cracks and eventually extends to the surface of the workpiece. Although the stress concentration of SiC particles occurred, particle shedding and interfacial debonding did not occur, and the chip is mainly composed of an aluminum matrix.

Figure 8(a1–b4) shows the formation of particle breakage and particle shedding when SiC particles are located at position C. Figure 8(a1–a4) shows the phenomenon of particle shedding under shear action. After abrasive contact with SiC particles, stress concentration occurs in the central region of SiC particles. The large stress is released through aluminum matrix tearing and interfacial debonding. Under the action of extrusion, the interface completely fails, and SiC particles fall off. Figure 8(b1–b4) shows the particle tumbling process of SiC particles. Under the action of cohesive force and extrusion pressure, the interface damage occurs due to stress concentration, and the interface damage expands continuously under the action of lapping force. However, the SiC particles do not entirely fall off but are deflected at an angle and pressed into the aluminum matrix, as shown in Figure 8(b4). It is worth noting that in this state, the SiC and aluminum matrix bonding area is small, the state is unstable, and it is prone to subsequent processing and fall-off.

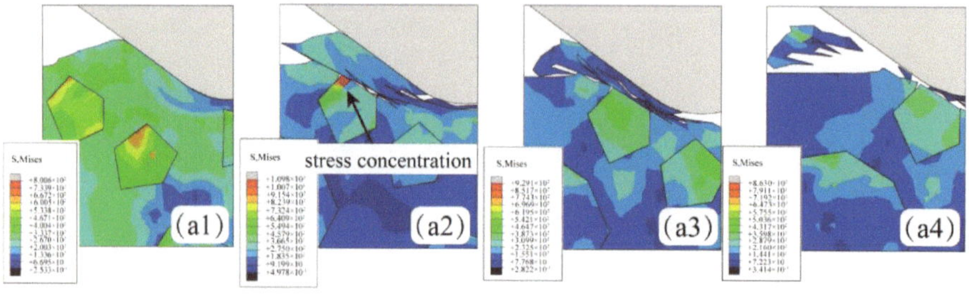

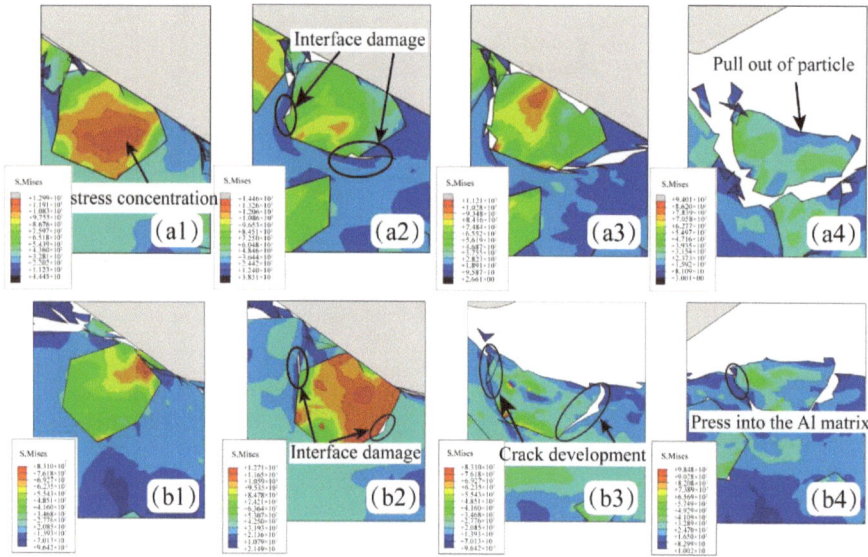

Figure 7. The abrasive is located above the SiC particle (**a1**–**a4**).

Figure 8. The abrasive is located in the middle of the SiC particle. (**a1**–**a4**) The SiC particles shed from the aluminum matrix. (**b1**–**b4**) The SiC particles were pressed into the aluminum matrix.

Figure 9(a1–a4) shows the formation process of surface defects when SiC particles are located at position A. The lapping depth is greater than the SiC particle size in this case. The cracks preferentially form in the aluminum matrix with a lower yield limit, expanding and connecting under continuous stress. Finally, the lapping chips composed of the aluminum matrix and SiC particles are formed. Figure 9(b1–b6) shows the process of SiC particles breaking under the effect of lapping force, which is more likely to occur when two SiC particles are close together. As shown in Figure 9(b1), the aluminum matrix deforms due to the low hardness and the pressing of SiC particles. The proximity of adjacent SiC particles to each other causes the surrounding aluminum matrix to be torn. The particles are in direct contact and produce a sizeable, concentrated stress at the contact position, as shown in Figure 9(b2). When the internal stress exceeds the yield limit of SiC particles, particle A produces a crack, as shown in Figure 9(b3,b4). As the extrusion degree intensifies, the crack in particle A propagates through particle A, and the particle is broken, as shown in Figure 9(b5,b6).

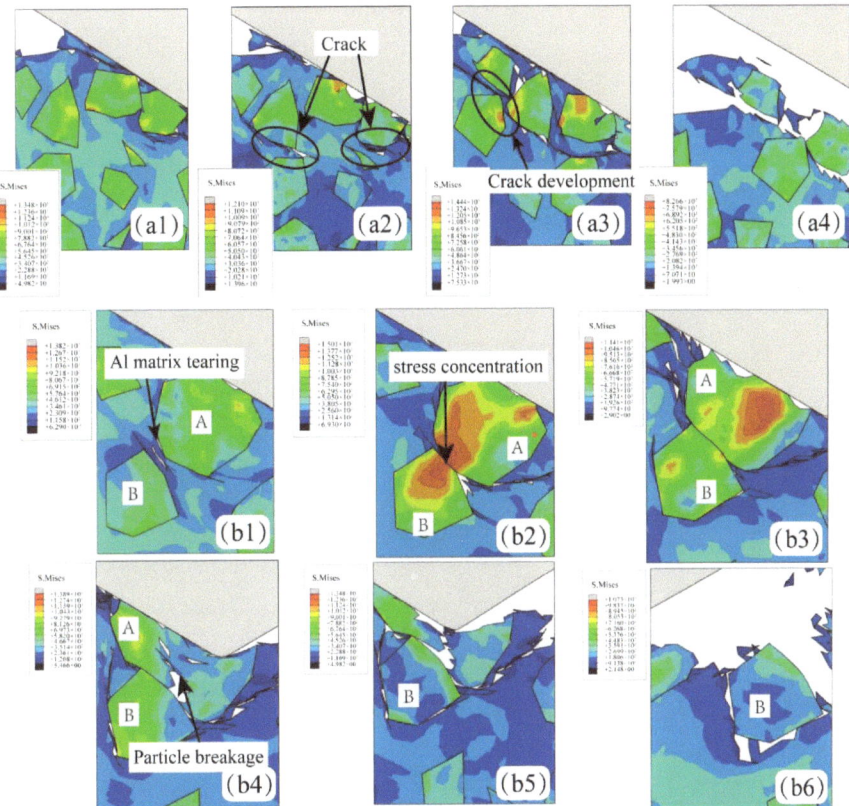

Figure 9. The abrasive is located below the SiC particle. (**a1–a4**) The SiC particles were removed together with the aluminum matrix.(**b1–b6**) The SiC particles were crushed. (A and B represent SiC particles).

4.3. Influence of Machining Parameters on SiCp/Al Lapping Surface Quality

Figure 10 shows the simulation comparison of the surface morphology of NVL SiCp/Al at different lapping depths a_p. When a_p is 4 μm, the lapping depth is small, and the abrasive removes lesser material. The lapping force is small, and the removal state remains stable, leading to better surface uniformity and surface quality after processing, as depicted in Figure 10a. As the a_p increases to 6 μm or 8 μm, the removal depth of the abrasive increases, and the phenomena of interfacial debonding and particle breakage increase. Moreover, some particles are more likely to fall off and experience interfacial debonding because of their weak bonding degree with the aluminum matrix. The SiC particles form pits on the surface after falling off. This phenomenon is particularly noticeable when the lapping depth increases to 8 μm, as shown in Figure 10b,c. When the a_p increases to 10 μm, particle fracture and shedding increase significantly, as shown in Figure 10d. In summary, excessive lapping depth causes SiC particle fracture, and the SiC particles fall out more obviously.

Figure 11 shows the simulation comparison of NVL SiCp/Al surface morphology when the lapping speed v_c changes. Because the two-dimensional single-abrasive lapping simulation cannot show all the velocity characteristics of abrasive, the lapping speed v_c in the simulation mainly represents the synthesis velocity of abrasive. When v_c is 12 mm/min, it can be observed that there are peaks, as shown in Figure 11a. When

v_c increases to 19 mm/min, the peak-shaped residue on the machined surface disappears, and the machined surface becomes flat, but there are still burrs on the surface, as shown in Figure 11b. When v_c is 25 mm/min, there are no obvious defects on the surface of the workpiece, and the phenomena of particle shedding and interfacial debonding are reduced, as shown in Figure 11c. When v_c increases to 31 mm/min, the phenomena of particle breakage, interfacial debonding, and particle shedding occur on the workpiece surface. In conclusion, With the increased lapping speed, the surface defect becomes better first and then worse. The surface defects are the least at v_c = 25 mm/min, and the particle fracture and the interfacial debonding are lesser.

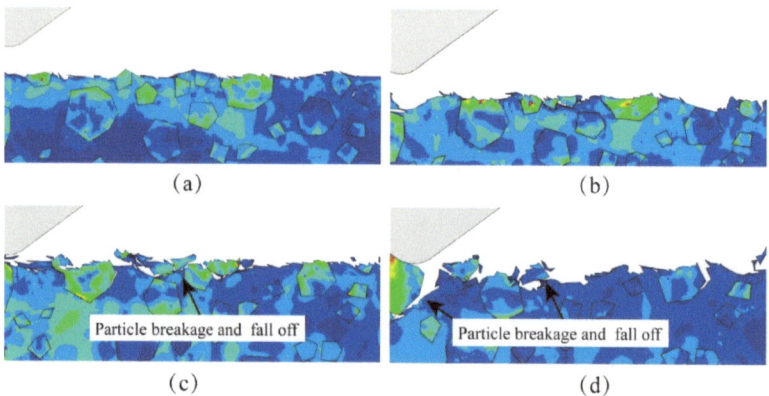

Figure 10. The simulation results of different lapping depths. (**a**) The lapping depth a_p is 4 µm. (**b**) The lapping depth a_p is 6 µm. (**c**) The lapping depth a_p is 8 µm. (**d**) The lapping depth a_p is 10 µm.

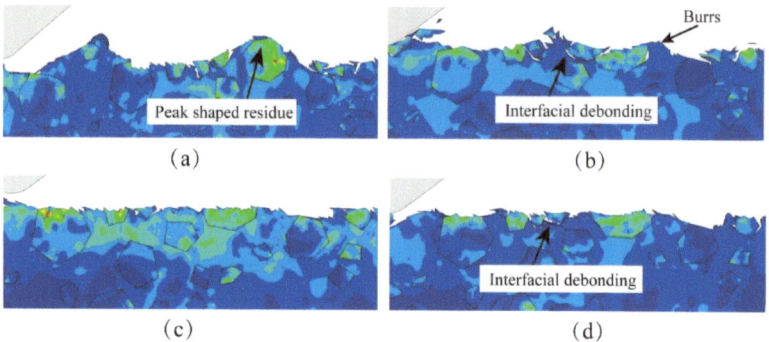

Figure 11. The simulation results of different lapping speeds. (**a**) The lapping depth v_c is 12 mm/min. (**b**) The lapping depth v_c is 19 mm/min. (**c**) The lapping depth v_c is 25 mm/min. (**d**) The lapping depth v_c is 31 mm/min.

Figure 12 shows the simulation comparison of NVL SiCp/Al surface morphology when the amplitude A changes. As seen in Figure 12a, when A is 1 µm, the surface after lapping is uneven, with more burrs, some fractured SiC particles, and obvious interface disbanding between particles and matrix. When amplitude A = 3 µm, large pits are formed after significant damage to SiC particles in a large area, as shown in Figure 12b. At an amplitude of A = 5 µm, although there are still burrs on the surface, the pits are significantly smaller compared to amplitudes A = 1 µm or A = 3 µm, as shown in Figure 12c. With the amplitude A = 7 µm, interfacial debonding and pits between particles and matrix reduces, as shown in Figure 12d.

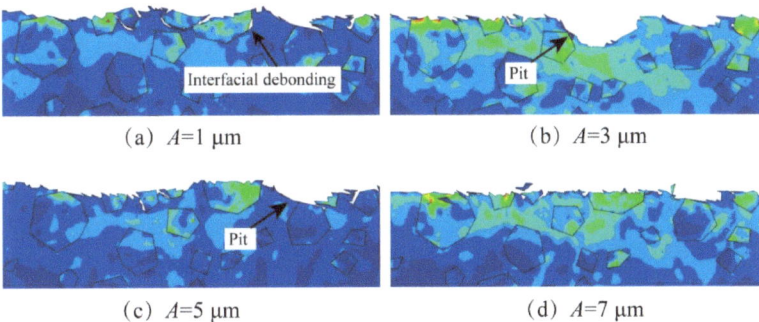

Figure 12. The simulation results of amplitude. (**a**) The lapping depth A is 1 μm. (**b**) The lapping depth A is 3 μm. (**c**) The lapping depth A is 5 μm. (**d**) The lapping depth A is 7 μm.

In conclusion, surface defects are more apparent when the amplitude is less than 7 μm. This is because the separation characteristics of lapping processing are not prominent, and the vibration processing characteristics could not be fully utilized. Additionally, the chips generated during the processing are difficult to discharge, resulting in the accumulation of chips in the abrasive front end, which affects the processing effect. The larger amplitude highlights the advantages of vibration processing. With the increase in amplitude, the motion trajectory of the abrasive changes, and the periodic contact process between the abrasive and the workpiece becomes significant, which is conducive to the discharge of lapping chips, and the surface quality after processing is also significantly improved.

Figure 13 shows the simulation comparison of lapping under different vibration frequencies, where the frequency f is 0 Hz, 150 Hz, 300 Hz, and 450 Hz, respectively. When f is set to 0 Hz (conventional lapping method) or 150 Hz, there are apparent large pits on the surface of the workpiece, which are formed after the SiC particles fall off, as shown in Figure 13a,d. When the frequency f is increased to 300 Hz, some small pits and burrs still exist on the surface. However, particle fracture and detachment occurrences are significantly reduced, improving surface topography, as shown in Figure 13c. Increasing the frequency f to 450 Hz eliminates noticeable pits on the workpiece surface, reduces shedding between the particle and matrix interface, and results in a flat and uniform surface with improved surface quality, as shown in Figure 13d. In conclusion, applying vibration in conjunction with CL processing proves beneficial in enhancing surface quality. The effect of low-frequency vibration on surface defects is weak. Increasing the frequency contributes to further improvements in surface quality and reduced occurrences of particle fracture and debonding.

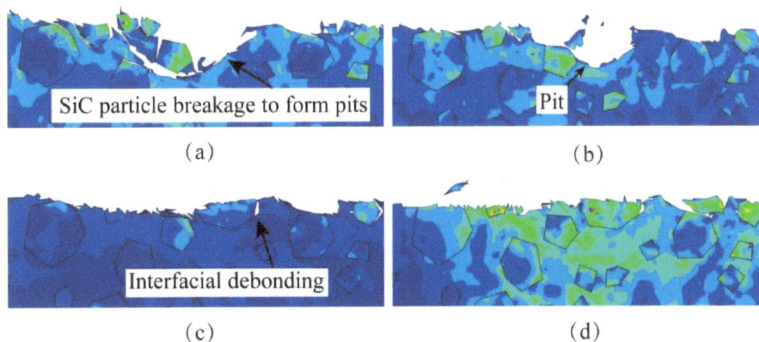

Figure 13. The simulation results of frequency. (**a**) The lapping depth f is 0 Hz. (**b**) The lapping depth f is 150 Hz. (**c**) The lapping depth f is 300 Hz. (**d**) The lapping depth f is 450 Hz.

5. Test of Vibration-Assisted Lapping Device and Lapping Experiment

In this section, firstly, the VLD test system is built, and its resolution, stroke, natural frequency, and motion trajectory are tested. Then, based on the designed VLD, an NVL SiCp/Al composite experimental system was built, and explored the effects of vibration frequency f, amplitude A, spindle speed n, feed speed v_f, and lapping depth a_p on the lapping force and workpiece surface roughness by SiCp/Al composite NVL processing experiment.

5.1. Performance Test of VLD

The VLD test system was built, and the VLD test experiment was carried out, which mainly included the natural frequency test, resolution test, output displacement test, and motion track acquisition. Figure 14 shows the performance test system of the VLD. The experimental instruments used in the test mainly include a piezoelectric actuator (PZT, COREMORROW Inc. Harbin, China), controller (Power PMAC), power amplifier (E-500, PI Inc. Shanghai, China), displacement sensor (2805, MicroSense Inc., Lowell, MA, USA), charge amplifier (DE 5300-013, MicroSense Inc.). The power amplifier amplifies the signal sent by the controller and acts as the drive signal of the piezoelectric actuator. The piezoelectric actuator outputs the displacement after receiving the signal; the displacement sensor collects the output displacement signal, transmits it to the controller, and outputs the displacement at the end.

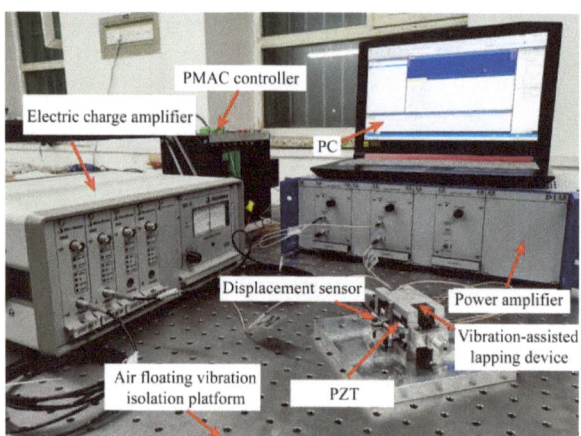

Figure 14. The test system of VLD.

The natural frequency, movement range, and resolution are the parameters that must be paid attention to. In order to avoid damage to the device, the VLD needs to operate at a frequency less than the first-order natural frequency. As shown in Figure 15a, the natural frequency of the VLD is 1460 Hz, which is slightly lower than the finite element simulation result of 1655 Hz. Due to machining errors and different contact conditions, there are usually some differences between the performance of the actual machining device and the simulation results under the ideal situation. The parameter range of VLD under working state should be determined by referring to the results of actual performance tests. Motion resolution significantly affects the control effect of the VLD. As shown in Figure 15b,c, a step signal is applied to two axes of the VLD, and the output displacement is measured by a charge sensor. As shown in Figure 15d,e, test results show that the resolution of both axes of the device can reach 40 nm. A triangular wave is used to test the output stroke of the VLD. Test the output displacement of the vibration-assisted lapping device with triangular waves and continuously increase the amplitude of triangular waves. Finally, the limit travel of the z1 axis and z2 axis of the vibration-assisted lapping device is 40 nm and 45 nm, respectively, which can realize fast-tracking of the input signal and meet the amplitude requirements in the machining process.

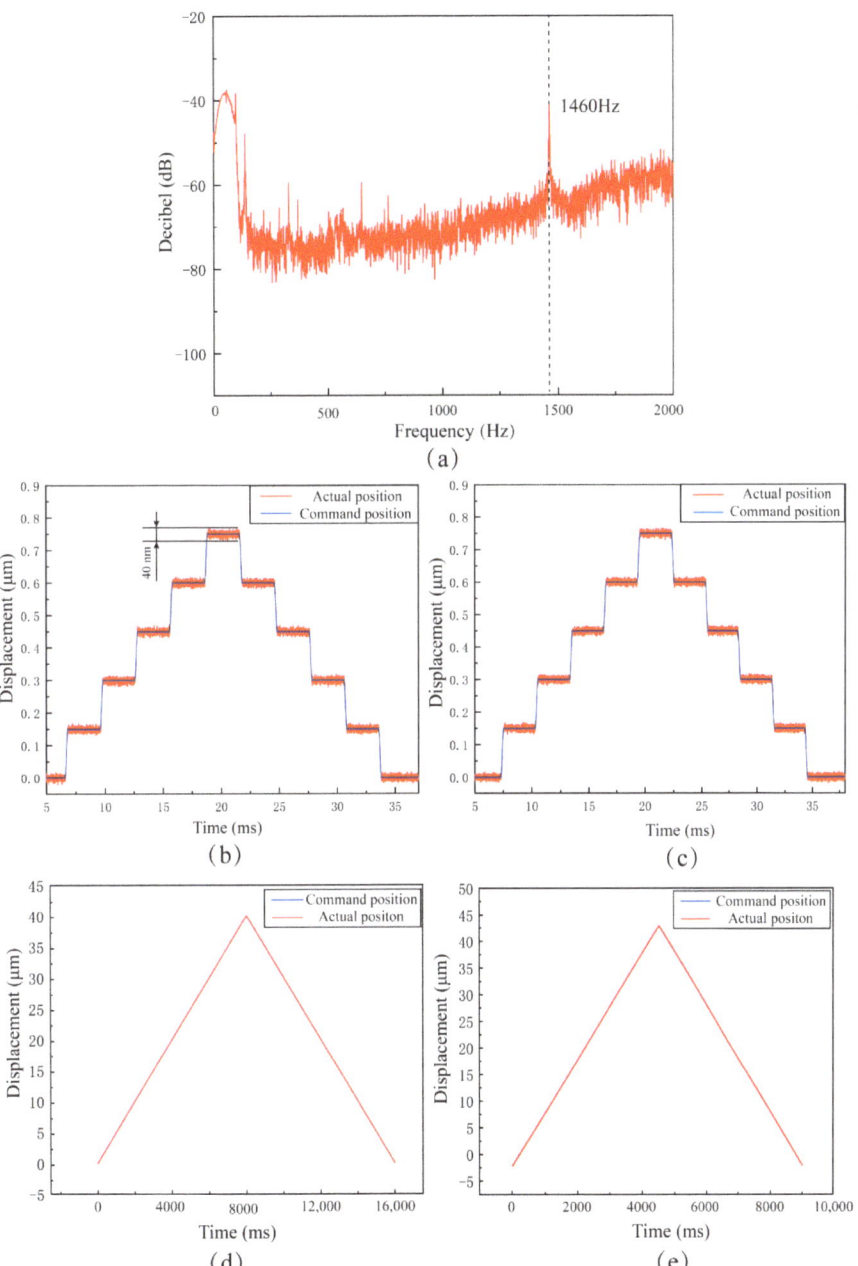

Figure 15. The test results of VLD: (**a**) natural frequency test results. (**b**) Z_1 axis resolution test results. (**c**) Z_2 axis resolution test results. (**d**) Z_1 axis displacement test results. (**e**) Z_2 axis displacement test results.

In the NVL process, the output of the device vibration track has a crucial impact on the lapping process. In order to determine the shape and stability of the device output track, the device trajectory output test is carried out. Sinusoidal signals with 45° and 90° phase differences are applied to the piezoelectric brake. Figure 16a,b shows the output trajectories of the device when the input signals are 45° and 90° phase differences, respectively. The output trajectories of the device under the phase differences of 45° and 90° are elliptical, and the trajectory outline is clear. It can meet the demand of motion trajectory output during processing.

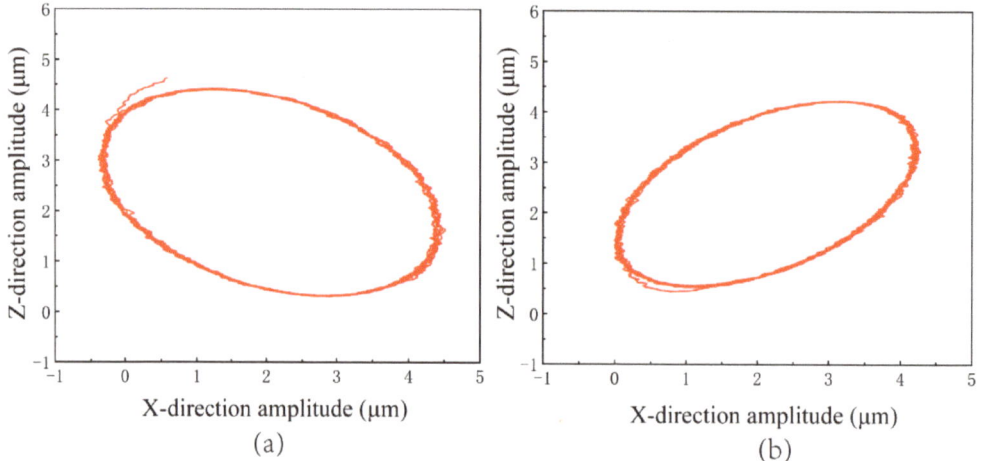

Figure 16. The movement trajectory of VLD under different phase differences: (**a**) 45° phase difference. (**b**) 90° phase difference.

5.2. Experiment on NVL of SiCp/Al Composites

Figure 17 shows the NVL experimental system of SiCp/Al. During the experiment, the dynamometer (9109, Kistler Inc. Winterthur, Switzerland) is installed on the machine guide rail platform, the VLD is installed on the dynamometer, the grinding head is installed in the front end of the VLD, and the machine guide rail does X-direction feed movement. A signal generator and power amplifier control the vibration parameters. SiCp/Al is $10 \times 10 \times 10$ mm, and the average size and volume fraction of SiC are 10 μm and 20%, respectively. Each workpiece is pretreated before the experiment. Each experimental parameter is repeated three times, and the experimental parameters are shown in Tables 3 and 4. The surface roughness of five random points on the polished surface was measured using a white-light interferometer (NewView 8000, Zygo Inc. Middlefield, CT, America).

Table 3. Experimental variable parameter setting of NVL.

	Experimental Variable				
Experiment Number	Spindle Speed n (r/min)	Feed Speed v_f (mm/min)	Lapping Depth a_p (μm)	Vibration Frequency f (Hz)	Vibration Amplitude A (μm)
Experiment 1	800	5	1	0	0
Experiment 2	800	5	1	850	8
Experiment 3	400, 600, 800, 1000, 1200	35	3	750	6
Experiment 4	800	5, 10, 20, 35, 55	3	750	6
Experiment 5	800	35	1, 2, 3, 4, 5	750	6
Experiment 6	800	35	3	450, 550, 650, 750, 850	6
Experiment 7	800	35	3	750	2, 4, 6, 8, 10

Table 4. Experimental controlled variable parameter setting of NVL.

Controlled Variable	
Diamond abrasive size (μm)	5
Lapping time t (s)	30
Initial surface roughness S_a (μm)	1.869
Diameter of lapping head (mm)	6

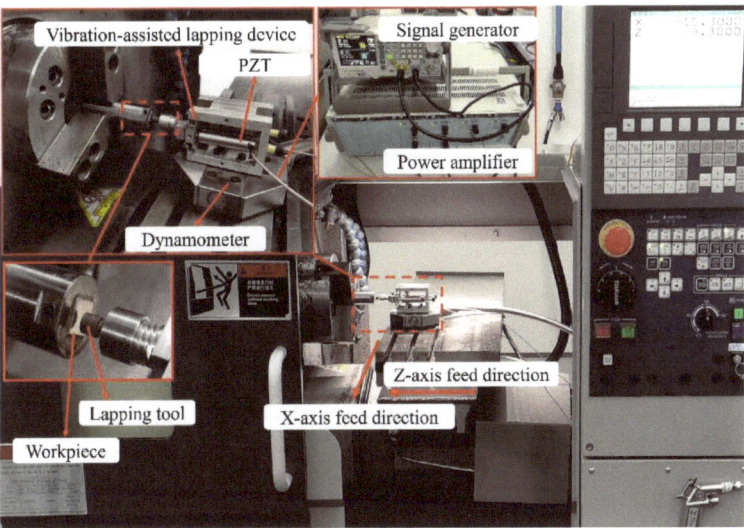

Figure 17. SiCp/Al composite material NVL experimental system.

5.3. Effect of Process Parameters

5.3.1. Comparison of NVL and CL

In order to study the difference between surface quality and lapping force after NVL and CL, the NVL experiment and CL experiment are conducted, respectively, and experimental parameters are shown in Experiment 1 and Experiment 2 in Table 3. Figure 18 shows the lapping force and the surface topography of the workpiece. As shown in Figure 18a,b, the abrasive of the CL method has a single motion path, and the removal process is continuous. There is a significant plow effect on surfaces using CL methods, and lapping debris in the lapping area is difficult to discharge, resulting in increased lapping force. The surface quality decreases from 1.869 μm to 0.321 μm. The NVL method has less lapping force, and the surface roughness is reduced from 1.869 μm to 0.045 μm, which is better than the CL method, as shown in Figure 18c,d.

Figure 19a shows the initial surface topography. The surface defects are mainly pits and scratches. Due to the difference in brittleness and plasticity between SiC particles and aluminum matrix, the SiC particles are more prone to brittle crushing and forming pits during processing. The microscopic surface of SiCp/Al processed by CL method is shown in Figure 19b. There are apparent scratches on the surface, but the pits disappear, which is because the aluminum matrix is coated on the surface after melting. The lapping chips are difficult to discharge, and lapping force and lapping heat increase. All these effects hurt the surface quality of the workpiece [32–34]. The microscopic morphology of SiCp/Al processed by the NVL method is shown in Figure 19c. The scratches on SiCp/Al surface are significantly reduced. The above phenomena indicate that the application of vibration reduces surface damage.

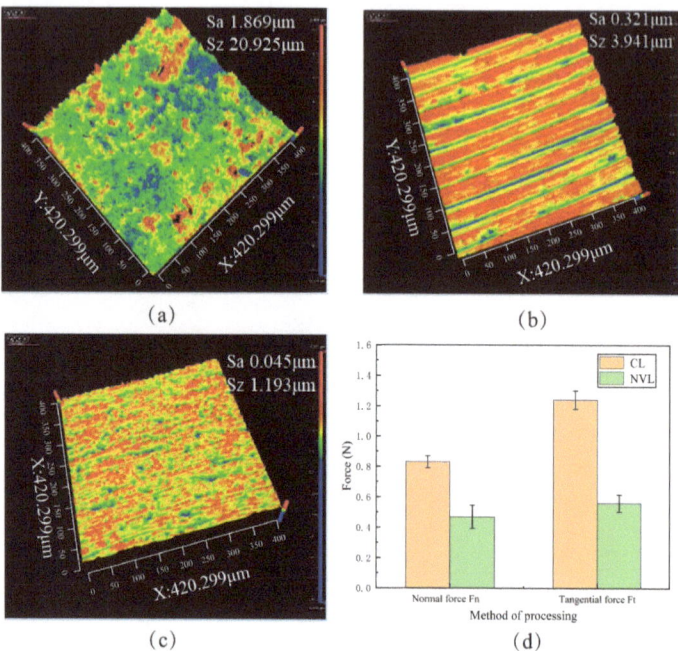

Figure 18. The surface roughness of the workpiece: (**a**) Original surface roughness. (**b**) Conventional lapping surface roughness. (**c**) Non-resonant vibration-assisted lapping surface roughness. (**d**) Lapping force of NVL and CL.

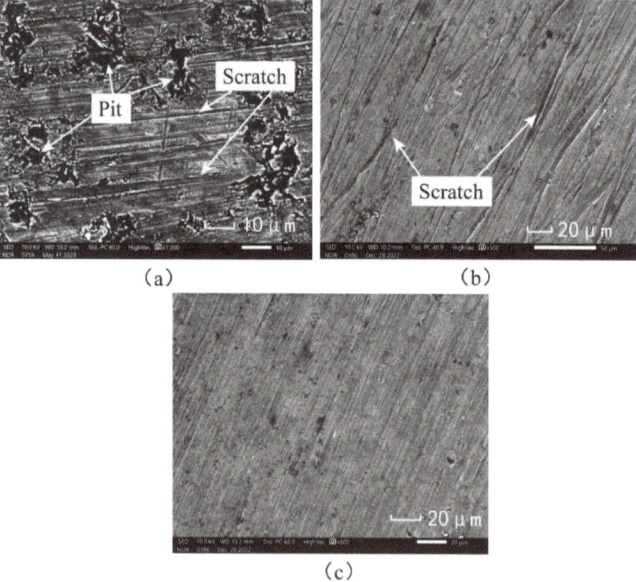

Figure 19. Surface Topography of the workpiece. (**a**) Original surface topography of the workpiece. (**b**) Conventional lapping surface morphology. (**c**) Non-resonant vibration-assisted lapping surface morphology.

5.3.2. Effect of Spindle Speed n

Figure 20a shows the influence of surface roughness with n. Figure 20b shows the law of lapping force variation in the spindle speed n range of 400 r/min to 1200 r/min. The experimental parameters are shown in Experiment 3 of Table 3. The surface roughness S_a shows a trend of first decreasing and then increasing, which is similar to the trend of the simulation analysis results, and the surface roughness $S_a = 0.085$ μm is the lowest when $n = 800$ r/min. The normal and tangential lapping force of NVL shows a decreasing trend. This phenomenon is because the change in spindle speed affects the tool wear, lapping force, and lapping trajectory. Because the spindle speed n is inversely proportional to the lapping depth and proportional to the lapping path length per unit time, the lapping force and friction resistance decrease with the increase of the spindle speed, and the surface quality is improved. When n exceeds 800 r/min, the wear of lapping tools is intensified due to excessive speed, and the temperature in the lapping area increases, resulting in the lapping chips becoming soft and sticking to the machined area. The secondary scratch of lapping chips increases the roughness of the machined surface.

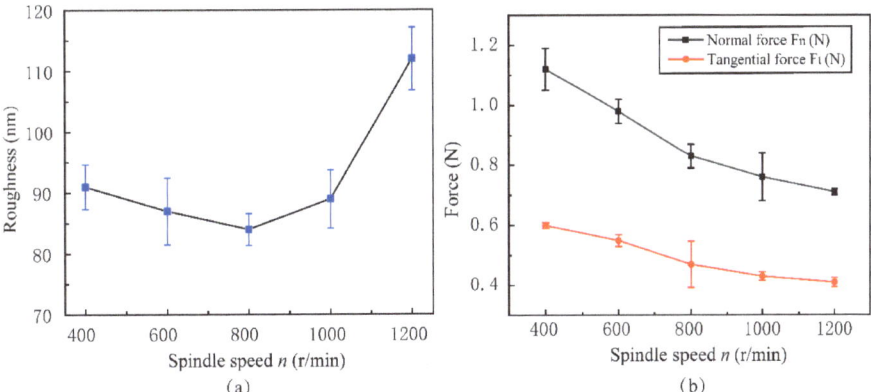

Figure 20. Effect of spindle speed on surface roughness and lapping force: (**a**) Roughness, (**b**) Force.

5.3.3. Effect of Feed Speed v_f

Figure 21 shows the influence of feed speed v_f on surface roughness and lapping force in the range of 5 mm/min to 55 mm/min. The experimental parameters are shown in Experiment 4 of Table 3. It can be seen from Figure 21a that the surface roughness is proportional to the feed speed, and the point corresponding to the best surface roughness is $v_f = 5$ mm/min because the high feed rate causes the workpiece surface to be unable to be fully lapping. At the same time, the material removal rate and lapping force of a single abrasive increases, resulting in more surface damage to SiCp/Al and an increase in SiC particle shedding and particle fracture on the machined surface, which affects the surface quality after machining. The difference between experimental and simulation results of feed speed v_f may be due to different parameter Settings, and the abrasive motion path of the NVL method is dependent on the spindle speed [24,35].

5.3.4. Effect of Lapping Depth a_p

Figure 22 shows the influence of surface roughness and lapping force in the range of lapping depth a_p from 1 μm to 5 μm. The experimental parameters are shown in Experiment 5 of Table 3. With the increase of a_p, surface roughness and lapping force increase, which is consistent with the simulation results. When the lapping depth $a_p = 1$ μm, the surface roughness reaches the minimum value of 60 nm, and the lapping force is also the minimum value. The reason for this phenomenon is that when the a_p is small, the SiC particle fracture and particle shedding phenomenon in the processing process is lesser, the lapping chips are also small, and the fine debris is more easily discharged from the processing area under the action of vibration. The contact area between the abrasive and workpiece will increase with the increased lapping depth, resulting in increased chip deformation force and friction, and SiC particles are prone to crushing and particle pulling out phenomenon. As the lapping depth is close to the amplitude of the lapping tool, the advantages of NVL are weakened, and the discharge of grind chips is difficult and accumulates between the workpiece and the lapping head, causing lapping force and heat increase, resulting in a reduction in surface quality.

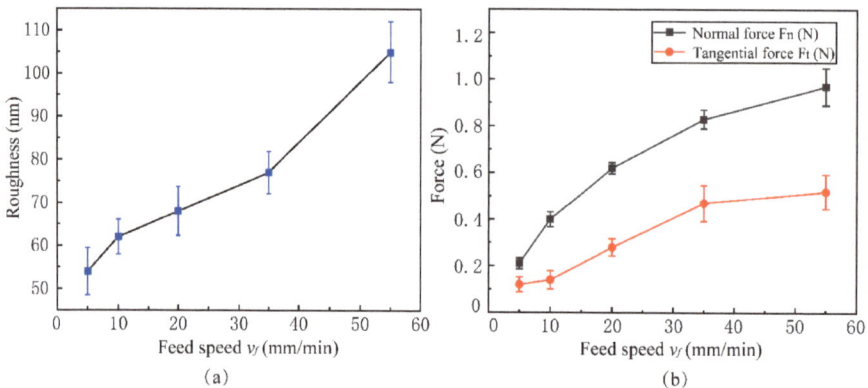

Figure 21. Effect of feed speed on surface roughness and lapping force: (**a**) Roughness, (**b**) Force.

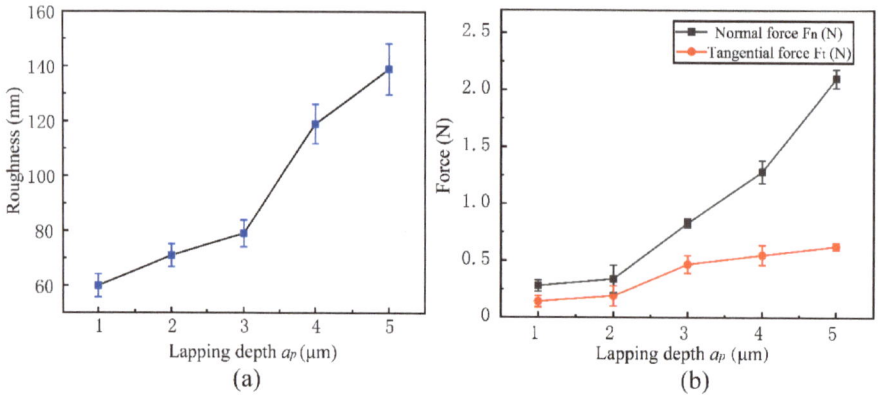

Figure 22. Effect of lapping depth on surface roughness and lapping force: (**a**) Roughness, (**b**) Force.

5.3.5. Effect of Vibration Frequency f

Figure 23 shows the variation of surface roughness and lapping force with vibrational frequency f between 450 Hz and 850 Hz. The experimental parameters are shown in Experiment 6 of Table 3. When the vibration frequency f = 850 Hz, the surface roughness reaches the minimum value of 81 nm, and the vibration frequency is inversely proportional to the lapping force. According to the simulation results of the frequency factor, the periodic separation process of abrasive and workpiece is not apparent at low-frequency vibration, and the suppression effect on surface defects is poor. High-frequency vibration enhances the effect of abrasive periodically impacting the workpiece, and the material removal volume within a single vibration cycle decreases. The number of chips discharged with vibration per unit time also increases, reducing the attachment and bonding of lapping chips, thereby delaying tool wear and reducing lapping force and surface roughness.

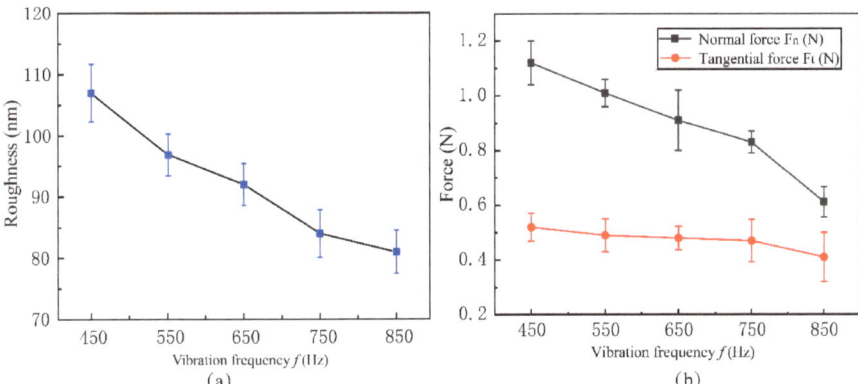

Figure 23. Effect of vibration frequency on surface roughness and lapping force: (**a**) Roughness, (**b**) Force.

5.3.6. Effect of Vibration Amplitude A

Figure 24 shows the influence of lapping force and surface roughness in the vibration amplitude A range from 2 μm to 10 μm. The experimental parameters were shown in experiment 7 of Table 3. When A = 8 μm, the lowest surface roughness value is 0.077 μm. Before the amplitude A reaches 8 μm, the surface quality is inversely proportional to the amplitude. This is because the vibration amplitude affects the contact separation process of the abrasive. The low amplitude is not conducive to the removal of lapping chips and the reduction of surface defects, which is consistent with the trend of simulation results. After the A exceeds 8 μm, the impact of the abrasive on the workpiece is too strong, increasing surface roughness. The lapping force decreases with the increase of A, as shown in Figure 24b, which is because the increase in amplitude makes the contact–separation effect more apparent, and the lapping chips can be discharged more smoothly, thus reducing the lapping force during the lapping process.

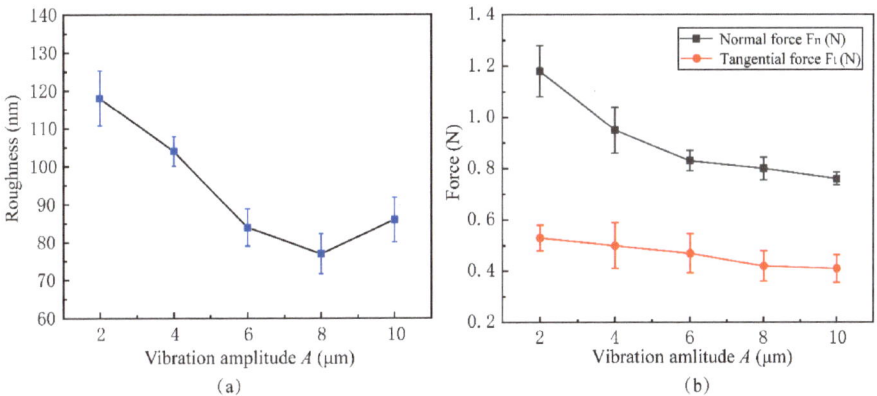

Figure 24. Effect of vibration amplitude on surface roughness and lapping force: (**a**) Roughness, (**b**) Force.

5.4. Analysis of Lapping Surface Micro-Morphology

From the simulation analysis in Section 4, there will be interfacial debonding, particle fracture, particle tumbling, and particle fracture of SiC particles in the lapping process, which will cause damage to the workpiece surface. The surface of the SiCp/Al workpiece after lapping is detected and analyzed by an electron microscope, and the results are compared with the phenomena in the simulation to verify the accuracy of the simulation.

Figure 25 shows the microscopic surface topography of SiC particles after lapping by electron microscope. After complete fracture, SiC particles are not separated from the matrix. They are still embedded in the matrix to form SiC particle clusters in Figure 25a because SiC particles generate huge internal stress concentrations under the action of the lapping tool. When the vibrating lapping head passes through SiC particles with severe stress concentration, SiC particles are completely fractured under impact and extrusion, and the fracture particles are squeezed again and then converge to form clusters. As shown in Figure 25b, SiC particles are crushed, and some particles remain in the aluminum matrix. As shown in Figure 25c, the SiC particles did not immediately break away from the surface after shedding with the matrix. However, they tumbled on the surface of the workpiece through the extrusion and pushing of the lapping head and are embedded in the aluminum matrix after leaving a period of plow marks on the surface of the workpiece. The remaining lapping chips will likely fall off in this state and eventually form pits. As shown in Figure 25d, under the action of lapping force, cracks occur in the SiC particles and further expand and finally penetrate the SiC particles.

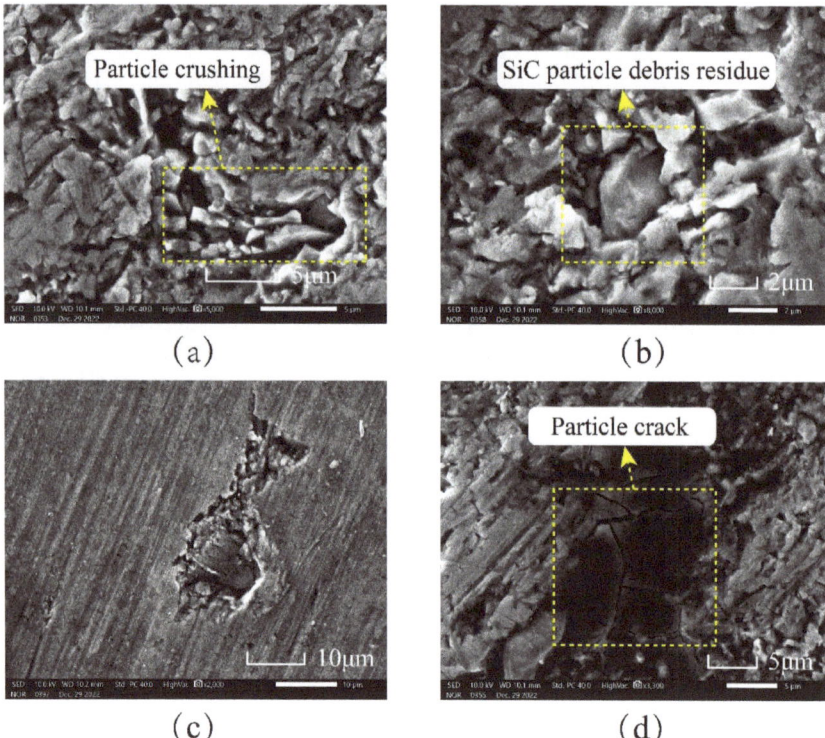

Figure 25. Morphology of SiC particles on workpiece surface after lapping: (**a**) SiC particle fracture; (**b**) the SiC particulate debris remained in the matrix; (**c**) SiC particle tumbling; (**d**) the SiC particles produce cracks.

6. Conclusions

A VLD is designed and tested. The NVL SiCp/Al process is studied using the VLD. The removal characteristics of the NVL SiCp/Al are analyzed. Single-factor experiments were carried out on the process parameters such as spindle speed n, feed speed

v_f, lapping depth a_p, vibration frequency f, and vibration amplitude A. The following conclusions can be drawn:

1. A VLD is designed. The motion trajectory of the VLD is analyzed, and its modal and stress distribution are simulated by the simulation method. The test results show that the first-order natural frequency is 1460 Hz, the resolution is 40 nm, and the two-axis stroke is 45 μm and 40 μm, respectively;
2. NVL SiCp/Al can reduce the lapping force. Spindle speed, vibration amplitude, and vibration frequency are inversely correlated with lapping force, and the lapping force reaches the minimum value at f = 900 Hz, A = 10 μm and n = 1200 r/min, respectively. The feed speed and lapping depth positively correlate to the lapping force;
3. NVL lapping SiCp/Al effectively reduces the surface roughness. The vibration amplitude and spindle speed decreased first, then increased, reaching the minimum values at A = 8 μm and n = 800 r/min, respectively. The lapping depth and feed speed were inversely correlated with the surface roughness. Reaches the minimum values of 0.054 μm and 0.059 μm at v_c = 5 mm/min and a_p = 1 μm, respectively. The vibration frequency is proportional to the surface roughness and reaches a minimum of 0.081 μm at f = 900 Hz. The changing trend of process parameters in machining experiments is similar to that in simulation.

Author Contributions: Conceptualization, X.F.; Methodology, H.Z.; Software, Y.G. (Yinghuan Gao); Investigation, J.W. and L.X.; Data curation, Y.X. and G.L.; Writing—original draft, H.Z.; Writing—review & editing, Y.G. (Yan Gu). All authors have read and agreed to the published version of the manuscript.

Funding: This work is supported by the Jilin Province Science and Technology Development Plan Project (Grant No. 20220201025GX).

Data Availability Statement: The data that support the findings of this study are available from the corresponding author upon reasonable request.

Conflicts of Interest: The authors declare no conflict of interest.

References

1. Chak, V.; Chattopadhyay, H.; Dora, T.L. A review on fabrication methods, reinforcements and mechanical properties of aluminum matrix composites. *J. Manuf. Process.* **2020**, *56*, 1059–1074. [CrossRef]
2. Wang, Z.Y.; He, Y.J.; Yu, T.B. Surface quality and milling force of SiCp/Al ceramic for ultrasonic vibration-assisted milling. *Ceram. Int.* **2022**, *48*, 33819–33834. [CrossRef]
3. Zheng, W.; Qu, D.; Qiao, G.H. Multi-phase modeling of SiC particle removal mechanism in ultrasonic vibration–assisted scratching of SiCp/Al composites. *Int. J. Adv. Manuf. Technol.* **2021**, *113*, 535–551. [CrossRef]
4. Liu, H.Z.; Wang, S.J.; Zong, W.J. Tool rake angle selection in micro-machining of 45 vol.% SiCp/2024Al based on its brittle-plastic properties. *J. Manuf. Process.* **2019**, *37*, 556–562. [CrossRef]
5. Gao, Q.; Guo, G.Y.; Wang, Q.Z. Study on micro-grinding mechanism and surface quality of high-volume fraction SiCp/Al composites. *J. Mech. Sci. Technol.* **2021**, *35*, 2885–2894. [CrossRef]
6. Yin, G.Q.; Wang, D.; Cheng, J. Experimental investigation on micro-grinding of SiCp/Al metal matrix composites. *Int. J. Adv. Manuf. Technol.* **2019**, *102*, 3503–3517. [CrossRef]
7. Brehl, D.E.; Dow, T.A. Review of vibration-assisted machining. *Precis. Eng.* **2008**, *32*, 153–172. [CrossRef]
8. Ding, W.F.; Huang, Q.; Zhao, B.; Cao, Y.; Tang, M.L.; Deng, M.M.; Liu, G.L.; Zhao, Z.C.; Chen, Q.L.; Deng, M. Wear characteristics of white corundum abrasive wheel in ultrasonic vibration-assisted gapping of AISI 9310 steel. *Ceram. Int.* **2023**, *49*, 12832–12839. [CrossRef]
9. Bie, W.B.; Zhao, B.; Zhao, C.Y.; Yin, L.; Guo, X.C. System design and experimental research on the tangential ultrasonic vibration-assisted grinding gear. *Int. J. Adv. Manuf. Technol.* **2021**, *116*, 597–610. [CrossRef]
10. Yin, J.T.; Zhao, J.; Song, F.Q.; Xu, X.Q.; Lan, Y.S. Processing Optimization of Shear Thickening Fluid Assisted Micro-Ultrasonic Machining Method for Hemispherical Mold Based on Integrated CatBoost-GA Model. *Materials* **2023**, *16*, 2683. [CrossRef]
11. Cao, Y.; Ding, W.F.; Zhao, B.; Wen, X.B.; Li, S.P.; Wang, Z.J. Effect of intermittent cutting behavior on the ultrasonic vibration-assisted grinding performance of Inconel718 nickel-based superalloy. *Precis. Eng.* **2022**, *78*, 248–260. [CrossRef]
12. Chen, Y.R.; Su, H.H.; Qian, N.; He, J.Y.; Gu, J.Q.; Xu, J.H.; Ding, K. Ultrasonic vibration-assisted grinding of silicon carbide ceramics based on actual amplitude measurement: Grinding force and surface quality. *Ceram. Int.* **2021**, *47*, 15433–15441. [CrossRef]

13. Jiang, J.L.; Sun, S.F.; Wang, D.X.; Yang, Y.; Liu, X.F. Surface texture formation mechanism based on the ultrasonic vibration-assisted gapping process. *Int. J. Mach. Tools Manuf.* **2020**, *156*, 103595. [CrossRef]
14. Gu, P.; Zhu, C.M.; Sun, Y.C.; Wang, Z.; Tao, Z.; Shi, Z.Q. Surface roughness prediction of SiCp/Al composites in ultrasonic vibration-assisted gapping. *J. Manuf. Process.* **2023**, *101*, 687–700. [CrossRef]
15. Zheng, W.; Wang, Y.J.; Zhou, M.; Wang, Q.; Ling, L. Material deformation and removal mechanism of SiCp/Al composites in ultrasonic vibration assisted scratch test. *Ceram. Int.* **2018**, *44*, 15133–15144. [CrossRef]
16. Li, Q.L.; Yuan, S.M.; Gao, X.X.; Zhang, Z.K.; Chen, B.C.; Li, Z.; Batako, A. Surface and subsurface formation mechanism of SiCp/Al composites under ultrasonic scratching. *Ceram. Int.* **2023**, *49*, 817–833. [CrossRef]
17. Wu, X.; Qin, H.B.; Zhu, X.J.; Feng, Y.; Zhou, R.F.; Ye, L.Z. Investigation of the longitudinal-flexural resonating spindle in ultrasound-assisted grinding of SiCp/Al composites. *Int. J. Adv. Manuf. Technol.* **2022**, *121*, 3511–3526. [CrossRef]
18. Zhao, J.; Huang, J.F.; Xiang, Y.C.; Wang, R.; Xu, X.Q.; Ji, S.M.; Hang, W. Effect of a protective coating on the surface integrity of a microchannel produced by microultrasonic machining. *J. Manuf. Process.* **2021**, *61*, 280–295. [CrossRef]
19. Zheng, L.; Chen, W.Q.; Huo, D.H. Review of vibration devices for vibration-assisted machining. *Int. J. Adv. Manuf. Technol.* **2020**, *108*, 1631–1651. [CrossRef]
20. Kang, M.S.; Gu, Y.; Lin, J.Q.; Zhou, X.Q.; Zhang, S.; Zhao, H.B.; Li, Z.; Yu, B.J.; Fu, B. Material removal mechanism of non-resonant vibration-assisted magnetorheological finishing of silicon carbide ceramics. *Int. J. Mech. Sci.* **2024**, *242*, 107986. [CrossRef]
21. Yuan, Y.J.; Yu, K.K.; Zhang, C.; Chen, Q.; Yang, W.X. Generation of Textured Surfaces by Vibration-Assisted Ball-End Milling. *Nanomanuf. Metrol.* **2023**, *6*, 19. [CrossRef]
22. Guo, Z.Y.; Zhang, W.C.; Zhao, P.C.; Liu, W.D.; Wang, X.H.; Zhang, L.F.; Hu, G.F. Development of a novel piezoelectric-driven non-resonant elliptical vibrator with adjustable characteristics. *Rev. Sci. Instrum.* **2023**, *94*, 065008. [CrossRef]
23. Wang, Z.J.; Luo, X.C.; Liu, H.T.; Ding, F.; Chang, W.L.; Yang, L.; Zhang, J.G.; Cox, A. A high-frequency non-resonant elliptical vibration-assisted cutting device for diamond turning microstructured surfaces. *Int. J. Adv. Manuf. Technol.* **2021**, *112*, 3247–3261. [CrossRef]
24. Zhou, J.K.; Lu, M.M.; Lin, J.Q.; Du, Y.S. Elliptic vibration assisted cutting of metal matrix composite reinforced by silicon carbide: An investigation of machining mechanisms and surface integrity. *J. Mater. Res. Technol.* **2021**, *15*, 1115–1129. [CrossRef]
25. Zheng, L.; Chen, W.Q.; Huo, D.H. Investigation on the tool wear suppression mechanism in non-resonant vibration-assisted micro milling. *Micromachines* **2020**, *11*, 380. [CrossRef] [PubMed]
26. Gu, Y.; Li, Z.; Lin, J.Q.; Zhou, X.Q.; Xu, Z.S.; Zhou, W.D.; Zhang, S.; Gao, Y.H. Enhanced machinability of aluminium-based silicon carbide by non-resonant vibration-assisted magnetorheological finishing. *J. Mater. Process. Technol.* **2024**, *324*, 118223. [CrossRef]
27. Gu, Y.; Fu, B.; Lin, J.Q.; Chen, X.Y.; Zhou, W.D.; Yu, B.J.; Zhao, H.B.; Li, Z.; Xu, Z.S. A novel wheel-type vibration-magnetorheological compound finishing method. *Int. J. Adv. Manuf. Technol.* **2023**, *125*, 4213–4235. [CrossRef]
28. Wang, G.L.; Lv, B.G.; Zheng, Q.C.; Zhou, H.B.; Liu, Z.Z. Polishing trajectory planning of three-dimensional vibration assisted finishing the structured surface. *AIP Adv.* **2019**, *9*, 015012. [CrossRef]
29. Prabhu, P.; Rao, M. Investigations on piezo actuated micro XY stage for vibration-assisted micro milling. *J. Micromech. Microeng.* **2021**, *31*, 065007. [CrossRef]
30. Zhao, J.; Ge, J.Y.; Khudoley, A.; Chen, H.Y. Numerical and experimental investigation on the material removal profile during polishing of inner surfaces using an abrasive rotating jet. *Tribol. Int.* **2024**, *191*, 109125. [CrossRef]
31. Storchak, M.; Rupp, P.; Möhring, H.; Stehle, T. Determination of Johnson–Cook constitutive parameters for cutting simulations. *Metals* **2019**, *9*, 473. [CrossRef]
32. Miao, Q.; Ding, W.F.; Gu, Y.L.; Xu, J.H. Comparative investigation on wear behavior of brown alumina and microcrystalline alumina abrasive wheels during creep feed grinding of different nickel-based superalloys. *Wear* **2019**, *426*, 1624–1634. [CrossRef]
33. Xiong, Q.; Nie, X.W.; Lu, J.B.; Yan, Q.S.; Deng, J.Y. Processing performance of vitrified bonded fixed-abrasive lapping plates for sapphire wafers. *Int. J. Adv. Manuf. Technol.* **2022**, *123*, 1945–1955. [CrossRef]
34. Tang, S.Y.; Sun, Y.L.; Lou, Y.S.; Xu, Y.; Lu, W.Z.; Li, J.; Zuo, D.W. Study on the influence of ambient temperature on surface/subsurface damage of monocrystalline germanium lapping wafer. *Procedia CIRP* **2018**, *71*, 435–439.
35. Gao, T.; Zhang, X.P.; Li, C.H.; Zhang, Y.B.; Yang, M.; Jia, D.Z.; Ji, H.J.; Zhao, Y.J.; Li, R.Z.; Yao, P.; et al. Surface morphology evaluation of multi-angle 2D ultrasonic vibration integrated with nanofluid minimum quantity lubrication grinding. *J. Manuf. Process.* **2020**, *51*, 44–61. [CrossRef]

Disclaimer/Publisher's Note: The statements, opinions and data contained in all publications are solely those of the individual author(s) and contributor(s) and not of MDPI and/or the editor(s). MDPI and/or the editor(s) disclaim responsibility for any injury to people or property resulting from any ideas, methods, instructions or products referred to in the content.

Article

Investigation of Gallium Arsenide Deformation Anisotropy During Nanopolishing via Molecular Dynamics Simulation

Bo Zhao [1,2], Xifeng Gao [1,2,*], Jiansheng Pan [1,2], Huan Liu [1,2] and Pengyue Zhao [1,2]

1. Center of Ultra-Precision Optoelectronic Instrumentation Engineering, Harbin Institute of Technology, Harbin 150001, China; 22b901011@stu.edu.cn (B.Z.); 22b301002@stu.edu.cn (J.P.); liuhuanxues@163.com (H.L.); pyzhao@hit.edu.cn (P.Z.)
2. Key Lab of Ultra-Precision Intelligent Instrumentation, Ministry of Industry Information Technology, Harbin 150080, China
* Correspondence: xifenggao@hit.edu.cn; Tel.: +86-0451-86413840

Abstract: Crystal orientation significantly influences deformation during nanopolishing due to crystal anisotropy. In this work, molecular dynamics (MD) simulations were employed to examine the process of surface generation and subsurface damage. We conducted analyses of surface morphology, mechanical response, and amorphization in various crystal orientations to elucidate the impact of crystal orientation on deformation and amorphization severity. Additionally, we investigated the concentration of residual stress and temperature. This work unveils the underlying deformation mechanism and enhances our comprehension of the anisotropic deformation in gallium arsenide during the nanogrinding process.

Keywords: molecular dynamics; crystal orientation; nanopolishing; surface quality

Citation: Zhao, B.; Gao, X.; Pan, J.; Liu, H.; Zhao, P. Investigation of Gallium Arsenide Deformation Anisotropy During Nanopolishing via Molecular Dynamics Simulation. *Micromachines* **2024**, *15*, 110. https://doi.org/10.3390/mi15010110

Academic Editors: Kun Li and Aiqun Liu

Received: 9 December 2023
Revised: 27 December 2023
Accepted: 28 December 2023
Published: 8 January 2024

Copyright: © 2024 by the authors. Licensee MDPI, Basel, Switzerland. This article is an open access article distributed under the terms and conditions of the Creative Commons Attribution (CC BY) license (https://creativecommons.org/licenses/by/4.0/).

1. Introduction

Gallium arsenide, as a III-V compound semiconductor, exhibits direct bandgap characteristics when compared to traditional elemental semiconductor materials such as silicon (Si). It finds extensive applications in the manufacturing of laser diodes [1] and offers reduced noise levels in high-frequency operating conditions compared to silicon devices [2]. Furthermore, gallium arsenide material demonstrates high carrier mobility [3] and optical coupling effects [4], making it well-suited for next-generation communication and advanced optical device fabrication [5,6]. However, during semiconductor processing, surface defects induced by fabrication processes have a significant impact on the electrical characteristics and service life of the final devices [7]. Existing research has indicated that the crystal orientation of gallium arsenide surfaces significantly influences the performance of the final processed devices, and selecting different crystal orientations during processing can result in substantial enhancements of semiconductor components [8,9].

Due to the significant impact of surface defects generated during semiconductor device processing on the final quality, scholars have conducted extensive experimental research. These experiments primarily include indentation tests and scratch tests [10,11], which employ experimental methods to observe structural surface defects and subsequently investigate surface morphology and crystal structure damage. Gao et al. [12] utilized molecular dynamics to examine GaAs laser bar cleavage. Their study highlighted the influence of scratching depth on scratch quality and provided optimal parameters for GaAs cleavage. Li et al. [13] conducted Vickers indentations on a GaAs single crystal, yielding defects like dislocations, microtwins, stacking faults, and amorphization. Proposed amorphization mechanisms include high-pressure and shear deformation; high-pressure induced amorphization and shear deformation induced amorphization indicate the transformation from crystalline to amorphous structure. Li et al. [14] investigated cracks induced by 0.049 N load indentations in GaAs, observing shear-related crack initiation, dislocation generation, lattice distortion, and amorphous band formation. Annealing eliminated the amorphous band, revealing a crack propagation via decohesion. Huang et al. [15] studied monocrystalline GaAs deformation during nanoscratching, revealing atomic-scale lattice bending in semiconductor materials. They discussed the lattice bending mechanism and found the residual stress could be responsible for the local lattice bending. Parlinska et al. [16] explored GaAs nanoindentations and nanoscratches using different indenters. The Berkovich indentations caused convergent dislocations, twins, and slip bands, while the 60° wedge indentations led to divergent bands and median cracks. They discussed the mechanism of deformation of the crystals and found that the deformation was mainly concentrated at the front of the indenter. They similarly found by TEM experiments that the crystal deformation was mainly concentrated at the front of the indenter. Wasmer et al. [17] employed nanoindentation and scratching to study gallium arsenide. They discovered twinning during indentation and slip bands and perfect dislocations during scratching. This phenomenon was attributed to differing strain rates, higher in scratching, promoting a perfect α dislocation propagation, while slower indentation velocities enable twinning nucleation from surface inhomogeneities. Wasmer et al. [18] employed scratch tests on GaAs {001} crystals with loads (5–100 mN) and a Berkovich indenter. They unveiled the plastic deformation stages, including dislocation cloud formation, median cracks nucleation, surface radial cracks, plastic flow, lateral cracks, and chip formation. These events exhibited a power-law dependence. Elastic recovery was approximately 15%, explained by the rheological factor X. Gao et al. [19] utilized scratching and cleavage operations to enhance GaAs cleavage planes in high-power semiconductor laser cavities. Scratching with a lower load and higher speed reduced damage, while the scratch capability index (SCI) indicated the cleavage plane quality. This approach can advance semiconductor laser chip manufacturing. They also discussed the relationship between scratch quality and load and found that the load has a significant effect on scratch quality. Yu et al. [20] employed nanoscratch tests on GaAs {100} using an atomic force microscope (AFM) with a SiO_2 tip. Decreasing the sliding velocity increased the scratch depth. High-resolution transmission electron microscopy (HRTEM) found no lattice damage. The material removal was attributed to dynamic interfacial bond breakage. High-speed sliding resulted in a faster GaAs surface material removal, ideal for SiO_2 polishing without surface damage. Chen et al. [21] used molecular dynamics (MD) simulations to investigate single-crystal copper nanoscratching. They observed depth-dependent subsurface changes, differing (100) and (111) plane behaviors, and identified stack faults. It was shown that the surface integrity was not only related to the scratch depth, but the surface grain orientation also had a non-negligible effect on the surface integrity. Fan et al. [22] employed oblique nanomachining to enhance GaAs machining quality. They observed an early dislocation avalanche and a favorable plasticity during cutting under certain tip conditions, particularly oblique cutting. Gao et al. [23] employed a novel method to study anisotropic stress in GaAs. They found a lower stress along {100} than {110} orientations. The (011) plane displayed potential as a preferential cleavage plane with improved quality. This research enhanced the understanding of cleavage mechanisms. The study discussed the stress field of the GaAs scribing process and showed that the maximum stress was concentrated at the tip of the indenter and appeared anisotropic in different directions. Wang et al. [24] employed AFM tip-based nanoscratching to create GaAs nanochannels, studying the material removal and subsurface damage. Depths below 11 nm favored cutting over plowing, inducing stacking faults, dislocations, nanocrystallization, and amorphization. Wu et al. [25] probed GaAs surface defects using a conductive atomic force microscope (C-AFM). Scratches showed a higher edge current, influenced by the load. Etching increased currents, with scratch-induced Schottky barrier height changes. Fang et al. [26] studied Si and GaAs nanomechanical properties via nanoindentation and nanoscratch. Results showed a decreased Young's modulus and hardness with a higher load, hold time, and cycles. GaAs exhibited a pop-in effect, and the wear behavior varied with the feed and load. The scratch technique used the material removal volume to evaluate hardness. The study found an effect of the applied load on the GaAs surface quality, which related to the surface hardness and Young's modulus.

However, due to the high cost of experimental research and the stringent requirements for experimental environments, many scholars are gradually adopting a combination of MD simulation with experimental research. The MD simulation studies of materials are widely used to investigate the mechanical behavior and deformation mechanisms of materials at the nanoscale. It has been widely applied in the study of atomic-scale surface deformation and crystal structure and is suitable for the study of properties that are difficult to measure with many traditional experimental methods [27–31]. Li et al. [32] employed molecular dynamics simulations to investigate the influence of cracking on GaAs deformation in different crystal orientations during processing. Their findings revealed cracking-induced alterations in atomic-level deformation behavior, attributed to the tensile stress distribution and fracture surface variations. The anisotropy induced by the surface grain orientation, which has an important effect on the surface defects, can also be seen by MD simulation. Xu et al. [33] used molecular dynamics simulations to investigate GaAs crystal anisotropy during nanoscratching. They found significant anisotropic effects on the deformation, residual stress, and surface properties, offering new insights into the material behavior. The study also confirmed that the anisotropy of the surface grain direction had an important influence on the distribution of residual stresses. Yi et al. [34] utilized molecular dynamics to examine GaAs nanoscratching in chemical mechanical polishing. Phase transformation and amorphization were the dominant deformation mechanisms. Anisotropic effects were observed, with varied scratching resistance and friction coefficients among different GaAs crystal orientations, providing insights into the mechanical wear in GaAs polishing. The study also confirmed that the anisotropy of the surface grain direction had an important influence on the scratching forces. Chen et al. [35] employed molecular dynamics simulations to explore surface and subsurface deformations in gallium arsenide during nanocutting. Dislocations, phase transformations, and anisotropic effects were investigated, providing insights into performance-affecting factors in GaAs machining. Li et al. [36] used molecular dynamics simulations to explore plowing-induced deformation in GaAs. They observed crack initiation, propagation, and dislocation-dominated plasticity, providing atomic-level insights into a novel deformation pattern in GaAs during plowing. The MD simulations also found that the deformation and high stress areas were mainly distributed at the front end of the indenter, which was consistent with the experimentally generated phenomena. Fan et al. [37] simulated the AFM tip-based hot machining of GaAs at temperatures of 600 K, 900 K, and 1200 K, revealing reduced cutting forces, increased friction, enhanced material removal rate, and ductile response with dislocations, along with chip densification during hot cutting. Fan et al. [38] studied nanoscale friction using MD simulations and AFM nanoscratch experiments on gallium arsenide. They examined the scratch depth effects, revealing a size-dependent behavior. The study found correlations between MD simulations and AFM experiments, indicating a specific scratch energy insensitivity to the tool geometry and scratch speed. However, the pile-up and kinetic coefficient of friction were influenced by the tool's tip geometry. Fan et al. [39] investigated a diamond wear during AFM-based nanomachining of GaAs via MD simulations. They observed the diamond tip's elastic–plastic deformation and transformation from a cubic to graphite structure, identifying graphitization as the dominant wear mechanism, introducing a novel method for quantifying the graphitization conversion rate. Chen et al. [40] investigated gallium arsenide's crack formation during nanocutting. They found a transition from dislocation to phase transformation at higher cutting speeds, with more cracks at greater depths. Deformation shifted from ductile to ductile–brittle, with cracks at the amorphous–single crystal boundaries. Tensile stress was concentrated at crack tips. Taper-cutting experiments revealed a 25 nm brittle–ductile transition depth, supported by transmission electron microscopy (TEM) showing microcracks and polycrystals in the subsurface, aligning with simulation findings. Li et al. [41] reviewed molecular dynamics simulations in tip-based nanomachining (TBN), covering material-specific models, TBN mechanisms, and future prospects, offering valuable insights for further research in this field. The study provided a systematic overview of the molecular dynamics study of TBN, showing that the molecular dynamics approach was applicable to the study of mechanical properties and surface defects. Fan et al. [42] used molecular dynamics to explain ductile plasticity in polycrystalline gallium arsenide during nanoscratching, emphasizing the dislocation nucleation at grain boundaries and its impact on material behavior. Rino et al. [43] studied structural phase transformation in crystalline gallium arsenide under a 22 GPa pressure, with a reverse transformation observed at 10 GPa, showing hysteresis. Molecular dynamics results matched experiments, estimating a 0.06 eV energy barrier. The simulation results showed that there was a clear relationship between the stresses and the changes in the crystal structure. Kodiyalam et al. [44] investigated pressure-induced structural transformation in gallium arsenide nanocrystals, with nucleation occurring at the surface, leading to inhomogeneous deformation and grain boundaries. It was also found that the region of high-pressure distribution had an important influence on the transformation of the crystal structure. Parasolov et al. [45] developed a nanoindentation model using molecular dynamics on GaAs. Above 100 K, nanoindentation led to increased point defects in GaAs atomic layers, attributed to thermal energy fluctuations and external stress. Gular et al. [46] conducted geometry optimization calculations on GaAs up to 25 GPa using a Stillinger–Weber potential. They determined a B3 to B1 phase transition at 17 GPa and evaluated various material properties, providing valuable insights for future GaAs pressure studies. The comprehensive analysis of MD simulations shows that anisotropy has an important effect on surface defects and crystal structure, and the MD simulation method is also applicable to the study of micromechanical properties in nanofabrication; the effect of crystal orientation will be further investigated in this study.

Existing studies have shown that in the processing of GaAs crystalline materials, the selection of the appropriate crystal orientation has an important impact on the performance of the final processed workpiece [8,9]. In this work, the machined surface of GaAs with different crystallographic orientations is modeled and the surface morphology and amorphous damage layer after nanopolishing are investigated; in addition, residual stresses as well as temperatures are analyzed in order to select a suitable crystallographic orientation for nanofabrication. This work utilizes MD simulations to investigate the processes of surface generation and subsurface damage. A nanoscale-polishing molecular dynamics model incorporating the microasperity structure of the actual processed surface is established. The surface topography, mechanical properties, and phase transition processes under {100}, {110}, and {111} crystal orientations are analyzed, validating the influence of anisotropy on the surface morphology and subsurface crystal phase transformation extent. Furthermore, by analyzing the differences in surface pile-up after nanoscale polishing for three crystal orientations, this work also examines the impact of the surface crystal orientation on the temperature distribution and residual stress distribution during the nanoscale polishing process, which may have practical implications for nanoscale polishing processes.

2. Methods

Simulation Methods

In comparison to the traditional nanoscale polishing model, the nanoscale polishing model employed in this study takes into account the microconvex structures present on the actual processed surface. The variables under investigation pertain to the crystallographic orientations of gallium arsenide (GaAs) surfaces during the nanoscale polishing process, specifically the {100}, {110}, and {111} crystallographic orientations. The nanoscale polishing model for GaAs crystals, as illustrated in Figure 1, can be conceptually divided into two main components: the equivalent spherical representation of the diamond polishing tool and the GaAs surface with its microconvex structures.

As depicted in Figure 1a, the equivalent diamond polishing particle had a diameter of 12 Å, consisting of 159,486 atoms, and possessed a lattice constant of 3.57 Å. The equivalent GaAs surface was composed of two parts: a substrate with dimensions of 300 Å × 220 Å × 50 Å and microconvex structures comprising one-quarter spheres at both ends and a central half-cylinder, all with a radius of 7 Å. The centers of the spherical structures at the two ends were located at (110 Å, 110 Å, 50 Å) and (190 Å, 110 Å, 50 Å), respectively. The position of the diamond particle was (−60 Å, 110 Å, 120 Å). The total number of gallium atoms was 104,963, and the total number of arsenic atoms was 103,420. The crystallographic structure of the GaAs crystal is depicted in Figure 1b, with a lattice constant of 5.654 Å.

The equivalent model for the gallium arsenide (GaAs) surface was divided into three distinct layers, as shown in Figure 1a: the Newtonian atomic layer situated at the top, where atomic motion follows Newton's second law and is calculated using the velocity Verlet algorithm [47]; the isothermal atomic layer in the middle, which regulates temperature changes based on the Berendsen thermostat [48]; and the fixed atomic layer at the bottom, where atomic positions and velocities are constrained to prevent atoms from escaping the boundary. In the multilayer structure, the thickness of the Newtonian layer was 100 Å (70 Å for the radius of the microconvex body and 30 Å for the basal portion), the thickness of the thermostatic layer was 10 Å, and the temperature of the boundary layer was 10 Å. In addition to the potential energy parameters, to ensure convergence, the model set boundary conditions as well as energy minimization constraints so that the model was in a steady state before nanopolishing. To enhance computational efficiency in the simulation, this work employed periodic boundary conditions for the nanoscale polishing process. Specifically, periodic boundary conditions were applied in the y-direction to exploit the system's symmetric properties, while nonperiodic boundary conditions were imposed in the x-direction (processing direction) and the z-direction (normal to the surface) to ensure a realistic representation of the system.

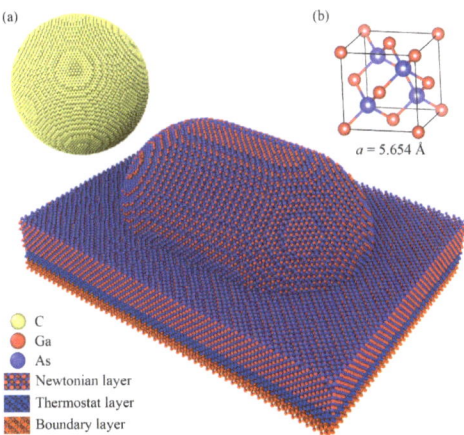

Figure 1. Molecular dynamics models of nanopolishing of gallium arsenide. (a) Model structure (b) GaAs crystal structure.

This study utilized the Large-scale Atomic/Molecular Massively Parallel Simulator (LAMMPS) [49] for molecular dynamics simulations and employed the open visualization tool (OVITO) [50] for the visualization and postprocessing of the simulation results. The detailed parameters of the model are presented in Table 1. The simulation workflow included the prepolishing energy minimization process using the conjugate gradient method [51]. The model's relaxation process was conducted under the NPT ensemble with a relaxation time of 100 ps. During this process, the model's temperature gradually stabilized at room temperature (293 K) using the Nose–Hoover thermostat, and the potential energy converged to -5.30×10^{-5} eV. The temperature and potential energy changes during the relaxation process are illustrated in Figure 2. Following the relaxation of the model, the ensemble was switched to NVE, and the simulation of nanoscale polishing was performed. During the relaxation phase of the model, the temperature gradually stabilized at 293 K, and the total potential energy of the model gradually stabilized at -5.30×10^{-5} eV. In this process, the polishing speed of diamond abrasive particles was set at 100 m/s in the (0,1,0) direction, with a polishing distance of 30 nm. Before the calculations for stresses, RDF, and temperature and after the nanopolishing simulation, the model was subjected to a relaxation process, which resulted in a more stable surface structure after processing. To observe the stable structure of the surface after the nanoscale polishing process, a second relaxation process was conducted for the model, also with a relaxation time of 100 ps.

Table 1. The MD simulation parameters.

Simulation Parameters	Value
Material of the workpiece	Gallium arsenide (GaAs)
Material of the nanopolishing grit	Diamond (C)
Dimension of the workpiece (nm)	$30 \times 22 \times 5$
Radius of the nanopolishing grit (nm)	6
Surface crystal orientation of the workpiece	{100}, {110}, and {111}
Potential function	Tersoff, ZBL
Nanogrinding speed (m/s)	100
Ambient temperature (K)	300
Nanogrinding distance (nm)	30
Timestep (fs)	1

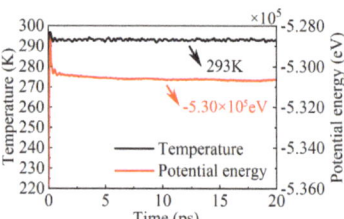

Figure 2. Temperature and potential energy of the relaxation process before nanopolishing.

During the process of nanoscale machining, the selection of the interatomic potential energy is of paramount importance. In the case of polishing gallium arsenide (GaAs) workpieces, the interatomic potential energy functions in Ga-Ga, Ga-As, and As-As atoms are described by the Tersoff potential [52] and the parameters refers to [53]. The expression of the Tersoff potential function is shown in Equation (1). For the interatomic potential energy function in carbon–carbon (C-C) atoms in diamond polishing particles, the Tersoff potential was employed. The interatomic potential energy functions between carbon (C) atoms in diamond polishing particles and gallium (Ga) or arsenic (As) atoms in GaAs workpieces are governed by the Ziegler–Biersack–Littmark universal screening function (ZBL) potential [54]. The expression of the ZBL potential is presented in Equation (2), where the parameter *inner* is the distance where the switching function begins, and *outer* is the global cutoff for the ZBL interaction. The parameters *inner* of Ga-C and As-C are 31.0 and 33.0, respectively. The parameters *outer* of Ga-C and As-C are 12.0.

$$\begin{cases} E = \frac{1}{2} \sum_i \sum_{i \neq j} V_{ij} \\ V_{ij} = f_c(r_{ij})[f_R(r_{ij}) + b_{ij} f_A(r_{ij})] \end{cases} \quad (1)$$

where V_{ij} is the Tersoff potential energy, f_R means the two-body term, f_A means the three-body term, f_C means the cutoff of the coefficient.

$$V_{ij} = \frac{1}{4\pi\varepsilon_0} \frac{Z_1 Z_2 e^2}{r_{ij}} \phi(r_{ij}/a) \quad (2)$$

where Z_1, Z_1 are the number of protons in the nucleus, e is the electron charge, ε_0 is the permittivity of vacuum, and $\phi(r_{ij}/a)$ is the universal screening function of ZBL potential.

When evaluating surface residual stresses in polished gallium arsenide (GaAs) workpieces, the von Mises stress was calculated. It was determined based on the atomic stress tensor, taking into account the combined effects of six stress components, as expressed in Equation (3). When considering temperature variations during the nanoscale polishing process, the temperature change was represented using the average kinetic energy expression [48], as shown in Equation (4).

$$\sigma_{vm}(i) = \left\{ \frac{1}{2} \left[\begin{array}{l} (\sigma_{xx}(i) - \sigma_{yy}(i))^2 + (\sigma_{yy}(i) - \sigma_{zz}(i))^2 \\ +(\sigma_{zz}(i) - \sigma_{xx}(i))^2 + 6(\sigma_{xy}^2(i) + \sigma_{yz}^2(i) + \sigma_{zx}^2(i)) \end{array} \right] \right\}^{1/2} \quad (3)$$

where $\sigma_{vm}(i)$ denotes the von Mises stress, and $\sigma(i)$ denotes an atomic stress tensor.

$$E_k = (3/2)kT \quad (4)$$

where E_k represents the average atomic kinetic energy, k denotes the Boltzmann constant which is 1.381×10^{-23} J/K, and T denotes the temperature.

3. Results and Discussion

3.1. Surface Quality

After nanoscale polishing on the gallium arsenide (GaAs) surface, significant atomic displacements were observed primarily due to the intense interaction between diamond polishing particles and the rapidly moving surface. The atomic displacements of the three crystallographic orientations were color-coded based on the total atomic displacement and displacement along the z-axis, as depicted in Figure 3(a1–c1). The regions of maximum atomic displacement for different crystallographic orientations were concentrated at the ends of microprotuberances formed as chip piles, with varying specific distributions. The {100} crystallographic orientation exhibited a maximum displacement concentration at the top of the polished chip pile, while the {110}, and {111} orientations, as a result of their detachment from the surface after microprotuberance polishing, showed an atomic displacement accumulation along the x-axis. A profile analysis of atomic displacements within the chip along the y-axis is shown in Figure 3(a2–c2). Along the z-axis, atomic displacements within the residual pile-up on the surface after polishing gradually decreased for the {100}, {110}, and {111} orientations. Due to the anisotropy caused by the crystallographic orientations, atoms from the {110} and {111} orientations exhibited the highest atomic displacements as they detached from the surface following their interaction with microprotuberances and diamond polishing particles, resulting in relatively smaller atomic displacements within the remaining portion compared to the {100} orientation.

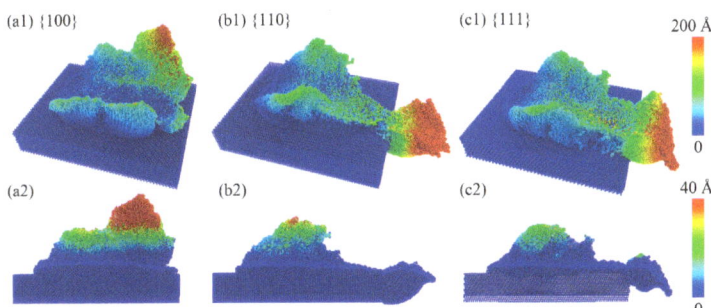

Figure 3. The total atomic displacement of the surface following nanoscale polishing. In panels (**a1–c1**), the cumulative atomic displacement is depicted, while panels (**a2–c2**) specifically represent atomic displacement in the z-axis direction.

After analyzing the atomic displacements following nanoscale polishing, postprocessing was carried out based on the z-coordinate positions of gallium arsenide polished surface atoms. As illustrated in Figure 4, observations from the z-direction top view and z-direction cross-sectional view revealed significant alterations in atomic distribution after nanoscale polishing. Notably, different crystallographic orientations of microprotrusions exhibited distinct impacts on the surface quality following polishing. Specifically, microprotrusions with a {100} crystallographic orientation did not disintegrate along with the diamond polishing particles during nanoscale polishing; instead, they accumulated on the preexisting surface, resulting in a maximum surface asperity height of 138 Å. In contrast, atoms from the {110} and {111} crystallographic orientations split into two parts after nanoscale polishing, with pile-up heights of 98 Å and 85 Å, respectively, exerting a lesser influence on the postpolishing surface quality.

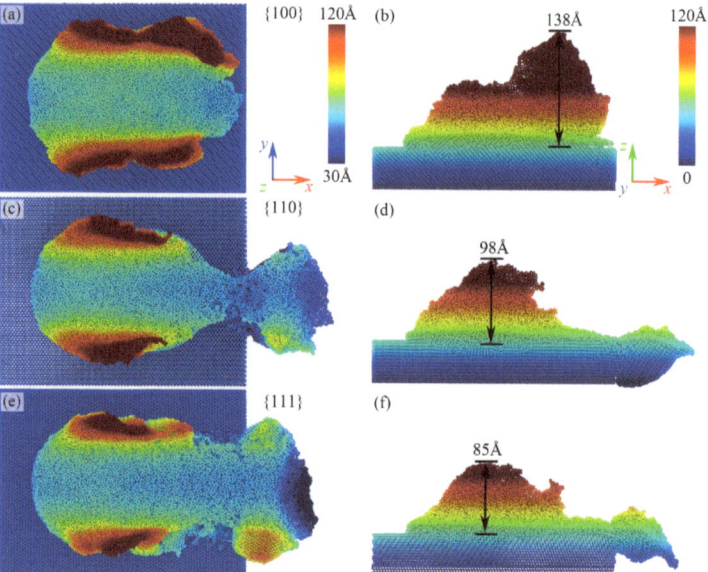

Figure 4. The surface quality (**a**,**c**,**e**) in the z-direction view, and the cross-sectional view (**b**,**d**,**f**) in the y-direction after nanoscale polishing.

3.2. Mechanical Property

In order to further investigate the influence of surface crystallographic anisotropy on mechanical properties, the normal force F_z and tangential force F_x during nanoscale polishing were separately calculated, as shown in Figure 5. The relationships between the normal force F_z and the polishing distance, as well as the tangential force F_x and the polishing distance, were analyzed. After the contact between diamond polishing particles and gallium arsenide microprotrusions, the contact force gradually increased with the moving distance, reaching a maximum value before gradually decreasing. Furthermore, it could be observed that F_z was numerically greater than F_x. For polishing distances less than 5 nm, there were no significant differences in F_z and F_x among the three crystallographic orientations. However, when the polishing distance was between 5 nm and 20 nm, significant differences in F_z and F_x appeared among the three crystallographic orientations, and after the diamond polishing particles left the microprotrusions, both F_z and F_x gradually decreased to their minimum values.

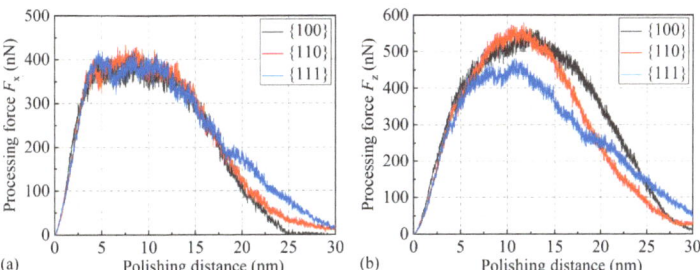

Figure 5. The relationship between (**a**) the tangential force, F_x, and (**b**) the normal force, F_z, with respect to the polishing distance.

After calculating the contact forces in three crystallographic orientations, the average contact forces within the nanoscale polishing range of 5–20 nm were determined, as illustrated in Figure 6. It is evident that the F_z and F_x components were highest for the {110} crystallographic orientation. Conversely, the F_x component was at its minimum for the {100} orientation, indicating reduced interatomic interactions. This observation aligned with the earlier analysis of the surface atomic displacement. It is plausible that atomic displacements in the x-direction were limited due to the nanopolishing, resulting in a lesser interaction for {100}. In contrast, the {110} and {111} orientations exhibited larger F_x values, implying stronger interatomic interactions, potentially leading to the detachment of some atoms as the microprotrusions decomposed due to the enhanced F_x. In terms of F_z, the {111} orientation experienced the lowest contact force, suggesting weaker interactions in the z-direction. This may be correlated with the earlier observation of the lowest atomic stacking height in the z-direction. A further investigation is required to understand its implications on the subsurface atomic structure.

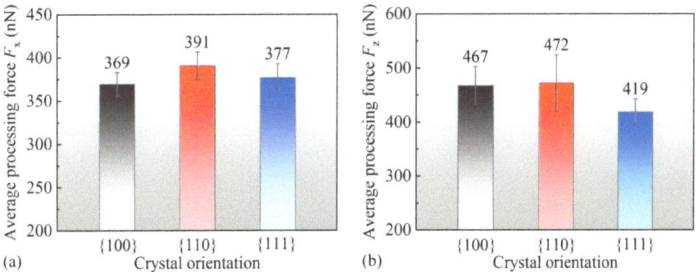

Figure 6. The mean contact force within a range of 5 nm to 20 nm in nanoscale polishing. (**a**) F_x; (**b**) F_z.

3.3. Amorphization Analysis

After nanoscale polishing, the anisotropy not only influences the mechanical properties but also results in differences in the crystal structure of the subsurface after processing. By examining the crystal structures of subsurface atoms, as depicted in Figure 7, it could be observed that following nanoscale polishing, most of the remaining atoms on the processed surface underwent an amorphization process, with only the outermost atoms of the microconvex polished surface retaining their original structure, namely the cubic diamond structure. From the cross-sectional view on the y-z plane, it is evident that the thickness of the subsurface damaged layer (SDL) varied after nanoscale polishing, with the {110} crystal orientation reaching a maximum thickness of 3.16 nm, the {100} crystal orientation measuring 2.89 nm, and the {111} crystal orientation having a minimum thickness of 2.19 nm. Consistent with the previous analysis of mechanical properties, the thickness of the SDL exhibited a similar trend to the variation in F_z values, potentially attributed to the varying degrees of z-directional interactions resulting from the crystal anisotropy.

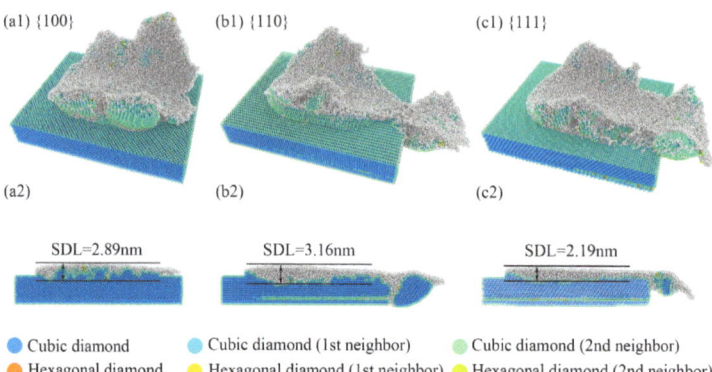

Figure 7. The thickness of the subsurface damage layer in the (**a1,b1,c1**) z-direction view, and (**a2,b2,c2**) y-side cross-sectional image after nanoscale polishing.

The thickness of the amorphous SDL layer exhibited anisotropy, indicating that the selection of different crystallographic orientations for processing had an influence on the crystal structure within the polished surface. In order to further investigate the subsurface damage process, an analysis was conducted using the radial distribution function (RDF), as shown in Figure 8. The RDF is a commonly used function for studying crystal structures, representing the distribution of atomic distances, with different peaks in the curve characterizing different crystal structures. As depicted in Figure 8a, RDF calculations were performed for the processed surface before, during, and after polishing. It can be observed that during the nanoscale polishing process, there was a decrease in the peak at 2.45 Å and an increase in the peak at 2.85 Å, indicating that some crystal structures with an atomic spacing of 2.45 Å were disrupted during the processing and transformed into structures with an atomic spacing of 2.85 Å, and this process was irreversible. For different crystal orientations, as shown in Figure 8b, it can be seen that the {110} orientation exhibited a higher peak at 2.85 Å compared to other orientations, indicating a lower proportion of amorphization atoms in the residual atoms after processing for the {110} orientation.

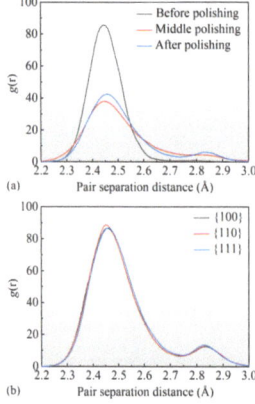

Figure 8. The RDF curve for the nanopolishing process. (**a**) Different stages of the machining process, (**b**) different crystal orientations.

3.4. Analysis of Temperature Distribution

Figure 9(a1–c1) represents the temperature distribution on the surface during the nanoscale polishing process. It can be observed that atoms with higher total displacements correspond to regions with higher temperatures, consistent with the previous analysis of atomic displacements. This observation indicates a relationship between the temperature distribution and the total atomic displacement during nanoscale polishing. Figure 9(a2–c2) displays a side view of the temperature distribution along the y-axis. It is evident that the temperature of {111} crystal facet debris reached a maximum of 600 K, while the surface atom temperature reached a minimum of 210 K. In contrast, the {100} facet debris exhibited a minimum temperature of 460 K, with the surface temperature reaching a maximum of 310 K. The opposite trends in temperature variation between different crystal facets during nanoscale grinding may be attributed to the fact that debris atoms carry away more heat, resulting in lower residual atomic temperatures on the surface.

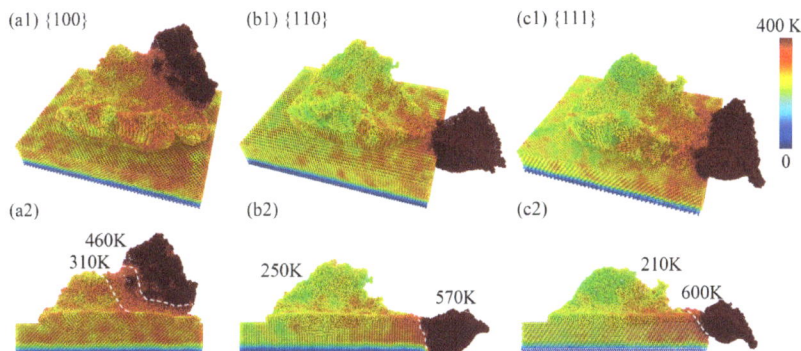

Figure 9. The temperature distribution of the surface quality after nanoscale polishing (**a1–c1**), and the lateral cross-sectional perspective (**a2–c2**).

In Figure 10, the average temperatures and errors in different regions after nanopolishing with different crystal orientations are presented, which are consistent with the trends in Figure 10. The average temperature of the {100} crystal orientation for the cutting chips was the lowest at 421.85 K, while the average temperature of surface atoms was the highest at 339.21 K. On the other hand, the {111} crystal orientation exhibited the highest average temperature for the cutting chips at 607.725 K, with the surface atoms having an average temperature of 284.99 K. From the statistical analysis of the average temperatures, it could be inferred that following nanopolishing, the cutting chip atoms in the {100} crystal orientation did not separate into two parts with the movement of the diamond polishing particles. This resulted in a higher temperature accumulation, indirectly providing a higher temperature environment for amorphization, possibly leading to a more pronounced degree of amorphization. In contrast, the {111} crystal orientation showed a lower average temperature for the remaining surface atoms after nanopolishing, which may have resulted in a lower degree of amorphization.

3.5. Analysis of Residual Stress

The degree of crystal amorphization is not only related to the temperature environment but also to the internal stress intensity within the region. In Figure 11, the visualization results of the von Mises stress distribution on the surface after nanoscale polishing are presented. It can be observed in Figure 11(a1–c1) that following nanoscale polishing, there existed a high-stress distribution region (in red) with values ranging from 3 to 4 GPa on the surface in contact with the polishing particles, as well as stress distribution regions of 1–2 GPa, corresponding to the previously mentioned amorphous surface atomic regions. In the y-directional cross-sectional view, it can be observed that the highest stress in the SDL layer on the surface after three crystallographic directions processing was approximately

3.9 GPa. The stress distribution in the {110} and {111} facets was generally similar to that on the surface, although specific numerical values require further calculation.

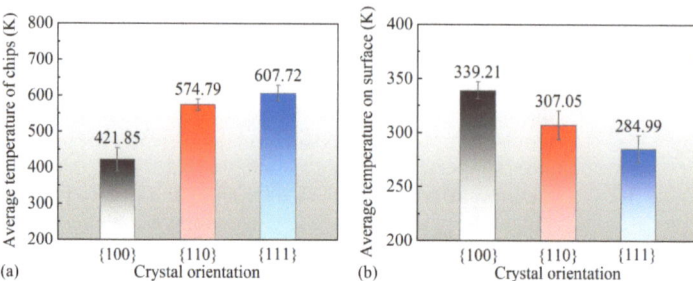

Figure 10. The average temperature profiles subsequent to nanoscale polishing, with distinct delineations for (**a**) the chip-removal atomic domain and (**b**) the residual-surface atomic domain.

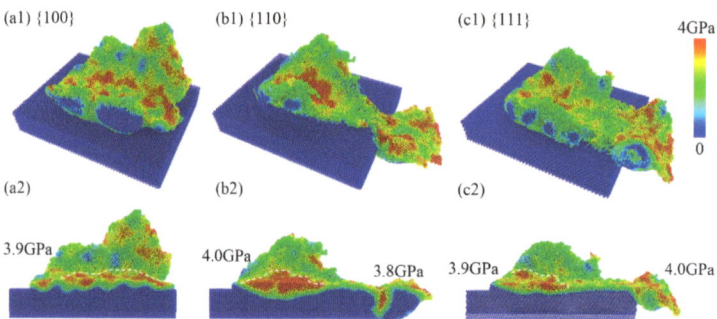

Figure 11. The stress distribution in the (**a1**–**c1**) region and the cross-sectional view on the (**a2**–**c2**) y-plane after nanoscale polishing.

As shown in Figure 11, for the calculation of residual stresses on the processed surface, Figure 12a illustrates the distribution of total residual stresses at a depth of 56 nm along the z-axis for different crystallographic orientations. It can be observed that the 110 orientation exhibited the highest stress, followed by the {111} orientation, while the {100} orientation showed the lowest residual stress. The residual stress distribution at various depths along the {100} orientation is depicted in Figure 12b, revealing a generally negative correlation between residual stress and depth. The minimum residual stress was observed at a depth of 54 nm, while stress levels increased closer to the surface, reaching higher values at a depth of 60 nm. Furthermore, an analysis of different stress components on the {100} orientation is presented in Figure 12c. It is evident that the shear stress component σ_{xx} in the xx direction was significantly larger than in other directions. Additionally, a comparison of the xx direction stress component σ_{xx} was made among different crystallographic orientations. Notably, the {111} orientation exhibited a smaller stress component value, whereas the {110} orientation displayed the largest residual stress component value. These findings may have implications for the stability of the crystal structure and surface quality following surface polishing.

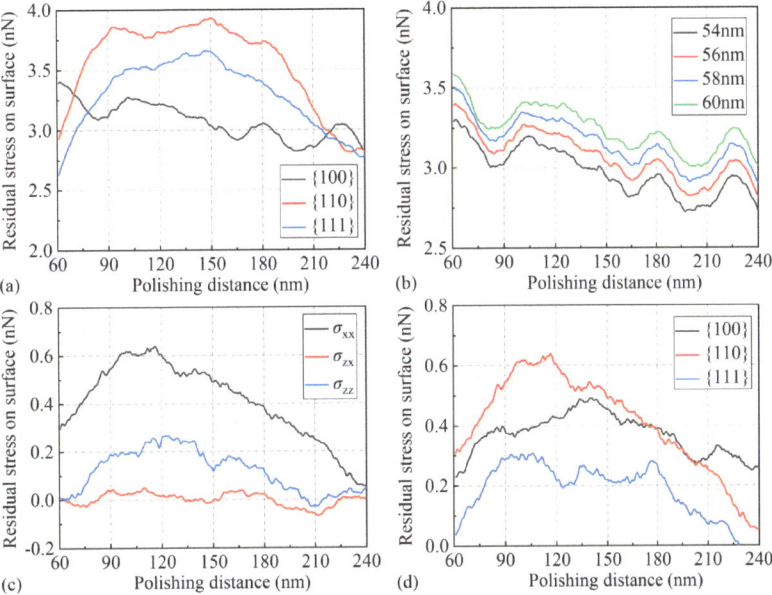

Figure 12. (a) The distribution of total stresses along different crystal orientations, (b) the relationship between residual stresses and depth distribution, (c) the stress components along different directions, and (d) the residual stress components along different crystal orientations in the processing direction, σ_{xx}, after nanoscale polishing.

4. Conclusions

In this work, MD simulations were employed to investigate the processes of surface generation and subsurface damage. A nanoscale polishing molecular dynamics model was established, taking into consideration the microconvex structures of the actual processed surface. The analysis encompassed surface morphology, mechanical properties, and amorphization processes under the {100}, {110}, and {111} crystal orientations, thus confirming the influence of anisotropy on surface morphology and subsurface crystalline amorphization extent. After analyzing the disparities in surface pile-up following nanoscale polishing in three crystal orientations, it was discerned that the {111} crystal orientation exhibited a lower residual atomic height and a lower normal contact force during the processing. Additionally, an investigation of subsurface crystalline amorphization revealed a thinner amorphous layer beneath the {111} crystal orientation. In the RDF analysis, it was observed that the proportion of atoms undergoing amorphization was slightly lower under the {110} crystal orientation compared to the other two orientations. Furthermore, this work examined the influence of the surface crystal orientation on the temperature distribution and residual stress distribution during the nanoscale polishing process. Regarding temperature, the {111} crystal orientation exhibited lower surface temperatures during the processing. In terms of stress, it was found that the tangential residual stress component, σ_{xx}, was larger compared to the normal residual stress component, σ_{zz}. Additionally, σ_{xx} under the {111} crystal orientation was lower. Considering the comprehensive analysis of postpolishing surface morphology, contact forces, SDL thickness, temperature, and stress distribution, it can be concluded that the microconvex structures under the {111} crystal orientation have a lesser impact on surface quality and subsurface amorphization after polishing, which may hold significance for practical nanoscale polishing processes.

Author Contributions: Conceptualization, B.Z.; methodology, P.Z.; validation, X.G.; investigation, H.L.; writing, B.Z.; supervision, J.P.; funding acquisition, P.Z. All authors have read and agreed to the published version of the manuscript.

Funding: This work was supported by the National Natural Science Foundation of China (52305569 and 52105547), the China Postdoctoral Science Foundation (2021M700995), the Natural Science Foundation of Heilongjiang Province, China (LBH-Z21063), and the Young Elite Scientists Sponsorship Program by CAST (2022QNRC001).

Institutional Review Board Statement: Not applicable.

Informed Consent Statement: Not applicable.

Data Availability Statement: The original contributions presented in the study are included in the article, further inquiries can be directed to the corresponding author.

Acknowledgments: We would like to express our gratitude to Academician Jiubin Tan for his guidance and support in the field of molecular dynamics.

Conflicts of Interest: The authors declare no conflict of interest.

References

1. Ahn, H.J.; Chang, W.I.; Kim, S.M.; Park, B.J.; Yook, J.M.; Eo, Y.S. 28 GHz GaAs pHEMT MMICs and RF front-end module for 5G communication systems. *Microw. Opt. Technol. Lett.* **2019**, *61*, 878–882. [CrossRef]
2. Chow, T.P.; Omura, I.; Higashiwaki, M.; Kawarada, H.; Pala, V. Smart power devices and ICs using GaAs and wide and extreme bandgap semiconductors. *IEEE Trans. Electron Devices* **2017**, *64*, 856–873. [CrossRef]
3. Hayati-Roodbari, N.; Wheeldon, A.; Hendler, C.; Fian, A.; Trattnig, R. Ohmic contact formation for inkjet-printed nanoparticle copper inks on highly doped GaAs. *Nanotechnology* **2021**, *32*, 225205. [CrossRef] [PubMed]
4. Gao, J.; Zhou, H.; Du, J.; Peng, W.; Lin, Y.; Xiao, C.; Yu, B.; Qian, L. Effect of counter-surface chemical activity on mechanochemical removal of GaAs surface. *Tribol. Int.* **2022**, *176*, 107928. [CrossRef]
5. Dinodiya, S.; Bhargava, A. A comparative analysis of pressure sensing parameters for two dimensional photonic crystal sensors based on Si and GaAs. *Silicon* **2022**, *14*, 4611–4618. [CrossRef]
6. Hao, D.; Zhang, W.; Liu, X.; Liu, Y. A low insertion loss variation trombone true time delay in GaAs pHEMT monolithic microwave integrated circuit. *IEEE Microw. Wirel. Components Lett.* **2021**, *31*, 889–892. [CrossRef]
7. Jordan, A.; Von Neida, A.; Caruso, R. The theoretical and experimental fundamentals of decreasing dislocations in melt grown GaAs and InP. *J. Cryst. Growth* **1986**, *79*, 243–262. [CrossRef]
8. Xu, M.; Wu, Y.; Koybasi, O.; Shen, T.; Ye, P. Metal-oxide-semiconductor field-effect transistors on GaAs (111) A surface with atomic-layer-deposited Al_2O_3 as gate dielectrics. *Appl. Phys. Lett.* **2009**, *94*, 212104. [CrossRef]
9. Bisson, S.; Kulp, T.; Levi, O.; Harris, J.; Fejer, M. Long-wave IR chemical sensing based on difference frequency generation in orientation-patterned GaAs. *Appl. Phys. B* **2006**, *85*, 199–206. [CrossRef]
10. Johnson, K.L.; Johnson, K.L. *Contact Mechanics*; Cambridge University Press: Cambridge, UK, 1987.
11. Hill, R. *The Mathematical Theory of Plasticity*; Oxford University Press: Oxford, UK, 1998; Volume 11.
12. Gao, R.; Jiang, C.; Walker, D.; Li, H.; Zheng, Z. Molecular dynamics study on mechanical cleavage mechanisms of GaAs and experimental verification. *Ceram. Int.* **2022**, *48*, 36076–36083. [CrossRef]
13. Li, Z.; Liu, L.; Wu, X.; He, L.; Xu, Y. Indentation induced amorphization in gallium arsenide. *Mater. Sci. Eng. A* **2002**, *337*, 21–24. [CrossRef]
14. Li, Z.; Liu, L.; He, L.; Xu, Y.; Wu, X. Shear-activated indentation crack in GaAs single crystal. *J. Mater. Res.* **2001**, *16*, 2845–2849. [CrossRef]
15. Wu, Y.; Huang, H.; Zou, J. Lattice bending in monocrystalline GaAs induced by nanoscratching. *Mater. Lett.* **2012**, *80*, 187–190. [CrossRef]
16. Parlinska-Wojtan, M.; Wasmer, K.; Tharian, J.; Michler, J. Microstructural comparison of material damage in GaAs caused by Berkovich and wedge nanoindentation and nanoscratching. *Scr. Mater.* **2008**, *59*, 364–367. [CrossRef]
17. Wasmer, K.; Parlinska-Wojtan, M.; Gassilloud, R.; Pouvreau, C.; Tharian, J.; Micher, J. Plastic deformation modes of gallium arsenide in nanoindentation and nanoscratching. *Appl. Phys. Lett.* **2007**, *90*, 03190. [CrossRef]
18. Wasmer, K.; Parlinska-Wojtan, M.; Graça, S.; Michler, J. Sequence of deformation and cracking behaviours of Gallium–Arsenide during nano-scratching. *Mater. Chem. Phys.* **2013**, *138*, 38–48. [CrossRef]
19. Gao, R.; Jiang, C.; Lang, X.; Dong, K.; Li, F. Experimental investigation of influence of scratch features on GaAs cleavage plane during cleavage processing using a scratching capability index. *Int. J. Precis. Eng. Manuf.-Green Technol.* **2021**, *8*, 761–770. [CrossRef]
20. Yu, B.; Gao, J.; Chen, L.; Qian, L. Effect of sliding velocity on tribochemical removal of gallium arsenide surface. *Wear* **2015**, *330*, 59–63. [CrossRef]

21. Chen, J.; Liang, Y.; Chen, M.; Bai, Q.; Tang, Y. A study of the subsurface damaged layers in nanoscratching. *Int. J. Abras. Technol.* **2009**, *2*, 368–381. [CrossRef]
22. Fan, P.; Katiyar, N.K.; Goel, S.; He, Y.; Geng, Y.; Yan, Y.; Mao, H.; Luo, X. Oblique nanomachining of gallium arsenide explained using AFM experiments and MD simulations. *J. Manuf. Process.* **2023**, *90*, 125–138. [CrossRef]
23. Gao, R.; Jiang, C.; Lang, X.; Zheng, Z.; Jiang, J.; Huang, P. Study on mechanical cleavage mechanism of GaAs via anisotropic stress field and experiments. *IEEE Trans. Semicond. Manuf.* **2022**, *35*, 633–640. [CrossRef]
24. Wang, J.; Yan, Y.; Jia, B.; Geng, Y. Study on the processing outcomes of the atomic force microscopy tip-based nanoscratching on GaAs. *J. Manuf. Process.* **2021**, *70*, 238–247. [CrossRef]
25. Wu, L.; Yu, B.; Fan, Z.; Zhang, P.; Feng, C.; Chen, P.; Ji, J.; Qian, L. Effects of normal load and etching time on current evolution of scratched GaAs surface during selective etching. *Mater. Sci. Semicond. Process.* **2020**, *105*, 104744. [CrossRef]
26. Fang, T.H.; Chang, W.J.; Lin, C.M. Nanoindentation and nanoscratch characteristics of Si and GaAs. *Microelectron. Eng.* **2005**, *77*, 389–398. [CrossRef]
27. Xuan, T.; Li, J.; Li, B.; Fan, W. Effects of the non-uniform magnetic field on the shear stress and the microstructure of magnetorheological fluid. *J. Magn. Magn. Mater.* **2021**, *535*, 168066. [CrossRef]
28. Mahmood, A.; Chen, S.; Chen, C.; Weng, D.; Wang, J. Molecular dynamics study of temperature influence on directional motion of gold nanoparticle on nanocone surface. *J. Phys. Chem. C* **2019**, *123*, 4574–4581. [CrossRef]
29. Li, B. Molecular Dynamics Simulations of Deformation Behavior of AlN in Nanoscratching. In Proceedings of the International Manufacturing Science and Engineering Conference, Online, 3 September 2020; American Society of Mechanical Engineers: New York, NY, USA, 2020; Volume 84256, p. V001T05A002.
30. Yin, Z.; Zhu, P.; Li, B.; Xu, Y.; Li, R. Atomic simulations of deformation mechanism of 3C-SiC polishing process with a rolling abrasive. *Tribol. Lett.* **2021**, *69*, 146. [CrossRef]
31. Chen, W.; Wang, W.; Liang, H.; Zhu, P. Molecular dynamics simulations of lubricant outflow in porous polyimide retainers of bearings. *Langmuir* **2021**, *37*, 9162–9169. [CrossRef]
32. Li, B.; Li, J.; Xu, J.; Xuan, T.; Fan, W. Effects of cracking on the deformation anisotropy of GaAs with different crystal orientations during scratching using molecular dynamics simulations. *Tribol. Int.* **2023**, *179*, 108200. [CrossRef]
33. Xu, X.; Fan, W.; Li, B.; Cao, J. Influence of GaAs crystal anisotropy on deformation behavior and residual stress distribution of nanoscratching. *Appl. Phys. A* **2021**, *127*, 690. [CrossRef]
34. Yi, D.; Li, J.; Zhu, P. Study of nanoscratching process of GaAs using molecular dynamics. *Crystals* **2018**, *8*, 321. [CrossRef]
35. Chen, C.; Lai, M.; Fang, F. Subsurface deformation mechanism in nano-cutting of gallium arsenide using molecular dynamics simulation. *Nanoscale Res. Lett.* **2021**, *16*, 117. [CrossRef] [PubMed]
36. Li, B.; Li, J.; Fan, W.; Xuan, T.; Xu, J. The dislocation-and-cracking-mediated deformation of single asperity GaAs during plowing using molecular dynamics simulation. *Micromachines* **2022**, *13*, 502. [CrossRef] [PubMed]
37. Fan, P.; Goel, S.; Luo, X.; Yan, Y.; Geng, Y.; He, Y.; Wang, Y. Molecular dynamics simulation of AFM tip-based hot scratching of nanocrystalline GaAs. *Mater. Sci. Semicond. Process.* **2021**, *130*, 105832. [CrossRef]
38. Fan, P.; Goel, S.; Luo, X.; Upadhyaya, H.M. Atomic-scale friction studies on single-crystal gallium arsenide using atomic force microscope and molecular dynamics simulation. *Nanomanuf. Metrol.* **2021**, *5*, 39–49. [CrossRef]
39. Fan, P.; Goel, S.; Luo, X.; Yan, Y.; Geng, Y.; Wang, Y. An atomistic investigation on the wear of diamond during atomic force microscope tip-based nanomachining of gallium arsenide. *Comput. Mater. Sci.* **2021**, *187*, 110815. [CrossRef]
40. Chen, C.; Lai, M.; Fang, F. Study on the crack formation mechanism in nano-cutting of gallium arsenide. *Appl. Surf. Sci.* **2021**, *540*, 148322. [CrossRef]
41. Li, Z.; Yan, Y.; Wang, J.; Geng, Y. Molecular dynamics study on tip-based nanomachining: A review. *Nanoscale Res. Lett.* **2020**, *15*, 201. [CrossRef]
42. Fan, P.; Goel, S.; Luo, X.; Yan, Y.; Geng, Y.; He, Y. Origins of ductile plasticity in a polycrystalline gallium arsenide during scratching: MD simulation study. *Appl. Surf. Sci.* **2021**, *552*, 149489. [CrossRef]
43. Rino, J.P.; Chatterjee, A.; Ebbsjö, I.; Kalia, R.K.; Nakano, A.; Shimojo, F.; Vashishta, P. Pressure-induced structural transformation in GaAs: A molecular-dynamics study. *Phys. Rev. B* **2002**, *65*, 195206. [CrossRef]
44. Kodiyalam, S.; Kalia, R.K.; Kikuchi, H.; Nakano, A.; Shimojo, F.; Vashishta, P. Grain boundaries in gallium arsenide nanocrystals under pressure: A parallel molecular-dynamics study. *Phys. Rev. Lett.* **2001**, *86*, 55. [CrossRef] [PubMed]
45. Prasolov, N.; Brunkov, P.; Gutkin, A. Molecular dynamics simulations of GaAs-crystal surface modifications during nanoindentation with AFM tip. *J. Phys. Conf. Ser.* **2017**, *917*, 092018. [CrossRef]
46. Güler, E.; Güler, M. Phase transition and elasticity of gallium arsenide under pressure. *Mater. Res.* **2014**, *17*, 1268–1272. [CrossRef]
47. Spreiter, Q.; Walter, M. Classical molecular dynamics simulation with the Velocity Verlet algorithm at strong external magnetic fields. *J. Comput. Phys.* **1999**, *152*, 102–119. [CrossRef]
48. Berendsen, H.J.; Postma, J.P.M.; van Gunsteren, W.F.; DiNola, A.; Haak, J.R. Molecular dynamics with coupling to an external bath. *J. Chem. Phys.* **1984**, *81*, 3684–3690. [CrossRef]
49. Van Gunsteren, W.; Berendsen, H.J. Algorithms for macromolecular dynamics and constraint dynamics. *Mol. Phys.* **1977**, *34*, 1311–1327. [CrossRef]
50. Stukowski, A. Visualization and analysis of atomistic simulation data with OVITO–the Open Visualization Tool. *Model. Simul. Mater. Sci. Eng.* **2009**, *18*, 015012. [CrossRef]

51. Štich, I.; Car, R.; Parrinello, M.; Baroni, S. Conjugate gradient minimization of the energy functional: A new method for electronic structure calculation. *Phys. Rev. B* **1989**, *39*, 4997. [CrossRef]
52. Erhart, P.; Albe, K. Analytical potential for atomistic simulations of silicon, carbon, and silicon carbide. *Phys. Rev. B* **2005**, *71*, 035211. [CrossRef]
53. Albe, K.; Nordlund, K.; Nord, J.; Kuronen, A. Modeling of compound semiconductors: Analytical bond-order potential for Ga, As, and GaAs. *Phys. Rev. B* **2002**, *66*, 035205. [CrossRef]
54. Ziegler, J.; Biersack, J.; Littmark, U. *The Stopping and Range of Ions in Solids*; Pergamon Press: New York, NY, USA, 1985.

Disclaimer/Publisher's Note: The statements, opinions and data contained in all publications are solely those of the individual author(s) and contributor(s) and not of MDPI and/or the editor(s). MDPI and/or the editor(s) disclaim responsibility for any injury to people or property resulting from any ideas, methods, instructions or products referred to in the content.

 *micromachines*

Article

Surface-Wetting Characteristics of DLP-Based 3D Printing Outcomes under Various Printing Conditions for Microfluidic Device Fabrication

Jeon-Woong Kang, Jinpyo Jeon, Jun-Young Lee, Jun-Hyeong Jeon and Jiwoo Hong *

School of Mechanical Engineering, Soongsil University, 369 Sangdo-ro, Dongjak-gu, Seoul 06978, Republic of Korea; kangjw159@gmail.com (J.-W.K.); junjynpyo@gmail.com (J.J.); ktino27@gmail.com (J.-Y.L.); bread951030@gmail.com (J.-H.J.)
* Correspondence: jiwoohong@ssu.ac.kr

Citation: Kang, J.-W.; Jeon, J.; Lee, J.-Y.; Jeon, J.-H.; Hong, J. Surface-Wetting Characteristics of DLP-Based 3D Printing Outcomes under Various Printing Conditions for Microfluidic Device Fabrication. *Micromachines* **2024**, *15*, 61. https://doi.org/10.3390/mi15010061

Academic Editor: Kun Li

Received: 15 November 2023
Revised: 21 December 2023
Accepted: 26 December 2023
Published: 28 December 2023

Copyright: © 2023 by the authors. Licensee MDPI, Basel, Switzerland. This article is an open access article distributed under the terms and conditions of the Creative Commons Attribution (CC BY) license (https://creativecommons.org/licenses/by/4.0/).

Abstract: In recent times, the utilization of three-dimensional (3D) printing technology, particularly a variant using digital light processing (DLP), has gained increasing fascination in the realm of microfluidic research because it has proven advantageous and expedient for constructing microscale 3D structures. The surface wetting characteristics (e.g., contact angle and contact angle hysteresis) of 3D-printed microstructures are crucial factors influencing the operational effectiveness of 3D-printed microfluidic devices. Therefore, this study systematically examines the surface wetting characteristics of DLP-based 3D printing objects, focusing on various printing conditions such as lamination (or layer) thickness and direction. We preferentially examine the impact of lamination thickness on the surface roughness of 3D-printed structures through a quantitative assessment using a confocal laser scanning microscope. The influence of lamination thicknesses and lamination direction on the contact angle and contact angle hysteresis of both aqueous and oil droplets on the surfaces of 3D-printed outputs is then quantified. Finally, the performance of a DLP 3D-printed microfluidic device under various printing conditions is assessed. Current research indicates a connection between printing parameters, surface roughness, wetting properties, and capillary movement in 3D-printed microchannels. This correlation will greatly aid in the progress of microfluidic devices produced using DLP-based 3D printing technology.

Keywords: three-dimensional printing technology; digital light processing; microfluidics; contact angle; contact angle hysteresis

1. Introduction

The advent of three-dimensional (3D) printing technology, which originated from the process of building up three-dimensional structures layer by layer with computer-aided design (CAD) drawings, has had a notable and beneficial influence on various aspects of everyday life [1,2] and in several industrial sectors, such as aerospace [3], automotive [4], and medical applications [5,6]. This can be attributed to its remarkable capacity for producing sophisticated structures, speedy development, and mass customization in contrast to conventional manufacturing methods [1,2].

There has recently been a significant increase in interest regarding the implications of 3D printing technology for the fabrication of microfluidic systems in place of traditional lithography methods, primarily using poly(dimethylsiloxane) (PDMS) [7,8]. This is because 3D printing allows automated, assembly-free 3D fabrication, offering rapid cost reduction as well as rapidly increasing resolution and throughput [7]. Several 3D printing techniques have been used in the field of microfluidics and its applications, including fused deposition modeling (FDM) [9], binder jet 3D printing [10], digital light processing (DLP) [11–13], stereolithography (SLA) [14,15], and selective laser sintering [16]. In particular, photopolymerization-based 3D printing technologies, such as SLA and DLP, have shown better print accuracy and print quality when printing chips and making 3D microchannels, even those with complex structures [17]. SLA and DLP are 3D printing techniques that utilize a laser beam (or UV light) and a digital light projector, respectively, to progressively expose liquid photosensitive material, causing it to gradually solidify into the desired object [17].

To create microfluidic systems with these 3D printing technologies and achieve the desired performance, fundamental research on the mechanical and physicochemical properties of photopolymerization-based printed products, based on the printing conditions, remains essential [18,19]. Several researchers have focused on the relationship between printing conditions and the mechanical properties of 3D-printed products in different printing methods [20–25]. For instance, Favero et al. [20] evaluated the effect of layer height on accuracy when a model was created with a 3D printer, based on an SLA scheme. Zhang et al. [21] examined the accuracies of DLP and SLA printers at various layer thicknesses and discovered the optimum layer thickness for these printing techniques. Liu et al. [22] explored the impact of printing-layer thickness on mechanical properties and optimized printing conditions through its modulation via fused deposition modeling-based 3D printing. For SLA-manufactured products, Saini et al. [23] investigated the effect of layer directions on mechanical properties, such as tensile, compression, flexural, impact, and fatigue characteristics. Ouassil et al. [24] studied the effect of printing speed on the porosity and tensile characteristics of fused filament 3D-printed materials. Jiang et al. [25] investigated how layer thickness affected the mechanical characteristics and molding accuracy of 3D-printed samples using a DLP scheme. The majority of these prior studies, however, have concentrated on the effects of printing parameters on the mechanical characteristics of 3D-printed objects.

In microfluidic systems, e.g., when the system size is reduced to the millimeter or micrometer scale, continuous or discontinuous flow (e.g., droplets) is primarily influenced by interfacial tension (or surface tension) rather than volumetric forces such as inertia and gravity [26,27]. Wettability characteristics, such as the contact angle and contact angle hysteresis (CAH, the difference between the advancing and receding contact angles), are widely recognized as crucial physical factors for understanding wetting and the capillary phenomena resulting from interfacial tension [27]. For instance, the physicochemical inhomogeneity of a solid surface causes CAH, which hinders the movement of discontinuous fluids [28,29], like a raindrop clinging to a window. Therefore, a thorough investigation of the wetting properties of 3D-printed objects as a function of printing conditions is essential and crucial to achieving the desired functionality of 3D-printed microfluidic systems.

In the present study, we methodically investigate the surface-wetting characteristics of DLP-based 3D printing outcomes under varied printing conditions to offer crucial and practical guidelines for fabricating microfluidic devices. We primarily examine the correlation between surface roughness and lamination thickness through microscopic and confocal microscopic image analysis. The variations in CA and CAH seen in aqueous and oily droplets are then investigated as a function of lamination direction and thickness. Finally, we analyze the dynamics of liquid flow when driven by capillary action inside simple microchannels that have been fabricated under different printing conditions to evaluate the effect of printing conditions on the functionality of a DLP 3D-printed microfluidic device.

2. Materials and Methods

To examine the effects of 3D printing conditions on the surface characteristics of products, we initially designed a cube of 10 mm and exported the design in the form of STL files using SolidWorks 2020, a professional 3D CAD program (Dassault Systèmes SolidWorks Corp., Waltham, MA, USA). Using Asiga Composer 1.3 (Asiga, Sydney, Australia), the design was segmented into layers and lamination conditions were set for layer thicknesses of 10 µm, 50 µm, and 100 µm. Cubic objects with different layer thicknesses were then manufactured utilizing a DLP-based 3D printer (Asiga MAX X27, Asiga, Australia) and printable resin (PlasClear V2, Asiga, Australia) (Figure 1). This printable resin is a type of

UV-curable resin from the diurethane dimethacrylate family. It is widely used in microfluidics due to its many advantageous properties, such as transparency and chemical stability against organic solvents [30,31]. The printing durations for layer thicknesses of 10 μm, 50 μm, and 100 μm were approximately 370, 84, and 51 min, respectively. The same objects were printed in a vertical direction to examine the impact of the lamination direction. To eliminate any remaining photocurable resin, we cleaned the printed products utilizing an ultrasonic cleaner (CPX8800H-E, Branson, MO, USA) and isopropyl alcohol. Subsequently, the printed products were post-cured through the application of UV light with a UV curing apparatus (Flash, Asiga, Australia) to diminish deformation and augment rigidity.

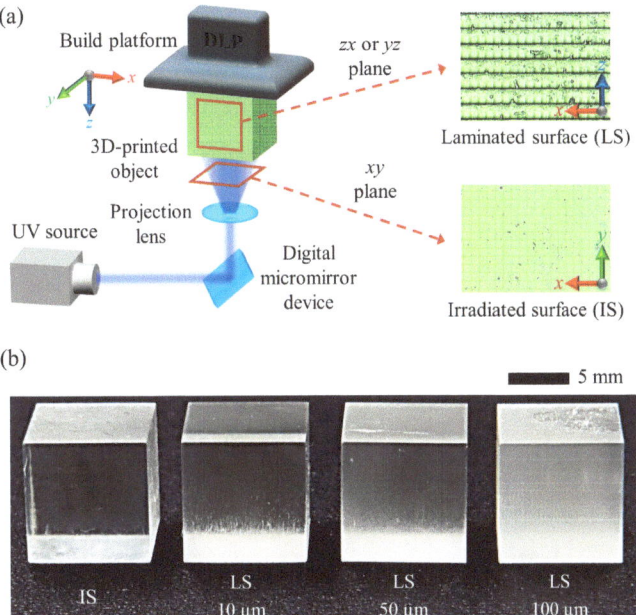

Figure 1. (**a**) Schematic of the DLP-based 3D printing process. (**b**) Actual photographs of the 3D-printed objects under different printing conditions. Here, IS and LS stand for irradiated and laminated surfaces, respectively, while the numerical values indicate the lamination thicknesses.

To qualitatively assess the surface roughness of 3D-printed objects created under various printing conditions, we captured microscopic images using an optical microscope (Eclipse Ci-L, Nikon, Tokyo, Japan) and a CCD camera (Fastcam mini UX100, Photron, Tokyo, Japan). A quantitative investigation of surface roughness as a function of printing conditions was also conducted using a confocal laser scanning microscope (LEXT OLS5000, Olympus, Tokyo, Japan) to enable the expected surface roughness and surface wetting characteristics to be correlated.

Aqueous and oil-liquid phases were prepared to assess the CA of droplets resting on the surfaces of 3D-printed objects created under varied printing conditions. Aqueous liquid phases included deionized (DI) water and a mixture of DI and Tween 20 (1 mM), while oil–liquid phases included mineral oil (Sigma-Aldrich, St. Louis, MO, USA), silicone oil with a viscosity of 50 cSt (Shinetsu, Tokyo, Japan), and hexadecane (Alfa Aesar, Haverhill, MA, USA). Here, Tween 20, otherwise known as polyoxyethylene (20) sorbitan monolaurate, is a water-soluble surfactant belonging to the polysorbate family [32]. It possesses the capability to modify the interaction between solids and liquids by reducing the surface tension of the liquid [33,34]. The surfactant was utilized in experiments to examine the effect of printing conditions on liquid flow in microchannels and to augment capillary action through the reduction of liquid surface tension. To improve the visibility of the aqueous droplets, a small amount (0.2 wt%) of a blue water-soluble dye was added to the aqueous liquid phase. Table 1 shows a summary of the physical properties of the aqueous and oil–liquid phases at room temperature. The viscosities and densities of the liquids were measured using a rotating viscometer (ViscoQC 300 L, Anton Paar, Graz, Austria) and an analytical balance (ME204T, Mettler Toledo, Columbus, OH, USA), respectively. Surface tension was also measured using the image processing of pendant droplets via the public-domain software ImageJ 1.53 (NIH Image, Bethesda, MD, USA). A micropipette was used to carefully dispense tiny aqueous and oil droplets with a volume of 5 µL onto the surfaces of 3D-printed objects, after which their images were captured using a DSLR camera (EOS 90D, Canon, Tokyo, Japan) equipped with a macro lens (MP-E 65 mm, Canon, Japan). The CA of the sessile droplets was measured from the acquired images using a low-bond axisymmetric drop shape analysis (LBADSA) approach via ImageJ software [35]. The LBADSA approach is based on fitting the Young–Laplace equation according to photographic images of axisymmetric sessile drops using a first-order perturbation method. This method is widely recognized for its ability to accurately measure the contact angle of a spherically shaped sessile droplet under low bond number conditions (using the ratio of the surface tension force to the gravitational force). It is freely available and is implemented as a plugin for the open-source program ImageJ [36].

Table 1. Physical properties of the aqueous and oil–liquid phases (23 ± 1 °C).

	Density (kg/m^3)	Viscosity (mPa·s)	Surface Tension (mN/m)
DI water	998	0.93	71.8
DI water + Tween 20 (1 mM)	998	1.01	37.5
Mineral oil	830	9.13	37.6
Silicone oil, 50 cSt	960	48	36.9
Hexadecane	774	3.46	27.3

To determine the CAH of droplets resting on the surfaces of 3D-printed objects, which is the difference between the advancing (ACA, θ_A) and receding contact angles (RCA, θ_R), we measured the ACA and RCA by inflating and deflating the droplet volumes, respectively (Figure 2). The ACA refers to the maximum CA just before the contact line moves forward on the surface as the droplet volume increases, while the RCA refers to the minimum CA just before the contact line moves backward on the surface as the droplet volume decreases. The ACA and RCA of the sessile droplets can be measured using the LBADSA approach via ImageJ software, which is similar to the CA measurement process.

To evaluate the effect of printing conditions on the functionality of a DLP 3D-printed microfluidic device, we designed a simple microfluidic device that enables the observation of the dynamics of liquid flow driven by capillary action (Figure 3). Figure 3b shows the geometry and dimensions of a microfluidic device. The microfluidic devices were fabricated utilizing a DLP-based 3D printer and using various lamination directions and thicknesses, such as those used for manufacturing the cubic objects. The microchannel features of the 3D-printed microfluidic devices for different lamination directions can be found in the Supplementary Materials. To observe capillary flows inside the microfluidic devices manufactured under different printing conditions, a droplet of 25 µL, consisting of a solution containing DI water and Tween 20 (1 mM), was initially deposited onto a reservoir of the microfluidic device using a micropipette. Subsequently, the displacement of liquid flow driven by capillary action was consecutively recorded using a DSLR camera (EOS 90D, Canon, Japan) equipped with a lens (AF-S NIKKOR 24–70 mm, Nikon, Japan). Finally, the captured images were subjected to digital image processing, using custom MATLAB® R2022a code to obtain data on the temporal progression of liquid flow. All experiments were completed with a minimum of three repetitions, and the reported data represent the average and standard deviations of the results.

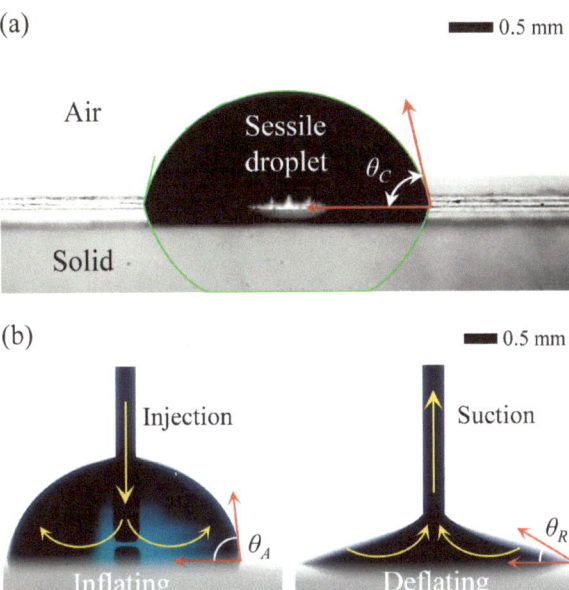

Figure 2. Measurements of: (**a**) the contact angle of a sessile droplet, using a low-bond axisymmetric drop shape analysis approach in ImageJ software; (**b**) changing the advancing and receding contact angles by inflating and deflating the droplet volume, respectively.

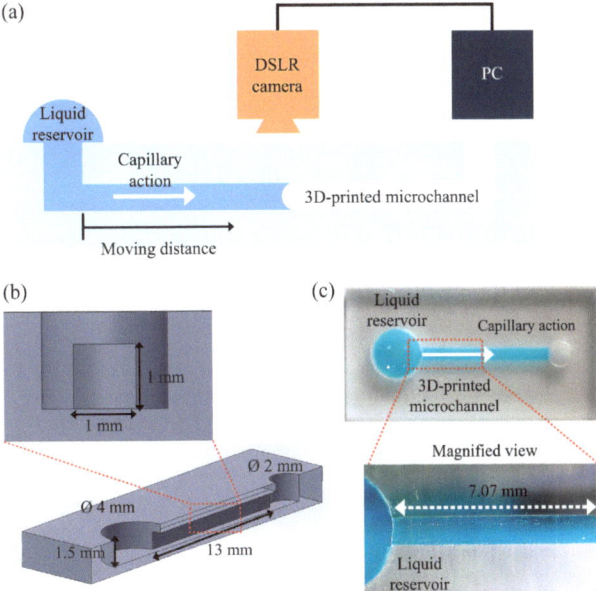

Figure 3. Simple capillary-driven microfluidic device used for evaluating the effect of printing conditions on the performance of DLP 3D-printing: (**a**) schematic, (**b**) drawing, and (**c**) actual photograph.

3. Results and Discussion

We initially examined the microscopic images of the surfaces of cubic objects created under various printing conditions to qualitatively investigate the relationship between printing conditions and surface roughness (Figure 4). The irradiated surface exhibits the lowest surface roughness compared to the other three surfaces while displaying a microscale lattice pattern of 27 μm in both width and height. This pattern results from the digital micromirror device (DMD) component, which comprises closely grouped small mirrors in a DLP printer. Each mirror of the DMD corresponds to a single pixel, resulting in a lattice pattern that matches the pixel resolution (27 μm) of the DLP printer used in this study. As expected, the surface roughness of the laminated surfaces was observed to increase as the layer thickness increased. In addition, when a layer was cured, the spot closer to the UV light source cured over a larger area than a spot that was farther away. Thus, even in the same layer, a spot closer to the light source had a greater height than other spots.

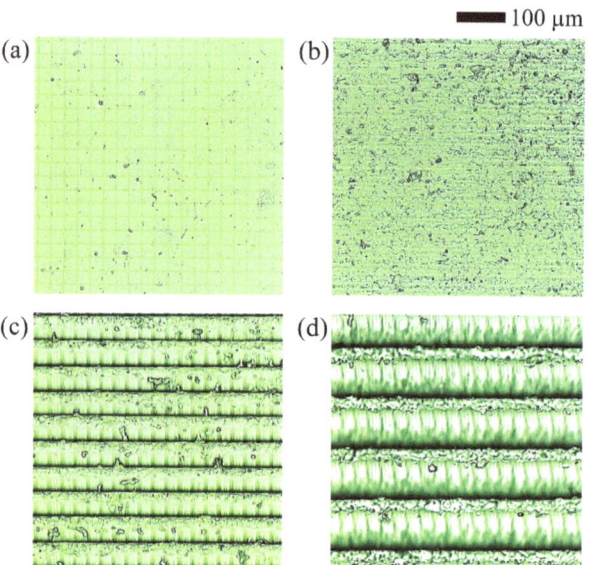

Figure 4. Microscopic images of (**a**) irradiated surfaces and (**b**–**d**) laminated surfaces of 3D-printed objects created under different printing conditions, specifically, horizontal lamination with layer thicknesses of 10 μm, 50 μm, and 100 μm, respectively.

We also conducted a quantitative assessment of the surface roughness of 3D-printed objects created under different printing conditions, using a confocal laser scanning microscope (LEXT OLS5000, Olympus, Japan), as shown in Figure 5. Confocal laser scanning microscopy is an optical imaging technique that improves the optical resolution and contrast of a micrograph and has the ability to capture multiple two-dimensional images at different depths within a sample, enabling the reconstruction of three-dimensional structures [37,38]. We employed the arithmetic mean roughness value (R_a) as the representative measure of surface roughness in this case, which was calculated by taking the average of all the profile values in the roughness profile. The arithmetic mean roughness values for irradiated and laminated surfaces with layer thicknesses of 10, 50, and 100 μm were approximately 0.06, 0.38, 2.57, and 7.03 μm, respectively. We quantitatively demonstrated a direct correlation between the increase in layer thickness and the increase in surface roughness. Due to the variations in roughness caused by different layer thicknesses, it is expected that the printing condition of layer thickness can affect surface wettability.

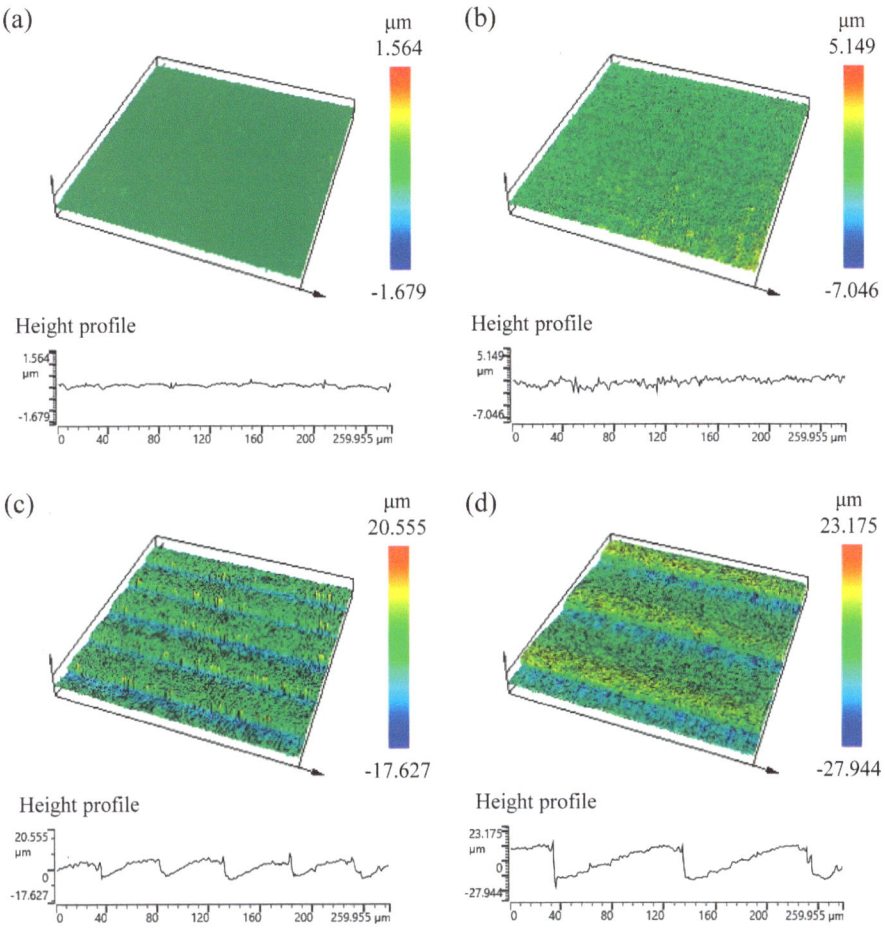

Figure 5. Confocal microscopic images and height profiles of (**a**) irradiated surfaces and (**b**–**d**) laminated surfaces of 3D-printed objects created under different printing conditions, specifically, horizontal lamination with layer thicknesses of 10 μm, 50 μm, and 100 μm, respectively.

Figure 6 shows the side and top views of aqueous and oil droplets resting on the irradiated and laminated surfaces of 3D-printed objects created with different layer thicknesses. All liquids, both aqueous and oily, except hexadecane, spread out axisymmetrically on an irradiated surface, giving each droplet the appearance of a sessile droplet. The aqueous droplets on the irradiated and laminated surfaces have an almost sessile droplet shape. However, the droplet shape is slightly distorted by the laminated grain as the layer thickness increases. In the case of oily liquids, the droplets spread unevenly, except on a few surfaces, due to the low surface tension. Based on these findings, we can infer that the thickness or direction of the lamination may affect the fluid dynamics seen on the 3D-printed surface. We will further investigate this assumption through a practical demonstration in the final examination of this study.

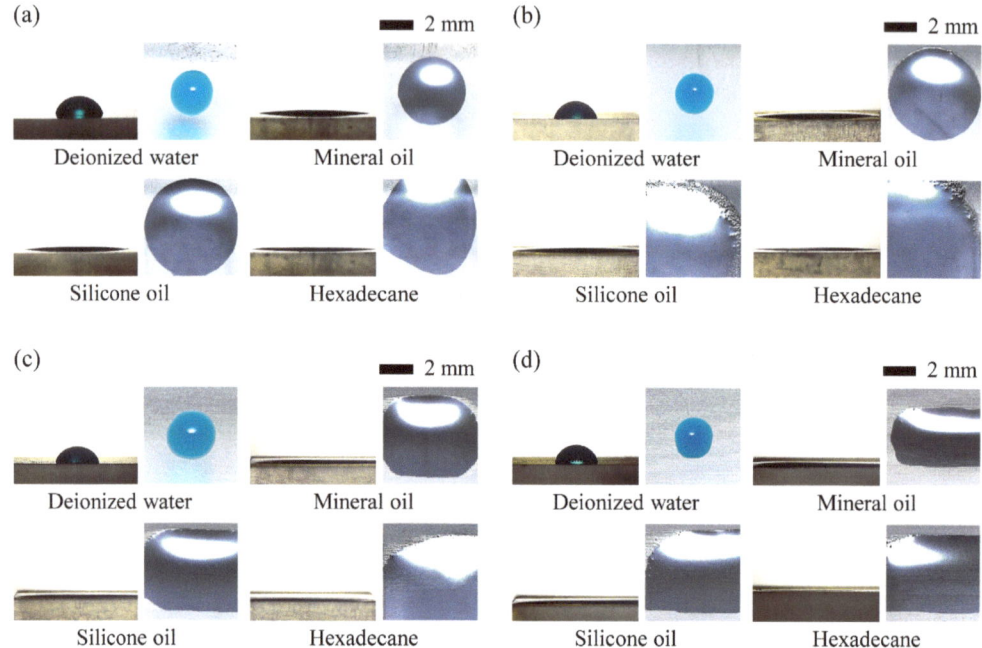

Figure 6. Side and top views of aqueous and oil droplets on: (**a**) irradiated surfaces; and (**b**–**d**) laminated surfaces with layer thicknesses of 10 μm, 50 μm, and 100 μm, respectively.

From the acquired image information in Figure 6, we measured the CAs of water and oil droplets on irradiated and laminated surfaces of different directions and thicknesses (Figure 7a). The CAs of DI water droplets on the horizontally laminated surfaces remained almost constant, regardless of the lamination thickness. However, the CA on the vertically laminated surfaces increased with increasing lamination thickness. The surface, especially at a lamination thickness of 100 μm, showed hydrophobic properties, as indicated by a CA of approximately 94°. This result may have occurred because when DI water droplets are placed on vertically laminated surfaces, they encounter greater resistance and pinning effects from the vertically laminated grain, which prevents them from spreading. As a result, the CA of these droplets is greater than that of droplets placed on horizontally laminated surfaces. However, in the case of droplets containing a mixture of DI water and Tween 20, the CA increases as the thickness increases, regardless of the lamination direction. The disparity in CA tendency between the two aqueous droplets may arise from the addition of surfactant, which reduces surface tension and significantly influences the solid–liquid interaction. Moreover, the CA is significantly greater on vertically laminated surfaces than on horizontally laminated surfaces. As a result, the fluidic behaviors in a 3D-printed microfluidic system are expected to be influenced by the lamination orientation. This will be demonstrated through a final assessment of this work. Finally, the oil droplets are unevenly spread in mineral oil placed on laminated surfaces with thicknesses of 50 and 100 μm and in silicone oil placed on laminated surfaces with thicknesses of 10, 50, and 100 μm, as shown in the top-view images in Figure 6. We were unable to collect any specific data on these behaviors due to the difficulty of determining and precisely measuring the oil droplet contact angle.

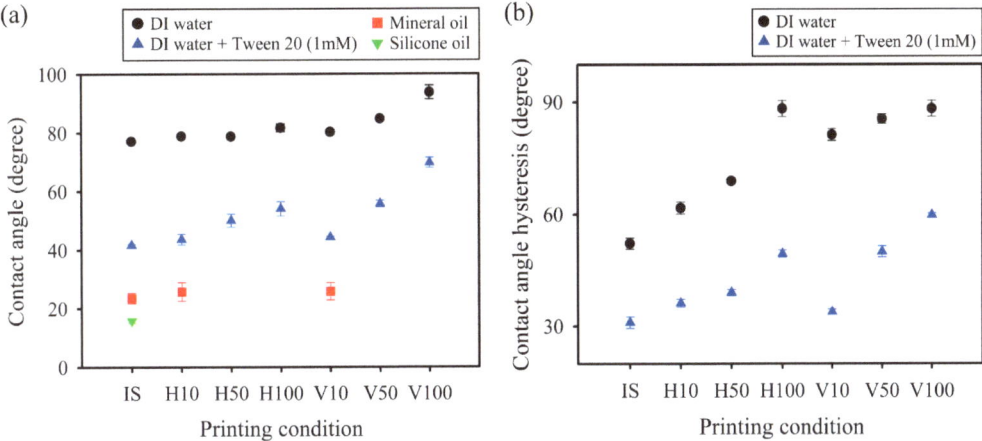

Figure 7. Aqueous and oil droplets on irradiated and laminated surfaces of varying directions and thicknesses: (**a**) contact angle and (**b**) contact angle hysteresis. IS, H, and V stand for irradiated, horizontally, and vertically laminated surfaces, respectively, while the numerical values indicate the lamination thickness.

The dependence of CAH of aqueous droplets on printing conditions was further investigated, as shown in Figure 7b. In contrast to the CA, it is evident that the lamination direction and thickness influenced the CAH of both aqueous droplets. This difference may be attributed to the fact that CAH is more affected by changes in surface roughness under different printing conditions than CA. Furthermore, the CAH increases with layer thickness, regardless of the lamination direction.

To assess the influence of printing conditions on the performance of a DLP 3D-printed microfluidic device, we observed the flow of a liquid consisting of a mixture of DI water and Tween 20 (1 mM), which was propelled by capillary action inside basic microfluidic devices under various printing conditions (Figure 8). A mixture of DI water and Tween 20 was used as the working fluid for the following reasons. The addition of Tween 20 surfactant resulted in a significant decrease in both CA and CAH, as previously shown in Figure 7b. A droplet consisting of a mixture of DI water and Tween 20 showed higher hydrophilic wettability on different printed surfaces than a droplet containing only DI water. In addition, a reduced CAH indicates a reduced pinning force, which hinders the movement of the liquid on the solid surface. As a result, we could experimentally investigate the effect of printing conditions on liquid flow in microchannels to improve capillary action by reducing the liquid surface tension and CAH. In a microfluidic channel created by vertical lamination with a thickness of 10 μm, the liquid rapidly enters the channel. However, when the layer thickness is increased to both 50 μm and 100 μm, the liquid barely flows. Conversely, in a microfluidic channel created via horizontal lamination, the liquid flow slows down as the layer thickness increases. The trends regarding whether the liquid flows and its speed may be correlated with its wetting characteristics, as shown in Figure 7. This correlation will be quantitatively analyzed and discussed later.

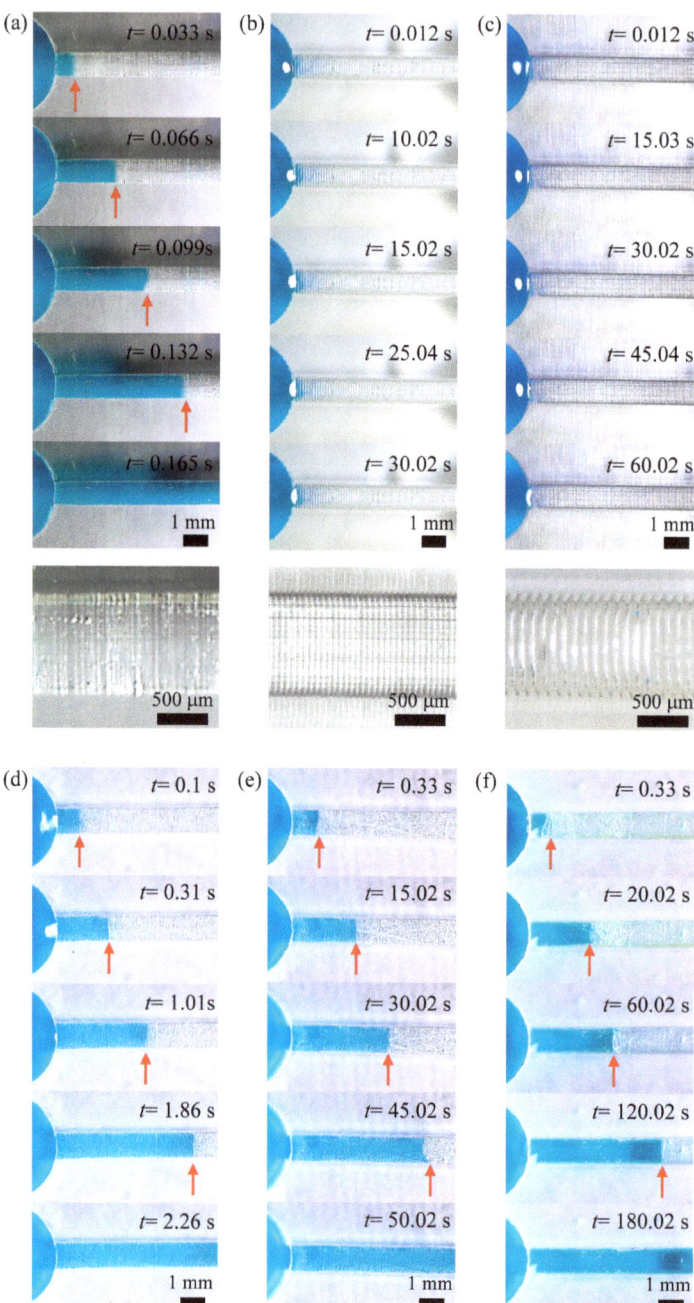

Figure 8. Consecutive images of capillary flow inside microfluidic devices fabricated under different printing conditions: images (**a–c**) correspond to printing conditions with a vertical lamination direction and layer thicknesses of 10 μm, 50 μm, and 100 μm, respectively. The bottom row exhibits microscopic photographs of the walls of microchannels created under different printing conditions; images (**d–f**) correspond to printing conditions with a horizontal direction and layer thicknesses of 10 μm, 50 μm, and 100 μm, respectively.

Using the acquired image data from Figure 8, we obtained quantitative information on the temporal evolution of capillary flows in microfluidic devices under different printing conditions (Figure 9). The microfluidic channel created by vertical lamination with a thickness of 10 µm showed the fastest capillary flow. In contrast, the microfluidic channel created by horizontal lamination exhibited slower capillary flow as the thickness increased. Based on the results presented in Figure 7, it is evident that the CA and CAH of droplets containing a mixture of DI water and Tween 20 decrease with decreasing layer thickness. Furthermore, an examination of Figures 7 and 8 shows that the CAH has a more pronounced effect on capillary flow in microfluidic devices than the CA. Contrary to our expectations, the capillary flow in the vertically laminated microchannels was faster than in the horizontal ones with a layer thickness of 10 µm. Possible reasons for this are the uncontrollable scratches found in the horizontally laminated channels and the small CAH in vertically laminated channels.

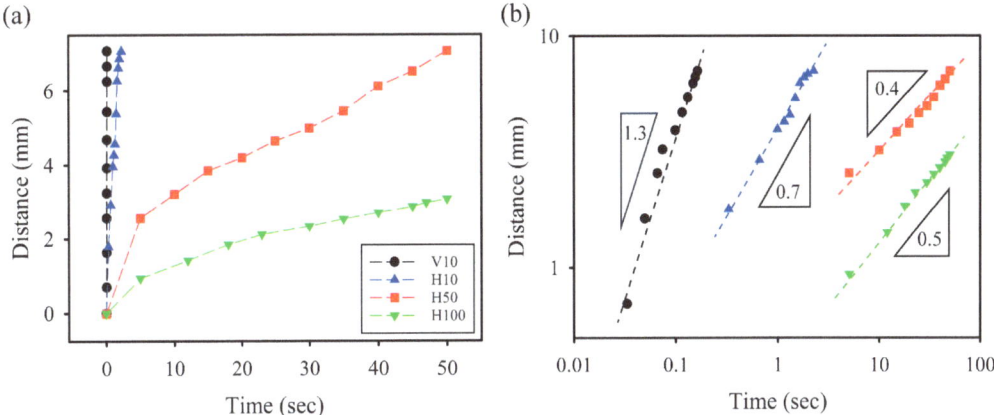

Figure 9. Temporal evolution of capillary flows inside microfluidic devices fabricated under different printing conditions: (**a**,**b**) are represented on a graph using linear and logarithmic scales, respectively. The dashed lines of (**b**) illustrate the power–law correlations between the moving distance of capillary flow and time. The triangle insets show the power–law exponents.

By adjusting the linear scales of the x and y axes in Figure 9a to a logarithmic scale, we established empirical relationships in the form of a power law between the distance traveled and the time needed for capillary flow (Figure 9b). In the case of relatively high capillary flow velocity—that is, in microfluidic channels constructed by laminating vertically and horizontally with a thickness of 10 µm—the distance traveled by the capillary flow tends to be roughly proportional to time. Conversely, in the case of relatively low capillary flow velocity—that is, microfluidic channels constructed by laminating horizontally with a thickness of 50 µm and 100 µm—the distance traveled by the capillary flow tends to be roughly proportional to the square root of time. These tendencies align closely with the power law correlation between the position of the advancing contact line and time, as discussed in previous literature on the dynamics of capillary flows in microchannels [39–42]. The position of the advancing contact line is known to be proportional to the time when the imbibed liquid passes through the inertial regime at short durations (<1 s, depending on the specific system) [42]. Conversely, the position of the advancing contact line is known to be the square root of time when long durations (<10 s, depending on the specific system) are reached, at which point the liquid enters the viscous or Lucas–Washburn regime [39,40].

The thickness of the horizontally laminated layers in 3D-printed objects has an effect on the surface roughness that can be evaluated by quantitative assessment with a confocal laser scanning microscope, as previously shown in Figure 5. It was also observed that the wetting properties, such as CA and CAH, of droplets containing a mixture of DI water and

Tween 20 increased with increasing layer thickness, as previously shown in Figure 7. The wetting properties, especially the CAH, influenced the flow of fluids by capillary action in the 3D-printed microchannels, based on the data in Figures 7 and 9. These results highlight the significance of assessing printing conditions before carrying out practical research when utilizing DLP printers to produce microfluidic devices.

4. Conclusions

This study investigates the surface wetting properties (e.g., CA and CAH) of 3D-printed items under different DLP printing conditions, including lamination direction and thickness. The thickness of the horizontally laminated layers in 3D-printed objects had an effect on the surface roughness, as seen in quantitative assessment with a confocal laser scanning microscope. The CA of the DI water droplets remained nearly constant on horizontally laminated surfaces but increased with increasing thickness on vertically laminated surfaces. Regardless of lamination direction, the CA increased with thickness for droplets containing a mixture of DI water and Tween 20. In contrast, the lamination direction and thickness influenced the CAH of both types of aqueous droplets. Moreover, the CA and CAH for oil droplets were limited to specific surfaces, due to uneven spreading. Thus, printing conditions showed a noteworthy effect on the efficiency of a microfluidic device produced through DLP 3D printing, as evidenced by the observed flow of liquid through microfluidic channels, driven by capillary action. This study offers essential and vital insights into grafting DLP-based 3D printing in various microfluidic applications, such as chemistry, materials science, medicine, biology, pharmaceuticals, and healthcare.

Supplementary Materials: The following supporting information can be downloaded at: https://www.mdpi.com/article/10.3390/mi15010061/s1, Figure S1: Schematic diagrams of (a) the microfluidic channels with vertical lamination and (b) the microfluidic channels with horizontal lamination.

Author Contributions: J.H. and J.-W.K. conceived the initial idea for this research. J.-W.K. and J.J. developed and conducted the experiments. J.-Y.L. and J.-H.J. performed image and data analysis. All authors were responsible for writing the paper. All authors have read and agreed to the published version of the manuscript.

Funding: This work was supported by the Soongsil University Research Fund (New Professor Support Research) of 2019.

Data Availability Statement: Data are contained within the article and supplementary materials.

Conflicts of Interest: The authors declare no conflicts of interest.

References

1. Huang, S.H.; Liu, P.; Mokasdar, A.; Hou, L. Additive manufacturing and its societal impact: A literature review. *Int. J. Adv. Manuf. Technol.* **2013**, *67*, 1191–1203. [CrossRef]
2. Mpofu, T.P.; Mawere, C.; Mukosera, M. The Impact and Application of 3D Printing Technology. *Int. J. Sci. Res.* **2014**, *3*, 2148–2152.
3. Najmon, J.C.; Raeisi, S.; Tovar, A. Review of additive manufacturing technologies and applications in the aerospace industry. In *Additive Manufacturing for the Aerospace Industry*; Froes, F., Boyer, R., Eds.; Elsevier: Amsterdam, The Netherlands, 2019; pp. 7–31.
4. Elakkad, A.S. 3D Technology in the Automotive Industry. *Int. J. Eng. Res.* **2019**, *8*, 248–251. [CrossRef]
5. Singh, S.; Ramakrishna, S. Biomedical applications of additive manufacturing: Present and future. *Curr. Opin. Biomed. Eng.* **2017**, *2*, 105–115. [CrossRef]
6. Ventola, C.L. Medical applications for 3D printing: Current and projected uses. *Pharm. Ther.* **2014**, *39*, 704.
7. Au, A.K.; Huynh, W.; Horowitz, L.F.; Folch, A. 3D-Printed Microfluidics. *Angew. Chem. Int. Ed.* **2016**, *55*, 3862–3881. [CrossRef]
8. Bhattacharjee, N.; Urrios, A.; Kang, S.; Folch, A. The upcoming 3D-printing revolution in microfluidics. *Lab Chip* **2016**, *16*, 1720–1742. [CrossRef] [PubMed]
9. Quero, R.F.; da Silveira, G.D.; da Silva, J.A.F.; de Jesus, D.P. Understanding and improving FDM 3D printing to fabricate high-resolution and optically transparent microfluidic devices. *Lab Chip* **2021**, *21*, 3715–3729. [CrossRef]
10. Achille, C.; Parra-Cabrera, C.; Dochy, R.; Ordutowski, H.; Piovesan, A.; Piron, P.; Van Looy, L.; Kushwaha, S.; Reynaerts, D.; Verboven, P.; et al. 3D Printing of Monolithic Capillarity-Driven Microfluidic Devices for Diagnostics. *Adv. Mater.* **2021**, *33*, 2008712. [CrossRef]
11. Olanrewaju, A.O.; Robillard, A.; Dagher, M.; Juncker, D. Autonomous microfluidic capillaric circuits replicated from 3D-printed molds. *Lab Chip* **2016**, *16*, 3804–3814. [CrossRef]

12. Parandakh, A.; Ymbern, O.; Jogia, W.; Renault, J.; Ng, A.; Juncker, D. 3D-printed capillaric ELISA-on-a-chip with aliquoting. *Lab Chip* **2023**, *23*, 1547–1560. [CrossRef] [PubMed]
13. Karamzadeh, V.; Sohrabi-Kashani, A.; Shen, M.; Juncker, D. Digital Manufacturing of Functional Ready-to-Use Microfluidic Systems. *Adv. Mater.* **2023**, *35*, 2303867. [CrossRef] [PubMed]
14. Au, A.K.; Lee, W.; Folch, A. Mail-order microfluidics: Evaluation of stereolithography for the production of microfluidic devices. *Lab Chip* **2014**, *14*, 1294–1301. [CrossRef] [PubMed]
15. Kanai, T.; Tsuchiya, M. Microfluidic devices fabricated using stereolithography for preparation of monodisperse double emulsions. *Chem. Eng. J.* **2016**, *290*, 400–404. [CrossRef]
16. Roy, N.K.; Behera, D.; Dibua, O.G.; Foong, C.S.; Cullinan, M.A. A novel microscale selective laser sintering (μ-SLS) process for the fabrication of microelectronic parts. *Microsyst. Nanoeng.* **2019**, *5*, 64. [CrossRef]
17. Quan, H.; Zhang, T.; Xu, H.; Luo, S.; Nie, J.; Zhu, X. Photo-curing 3D printing technique and its challenges. *Bioact. Mater.* **2020**, *5*, 110–115. [CrossRef]
18. Hsieh, M.T.; Ha, C.S.; Xu, Z.; Kim, S.; Wu, H.F.; Kunc, V.; Zheng, X. Stiff and strong, lightweight bi-material sandwich plate-lattices with enhanced energy absorption. *J. Mater. Res.* **2021**, *36*, 3628–3641. [CrossRef]
19. Xu, Z.; Ha, C.S.; Kadam, R.; Lindahl, J.; Kim, S.; Wu, H.F.; Kunc, V.; Zheng, X. Additive manufacturing of two-phase lightweight, stiff and high damping carbon fiber reinforced polymer microlattices. *Addit. Manuf.* **2020**, *32*, 101106. [CrossRef]
20. Favero, C.S.; English, J.D.; Cozad, B.E.; Wirthlin, J.O.; Short, M.M.; Kasper, F.K. Effect of print layer height and printer type on the accuracy of 3-dimensional printed orthodontic models. *Am. J. Orthod. Dentofac. Orthop.* **2017**, *152*, 557–565. [CrossRef]
21. Zhang, Z.C.; Li, P.L.; Chu, F.T.; Shen, G. Influence of the three-dimensional printing technique and printing layer thickness on model accuracy. *J. Orofac. Orthop.* **2019**, *80*, 194–204. [CrossRef]
22. Liu, Y.; Bai, W.; Cheng, X.; Tian, J.; Wei, D.; Sun, Y.; Di, P. Effects of printing layer thickness on mechanical properties of 3D-printed custom trays. *J. Prosthet. Dent.* **2021**, *126*, 671.e1–671.e7. [CrossRef] [PubMed]
23. Saini, J.S.; Dowling, L.; Kennedy, J.; Trimble, D. Investigations of the mechanical properties on different print orientations in SLA 3D printed resin. *Proc. Inst. Mech. Eng. C J. Mech. Eng. Sci.* **2020**, *234*, 2279–2293. [CrossRef]
24. Ouassil, S.-E.; Magri, A.E.; Vanaei, H.R.; Vaudreuil, S. Investigating the effect of printing conditions and annealing on the porosity and tensile behavior of 3D-printed polyetherimide material in Z-direction. *J. Appl. Polym. Sci.* **2023**, *140*, e53353. [CrossRef]
25. Jiang, T.; Yan, B.; Jiang, M.; Xu, B.; Gao, S.; Xu, Y.; Yu, Y.; Ma, T.; Qin, T. Study of Forming Performance and Characterization of DLP 3D Printed Parts. *Materials* **2023**, *16*, 3847. [CrossRef] [PubMed]
26. Squires, T.M.; Quake, S.R. Microfluidics: Fluid physics at the nanoliter scale. *Rev. Mod. Phys.* **2005**, *77*, 977. [CrossRef]
27. De Gennes, P.G.; Brochard-Wyart, F.; Quéré, D. *Capillarity and Wetting Phenomena: Drops, Bubbles, Pearls, Waves*; Springer: New York, NY, USA, 2004.
28. Johnson, R.E., Jr.; Dettre, R.H.; Brandreth, M.D. Dynamic contact angle and contact angle hysteresis. *J. Colloid Interface Sci.* **1977**, *62*, 205–212. [CrossRef]
29. Eral, H.B.; 't Mannetje, D.J.C.M.; Oh, J.M. Contact angle hysteresis: A review of fundamentals and applications. *Colloid Polym. Sci.* **2013**, *291*, 247–260. [CrossRef]
30. Richardson, A.K.; Irlam, R.C.; Wright, H.R.; Mills, G.A.; Fones, G.R.; Stürzenbaum, S.R.; Cowan, D.A.; Neep, D.J.; Barron, L.P. A miniaturized passive sampling-based workflow for monitoring chemicals of emerging concern in water. *Sci. Total Environ.* **2022**, *839*, 152620. [CrossRef]
31. Irlam, R.C.; Hughes, C.; Parkin, M.C.; Beardah, M.S.; O'Donnell, M.; Brabazon, D.; Barron, L.P. Trace multi-class organic explosives analysis in complex matrices enabled using LEGO®-inspired clickable 3D-printed solid phase extraction block arrays. *J. Chromatogr. A* **2020**, *1629*, 461506. [CrossRef]
32. Kerwin, B.A. Polysorbates 20 and 80 Used in the Formulation of Protein Biotherapeutics: Structure and Degradation Pathways. *J. Pharm. Sci.* **2008**, *97*, 2924–2935. [CrossRef]
33. Bąk, A.; Podgórska, W. Interfacial and surface tensions of toluene/water and air/water systems with nonionic surfactants Tween 20 and Tween 80. *Colloids Surf. A-Physicochem. Eng. Asp.* **2016**, *504*, 414–425. [CrossRef]
34. Kothekar, S.C.; Ware, A.M.; Waghmare, J.T.; Momin, S.A. Comparative Analysis of the Properties of Tween-20, Tween-60, Tween-80, Arlacel-60, and Arlacel-80. *J. Dispers. Sci. Technol.* **2007**, *28*, 477–484. [CrossRef]
35. Stalder, A.F.; Melchior, T.; Müller, M.; Sage, D.; Blu, T.; Unser, M. Low-bond axisymmetric drop shape analysis for surface tension and contact angle measurements of of sessile drops. *Colloids Surf. A-Physicochem. Eng. Asp.* **2010**, *364*, 72–81. [CrossRef]
36. Stalder, A.F. Biomedical Imaging Group of École Polytechnique Fédérale de Lausanne, Drop Shape Analysis—Free Software for High Precision Contact Angle Measurement. Available online: http://bigwww.epfl.ch/demo/dropanalysis (accessed on 13 December 2009).
37. Tata, B.V.R.; Raj, B. Confocal laser scanning microscopy: Applications in material science and technology. *Bull. Mat. Sci.* **1998**, *21*, 263–278. [CrossRef]
38. Al-Nawas, B.; Grötz, K.A.; Götz, H.; Heinrich, G.; Rippin, G.; Stender, E.; Duschner, H.; Wagner, W. Validation of three-dimensional surface characterising methods: Scanning electron microscopy and confocal laser scanning microscopy. *Scanning* **2001**, *23*, 227–231. [CrossRef] [PubMed]
39. Lucas, R. Rate of capillary ascension of liquids. *Kolloid-Zeitschrift* **1918**, *23*, 15–22. [CrossRef]
40. Washburn, E.W. The dynamics of capillary flow. *Phys. Rev.* **1921**, *17*, 273–283. [CrossRef]

41. Rideal, E.K. CVIII. On the Flow of Liquids under Capillary Pressure. *Lond. Edinb. Dublin Philos. Mag. J. Sci.* **1922**, *44*, 1152–1159. [CrossRef]
42. Lade, R.K.; Jochem, K.S.; Macosko, C.W.; Francis, L.F. Capillary Coatings: Flow and Drying Dynamics in Open Microchannels. *Langmuir* **2018**, *34*, 7624–7639. [CrossRef]

Disclaimer/Publisher's Note: The statements, opinions and data contained in all publications are solely those of the individual author(s) and contributor(s) and not of MDPI and/or the editor(s). MDPI and/or the editor(s) disclaim responsibility for any injury to people or property resulting from any ideas, methods, instructions or products referred to in the content.

Review

Radiation Synthesis of High-Temperature Wide-Bandgap Ceramics

Victor Lisitsyn [1,*], Aida Tulegenova [2,3,*], Mikhail Golkovski [4], Elena Polisadova [1], Liudmila Lisitsyna [5], Dossymkhan Mussakhanov [6] and Gulnur Alpyssova [7]

[1] Department of Materials Science, Engineering School, National Research Tomsk Polytechnic University, 30, Lenin Ave., Tomsk 634050, Russia; elp@tpu.ru
[2] Institute of Applied Science & Information Technology, Almaty 050042, Kazakhstan
[3] National Nanotechnology Laboratory of Open Type (NNLOT), Al-Farabi Kazakh National University, 71, Al-Farabi Ave., Almaty 050040, Kazakhstan
[4] Budker Institute of Nuclear Physics, SB RAS, 11, Lavrentiev Ave., Novosibirsk 630090, Russia; golkovski@mail.ru
[5] Department of Physics, Chemistry and Theoretical Mechanics, Tomsk State University of Architecture and Building, 2, Solyanaya Sq., Tomsk 634003, Russia; lisitsyna@mail.ru
[6] Department of Technical Physics, L.N. Gumilyov Eurasian National University, Astana 010000, Kazakhstan; dos_f@mail.ru
[7] Department of Radiophysics and Electronics, Karaganda Buketov University, Karaganda 100028, Kazakhstan; gulnur-0909@bk.ru
* Correspondence: lisitsyn@tpu.ru (V.L.); tulegenova.aida@gmail.com (A.T.); Tel.: +79-138242469 (V.L.); +770-79199951 (A.T.)

Citation: Lisitsyn, V.; Tulegenova, A.; Golkovski, M.; Polisadova, E.; Lisitsyna, L.; Mussakhanov, D.; Alpyssova, G. Radiation Synthesis of High-Temperature Wide-Bandgap Ceramics. *Micromachines* **2023**, *14*, 2193. https://doi.org/10.3390/mi14122193

Academic Editor: Kun Li

Received: 30 September 2023
Revised: 26 November 2023
Accepted: 27 November 2023
Published: 30 November 2023

Copyright: © 2023 by the authors. Licensee MDPI, Basel, Switzerland. This article is an open access article distributed under the terms and conditions of the Creative Commons Attribution (CC BY) license (https://creativecommons.org/licenses/by/4.0/).

Abstract: This paper presents the results of ceramic synthesis in the field of a powerful flux of high-energy electrons on powder mixtures. The synthesis is carried out via the direct exposure of the radiation flux to a mixture with high speed (up to 10 g/s) and efficiency without the use of any methods or means for stimulation. These synthesis qualities provide the opportunity to optimize compositions and conditions in a short time while maintaining the purity of the ceramics. The possibility of synthesizing ceramics from powders of metal oxides and fluorides (MgF_2, BaF_2, WO_3, Ga_2O_3, Al_2O_3, Y_2O_3, ZrO_2, MgO) and complex compounds from their stoichiometric mixtures ($Y_3Al_5O_{12}$, $Y_3Al_xGa_{(5-x)}O_{12}$, $MgAl_2O_4$, $ZnAl_2O_4$, $MgWO_4$, $ZnWO_4$, $Ba_xMg_{(2-x)}F_4$), including activators, is demonstrated. The ceramics synthesized in the field of high-energy electron flux have a structure and luminescence properties similar to those obtained by other methods, such as thermal methods. The results of studying the processes of energy transfer of the electron beam mixture, quantitative assessments of the distribution of absorbed energy, and the dissipation of this energy are presented. The optimal conditions for beam treatment of the mixture during synthesis are determined. It is shown that the efficiency of radiation synthesis of ceramics depends on the particle dispersion of the initial powders. Powders with particle sizes of 1–10 μm, uniform for the synthesis of ceramics of complex compositions, are optimal. A hypothesis is put forward that ionization processes, resulting in the radiolysis of particles and the exchange of elements in the ion–electron plasma, dominate in the formation of new structural phases during radiation synthesis.

Keywords: radiation synthesis; refractory dielectric materials; luminescence; high-power electron flux; ceramics

1. Introduction

Luminophores and scintillators are subject to stringent requirements regarding their response to the influence of the surrounding environment. These materials should not undergo any changes in their properties during the extended operation of the devices in which they are employed, even when operating conditions change. At the same time, they need to be sensitive to ultraviolet radiation and harsh radiation (high-energy particle flux, gamma, and X-ray radiation). Materials based on metal oxides and fluorides are used to achieve high resistance to potential external factors. High sensitivity to UV and harsh radiation is achieved by introducing activators and intrinsic structural defects (luminescent centers) that effectively convert absorbed radiation energy into light, which is detected by light-sensitive receivers.

Optical materials based on metal oxides and fluorides have found wide-ranging applications as phosphors for LEDs [1,2], scintillators [3,4], and long-lasting afterglow markers [5]. These materials are also employed as temperature sensors [6] and in transparent optical media [7]. The ability to visualize radiation flux fields using phosphors and scintillators has led to the development of entire branches of medicine. Effective methods of medical diagnostics, such as X-ray and tomography, are evolving and gaining widespread use [8–10]. Each application area requires materials with specific properties. There is a growing demand for the development of new materials with novel characteristics, complex elemental compositions, and structures. Metal oxides and fluorides are promising for use in the mentioned application areas, offering a multitude of material options to meet the increasing practical needs.

The synthesizing of refractory dielectric materials remains a complex task. It requires not only high temperatures but also the creation of new materials from basic elements with significantly different melting temperatures. Therefore, complex multi-step technological techniques are employed for synthesis, creating conditions that promote the exchange of elements in the initial materials through the addition of supplementary substances and mechanical manipulation.

Numerous research studies have been conducted aimed at developing and improving methods for synthesizing materials based on refractory oxides and metal fluorides. In works [11–13], a brief description of the synthesis methods employed and their comparison is provided.

The thermal methods have gained the widest acceptance [14,15]. In the thermal synthesis method, high-quality initial materials, typically in the form of finely dispersed powders with a specified stoichiometric composition, are carefully mixed and heated to temperatures below the melting point of the most easily melted component. Over an extended period of time in the heated mixture, partial sintering and element exchange occur. To expedite the process of diffusional element exchange between particles of varying composition in the mixture prior to heating, a flux is added, a substance with a melting temperature lower (approximately 70%) than that of the most easily melted component in the mixture. This serves to replace solid-state diffusion processes with significantly higher rates of liquid-phase diffusion. After cooling, the obtained ceramic is crushed into micrometer-sized particles. Subsequently, multiple high-temperature annealing cycles are conducted (at a temperature of approximately 80% of the melting temperature) for 40–50 h to complete the formation of the desired phase and the evaporation of the flux. Consequently, the process of forming a new phase through thermal methods is time-consuming, and it may be challenging to eliminate all substances introduced during synthesis. Nevertheless, this method is the most widely used and allows for the production of a high-quality final product with high reproducibility due to the careful adherence to technological regulations.

There are also alternative methods of synthesis [16–19]. For example, the sol–gel method involves the formation of molecules with the desired composition through chemical reactions from precursors. These reactions are carried out in an aqueous solution of nitrates of cations in stoichiometric proportions. Then, a lengthy multi-stage process of removing excess elements follows, along with multiple high-temperature annealing cycles to complete the formation of particle structure and size. The advantage of this method is the ability to obtain powders with a specified particle size distribution. However, the synthesis process is challenging to control and time-consuming.

Exploring the possibility of synthesizing refractory dielectric materials using the combustion method, synthesis in a burner flame [20–22] is also considered. The mixture prepared for synthesis is blended with combustible materials and heated to the ignition temperature of the fuel. The desired structure is formed in the high-temperature flame of the combustible. The synthesis time takes only a few minutes, which is the primary advantage of this method. However, post-synthesis cleaning of the obtained powder from residual combustibles is necessary.

In recent years, much attention has been given to the synthesis of refractory ceramics using the spark plasma sintering method [23]. Large currents, representing discharges within the material, are passed through the prepared mixture. The material rapidly melts, leading to the formation of new phases. Substances to enhance conductivity can be added to the mixture to increase current flow. Synthesis can be performed under high pressure and in any atmosphere. The synthesis is completed within minutes, and it is possible to obtain transparent ceramics. This method holds promise, and research is ongoing to explore the most optimal applications.

The influence of intense radiation fluxes during synthesis may promote the necessary solid-state reactions between elements in the environment and enhance the efficiency of new structure formation [24]. Exceeding certain power thresholds in radiation flux can lead to changes in the character of exchange reactions between particles in the environment, with reactions potentially involving short-lived radiolysis products. In [25–27], it is shown that high-power, high-energy electron fluxes can be used for synthesizing refractory dielectric materials with high efficiency.

It is established that the synthesis of ceramics from metal oxides and fluorides in the presence of a powerful high-energy electron flux takes less than 1 s, without the use of any synthesis-facilitating substances, and with high productivity. Further research is needed to understand the dependence of synthesis results, the efficiency of new phase formation, their properties, on the technological parameters of radiation synthesis, the history of initial substances, and the limits of universality of the developed method. It is presumed that processes related to the ionization of dielectric materials dominate during ceramic formation. Research into the processes occurring in the mixture under the influence of a powerful ionizing radiation flux is essential.

This work presents a compilation of existing results on the synthesis of luminescent inorganic materials based on metal oxides and fluorides and their analysis, as conducted by the authors up to the present time.

2. Ceramic Synthesis

The radiation synthesis method is fundamentally new. At the outset of the synthesis work, the radiation treatment modes, requirements for initial materials, and potential mutual influences between the radiation processing conditions and the properties of the initial materials were completely unknown. Below, we present, describe, and discuss the achieved results of synthesizing a group of ceramic samples differing in their physicochemical properties.

2.1. Ceramic Synthesis Methodology

Ceramic synthesis was achieved by directly exposing an electron beam to an initial mixture of a specified composition at the ELV6 electron accelerator at the Budker Institute of Nuclear Physics, Siberian Branch of the Russian Academy of Sciences. High-energy electron beams with energies of 1.4, 2.0, and 2.5 MeV were utilized for the synthesis. The resulting beam, extracted through a differential pumping system, exhibited a Gaussian-shaped profile with an area of 1 cm^2 on the target surface. Synthesis occurred when the threshold power density of the energy flux was exceeded. Since the penetration depth of electrons into the target increases with their energy, the power density of the beam was adjusted to account for changes in the absorbed energy distribution profile. For the synthesis of yttrium aluminum garnet (YAG) ceramics at 1.4 MeV, the optimal electron beam power densities were in the range of 20–25 kW/cm^2, while for 2.5 MeV, it was 37 kW/cm^2 [27].

To synthesize ceramics of the required composition, a mixture of oxide and fluoride powders was prepared in a stoichiometric ratio. To activate the process, oxide powders of rare-earth elements and tungsten were added in quantities ranging from 0.2% to 1.0% of the total mixture mass. Powders of various compositions, prehistory, and with different degrees of dispersion were used. The bulk density (mass of 1 cm^3 of powder) ranged from 0.2 g/cm^3 for nanopowders to 2.5 g/cm^3 for crushed crystals. The mixture was loaded into a massive copper crucible with a surface area of 10×5 cm^2 and depths of 7, 10, and 14 mm. The crucible's depth for the synthesis of specific ceramics was chosen to ensure the complete absorption of the electron beam of a specified energy by the mixture. The thickness of the mixture layer should exceed the linear range of penetration to prevent the crucible material from entering the forming ceramic. The linear penetration depth (in centimeters) is calculated from the known mass range of penetration σ (g/cm^2) and bulk density ρ (g/cm^3) according to the relationship: $l = \sigma/\rho$. The crucible was positioned under the accelerator's exit aperture on a substantial metal table. Two different methods of radiation treatment were used depending on the experimental goals:

1. The electron beam was scanned at a frequency of 50 Hz in the transverse direction of the crucible, while the crucible was moved relative to the scanning beam at a speed of 1 cm/s. This mode is referred to as "R1" in the text.
2. The electron beam was directed at the crucible, which was shifted without scanning at a speed of 1 cm/s relative to the electron beam. This mode is denoted as "R2."

The total exposure time of the electron beam on the treated surface of the mixture in the crucible was always 10 s due to the design features of the accelerator and the subject table. To achieve equal absorbed doses using both methods, the beam power in the "R2" mode was set to be 5 times lower than in the "R1" mode. Optimal values for the electron beam current densities for synthesizing ceramics from the mixture of selected initial powders were determined experimentally. For all the compositions investigated in this study, the sufficient electron beam power density with an energy of 1.4 MeV fell within the range of 10–25 kW/cm^2 in the "R1" mode or 2–7 kW/cm^2 in the "R2" mode. When using electron beams with energies of 2.0 and 2.5 MeV, the beam power density increased by 1.3 and 1.5 times, respectively, to achieve equal absorbed doses [27].

The synthesis of ceramics was achieved solely through the energy flux of radiation, using only the materials of the mixture, without the addition of other materials facilitating the process. The synthesis resulted in obtaining specimens in the form of plates with dimensions similar to the crucible: 10×5 cm^2 rods, as shown in Figure 1, or a series of specimens in the form of solidified droplets, rods. Samples from the same series, obtained in one experiment in one crucible, have identical structural and luminescent properties. Since the beam had a size in the target plane of 1 cm^2, each elementary surface area of the mixture was subjected to a 1 s exposure. In mode R2, the flux magnitude in each irradiated elementary area increased and then decreased according to a Gaussian law as the crucible moved relative to the beam. In mode R1, to achieve the same dose of radiation, the integral exposure time remained the same, but the radiation exposure to the elementary surface area of the target was carried out by a series of increasing and decreasing Gaussian-shaped envelope pulses with a duration of 2 ms and a period of 10 ms. Consequently, the synthesis was accomplished in less than 2 ms.

Figure 1. Photographs of YAG:Ce samples synthesized under the influence of an electron beam with E = 1.4 MeV: (**a**) P = 22 kW/cm^2, R1; b,b_1 P = 5 kW/cm^2, R2 in crucibles; (**b**) removed from the crucible; (**b_1**) traces of the impact of beams with P = 27 and 22 kW/cm^2, R1 mode, on a steel plate (c_1,c_2).

2.2. YAG:Ce Ceramics

For the synthesis of YAG:Ce ceramics, a mixture with a stoichiometric composition was prepared, consisting of 57 wt% Y_2O_3 and 43 wt% Al_2O_3. Cerium oxide (Ce_2O_3) was added for activation in an amount ranging from 0.2 to 2 wt% of the total weight. All initial materials had a purity level of no less than 99.5% and were thoroughly mixed. The YAG:Ce ceramic synthesis mixture consisted of Y_2O_3, Al_2O_3, Ga_2O_3 oxides, with 1 wt% Ce_2O_3, and a specified Al/Ga ratio. YAG:Ce ceramics were also synthesized with up to 15% substitution of yttrium by gadolinium (Gd). When using an electron beam with E = 1.4 MeV for synthesis in the R1 mode, power density values in the range of 10–25 kW/cm^2 were employed. A typical appearance of YAG:Ce ceramic plates and rods in the crucible obtained as a result of the synthesis is shown in Figure 1a,b.

When processing with an electron beam at E = 1.4 MeV and P = 22 kW/cm^2, a ceramic plate with dimensions of 40 × 90 × 7 mm^3 is formed in the crucible under condition R1. Exposure to an electron beam with E = 1.4 MeV and P = 5 kW/cm^2 in mode R2 leads to the formation of an YAG ceramic rod inside the mixture with an open upper surface. The rod has a thickness of approximately 5 mm. Often, a series of ceramic samples in the form of solidified droplets are formed in the crucible. The number of samples in one series in a single crucible can vary from one to ten, with sizes ranging from 3 to 30 mm, depending on the properties of the initial materials, irradiation modes, and bulk density. The samples have a solid outer shell with a thickness of up to 2 mm and a porous structure inside. The total weight of a series of samples in the crucible when exposed to an electron beam with E = 1.4 MeV usually ranges from 20 to 30 g. The maximum sample mass obtained under our synthesis conditions at E = 2.5 MeV was 83 g.

In the same figure, a photograph is provided showing traces of the impact on a steel plate (c_1 and c_2) by a flux of electrons with E = 1.4 MeV, P = 27, and 20 kW/cm^2, R1. When exposed to a flux with P = 27 kW/cm^2, the thin sub-millimeter layer on the surface of the metal plate exhibits melting, whereas at 20 kW/cm^2, a thin film appears. At lower P values, only an oxidation layer forms on the plate. Thus, the response of the dielectric and metallic materials to radiation treatment is entirely different.

The structure of the ceramics was studied using a D8 ADVANCE Bruker diffractometer with a CuKα radiation source. For qualitative phase analysis and diffraction pattern indexing, the following data from the PDF-2 database (ICDD, 2007) were used: PDF 01-089-6659 "Yttrium Gallium Aluminum Oxide ($Y_3Ga_2Al_3O_{12}$)," PDF 00-033-0040 "Aluminum Yttrium Oxide ($Al_5Y_3O_{12}$)," PDF 01-070-1677 "Yttrium Aluminium Oxide (YAlO$_3$)," and PDF 00-046-1212 "Aluminum Oxide (Al_2O_3)." The diffraction patterns of samples from four typical series, namely, YAG: Ce ceramics (1 and 2) consisting of Y_2O_3, Al_2O_3, Ga_2O_3 with 1 wt% Ce_2O_3, varying in the Al^{3+}/Ga^{3+} ratio; and YAG: Ce ceramics (3,4) with different morphologies of the initial materials, are shown in Figure 2. The presented research results indicate that the dominant structural type for all studied samples is yttrium aluminum garnet (YAG). Samples 1 and 2 are single-phase. Samples 3 and 4 contain yttrium aluminum perovskite (YAP) as a secondary phase with a content of approximately 7% and 11%, respectively.

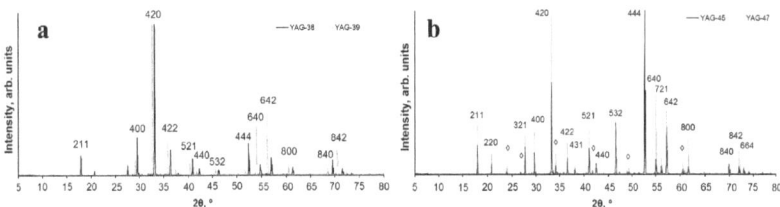

Figure 2. Diffraction patterns of YAG samples: (**a**) 1 (solid line) and 2 (dotted line); (**b**) 3 (solid line) and 4 (dotted line). Reflexes belonging to accompanying phases are marked with the symbol ◊.

YAG crystallizes in the cubic symmetry, possesses an elementary I-cell, and belongs to the space group $Ia\text{-}3d$. Ce^{3+} ions partially substitute for Y^{3+}. In samples 3 and 4, Al^{3+} ions occupy both tetrahedral and octahedral structural positions, whereas in samples 1 and 2, the octahedral position is primarily occupied by Al^{3+}/Ga^{3+} cations in approximately a 50/50 ratio, with the tetrahedral position predominantly occupied by Al^{3+} ions in sample 1 and Ga^{3+} ions in sample 2. Similar observations were made in [28].

Due to the difference in ionic radii between Al and Ga [29], the unit cell volume of $Y_3AlGa_4O_{12}$ (sample 2) significantly exceeds the volume calculated for the unit cell of $Y_3Al_4GaO_{12}$ (sample 1). This is reflected in Figure 2, where the diffraction peaks of sample 2 are shifted towards smaller 2θ angles. Samples 3 and 4 are nearly identical in their phase composition, containing YAG and YAP in approximately the same ratio, with the unit cell parameters for $Y_3Al_5O_{12}$ and $YAlO_3$ being very close. It can be concluded that the phase composition of the resulting sample depends on the composition and morphology of the initial material.

A cycle of research on the luminescent properties of synthesized YAG: Ce ceramics has been conducted, including photoluminescence (PL) and cathodoluminescence (CL) spectra under stationary conditions using the Cary Eclipse spectrofluorimeter for excitation and time-resolved cathodoluminescence spectra utilizing a pulsed electron accelerator with an energy of 250 keV [30]. In a generalized form, the spectral–kinetic properties of the obtained YAG: Ce ceramic are depicted in Figure 3b.

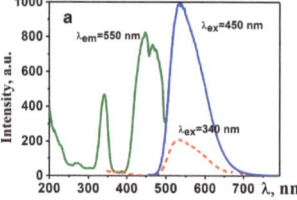 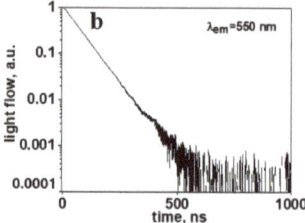

Figure 3. Excitation, luminescence (**a**), and decay kinetics spectra (**b**) of synthesized YAG: Ce ceramics.

Luminescence is efficiently excited by UV radiation in the range of 350 to 450 nm. The maximum of the broad luminescence band occurs at 550–560 nm when excited at λ = 450 nm. In samples of synthesized YAG: Ce ceramics, the luminescence spectrum exhibits typical characteristics of YAG: Ce phosphors with a dominant band around 550 nm and a characteristic relaxation time of 60–65 ns.

As evident from the presented results, the qualitative characteristics of luminescence (spectra, dynamics of their relaxation) are similar to those known for YAG: Ce phosphors [31,32]. In the synthesized YAG: Ce ceramic, crystallites with a structure characteristic of YAG:Ce crystals are formed.

2.3. Spinel

For the synthesis of $MgAl_2O_4$ ceramics, a mixture with a stoichiometric composition was prepared, consisting of 28.4% MgO and 71.6% Al_2O_3. For the synthesis of $ZnAl_2O_4$ ceramics, the batch comprised 44.4% ZnO and 55.6% Al_2O_3. Oxides RE (Eu, Er) were added for activation in an amount of 0.5% by weight. All initial materials had a purity level of not less than 99.5% and were thoroughly mixed.

During the radiation treatment of the mixture in the crucible, a zigzag-shaped ceramic plate forms across the entire area. Examples of the synthesized samples in the crucible are shown in the photographs in Figure 4. The samples resemble solidified melt. The formation of spinel ceramic samples occurs over a wide range of power densities used, ranging from 15 to 25 kW/cm^2, in the R1 mode. The samples are brittle, and their brittleness decreases with an increase in power density.

Figure 4. Photographs of MgAl$_2$O$_4$ (**a**) and ZnAl$_2$O$_4$ (**b**) sample synthesized at 27 kW/cm^2.

The structural properties of MAS were calculated and analyzed using X'pert highscore and Origin software (OriginPro 2018 (64 bit) SR1 b9.5.1.195). Diffractograms of the samples of magnesium aluminate spinel, pure and cerium- and erbium-ion activated, are presented in Figure 5.

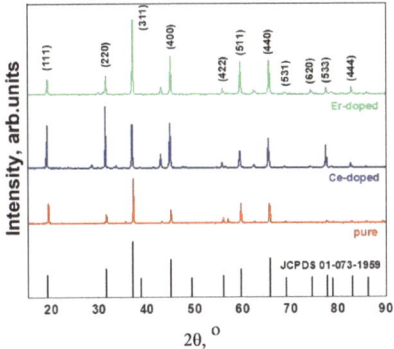

Figure 5. XRD spectra of MgAl$_2$O$_4$ spinel, pure and doped Ce and Er ions.

XRD results have shown that the synthesized MgAl$_2$O$_4$ by radiation method have cubic structure and are in the crystalline spinel MgAl$_2$O$_4$. In addition to the spinel main phase, there are weak peaks of the phase MgO. The weak peaks of phase MgO and no peaks related to Al$_2$O$_3$ indicate the rather high purity of the obtained MAS. The clear diffraction peaks demonstrate the good crystallinity of the synthesized MAS. The average crystallite size for samples with various impurities is approximately 48 nm. The structural characteristics are in good agreement with known data [33,34].

The luminescence properties of ceramic samples were investigated under various excitation conditions. Photoluminescence (PL) under excitation with λ_{ex} = 330 nm was measured using an Agilent Cary Eclipse spectrofluorimeter, while cathodoluminescence (CL) excitation was achieved using a pulsed electron accelerator with an energy of 250 keV [35]. The spectral characteristics of the synthesized MAS luminescence are presented in Figure 6.

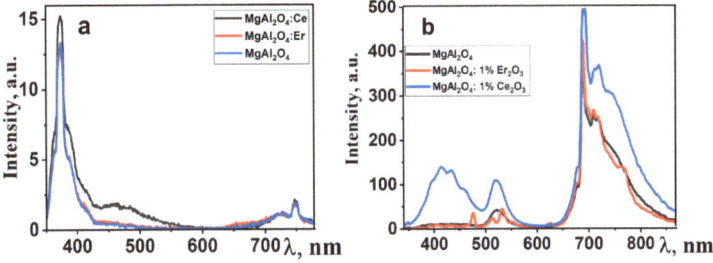

Figure 6. Photoluminescence (**a**) and cathodoluminescence (**b**) images of polycrystalline spinel samples doped with rare-earth elements.

In the photoluminescence spectra (Figure 6a), an intense emission band is observed in the UV region, along with broad emission in the "red" region of the spectrum. Under electronic excitation, the dominant emission in the spectrum is the "red" emission, both for pure and activated spinel samples. Additionally, there are bands with maxima at around 410 and 520 nm, and their intensity depends on the type of doping impurity. The emission in the region with λ_{max} = 380 nm may be associated with F-centers in the spinel structure. The emission band in the 700–760 nm range is attributed to oxygen vacancies. A characteristic feature of spinel is the emission of chromium ions, which appears as a set of narrow bands in the 700 nm region [36]. As evident from the presented results, the qualitative characteristics of luminescence are similar to well-known ones [37,38]. In this manner, in the synthesized $MgAl_2O_4$ ceramic, crystallites with a spinel-like structure are formed.

2.4. Fluorides

Fluorides of alkaline earth metals (MeF_2) and their solid solutions have been utilized as scintillation materials [39–41]. The synthesis of ceramics based on alkaline earth metal fluorides (MeF_2) is of interest for several reasons. The activation of crystals by multivalent ions makes them sensitive to the effects of ionizing radiation. However, introducing the most interesting ions for increasing sensitivity, such as W and U, is hindered by the fact that in a fluoride environment at temperatures of 20–60 °C, these ions form volatile compounds and are removed from the medium. Therefore, additives capable of retaining W and U ions in the melt are introduced into the crucible for the synthesis of such compounds. Incorporating hexavalent ions W and U into the lattice is complicated by the need to balance the charge difference, which can be resolved by using solid solution matrices, such as $BaMgF_4$. Finally, the melting temperatures of MeF_2 are 1.5–2 times lower than those of metal oxides. This is of interest in understanding synthesis processes in a radiation field. It can be expected that at a high rate of radiation synthesis, the activator ions will not have time to exit the ceramic formation zone [42].

Activated W ceramic samples of BaF_2, MgF_2, and $BaMgF_4$ were synthesized. For the ceramic synthesis, tungsten oxide was added to the mixture in amounts of 1–2 weight percent for activation. All initial materials had a purity level of no less than 99.5% and were thoroughly mixed.

When using an electron beam with E = 1.4 MeV under the R1 mode, power densities in the range of 10–20 kW/cm^2 were employed for synthesis. A typical appearance of activated W ceramic samples of BaF_2, MgF_2, and $BaMgF_4$ synthesized under the influence of an electron beam with E = 1.4 MeV and P = 15 kW/cm^2 in the R1 mode is shown in Figure 7.

Figure 7. Photographs of activated W samples of BaF_2 (**a**), MgF_2 (**b**), and $BaMgF_4$ (**c**) ceramics synthesized under the influence of an electron beam with E = 1.4 MeV, P = 15 kW/cm^2, R1.

During the radiation treatment of the mixture in the crucible, ceramic samples solidify in the form of a melt. Ceramic formation occurs over a wide range of used power densities from 10 to 20 kW/cm^2, in the R1 mode.

The results of phase analysis of the diffractograms for the synthesized fluoride samples of Ba and Mg using a power density of P = 15 kW/cm^2 are presented in Table 1.

Table 1. The results of studying the phase composition of magnesium fluoride samples synthesized at a power density of 20 kW/cm^2.

Main Phase	Secondary Phase
MgF$_2$	It is possible that WO$_3$ (tungsten trioxide) is present, with an impurity phase of 2–3%
BaF$_2$	Impurity phase of 2–3%
BaMgF$_4$	Impurity phase of 2–3%

The synthesis result depends on irradiation modes and the history of the initial substances. With changes in power density, the phase composition of the synthesized samples can undergo some variations: in MgF$_2$, MgO phase is detected, and in BaMgF$_4$, BaF$_2$ is preserved. The appearance of the MgO phase can be explained by oxidation: the synthesis is conducted in an open atmosphere. Nevertheless, XRD results show that during radiation treatment, ceramics with the mentioned phases, including complex ones, are formed.

A series of studies on the luminescent properties of W-activated samples of synthesized ceramics BaF$_2$, MgF$_2$, BaMgF$_4$ was carried out: photoluminescence (PL) excitation and emission spectra in stationary conditions using the SM2203 SOLAR spectrometer. In a generalized form, the spectra are presented in Figure 8. The figures present the results of spectral measurements of fluorescence from individual sections of the sample to demonstrate the homogeneity of properties. The inserts in the spectra show the excitation spectra of luminescence. The spectra highlighted in different colors in the graph (Figure 8) indicate that the spectra of the same sample from different regions are highlighted in different colors.

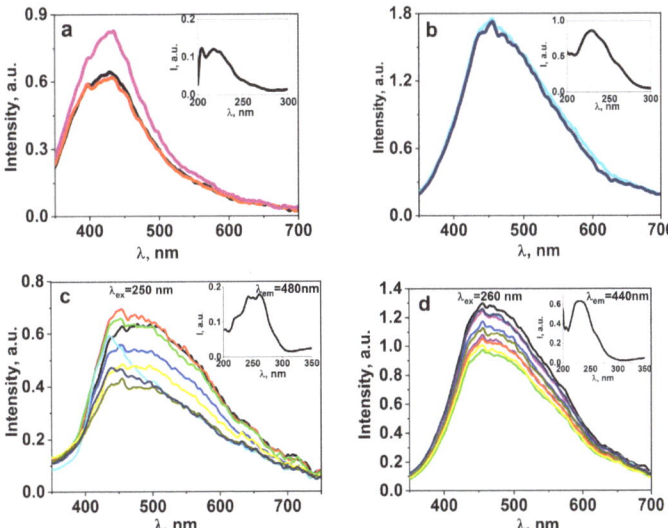

Figure 8. The FL spectra of BaMgF$_4$ (**a**) samples under excitation at 220 nm, as well as BaMgF$_4$:W (**b**), BaF$_2$:WO$_3$ (**c**), and MgF$_2$:WO$_3$ (**d**) samples under excitation in the 260 nm range.

Research on the spectral properties of synthesized ceramic samples has revealed two significant differences between activated and non-activated samples. In activated ceramic samples, the position of the luminescence band is shifted towards the longer-wavelength region. In non-activated samples, luminescence is excited by UV radiation up to 260 nm, while in activated samples, it extends up to 300 nm.

Kinetic decay curves were measured following excitation of cathodoluminescence using electron beam pulses with an energy of 250 keV and a duration of 10 ns. The measurement results are presented in Figure 9. In all activated ceramic samples, the appearance of a short-lived component is observed.

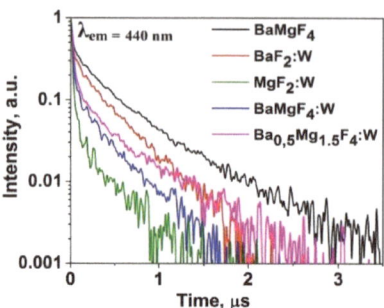

Figure 9. Kinetic decay curves of CL in ceramic samples.

The combination of the obtained research results on synthesized ceramic samples based on alkaline earth metal fluorides allows us to draw the following conclusion: during synthesis under the influence of intense radiation fluxes, tungsten can be successfully incorporated into the ceramic lattice. Within a short synthesis period, tungsten does not have sufficient time to exit the synthesis zone and remains trapped within the lattice.

2.5. Tungstates

Tungstates ($MeWO_4$, Me: Mg, Ca, Cd, Zn) are promising materials for use as scintillation materials [43–46]. To advance our understanding of the physicochemical processes during radiation synthesis in dielectric materials, tungstates are of interest due to the necessity to synthesize them from a mixture with significantly different melting temperatures: WO_3 (1473 °C), ZnO (1975 °C), MgO (2570 °C), CaO (2572 °C), CdO (1559 °C).

We have successfully synthesized ceramic samples of $MeWO_4$, where Me represents Mg, Zn, Ca, from the following mixture compositions: $MgWO_4$ (MgO 14.7%, WO_3 85.3%); $ZnWO_4$ (ZnO 26%, WO_3 74%); $CaWO_4$ (CaO 19.5%, WO_3 80.5%). After mixing, the mixture was poured into a crucible and compacted to even out the surface. All initial materials had a purity level of no less than 99.5% and were thoroughly mixed. For the synthesis, electron irradiation with E = 1.4 MeV at power densities ranging from 10 to 25 kW/cm^2 in R1 mode was employed. The typical appearance of the synthesized ceramic samples after exposure to an electron beam with E = 1.4 MeV, P = 18 kW/cm^2, R1 mode, is shown in Figure 10.

Figure 10. Photographs of ceramic samples (**a**) $ZnWO_4$, (**b**) $MgWO_4$, (**c**) $CaWO_4$, synthesized under the influence of an electron beam with E = 1.4 MeV, P = 18 kW/cm^2, R1.

In the process of synthesis, ceramic samples were formed in the shape of plates with dimensions similar to that of a crucible. It should be emphasized that ceramic formation occurs from a mixture composed of initial metal oxides with significantly different melting temperatures under identical conditions of radiation treatment.

The surface structure of the synthesized samples of $ZnWO_4$, $MgWO_4$, and $CaWO_4$ tungstates was investigated using a scanning electron microscope, Mira 3 (TESCAN). Since the examined samples are dielectrics, they were coated with a conductive carbon layer using a Quorum Q150R ES sputtering system. The investigation was conducted at an accelerating voltage of 25 kV.

SEM images of the measured samples are presented in Figure 11. On the surface of the ZnWO$_4$ samples, large formations of spherical shape with dimensions on the order of 180 µm, which can reach sizes of up to 600 µm, are visible. Upon increasing the image resolution, a porous microstructure with elongated elements ranging in size from 7 to 20 µm and a thickness of approximately 7 µm is observed. These elements may be microcrystals of the synthesized material.

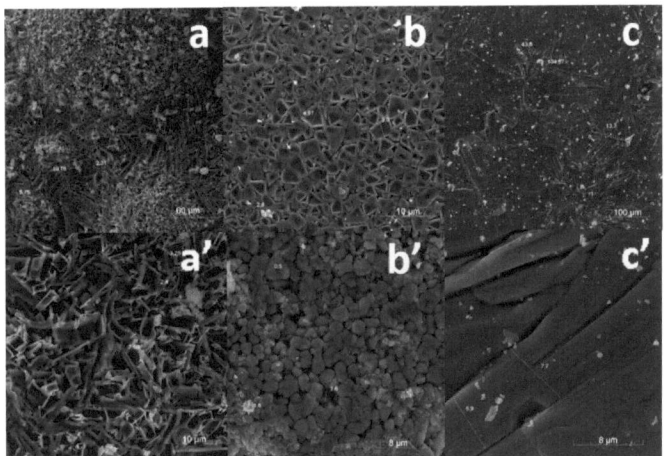

Figure 11. SEM images of the surface of ceramic samples (a,a′)ZnWO$_4$, (b,b′) MgWO$_4$, and (c,c′) CaWO$_4$ synthesized under the influence of an electron beam with E = 1.4 MeV, P = 18 kW/cm^2, R1.

On the surface of the MgWO$_4$ sample, a dense compact microstructure with inclusions (approximately 50 µm) and small holes (10–20 µm) is observed. With an increase in image resolution, tightly packed microcrystals of various shapes with average sizes ranging from 2 to 5 µm become visible.

On the surface of the CaWO$_4$ sample, densely packed microcrystals of elongated shape with an average length of 100 µm and a width of 7 µm are observed.

The presented results suggest the possibility of forming a crystalline structure in the synthesized samples.

2.6. Initial Materials for the Synthesis of Metal Oxides and Fluorides

The necessity of establishing the possibility of synthesizing ceramics from the materials used for the synthesis of the aforementioned functional materials is entirely evident. We conducted experiments on ceramic synthesis using powders of all the oxides and fluorides of the following metals: Y$_2$O$_3$, Al$_2$O$_3$, Ga$_2$O$_3$, MgO, WO$_3$, ZnO, ZrO$_2$, MgF$_2$, BaF$_2$. To synthesize the powders, they were placed in crucibles, and the conditions of radiation treatment were experimentally selected, including the mode and power density. Additionally, experiments were conducted for the synthesis of activated ceramics, where an activator in the form of its oxide was added to the powder. The introduction of the activator allowed us to demonstrate its potential incorporation into the crystalline lattice through luminescence techniques.

In Figure 12, photographs of ceramic samples made from the primary types of substances used in the synthesis of functional materials are presented. It was found that the synthesis results, primarily the efficiency of converting initial materials into ceramics, strongly depend on their prior history. We understand the prehistory not only in terms of the quality of the chemical composition but also in terms of dispersity, the sizes and quantity of particles, and their distribution by size. The available information about the obtained materials was insufficient for selection. Therefore, substances with different histories were used for synthesis. Figure 12 presents the results of the radiation synthesis of ceramic samples from the initial substances that provided the highest efficiency. More details about the dependence of synthesis efficiency on the history of the initial materials will be discussed below.

Figure 12. Photographs of synthesized ceramic samples under the influence of a 1.4 MeV electron beam in mode R1: Y_2O_3; Al_2O_3; MgO; ZrO_2 (P = 25 kW); ZnO (P = 22 kW); Ga_2O_3 (P = 18 kW); WO_3 (P = 17 kW); MgF_2 (P = 15 kW).

Among all the selected initial substances used for synthesis, except for WO_3, it was possible to obtain ceramic plates with dimensions comparable to the crucible sizes. Introducing additives into the mixture to obtain activated samples in the required quantity up to 2 wt% does not affect the synthesis results. The conditions of radiation treatment for ceramic synthesis, particularly power density, were primarily selected through experimental means.

One can draw the following conclusion from the above:

A series of experiments has been conducted to investigate the possibility of synthesizing ceramics based on oxides and metal fluorides by direct exposure to high-energy electron beams on a stoichiometric mixture.

It has been determined that under the influence of electron beams with energies in the range of E = 1.4–2.5 MeV and power densities of P ~10–25 kW/cm^2, it is possible to form ceramics, including activated ceramics, with a crystalline structure of YAG ($Y_3Al_5O_{12}$, $Y_3Al_xGa_{5-x}O_{12}$), spinels ($MgAl_2O_4$, $ZnAl_2O_4$), and metal fluorides (BaF_2, MgF_2, $BaMgF_4$). The synthesis of tungstates $MgWO_4$ and $ZnWO_4$ is also achievable under these conditions.

Under the given radiation exposure conditions, the synthesis of samples occurs at a rate of 1 cm^2/s. The synthesis time for an elementary section of ceramics from the mixture does not exceed 2 ms.

The synthesis of ceramics is achieved without the use of any additional substances that facilitate the process. Radiation exposure promotes efficient mixing of the mixture components.

During radiation synthesis, it is possible to introduce ions of activators and modifiers into the lattice, which is difficult to achieve using other methods. At high synthesis rates, ions do not have time to leave the reaction zone.

The combination of processes in the mixture during radiation synthesis differs significantly from those initiated by heating during thermal treatments. The synthesis of ceramics from the investigated dielectric materials, which have different melting temperatures, occurs within similar ranges of the applied conditions. The results of radiation exposure in these materials are significantly different from those in metals.

3. The Dependence of Radiation Synthesis Efficiency on the Prior History of the Initial Materials

Establishing the relationship between the efficiency of material synthesis and the state and properties of the initial substances is a multifactorial task. Therefore, these studies receive significant attention. Such research is ongoing because the technologies for obtaining initial materials are constantly improving, and their scope is expanding.

It is evident that the requirements for the initial materials vary depending on the use of different technologies. For instance, in thermal synthesis, the efficiency of synthesis (time, quality) is higher when the precursor particle sizes are smaller. Smaller particle sizes increase the likelihood of element exchange between particles. In radiative synthesis, the formation of new structures occurs in the electron-ion plasma created by a powerful stream of high-energy radiation. Therefore, it is necessary to investigate the dependence of the results of radiative synthesis on the properties of the initial substances. In [28], when studying the dependence of the efficiency of radiation synthesis on the history of the initial material, it was found that the synthesis outcome is significantly influenced by the dispersed composition of the substances used for synthesis. This dependence has a greater impact on the synthesis outcome than the degree of their purity. We conducted a series of studies aimed at investigating the dependence of the efficiency of radiation synthesis of ceramics based on metal oxides and fluorides on the particle sizes of the initial powders.

3.1. Experimental Assessment of Synthesis Efficiency

To discuss the dependence of synthesis results on the history of powders and radiation treatment conditions, the concept of "synthesis efficiency" is proposed to be introduced. By "synthesis efficiency" here and in the following context, we mean the ratio of the mass of the obtained series of samples in one crucible in a single experiment to the mass of the mixture in the crucible. It should be emphasized that this assessment provides fairly valuable information for establishing the relationship between the synthesis outcome and the conditions of radiation exposure and the history of the initial materials. However, it is important to remember that the obtained samples may be coated with remnants of the mixture in the crucible, which can sometimes be challenging to remove, especially when using the R2 mode. A layer of mixture thicker than the penetration depth of electrons is always added to the crucible to ensure that the crucible material does not affect the results. Finally, a portion of the mixture is dispersed in the radiation field due to the charging of dielectric particles and rapid heating of the air in the mixture. Therefore, the values of the synthesis reaction yield should be considered as approximate. Nevertheless, they provide a good understanding of the efficiency of synthesis of different materials and its dependence on the properties of the initial materials and radiation treatment modes.

Table 2 presents examples of the results of evaluating the efficiency of radiation synthesis of ceramics. To date, a total of 538 experiments have been conducted for the synthesis of ceramics with various compositions and different initial materials. The table primarily showcases the evaluation results of compositions with the highest synthesis efficiency. It is evident that these results can be achieved through radiation synthesis. To provide a comparison, the evaluation results of synthesis with low values are also included to establish the causes of the influence of the precursor's history.

As noted previously, the impact of a high-energy electron beam on the mixture in the crucible results in the formation of a series of samples ranging from one (yielding a plate) to 10–15. All samples within a given series exhibit similar properties [47], with differences in synthesis efficiency not exceeding 10%. It is worth noting that synthesizing ceramics from a mixture of the same composition and powders with the same history, under identical conditions of radiation treatment, also leads to the formation of specimens with similar properties and an efficiency difference not exceeding 20%. The text and tables include serial numbers for the samples according to the system used by the authors to track synthesis results.

As indicated by the results presented in Table 2, there is a significant dispersion in the values of reaction yield and mass loss. For instance, the synthesis reaction yields of MgO (2), Al_2O_3 (nano), and ZrO_2 (2) ceramics range from 0% to 5%. Meanwhile, under similar radiation exposure conditions, the synthesis reaction yields for MgO (K11) are 90.9%, Al_2O_3 (F-800) is 94.9%, and ZrO_2 (1) is 80%. Based on the available information, these two groups of materials differ in the dispersion of the initial substances used for synthesis.

We conducted studies on the morphology of Y_2O_3 powders of grades ITO/I, ITO/B, and Al_2O_3 grades F600–F1200, as well as nanooxide, using an optical microscope µVizo (LOMO). Microphotographs of the initial aluminum oxide powders are presented in Figure 13.

Table 2. Efficiency of radiation ceramic synthesis.

Continuity No	Sample	Description	Power P, kW	Mixture Weight, g	Mass Loss, %	Yield Sample/Mixture, %
407	Al_2O_3	F-800	26	40.53	1.8	94.9
493	Al_2O_3	nano	25	43.33	17.2	4.5
519	BaF_2	(K14)	15	88.12	0.8	76
516	MgF_2	(K13)	15	42.52	0.4	99.3
485	MgO	(K11)	25	26.94	5.7	90.9
486	MgO	G	35	39.94	90	4.6
388	MgO	MgO (1)	8	22.22	1.76	20.3
397	MgO	MgO (2)	8	14.19	46.65	0.0
528	WO_3	K12	17	95.4	4.95	77.2
439	WO_3	WO_3 (1)	24	100.38	35.02	46.1
490	Y_2O_3		25	56.01	1.3	91.9
495	ZrO_2	ZrO_2 (2)	5	56.32	52.1	0
146	ZrO_2	ZrO_2 (1)				80
474	Ga_2O_3		17	53.34	0.8	95.7
525	$MgAl_2O_4$: Eu	Al_2O_3 (K7), MgO (K11)	25	36.58	0.4	99
384	$MgAl_2O_4$: Eu	MgO (1), Al_2O_3 (F-800)	25	41.31	51.9	23.1
377	$MgAl_2O_4$: Eu	MgO K11, Al_2O_3 (F-800)	25	33.95	1.1	96.9
383	$MgAl_2O_4$: Eu	MgO (1), Al_2O_3 (F-800)	25	39.23	53.5	34.7
525	$MgAl_2O_4$: Eu	MgO (1), Al_2O_3 (F-800)	25	41.31	51.9	23.1
521	$BaMgF_4$	BaF_2 (K14) MgF_2 (K13)	15	81.57	0.5	92.8
335	$Y_3Al_5O_{12}$: Ce	Al_2O_3 (alund) Y_2O_3 (ITO-B)	25	35.2	0.7	97.8
338	$Y_3Al_5O_{12}$: Ce	Al_2O_3 (nano) Y_2O_3 (ITO-B)	25	33.38	32.2	53.02
505	$Y_3Al_5O_{12}$: Ce	Al_2O_3 (K7), Y_2O_3 (K6)		25	45.9	99.1
504	ZnAl GaO_4	ZnO K9, Al_2O_3 (K7) Ga_2O_3 (K9)	22	42.08	0.9	99.3
526	$ZnAl_2O_4$	ZnO K9, Al_2O_3 (K7)	25	43.3	1.1	98.3
512	$MgWO_4$	MgO (K11)	18	73.53	0.7	97.4
444	MgWO	MgO (1)	15	45.72	30.3	42.20

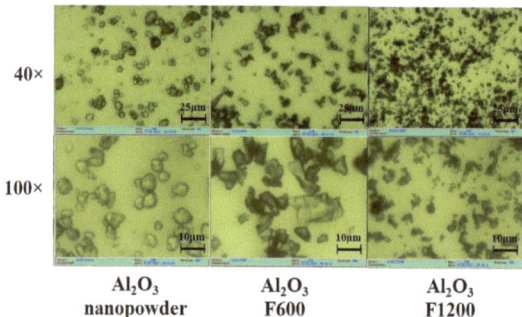

Figure 13. Microphotographs of the initial aluminum oxide powders on zoom.

The morphology of powders significantly varies. Nanopowder particles exhibit a clearly non-crystalline appearance, typical of agglomerated nanoparticles. The particles of F800 and F1200 powders appear as fragmented crystals with distinct fractures. The sizes of agglomerated nanoparticles and microcrystals are comparable. Therefore, radiation synthesis using nanoparticles is inefficient. A high synthesis efficiency is achieved when particles with sizes of 5–10 µm are used. It is worth noting that there is no such difference in thermal synthesis. The low efficiency of radiation synthesis may be attributed to significant differences in the processes of electronic excitation decay in bulk and nanocrystals. In nanocrystals, radiation-induced electronic excitations localize and decay on the surfaces of nanoparticles without producing radiolysis products.

It is hypothesized that synthesis efficiency is also low for powders of larger sizes. As indicated in the table, the synthesis of ceramics from MgO (G) powder is extremely low. Unfortunately, there is no information about the particle sizes for this material, but direct observations using an optical microscope have shown the presence of a large number of large, sub-millimeter-sized particles. Therefore, to establish the relationship between synthesis efficiency, one needs to know not only the particle sizes but also the full spectrum of particle sizes.

3.2. Dispersion of Initial Materials

To establish the validity of this assumption, we conducted a series of measurements of the dispersion of all currently used initial materials. Measurements were performed using the laser diffraction method with a laser particle size analyzer Shimadzu SALD-7101. Below, we will consider the results of the dispersion study of the most commonly used materials in the synthesis of MgO, Al_2O_3, Y_2O_3 powders, and the possible influence of dispersion on the radiation synthesis of ceramics.

Figure 14 shows the results of the dispersion analysis of MgO, Al_2O_3, and Y_2O_3 powders with different histories. The method used allows us to obtain relationships between the quantity of particles and their volume as a function of particle size. This may provide the opportunity to establish the nature of processes that determine the relationship between radiation efficiency and particle sizes and optimize the synthesis process by selecting the initial materials.

As can be seen from the presented measurement results of particle distribution, in all examined powders, the distribution of particles by size in terms of volume and quantity does not match. The volume of larger particles is usually much greater than the volume of smaller particles. This conclusion is based not only on the data shown in the figure but also on all 24 measured powders of MgO, Al_2O_3, Y_2O_3 with different histories. As seen from Table 2, the sample's mass is almost equal to the mass of the mixture. Therefore, synthesis primarily occurs from the larger particles in the mixture. It has been shown earlier that the efficiency of radiation synthesis is low for nanoparticles. Thus, there exists an optimal range of particle sizes for radiation synthesis, within which the formation of ceramics is most efficient. It is evident that radiolysis of larger particles is less probable than that of smaller ones.

Ceramics synthesis of the listed complex compositions from mixtures of MgO, Al_2O_3, Ga_2O_3, Y_2O_3, ZnO, WO_3, BaF_2, and MgF_2 has been realized. As expected, the efficiency of ceramics synthesis of yttrium aluminum garnets, spinels, tungstates, and alkali earth metal fluorides ($BaMgF_4$) depends on the dispersity of the initial powders in the mixture. The ceramics selected here are wide-bandgap dielectrics with a high degree of ionic bonding, where the process of electronic excitation decay into pairs of defects is efficient. The exit of decay products beyond the particle depends on its size. Large ceramic samples with sizes up to 10 cm^2 and thicknesses of 0.6–0.8 cm are obtained successfully from mixtures of powders with grain sizes of 1–10 µm. Mixtures of powders with different sizes (nano and microparticles) result in smaller samples with significantly lower yields. For example, the efficiency of $Y_3Al_5O_{12}$:Ce ceramics synthesis from a mixture of Al_2O_3 (K7) and Y_2O_3 (K6) is 99.1%, whereas from a mixture of Al_2O_3 (nano) and Y_2O_3 (ITO-V), the efficiency is 53%. It should be noted that the efficiency of synthesis of the same ceramics from a mixture of Al_2O_3 (BV) and Y_2O_3 (ITO-V) does not exceed 50%. In this mixture, the Al_2O_3 (BV) powder consists of particles with sizes of 10–50 μm (Figure 14). Thus, the efficiency of $Y_3Al_5O_{12}$:Ce ceramics synthesis is maximum when using mixtures of powders with similar dispersion.

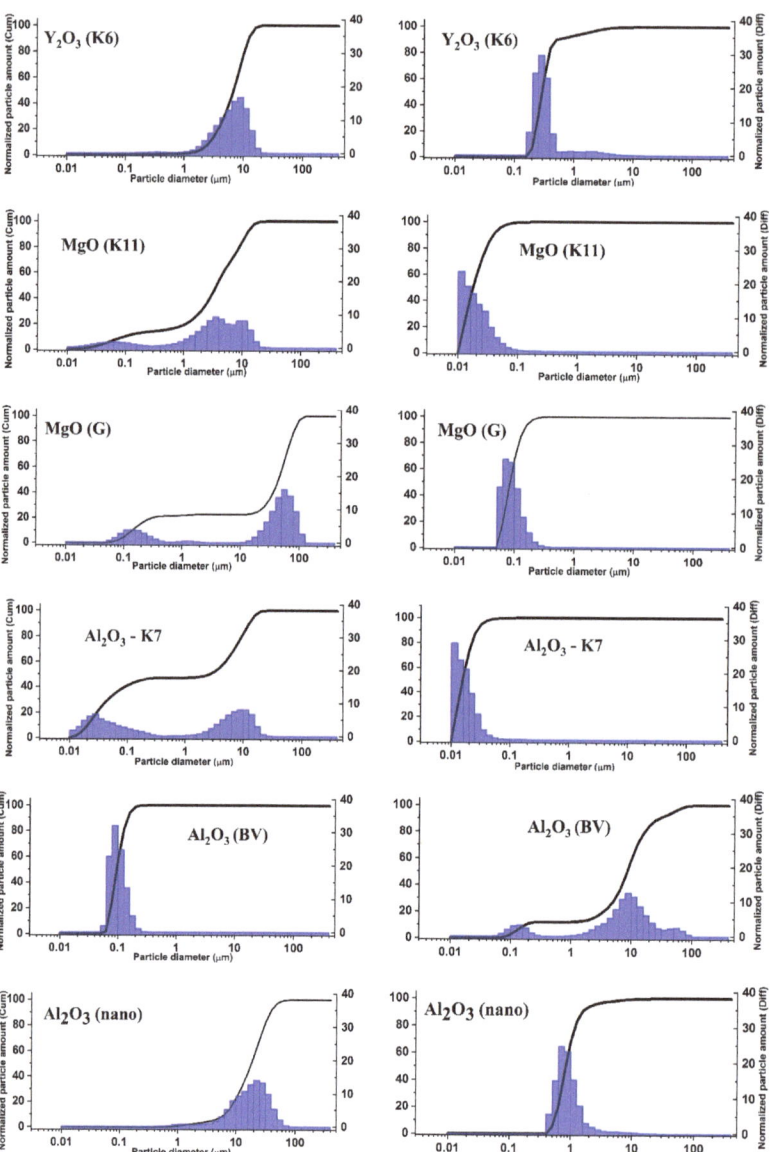

Figure 14. Dependence of the quantity of particles and volume on their sizes in the initial powders of MgO, Al_2O_3, and Y_2O_3. (**Left**) dispersion by volume; (**right**) dispersion by quantity of particles.

A similar conclusion can be drawn from the comparison of the dispersion of initial powders for the synthesis of $MgAl_2O_4$:Eu ceramics. The synthesis was performed using two variants of the initial mixture. Aluminum oxides of grades K7 and F-800 have very similar particle size distributions. The main volume consists of Al_2O_3 particles with sizes in the range of 3–12 μm. The sizes of MgO (K11) particles are in the range of 1–15 μm, and Al_2O_3 (K7) particles are in the range of 3–12 μm. The efficiency of $MgAl_2O_4$:Eu ceramics synthesis is 99%. From a mixture of Al_2O_3 (F-800) powders with particle sizes ranging from 3 to 12 μm and MgO (1) powders with particle sizes ranging from 0.4 to 2 μm, the synthesis occurs with an efficiency of 23%.

The reason for the dependence of the efficiency of ceramics synthesis of complex compositions on particle sizes is the difference in the dispersion of the initial compositions' powders. When their sizes are unequal, local non-stoichiometry may occur because large particles are surrounded by many small ones. Effective element exchange between such particles cannot occur in a short time. As seen from Table 2 and Figure 14, the synthesis efficiency is optimal when the dispersion of the mixture components is close.

It is worth noting that efficient radiation synthesis of three-component compositions is also possible. A high-efficiency (99.3%) synthesis of $ZnAlGaO_4$ ceramics was obtained from a mixture of ZnO, Al_2O_3, and G_2O_3 powders.

This work does not consider the possible dependence of synthesis efficiency on the purity of the initial substances. As demonstrated by the results of activated ceramics synthesis, introducing activator oxides into the mixture up to concentrations of 1% does not affect efficiency. Apparently, other possible impurities do not influence the synthesis results as much as the difference in dispersion does.

4. Energy Losses of an Electron Beam in a Material

In the modes of radiation treatment used for ceramic synthesis, the electron beam interacts with the mixture for a short duration. The beam has a Gaussian distribution in cross-section. Ceramic synthesis was carried out using electron beams of varying energy and power in two different modes: R1 (with scanning) and R2 (without scanning). The absorption of energy by the incident electrons in the material is non-uniform in depth. All of these characteristics of radiation treatment can influence the synthesis outcome. Therefore, a thorough analysis of the energy transfer processes from the electron beam to the material is required.

The energy loss of electrons with energies ranging from 0.5 to 5 MeV when passing through matter predominantly occurs due to ionization processes. The fraction of radiation losses is negligible. The mass depth of run of electrons, in close agreement with experimentally measured values, is calculated using empirical formulae [47,48]. With high precision, it can be calculated based on known values of substance density, average number of electrons in atoms, and the average ionization potential. Modeling the passage of electrons through a substance using the Monte Carlo method allows not only for determining the distribution of energy losses along the path of electron propagation; it also enables a rigorous estimation of energy loss distribution in the transverse cross-section relative to the beam propagation. Such estimations are important in cases where beams of limited cross-section are used, for example, of Gaussian shape. It is precisely such beams that were applied in the current study for ceramic synthesis. The high reliability and quantitative correspondence to experimental results are ensured by Monte Carlo calculations using the CASINO program [49–51].

The distribution of absorbed energy in the material when exposed to a non-uniform cross-sectional high-energy electron beam can be accurately calculated using the CASINO V2.5 program.

4.1. Energy Loss Modeling

Figure 15 shows profiles of energy loss distribution for electrons passing through a mixture with a bulk density of 1.2 g/cm^3 made of Y_2O_3 (57%) and A_2O_3 (43%) powders, which were used for synthesizing $Y_3Al_5O_{12}$ ceramics. The calculations were performed using the Monte Carlo method for beams with Gaussian flux distribution across the cross-section and a total of 10,000 incident electrons at energies of 1.4, 2.0, and 2.5 MeV, which were utilized in the experiments.

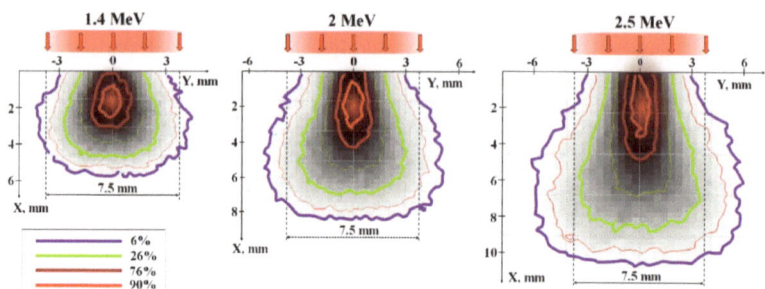

Figure 15. Energy loss distribution of electrons with E = 1.4, 2.0, 2.5 MeV in a mixture with a bulk density of 1.2 g/cm^3 for the synthesis of $Y_3Al_5O_{12}$ ceramics. Colored lines of equal losses are presented in units relative to the losses at the center.

When an incoming beam of electrons penetrates a material, it scatters off atoms and ions within the substance, transferring its energy to ionization and the generation of secondary electrons. As a result of these processes, the spatial structure of energy transfer within the beam changes as it progresses through the material. A portion of the energy is imparted to the material beyond the cross-section of the beam. Energy losses are concentrated along the axis of the beam. Approximately 50% of the total energy loss of the beam occurs within the region along the beam axis, with a cross-section of 0.3–0.4 relative to the beam's surface area. This results in a characteristic distribution of energy losses along the beam axis.

Figure 16a shows calculated profiles of energy loss distribution (dE/dx) of electrons within the material as a function of depth, assuming an equal number of incident electrons with energies of 1.4, 2.0, and 2.5 MeV. Here, energy losses of the electron beam as it passes through the material are understood as the magnitude of losses at depth X across the entire region perpendicular to the beam axis. The maximum absorbed energy is found at a certain depth from the surface, depending on the electron energies. The positions of the peaks in the dE/dx energy loss for the beams correspond to 2.8, 3.7, and 4.6 mm for the specified electron energies. The magnitude of energy losses at the peaks is 30–40% higher than at the surface of the target.

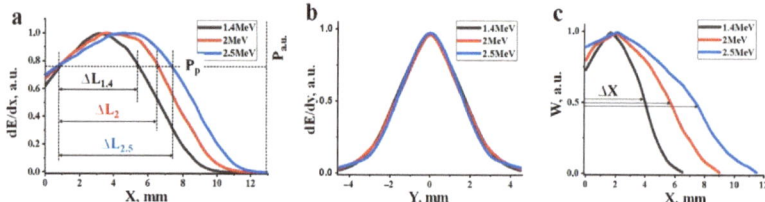

Figure 16. Profiles of energy loss distributions dE/dx (**a**), dE/dy (**b**) for electrons with energies of 1.4, 2.0, 2.5 MeV in a target, and the absorbed energy density W (**c**).

Figure 16b presents calculated profiles of the energy loss distribution of electrons (dE/dy) in the direction perpendicular to the beam axis. From the results presented in Figure 15, it is evident that the profiles of dE/dy change with depth. These changes are different for electron beams of different energies. The profiles of dE/dy shown in the figure correspond to depths that correspond to the maximum density of absorbed energy (Wr) of the electron beam. Here, the absorbed energy density (W) is defined as the energy loss per unit volume of material. The dE/dy profiles coincide, indicating that under the calculation conditions with an equal number of incident electrons, the values of Wr are the same.

Energy loss densities along the axis of the passing beam are always higher than off-axis and exhibit a curve with a maximum. Figure 16c shows calculated profiles of the dependence of W on the depth of electron penetration into a material with a specified composition and bulk density. These dependencies appear as curves with maxima at 1.8, 2.1, and 2.3 mm for electrons with energies of 1.4, 2.0, and 2.5 MeV, respectively. The length of the region (Δx) with equal absorbed energy density along the beam axis increases by a factor of 2 on average as the electron energy increases from 1.4 to 2.5 MeV. It is worth noting that the maximum of energy loss (dE/dx) occur at depths of 2.8, 3.7, and 4.6 mm for energies E equal to 1.4, 2.0, and 2.5 MeV, respectively. However, the length of the region with equal absorbed energy along the beam axis increases by only 25% on average. This is due to the expansion of the energy loss region towards the end of the range. Consequently, the maximum of energy loss with depth (Figure 16a) (dE/dx) and energy loss densities (W) (Figure 16c) do not coincide.

Clearly, regions of maximum energy loss densities should govern subsequent processes. It is within these regions that ionization density is maximized, and the material reaches its highest temperature. In these regions, when the energy loss exceeds a certain threshold, the crystalline structure of yttrium and aluminum oxides transforms into yttrium aluminum garnet. Primarily, synthesis should take place at depths corresponding to the maximum energy loss densities along the beam axis, not just energy loss.

With changes in the power of the electron beam (P), the distribution profile remains unchanged, but the absolute values of energy losses (dE/dx and dE/dy) change proportionally. Synthesis occurs in the material when the energy losses (dE/dx and dE/dy) in a specific region of the material with coordinates X-Y exceed a certain threshold power (Pp) of the beam. The threshold (Pp) at which synthesis can occur depends solely on the composition of the initial mixture, i.e., the composition of the synthesized material. Synthesis may not occur at the surface of the target and at great depths but rather within a depth range (ΔL) where energy losses exceed the required threshold for synthesis. The depth range (ΔL) where synthesis can occur increases with increasing electron energy (E). Synthesis takes place within depths where energy loss densities (W) exceed the required threshold for synthesis. As the power of the electron beam (P) increases, the length and diameter of the region with maximum energy losses (W) along the beam axis increases. Therefore, synthesis can occur in a larger volume where energy loss densities exceed the synthesis threshold (Pp).

In summary, the volumetric energy loss density undergoes significant changes in both the longitudinal and transverse directions to the electron beam propagation in the material. As the power density of the incident electron beam increases, energy losses proportionally increase. In regions of maximum energy loss density, material synthesis is most likely to occur. With higher electron beam power, the volume in which synthesis can occur increases, with the upper limit potentially reaching the surface of the target and the lower limit extending to depths equal to the extrapolated electron range (X_e). The extrapolated range depth is defined as the value on the depth-of-penetration axis (X_e) at which the axis intersects the tangent to the energy loss curve at its descent at the inflection point of the function. The extrapolated electron range in the mixturet for YAG:Ce ceramic synthesis (Figure 15) is 9, 10, and 11 mm for electrons with energies of 1.4, 2.0, and 2.5 MeV, respectively.

4.2. Experimental Verification of Energy Loss Distribution

A research cycle was conducted to investigate the dependence of the efficiency of YAG:Ce ceramic radiation synthesis on electron energy and beam power. The concentration of the Ce activator introduced for activation was 0.5%. Such a quantity of activator does not affect the fundamental energy loss patterns but allows confirmation through luminescent methods that Ce is incorporated into the ceramic's crystalline structure. YAG:Ce ceramic synthesis was performed in copper crucibles with depths of 10 or 14 mm, exceeding the full electron range from 1.4 to 2.5 MeV, and dimensions of 50×100 mm^2. The ceramic synthesis was carried out in R1 and R2 modes.

Since the distribution of absorbed energy by the electron beam in the material is non-uniform, it is necessary to define criteria for selecting irradiation conditions for a correct comparison of synthesis results using electron beams with different energies. The maximum energy loss densities in the irradiated material region should be similar. Our previous studies have shown that when using an electron beam with E = 1.4 MeV, the synthesis of YAG:Ce ceramics in R2 mode is successful when delivering 4 kJ/s·cm^3 of energy to the central region with a bulk density of 1.2 g/cm^3, which corresponds to 50% of the absorbed energy. Such energy absorption density is achieved with a beam power of 5 kW/cm^2 under the used irradiation conditions. When irradiated with electrons of higher energies, 50% of the absorbed energy occurs in a larger volume along the beam path. Based on the study of the dependence of absorbed energy distribution on electron energy (Figures 15 and 16), we have shown that the electron beam power should be 1.4 times higher for electrons with E = 2.0 MeV and 1.8 times higher for electrons with E = 2.5 MeV. The adjustment of synthesis modes was carried out experimentally.

Photographs of ceramic samples in crucibles synthesized under the influence of electron fluxes with E = 1.4 MeV, E = 2.5 MeV at different power densities P are shown in Figure 17. The synthesis was carried out in the "without scanning" mode, which makes it possible to visually compare the results of the analysis of energy losses and synthesis.

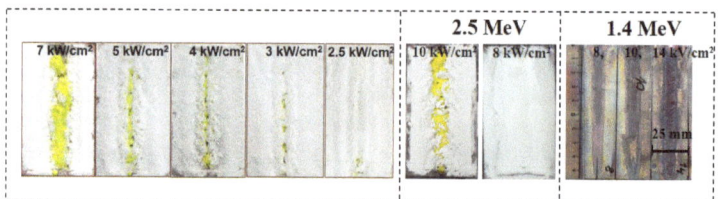

Figure 17. Photographs of ceramic samples synthesized under the exposed of electron fluxes with E = 1.4 MeV (P = 4—2.5 kW/cm^2), E = 2.5 MeV (P= 10 and 8 kW/cm^2), and traces of the impact of electron flows with E = 1.4 MeV (P = 8, 10, 14 kW/cm^2) on the copper plate.

Figure 17 shows photographs of YAG: Ce ceramics samples synthesized under the influence of electron flux with E= 1.4 MeV at various power levels. The samples were synthesized in a R2 mode, which visually presents the morphology of the formed samples. The electron beam was directed along the target, and the sample was formed in the shape of a strip against the background of the rest of the mixture.

At a power density of 7 kW/cm^2 against the background of the mixture, a strip of YAG: Ce ceramics, characterized by its distinctive yellow color, was formed. Upon reducing the power density to 5–2.5 kW/cm^2, the ceramics strip has become narrower.

From the provided images, it can be observed that during an exposure time of 10 s, ceramic samples in the form of characteristic yellow YAG:Ce rods are formed in the crucible by electron beams. At higher power (P), the rod-shaped samples are positioned on or near the irradiated surface. As the power (P) decreases, the formed samples can be concealed under a layer of mixture material. The depth at which the formed sample is situated within the mixture increases with higher energy (E). This trend corresponds to the conclusion made earlier regarding the dependence of the position of the region of maximum electron beam energy loss on E and P.

In the same figure, photographs of the traces of electron beam impact with E = 1.4 MeV in R2 mode on a thick copper plate are presented. In the experiment, the upper surface of the plate was placed at the same distance from the accelerator's exit aperture as the external surface of the mixture during synthesis. The images clearly show that the width of the impact trace in the center of the image reaches 7–10 mm, with much higher power density at the center. At a power density (P) of 8 kW/cm^2, only signs of oxidation are visible in the image, while at 12 kW/cm^2, melting of the outer surface is observed. It is worth noting that the melting temperature of copper is 1085 °C, while the oxides being synthesized, such as A_2O_3 (2044 °C) and Y_2O_3 (2410 °C), have significantly higher melting temperatures. YAG:Ce ceramic synthesis, on the other hand, occurs under the same conditions of electron beam exposure at P < 4 kW/cm², which can be explained by the differences in the dissipation of absorbed energy between metals and dielectrics.

Another noteworthy effect is observed. In the images shown in Figure 17, it can be seen that YAG:Ce ceramic synthesis occurs almost uniformly along the entire length of the crucible, which is moved relative to the electron beam. The trace of the electron beam impact on the copper plate has a variable width. As the plate is moved (or over time after the start of electron beam exposure), the trace expands and then remains constant over a longer length. This is explained by the fact that during radiation processing, the entire copper plate heats up to a certain threshold determined by the time it takes to establish equilibrium between the supplied and dissipated energy.

Figure 18 presents photographs of samples synthesized in R2 mode, removed from the crucibles. The first three were completely covered by mixture material, while the last two were open. All the samples are in the form of rods with different diameters. Sample 1 was synthesized with E = 1.4 MeV, P = 2.5 kW/cm², 2 with E = 2.0 MeV, P = 4 kW/cm², and 3 with E = 2.5 MeV, P = 8 kW/cm². Sample 4 was only slightly covered by mixture material from the top (E = 2.0 MeV, P = 6 kW/cm²), and 5 was nearly completely exposed (E = 2.5 MeV, P = 10 kW/cm²).

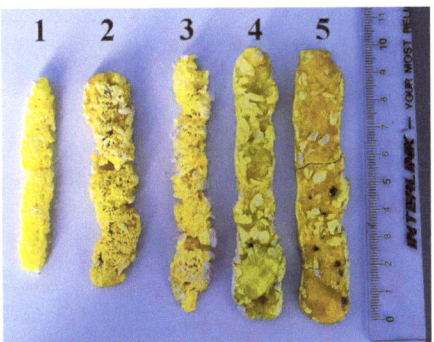

Figure 18. Photographs of YAG:Ce ceramic samples synthesized under the influence of electron beams with different values of E and P are as follows: 1—E = 1.4 MeV, P = 2.5 kW/cm²; 2—E = 2.0 MeV, P = 4 kW/cm²; 3—E = 2.5 MeV, P = 8 kW/cm²; 4—E = 2.0 MeV, P = 6 kW/cm²; 5—E = 2.5 MeV, P = 10 kW/cm².

Samples formed inside the mixture, at relatively low power density P, have a smaller length and a porous surface. Ceramic samples that have reached the surface of the mixture during ceramic formation have a solid continuous surface but are porous inside. It should be noted that the bright spots in the sample photographs are traces of the mixture, which are difficult to remove without damaging the sample.

As P decreases, the diameter of the forming sample decreases, and the solid rod turns into a dashed one. The smallest ceramic samples, in the form of rare, dashed particles with dimensions of approximately 3 mm in diameter and up to 10 mm in length, were obtained when exposed to electron beams with E = 1.4 MeV, P = 1.5 kW/cm². These samples are loose and disintegrate under slight pressure. Nevertheless, they exhibit the characteristic yellow color of YAG:Ce ceramics.

The experimentally obtained results of ceramic synthesis in mode R2 closely correspond to the patterns described above, obtained during the modeling of energy loss processes.

4.3. Redistribution of Absorbed Energy in the Developing Sample

As stated above, the energy deposition of high-energy electrons is unevenly distributed within the volume of the irradiated material. The maximum absorbed energy is located at a certain depth below the surface, depending on the electron energy. The positions of the energy loss maxima (dE/dx) for beams in the synthesis of YAG come at 2.8, 3.7, and

4.6 mm for energies of 1.4, 2.0, and 2.5 MeV, respectively. The energy loss in the maxima is 30–40% higher than at the surface and decreases to zero at the depth of the full range.

The energy transferred to the Irradiated material is converted into heat with non-uniform distribution within the material. In metals, the absorbed energy is converted into heat in less than 10^{-12} s, while in dielectric materials, including those we are studying, half of the energy is transferred within this time, and the other half within less than 10^{-6} s. After the electron beam's impact ceases, thermal energy redistributes within the target. Heat from the hottest regions of the target shifts towards the colder layers, eventually equalizing the temperature throughout the target [52,53].

The characteristic length of the temperature front displacement (*l*) during a selected time (*t*) is determined by the following equation:

$$l = \left(\frac{\lambda}{\rho C} \cdot t\right)^{\frac{1}{2}}, \quad (1)$$

where λ is thermal conductivity, C is specific heat capacity, and ρ is the material density (of the mixture). For the YAG ceramic mixture, $\lambda = 0.15 - 0.16$ W/mK [54], $C \approx 0.59$ J/gK, and $\rho = 1.15$ g/cm^3. During the 1-s synthesis, the displacement distance of the temperature front in the dielectric powder, such as YAG, is 0.28 mm. Therefore, within 1 s of radiation exposure, the temperature distribution remains relatively homogeneous when synthesizing a 1 cm long sample. Consequently, the resulting ceramic sample fully reflects the spatial structure of the absorbed energy distribution.

In contrast, heat energy redistributes differently in metals. The thermal conductivity of copper (401 W/mK) and steel (55 W/mK) is much higher than that of the materials used in the synthesis of the mixture, resulting in temperature front displacements of 50 mm and 20 mm, respectively, during the synthesis time. Thus, when electrons of the used energies (1.4–2.5 MeV) impact a copper or steel target, the temperature front can move a significant distance, sufficient to equalize the temperature within the volume. Since the electron penetration depth in copper and steel does not exceed 5 mm, all the heat reaches the surface, and any visible structural disturbances are concentrated there. This explains the difference in the formation of the radiation impact structure in ceramics and metals. In the ceramics of metal oxides and fluorides, the formation of a bulk structure is observed, whereas in metals at the energies of electrons employed, only surface disruption is observed (Figure 17).

Radiation synthesis, as shown in Section 2, is achieved in the studied materials by exposing the target to a flux of electrons with E = 1.4 MeV and power densities in the R1 mode ranging from 10 to 25 kW/cm^2. Research has shown that ceramic synthesis occurs when the power density exceeds a characteristic threshold P_c for that material. Therefore, for each synthesized material, power densities were experimentally selected such that the synthesis reaction yield and morphology were sufficient for conducting experiments in the study and application of ceramics. It can be considered established that the synthesis of YAG ceramics is achieved using power densities above 2.5 kW/cm^2 in the R2 mode (Figures 1 and 17) or above 12 kW/cm^2 in the R1 mode. Ranges have been identified within which aluminum magnesium spinel ceramics and alkaline earth metal fluorides ceramics are synthesized. To develop an understanding of the processes underlying radiation synthesis and to address technological issues, an assessment of the P_c power density values is necessary. Estimating P_c is complicated by two main factors. First, the distribution of absorbed electron energy is non-uniform within the volume, meaning that synthesis always occurs in the region of maximum energy loss, and synthesis thresholds are non-linearly related to the energy of the electrons used. Second, the synthesis threshold Pc may depend on the history of the materials used for synthesis. It is necessary to experimentally demonstrate the possible dependence of synthesis on the material's history and understand the reasons behind it.

To date, the following approach seems reasonable for determining P_c. During synthesis in the R2 mode, a rod-shaped specimen is formed in the mixture. As the power density P decreases, the samples cross-section decreases to a certain value. Below a certain P, the specimen becomes discontinuous, but its cross-sectional dimensions remain the same (Figure 18). Further reduction in P does not lead to ceramic formation. This value should be considered as the threshold.

5. Discussion

It has been experimentally demonstrated that by directly exposing a powerful flux of high-energy electrons with energies ranging from 1.4 to 2.5 MeV and power densities up to 40 kW/cm^2, the synthesis of ceramics from powders of Mg fluorides, Ba, Al oxides, Y, Ga, Zn, Zr, Mg, and W is possible. This synthesis also extends to ceramics with new phase compositions, such as yttrium aluminum garnet, spinels, tungstates, and solid solutions of alkali earth metal fluorides. It has been established that ceramic synthesis can be achieved in all examined combinations of initial powders within the range of power densities from 10 to 25 kW/cm^2 at E = 1.4 MeV (in "scanning" processing mode). It has been demonstrated that ceramic synthesis can be achieved under the specified conditions from powders with significantly different melting temperatures, ranging from 1263 °C (MgF$_2$) to 2825 °C (MgO), including mixtures of powders. The synthesis of ceramics based on metal fluorides (T_{mt} 1300–1400 °C), metal oxides such as YAG (T_{mt} 2044–2410 °C), and MgAl$_2$O$_4$ spinel (T_{mt} 2044–2825 °C) is achievable when utilizing power densities exceeding 10 kW/cm^2, even though the melting temperatures of the initial materials vary.

Radiation synthesis is successfully achieved from powder mixtures with particle sizes between 1–15 µm but is less effective when using nano- or sub-millimeter-sized particles. The combination of these observed regularities does not fit within the framework of thermal processes in radiation synthesis. It is hypothesized that ionization effects in the irradiated materials play a dominant role in radiation synthesis. When exposed to electron flux, the target material mix for synthesis is heated, and thermal processes contribute to the development of radiation-induced processes.

The high-speed synthesis of materials from refractory metal oxides in the presence of powerful fluxes of high-energy electrons suggests the existence of high-efficiency mutual element exchange between the mixture particles for the formation of new phases. Clearly, element exchange between crystalline particles is not possible within 1 s and is unlikely in the liquid phase after the instant melting of all particles. Element exchange between particles can occur within 1 s in an electron-ion plasma. It is known that relaxation times in the created ion–electron plasma have a magnitude of approximately 1–10 µs [55].

One can hypothesize that at high excitation densities in dielectric materials, the formation of radicals with high reactivity may occur, which will be capable of facilitating the creation of new phases corresponding to the specified stoichiometric composition.

5.1. Elementary Processes

The radiation fluxes used for synthesis provide a high density of electron–hole pairs. For example, in YAG at an electron energy of 1.4 MeV, a power density of 20 kW/cm^2, an electron range of 0.2 cm, and a forbidden bandgap width of 8 eV in the metal oxides used for synthesis, approximately 4×10^{22} electron–hole pairs are created in 1 s in 1 cm^3. This volume contains 4.7×10^{21} molecules in 6×10^{20} elementary cells. Thus, the concentration of electronic excitations capable of decaying into pairs of structural defects exceeds the number of molecules and cells by 1–2 orders of magnitude. The lifetime of electronic excitations before decay is within the range of $\tau_{ep} = 10^{-9}$ to 10^{-10} s. Consequently, at any moment of electron flux exposure, there are approximately 10^{13} electronic excitations within mutual distances of about 400 nm, comparable to the sizes of micro-particles, in the mixture. Therefore, there is a high probability of their localization on the surface, followed by decay under significantly different conditions through non-radiative recombination, leading to the formation of structural defects and radicals. Primary short-lived radiation defects transform into relatively long-lived pairs and radicals within the time range of 10^{-9} to 10^{-3} s. Therefore, the concentration of radicals can reach 10^{19} cm^{-3}. At such a concentration, the medium under the influence of a powerful radiation flux can be considered an ion–electron plasma. Therefore, under the described conditions of radiation processing, the decay products of electronic excitations and radiolysis are capable of entering into mutual reactions, leading to the formation of new structural units.

It is known that electronic excitations in alkali halide crystals readily decay into pairs of structural defects with high efficiency. The yield of the excitation decay reaction increases with temperature and can reach 0.8. High efficiency in the decay into pairs of structural defects is also observed in crystals of alkaline earth metal fluorides. There are grounds to expect such high yields in other dielectric materials with a high degree of ionic bonding [55].

The processes of decay of electronic excitations into pairs of structural defects in metal oxide crystals have not been thoroughly studied. However, the phenomenon of coloration, i.e., the creation of color centers, is known. Even the formation of short-lived color centers in corundum tubes of gas discharge lamps is known. At high electron excitation densities, their decay into structural pairs implies the decomposition of the material into ions and radicals.

Thus, the entire set of processes for dissipating radiation energy in dielectric materials can be schematically represented (Figure 19) and described as follows: 99% of the energy of the electron flux with energies of 1.4–2.5 MeV is spent on material ionization, where electrons transition from the valence band (VB) to the conduction band (CB). The creation of an electron–hole pair (EHP) requires energy equal to 2–3 times the width of the bandgap; e.g., the creation time of EHP is no more than 10^{-15} s.

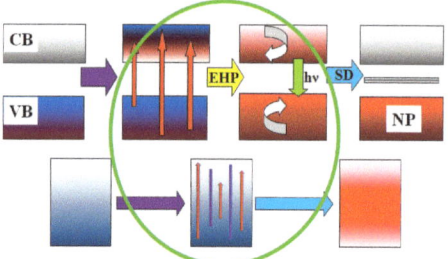

Figure 19. A schematic representation of the relaxation of excitation energy in dielectrics (at the top) and metals (at the bottom).

Then:

Relaxation to the lowest states (e_0 to e_r) occurs with the transfer of 0.5–0.7 energy to the lattice, and the relaxation time is less than $\tau = 10^{-12}$ s.

Decay of electronic excitations occurs, either radiative or non-radiative, into pairs of short-lived defects (SD), for example, Frenkel pairs and radicals (e_r F + H). The time range for these processes is from 10^{-12} s to 10^{-9} s, and some of the energy is transferred to the lattice. The decay of electronic excitations into SD pairs and their transformation into stable states is facilitated by the high temperature of the substance.

Recombination or transformation of primary pairs into stable complexes (F + H to F + V). The formation of new phases (NP) takes place in a time range from 10^{-9} to 10^{-3} s. Some of the energy is transferred to the emerging phase.

Cooling of the material (transfer of energy to the surrounding medium through radiation, thermal conductivity, convection) occurs over timescales longer than 1 s.

In metals, electronic excitations created under the influence of high-energy radiation disappear non-radiatively and without forming defects in timescales shorter than 10^{-12} s. The energy released during this process is immediately transferred to the lattice, heating the material.

Therefore, the main distinction between the energy dissipation processes of electronic excitations in dielectric and metallic materials lies in the existence of short-lived radiolysis products in dielectric materials. In metals, there are no processes associated with the decay of electronic excitations into pairs of defects or the formation of radicals—mobile intermediate components capable of participating in structural transformations during their existence.

The transformation of structural phases after exposure to a radiation flux occurs with the heating of the irradiated material. Figure 20 schematically represents the dependence of the amount of energy transferred to the material's heating during the relaxation of electronic excitations and structural phase transformations. At least half of all absorbed radiation energy is transferred to the lattice during the relaxation of created electronic excitations to their lowest states in a time period of $\tau = 10^{-12}$ s. The creation of primary short-lived defects occurs with the transfer of a portion of the energy from relaxed electronic excitations, excitons, to the lattice within a time period of $\tau = 10^{-9}$ s. This portion can constitute approximately half of the exciton's energy. Primary defects transform into stable radicals, which form a new phase within timescales of $\tau = 10^{-3}$ s. Transformations of stable radicals are also possible within timescales of $\tau = 10^3$ s.

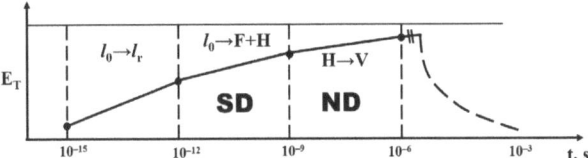

Figure 20. Schematic representation of the transfer of excitation energy to heating in a dielectric.

When exposed to a stationary electron flux, the target must heat up over time to temperatures at which thermal equilibrium is reached between the processes of heat input and heat exchange with the surrounding environment. Under the conditions employed in radiation synthesis, heat exchange occurs between the heated target, the charge, and the surrounding environment, including thermal radiation, convective exchange with the air, and heat transfer into the copper crucible. As evident from the graph (Figure 17), when the copper plate is displaced relative to the electron beam (Mode R2) at a velocity of 1 cm/s, the width of the trace increases, but it remains unchanged after 1–2 s. Consequently, thermal equilibrium between energy input and heat dissipation is reached within 1 s of electron beam exposure. A similar effect occurs when irradiating the charge of synthesized ceramics. The heating of the charge of synthesized ceramics precedes the displacement of the beam. Therefore, synthesis takes place in an environment that has already reached a maximum temperature. Since the efficiency of the decay of electronic excitations into structural pairs increases with temperature, the efficiency of radiolysis also increases and becomes constant, characteristic of the specific material.

5.2. On the Threshold of Synthesis

To implement radiation synthesis, it is necessary to create such a density of radicals in the particles of the substances used that can facilitate the exchange of elements between the particles. The lifetime of the radicals probably spans a wide range from 10^{-9} to 10^{-3} s. Therefore, the choice of the power density of the electron beam that provides a sufficient density of radicals for synthesis can only be determined experimentally at this time. Clearly, for each material or combination of materials, there exist characteristic thresholds for the power density of the electron beam above which synthesis can be realized. However, estimating these synthesis threshold values is only possible under specific conditions of radiation processing: electron energy, spatial distribution of the beam flux, irradiation modes (with scanning or without scanning).

In the conditions we are using for high-energy electron beam impact on the synthesis layer, there is a pronounced non-uniform distribution of absorbed energy, leading to the creation of a non-uniform density of electronic excitations. The probability of radiation synthesis is higher in the region of maximum electronic excitation density. In this region, synthesis will primarily occur as the power density of the electron beam increases. As the power density of the beam increases, this region expands (Figure 17).

Therefore, according to the presented understanding of ceramic synthesis processes under the influence of high-energy electron beams, there should be a threshold power density P_c for each material above which the process of forming a new phase or morphology becomes possible. Knowledge of this threshold is crucial for understanding the dependence of synthesis efficiency on the energy and morphological properties of the initial materials. At present, it seems reasonable to adopt the following approach to determine P_c: during synthesis in the R2 mode, a rod-shaped sample is formed in the layer. As the power density P decreases, the cross-sectional area of the sample decreases until it reaches a certain value. Below a certain P, the sample becomes discontinuous, but the dimensions of its cross-section remain the same (Figure 18). In the synthesis of YAG samples, the minimum diameter of the sample reaches 4–5 mm, with a minimum length of 10 mm. With further reduction in P, ceramic formation is not observed. This value should be considered as the threshold.

6. Conclusions

The synthesis of ceramics through the direct impact of a powerful flux of high-energy electrons on mixtures of dielectric powders with high melting temperatures has been achieved. These powders include MgF_2 (1263 °C), BaF_2 (1368 °C), WO_3 (1473 °C), Ga_2O_3 (1725 °C), ZnO (1975 °C), Al_2O_3 (2044 °C), Y_2O_3 (2410 °C), ZrO_2 (2715 °C), MgO (2825 °C), $Y_3Al_5O_{12}$, $Y_3Al_xGa_{5-x}O_{12}$, $MgAl_2O_4$, $ZnAl_2O_4$, $MgWO_4$, $ZnWO_4$, $Ba_xMg_{(2-x)}F_4$, $Ba_xMg_{(2-x)}F_4$: W, $Y_3Al_5O_{12}$: Gd, Ce, Cr, Eu, Er, and $MgAl_2O_4$: Ce, Cr, Eu, Er. The formation of ceramics from these compositions suggests the method's universality. It can be asserted at this time that radiation synthesis has produced ceramics based on YAG ($Y_3Al_5O_{12}$), spinel ($MgWO_4$), and solid solution ($BaMgF_4$) with the characteristic properties of these materials. Expanding the range of synthesized materials and establishing criteria for proving the implementation of radiation ceramic synthesis are necessary.

The main properties of radiation synthesis are as follows. Synthesis is achieved:

- Solely through radiation energy;
- Exclusively from mixture materials;
- Without the addition of any other materials to facilitate synthesis;
- In less than 1 s.

The high synthesis speed allows for the rapid execution of necessary improvements, conducting series of experimental studies to optimize synthesis, which is particularly important for multi-component luminescent materials.

The set of processes enabling the radiation synthesis of dielectric and metallic materials differs. Processes in dielectric and metallic materials vary due to relaxation of excited states after ionization by the radiation flux. It is essential to develop an understanding of radiation-induced processes, radiolysis, structural (phase) transformations, and the formation of new materials with new properties stimulated by high radiation flows.

Radiation exposure efficiently facilitates the mutual transfer of elements between charge particles of mixture materials and the formation of a new phase with the same morphology, as well as the introduction of additional elements from added particles (activation). Powerful radiation exposure ensures high-efficiency mixing of particle elements from the initial materials used.

Ceramic synthesis occurs when a certain threshold power density of the beam P_c is exceeded. Knowledge of this threshold is crucial for understanding the dependence of synthesis efficiency on the energetic and morphological properties of the initial substances. Experimental determination of this threshold and its relationship with the history of the initial substances is necessary.

Synthesis efficiency depends on the history of the initial materials, especially their dispersity. The best results are obtained from powders with particle sizes ranging from 3 to 10 µm.

Samples synthesized in a powerful electron beam have a solid shell and a porous internal structure. The structural and luminescent properties of the external and internal layers of ceramics are not different. The presence of pores affects the transparency of samples, causing light scattering. To obtain transparent ceramics, a method to reduce porosity needs to be found.

Author Contributions: Conceptualization, V.L. and L.L.; methodology, M.G.; software, E.P.; validation, L.L., G.A. and D.M.; formal analysis, V.L.; investigation, V.L.; resources, M.G.; data curation, E.P.; writing—original draft preparation, V.L.; writing—review and editing, A.T.; visualization, L.L.; super-vision, V.L.; project administration, V.L.; funding acquisition, A.T. All authors have read and agreed to the published version of the manuscript.

Funding: This research was funded by the Science Committee of the Ministry of Science and Higher Education of the Republic of Kazakhstan (Grant No. AP 19577213). This research was funded by the Russian Science Foundation of the Russian Federation. (Grant No. 23-73-00108).

Institutional Review Board Statement: Not applicable.

Informed Consent Statement: Not applicable.

Data Availability Statement: The data presented in this study are available on request from the corresponding author.

Acknowledgments: This research was supported by the Budker Institute of Nuclear Physics, Siberian Branch of the Russian Academy of Sciences (BINP SB RAS). The authors are grateful to the al-Farabi Kazakh National University, L.N. Gumilyov Eurasian National University, and Karaganda Buketov University and to Rustam Abelevich Salimov, the Chief Researcher of the Institute of Applied Physics of the Siberian Branch of the Russian Academy of Sciences, for his constant attention and support in conducting research on the synthesis of high-temperature inorganic ceramics.

Conflicts of Interest: The authors declare no conflict of interest.

References

1. Liu, X.; Qian, X.; Zheng, P.; Chen, X.; Feng, Y.; Shi, Y.; Zou, J.; Xie, R.; Li, J. Composition and Structure Design of Three-Layered Composite Phosphors for High Color Rendering Chip-on-Board Light-Emitting Diode Devices. *J. Adv. Ceram.* **2021**, *10*, 729–740. [CrossRef]
2. Xia, Z.; Meijerink, A. Ce^{3+}-Doped Garnet Phosphors: Composition Modification, Luminescence Properties and Applications. *Chem. Soc. Rev.* **2017**, *46*, 275–299. [CrossRef] [PubMed]
3. Li, J.; Sahi, S.; Groza, M.; Pan, Y.; Burger, A.; Kenarangui, R.; Chen, W. Optical and Scintillation Properties of Ce^{3+}-Doped LuAG and YAG Transparent Ceramics: A Comparative Study. *J. Am. Ceram. Soc.* **2017**, *100*, 150–156. [CrossRef]
4. Lecoq, P. Development of New Scintillators for Medical Applications. *Nucl. Instrum. Methods Phys. Res. Sect. A Accel. Spectrometers Detect. Assoc. Equip.* **2016**, *809*, 130–139. [CrossRef]
5. Sharma, S.K.; James, J.; Gupta, S.K.; Hussain, S. UV-A,B,C Emitting Persistent Luminescent Materials. *Materials* **2022**, *16*, 236. [CrossRef]
6. Sun, H.; Gao, Q.; Wang, A.; Liu, Y.; Wang, X.; Liu, F. Ultraviolet-B Persistent Luminescence and Thermoluminescence of Bismuth Ion Doped Garnet Phosphors. *Opt. Mater. Express OME* **2020**, *10*, 1296–1302. [CrossRef]
7. Xiao, Z.; Yu, S.; Li, Y.; Ruan, S.; Kong, L.B.; Huang, Q.; Huang, Z.; Zhou, K.; Su, H.; Yao, Z.; et al. Materials Development and Potential Applications of Transparent Ceramics: A Review. *Mater. Sci. Eng. R Rep.* **2020**, *139*, 100518. [CrossRef]
8. Yamamoto, S.; Nitta, H. Development of an Event-by-Event Based Radiation Imaging Detector Using GGAG: A Ceramic Scintillator for X-Ray CT. *Nucl. Instrum. Methods Phys. Res. Sect. A Accel. Spectrometers Detect. Assoc. Equip.* **2018**, *900*, 25–31. [CrossRef]
9. Ji, T.; Wang, T.; Li, H.; Peng, Q.; Tang, H.; Hu, S.; Yakovlev, A.; Zhong, Y.; Xu, X. Ce^{3+}-Doped Yttrium Aluminum Garnet Transparent Ceramics for High-Resolution X-Ray Imaging. *Adv. Opt. Mater.* **2022**, *10*, 2102056. [CrossRef]
10. Dujardin, C.; Auffray, E.; Bourret-Courchesne, E.; Dorenbos, P.; Lecoq, P.; Nikl, M.; Vasil'ev, A.N.; Yoshikawa, A.; Zhu, R.-Y. Needs, Trends, and Advances in Inorganic Scintillators. *IEEE Trans. Nucl. Sci.* **2018**, *65*, 1977–1997. [CrossRef]
11. Pan, Y.; Wu, M.; Su, Q. Comparative Investigation on Synthesis and Photoluminescence of YAG:Ce Phosphor. *Mater. Sci. Eng. B* **2004**, *106*, 251. [CrossRef]
12. Smet, P.F.; Parmentier, A.B.; Poelman, D. Selecting Conversion Phosphors for White Light-Emitting Diodes. *J. Electrochem. Soc.* **2011**, *158*, R37. [CrossRef]
13. Ye, S.; Xiao, F.; Pan, Y.X.; Ma, Y.Y.; Zhang, Q.Y. Phosphors in Phosphor-Converted White Light-Emitting Diodes: Recent Advances in Materials, Techniques and Properties. *Mater. Sci. Eng. R Rep.* **2010**, *71*, 1–34. [CrossRef]

14. Wang, X.; Li, J.; Shen, Q.; Shi, P. Flux-Grown $Y_3Al_5O_{12}$:Ce^{3+} Phosphors with Improved Crystallinity and Dispersibility. *Ceram. Int.* **2014**, *40*, 15313–15317. [CrossRef]
15. Yang, Y.; Wang, X.; Liu, B.; Zhang, Y.; Lv, X.; Li, J.; Wei, L.; Yu, H.; Hu, Y.; Zhang, H. Molten Salt Synthesis and Luminescence of Dy^{3+}-doped $Y_3Al_5O_{12}$ Phosphors. *Luminescence* **2020**, *35*, 580–585. [CrossRef] [PubMed]
16. Pereira, P.F.S.; Matos, M.G.; Ávila, L.R.; Nassor, E.C.O.; Cestari, A.; Ciuffi, K.J.; Calefi, P.S.; Nassar, E.J. Red, Green and Blue (RGB) Emission Doped Y3Al5O12 (YAG) Phosphors Prepared by Non-Hydrolytic Sol–Gel Route. *J. Lumin.* **2010**, *130*, 488–493. [CrossRef]
17. Abdullin, K.A.; Kemel'bekova, A.E.; Lisitsyn, V.M.; Mukhamedshina, D.M.; Nemkaeva, R.R.; Tulegenova, A.T. Aerosol Synthesis of Highly Dispersed $Y_3Al_5O_{12}$:Ce^{3+} Phosphor with Intense Photoluminescence. *Phys. Solid State* **2019**, *61*, 1840–1845. [CrossRef]
18. Grazenaite, E.; Garskaite, E.; Stankeviciute, Z.; Raudonyte-Svirbutaviciene, E.; Zarkov, A.; Kareiva, A. Ga-Substituted Cobalt-Chromium Spinels as Ceramic Pigments Produced by Sol–Gel Synthesis. *Crystals* **2020**, *10*, 1078. [CrossRef]
19. Dai, P.; Ji, C.; Shen, L.; Qian, Q.; Guo, G.; Zhang, X.; Bao, N. Photoluminescence Properties of YAG:Ce^{3+},Pr^{3+} Nano-Sized Phosphors Synthesized by a Modified Co-Precipitation Method. *J. Rare Earths* **2017**, *35*, 341–346. [CrossRef]
20. Serrano-Bayona, R.; Chu, C.; Liu, P.; Roberts, W.L. Flame Synthesis of Carbon and Metal-Oxide Nanoparticles: Flame Types, Effects of Combustion Parameters on Properties and Measurement Methods. *Materials* **2023**, *16*, 1192. [CrossRef]
21. Huczko, A.; Kurcz, M.; Baranowski, P.; Bystrzejewski, M.; Bhattarai, A.; Dyjak, S.; Bhatta, R.; Pokhrel, B.; Kafle, B.P. Fast Combustion Synthesis and Characterization of YAG:Ce $^{3+}$ Garnet Nanopowders: Fast Combustion Synthesis of YAG:Ce^{3+} Garnet Nanopowders. *Phys. Status Solidi B* **2013**, *250*, 2702–2708. [CrossRef]
22. Ohyama, J.; Zhu, C.; Saito, G.; Haga, M.; Nomura, T.; Sakaguchi, N.; Akiyama, T. Combustion Synthesis of YAG:Ce Phosphors via the Thermite Reaction of Aluminum. *J. Rare Earths* **2018**, *36*, 248–256. [CrossRef]
23. Le Godec, Y.; Le Floch, S. Recent Developments of High-Pressure Spark Plasma Sintering: An Overview of Current Applications, Challenges and Future Directions. *Materials* **2023**, *16*, 997. [CrossRef] [PubMed]
24. Boldyrev, V.V.; Zakharov, Y.A.; Konyshev, V.P.; Pinaevskaya, E.N.; Boldyreva, A. On Kinetic Factors Which Determine Specific Mechano-Chemical Processes in Inorganic Systems. *Kinetika i Kataliz* **1972**, *13*, 1411–1421.
25. Lisitsyn, V.; Tulegenova, A.; Kaneva, E.; Mussakhanov, D.; Gritsenko, B. Express Synthesis of YAG:Ce Ceramics in the High-Energy Electrons Flow Field. *Materials* **2023**, *16*, 1057. [CrossRef] [PubMed]
26. Lisitsyn, V.M.; Musakhanov, D.A.; Korzhneva, T.G.; Strelkova, A.V.; Lisitsyna, L.A.; Golkovsky, M.G.; Zhunusbekov, A.M.; Karipbaev, J.T.; Kozlovsky, A.L. Synthesis and Characterization of Ceramics BaxMg$_{2-x}$F4 Activated by Tungsten. *Glass Phys. Chem.* **2023**, *49*, 288–292. [CrossRef]
27. Lisitsyn, V.; Mussakhanov, D.; Tulegenova, A.; Kaneva, E.; Lisitsyna, L.; Golkovski, M.; Zhunusbekov, A. The Optimization of Radiation Synthesis Modes for YAG:Ce Ceramics. *Materials* **2023**, *16*, 3158. [CrossRef]
28. Xia, M.; Gu, S.; Zhou, C.; Liu, L.; Zhong, Y.; Zhang, Y.; Zhou, Z. Enhanced Photoluminescence and Energy Transfer Performance of $Y_3Al_4GaO_{12}$: Mn^{4+},Dy^{3+} Phosphors for Plant Growth LED Lights. *RSC Adv.* **2019**, *9*, 9244–9252. [CrossRef] [PubMed]
29. Shannon, R.D. Revised Effective Ionic Radii and Systematic Studies of Interatomic Distances in Halides and Chalcogenides. *Acta Cryst. A* **1976**, *32*, 751–767. [CrossRef]
30. Karipbayev, Z.T.; Lisitsyn, V.M.; Mussakhanov, D.A.; Alpyssova, G.K.; Popov, A.I.; Polisadova, E.F.; Elsts, E.; Akilbekov, A.T.; Kukenova, A.B.; Kemere, M.; et al. Time-Resolved Luminescence of YAG:Ce and YAGG:Ce Ceramics Prepared by Electron Beam Assisted Synthesis. *Nucl. Instrum. Methods Phys. Res. Sect. B Beam Interact. Mater. At.* **2020**, *479*, 222–228. [CrossRef]
31. Dorenbos, P. 5d-Level Energies of Ce^{3+} and the Crystalline Environment. IV. Aluminates and "Simple" Oxides. *J. Lumin.* **2002**, *99*, 283–299. [CrossRef]
32. Ueda, J.; Tanabe, S. (INVITED) Review of Luminescent Properties of Ce^{3+}-Doped Garnet Phosphors: New Insight into the Effect of Crystal and Electronic Structure. *Opt. Mater. X* **2019**, *1*, 100018. [CrossRef]
33. Li, J.-G.; Ikegami, T.; Lee, J.-H.; Mori, T. Fabrication of Translucent Magnesium Aluminum Spinel Ceramics. *J. Am. Ceram. Soc.* **2004**, *83*, 2866–2868. [CrossRef]
34. Chen, C.-F.; Doty, F.P.; Houk, R.J.T.; Loutfy, R.O.; Volz, H.M.; Yang, P. Characterizations of a Hot-Pressed Polycrystalline Spinel:Ce Scintillator: Characterizations of Spinel:Ce Scintillator. *J. Am. Ceram. Soc.* **2010**, *93*, 2399–2402. [CrossRef]
35. Korepanov, V.I.; Lisitsyn, V.M.; Oleshko, V.I. Application of High-Current Nanosecond Electron Beam for Control of Solids Parameters. *Izv. Vyss. Uchebnykh Zaved. Fiz.* **2000**, *43*, 22–30.
36. Polisadova, E.F.; Vaganov, V.A.; Stepanov, S.A.; Paygin, V.D.; Khasanov, O.L.; Dvilis, E.S.; Valiev, D.T.; Kalinin, R.G. Pulse cathodoluminescence of the impurity centers in ceramics based on the $MgAl_2O_4$ spinel. *J. Appl. Spectrosc.* **2018**, *85*, 407–412. [CrossRef]
37. Ganesh, I. A Review on Magnesium Aluminate ($MgAl_2O_4$) Spinel: Synthesis, Processing and Applications. *Int. Mater. Rev.* **2013**, *58*, 63–112. [CrossRef]
38. Choi, B.S.; Jeong, O.G.; Park, J.C.; Kim, J.W.; Lee, S.J. Photoluminescence Properties of Non-Rare Earth $MgAL_2O_4$:Mn^{2+} Green Phosphor for LEDs. *J. Ceram. Process. Res.* **2016**, *17*, 778–781.
39. Hayashi, E.; Ito, K.; Yabashi, S.; Yamaga, M.; Kodama, N.; Ono, S.; Sarukura, N. Vacuum Ultraviolet and Ultraviolet Spectroscopy of $BaMgF_4$ Codoped with Ce^{3+} and Na^+. *J. Lumin.* **2006**, *119–120*, 69–74. [CrossRef]
40. Kodama, N.; Hoshino, T.; Yamaga, M.; Ishizawa, N.; Shimamura, K.; Fukuda, T. Optical and Structural Studies on $BaMgF_4$:Ce^{3+} Crystals. *J. Cryst. Growth* **2001**, *229*, 492–496. [CrossRef]

41. Kore, B.P.; Kumar, A.; Erasmus, L.; Kroon, R.E.; Terblans, J.J.; Dhoble, S.J.; Swart, H.C. Energy Transfer Mechanisms and Optical Thermometry of BaMgF$_4$:Yb^{3+},Er^{3+} Phosphor. *Inorg. Chem.* **2018**, *57*, 288–299. [CrossRef] [PubMed]
42. Lisitsyn, V.; Lisitsyna, L.; Dauletbekova, A.; Golkovskii, M.; Karipbayev, Z.; Musakhanov, D.; Akilbekov, A.; Zdorovets, M.; Kozlovskiy, A.; Polisadova, E. Luminescence of the Tungsten-Activated MgF$_2$ Ceramics Synthesized under the Electron Beam. *Nucl. Instrum. Methods Phys. Res. Sect. B Beam Interact. Mater. At.* **2018**, *435*, 263–267. [CrossRef]
43. Mikhailik, V.B.; Kraus, H.; Miller, G.; Mykhaylyk, M.S.; Wahl, D. Luminescence of CaWO$_4$, CaMoO$_4$, and ZnWO$_4$ Scintillating Crystals under Different Excitations. *J. Appl. Phys.* **2005**, *97*, 083523. [CrossRef]
44. Itoh, M.; Fujita, N.; Inabe, Y. X-Ray Photoelectron Spectroscopy and Electronic Structures of Scheelite- and Wolframite-Type Tungstate Crystals. *J. Phys. Soc. Jpn.* **2006**, *75*, 084705. [CrossRef]
45. Nagirnyi, V.; Feldbach, E.; Jönsson, L.; Kirm, M.; Kotlov, A.; Lushchik, A.; Nefedov, V.A.; Zadneprovski, B.I. Energy Transfer in ZnWO4 and CdWO$_4$ Scintillators. *Nucl. Instrum. Methods Phys. Res. Sect. A Accel. Spectrometers Detect. Assoc. Equip.* **2002**, *486*, 395–398. [CrossRef]
46. Pereira, P.F.S.; Gouveia, A.F.; Assis, M.; De Oliveira, R.C.; Pinatti, I.M.; Penha, M.; Gonçalves, R.F.; Gracia, L.; Andrés, J.; Longo, E. ZnWO$_4$ Nanocrystals: Synthesis, Morphology, Photoluminescence and Photocatalytic Properties. *Phys. Chem. Chem. Phys.* **2018**, *20*, 1923–1937. [CrossRef] [PubMed]
47. Tatsuo, T.; Pedro, A.; Kunihiko, S.; Rinsuke, I. Energy Deposition through Radiative Processes in Absorbers Irradiated by Electron Beams. *Nucl. Instrum. Methods Phys. Res. Sect. B Beam Interact. Mater. At.* **1994**, *93*, 447–456. [CrossRef]
48. Tabata, T.; Andreo, P.; Shinoda, K. An Algorithm for Depth–Dose Curves of Electrons Fitted to Monte Carlo Data. *Radiat. Phys. Chem.* **1998**, *53*, 205–215. [CrossRef]
49. Hovington, P.; Drouin, D.; Gauvin, R. CASINO: A New Monte Carlo Code in C Language for Electron Beam Interaction —Part I: Description of the Program. *Scanning* **1997**, *19*, 1–14. [CrossRef]
50. Drouin, D.; Hovington, P.; Gauvin, R. CASINO: A New Monte Carlo Code in C Language for Electron Beam Interactions—Part II: Tabulated Values of the Mott Cross Section. *Scanning* **1997**, *19*, 20–28. [CrossRef]
51. Drouin, D.; Couture, A.R.; Joly, D.; Tastet, X.; Aimez, V.; Gauvin, R. CASINO V2.42—A Fast and Easy-to-use Modeling Tool for Scanning Electron Microscopy and Microanalysis Users. *Scanning* **2007**, *29*, 92–101. [CrossRef] [PubMed]
52. Lushchik, C.B.; Vitol, I.K.; Élango, M.A. Decay of Electronic Excitations into Radiation Defects in Ionic Crystals. *Sov. Phys. Usp.* **1977**, *20*, 489–505. [CrossRef]
53. Lisitsyn, V.M.; Korepanov, V.I.; Yakovlev, V.Y. Evolution of Primary Radiation Defects in Ionic Crystals. *Russ. Phys. J.* **1996**, *39*, 1009–1028. [CrossRef]
54. Volchenko, T.S.; Yalovets, A.P. Calculation of the Effective Thermal Conductivity of Powders Formed by Spherical Particles in a Gaseous Atmosphere. *Tech. Phys.* **2016**, *61*, 324–336. [CrossRef]
55. Cremers, D.A.; Radziemski, L.J. *Handbook of Laser-Induced Breakdown Spectroscopy*, 2nd ed.; John Wiley & Sons, Ltd.: Chichester, UK, 2013; ISBN 978-1-118-56736-4.

Disclaimer/Publisher's Note: The statements, opinions and data contained in all publications are solely those of the individual author(s) and contributor(s) and not of MDPI and/or the editor(s). MDPI and/or the editor(s) disclaim responsibility for any injury to people or property resulting from any ideas, methods, instructions or products referred to in the content.

Article

A Fast Transient Adaptive On-Time Controlled BUCK Converter with Dual Modulation

Mengyuan Sun [1,†], Chufan Chen [1,†], Leiyi Wang [1,2], Xinling Xie [1], Yuhang Wang [1] and Min Xu [1,2,*]

1. State Key Laboratory of ASIC and System, School of Microelectronics, Fudan University, Shanghai 200433, China; 20212020003@fudan.edu.cn (M.S.); 21212020002@m.fudan.edu.cn (C.C.); lywang2018@stu.suda.edu.cn (L.W.); 21212020013@m.fudan.edu.cn (X.X.); yuhangwang21@m.fudan.edu.cn (Y.W.)
2. Shanghai Integrated Circuit Manufacturing Innovation Center Co., Ltd., Shanghai 200433, China
* Correspondence: xu_min@fudan.edu.cn
† These authors contributed equally to this work.

Abstract: This paper proposed a fully integrated adaptive on-time (AOT) controlled buck converter with fast load transient. An adaptive on-time generator is presented to stabilize the output frequency. To enhance the light load efficiency, the converter could transfer from the pulse width modulation (PWM) to pulse skip modulation (PSM) as the load current decreases. The buck converter can switch between these two modulation modes adaptively with the assistance of a zero current detection circuit. Implemented in the TSMC 0.18 μm BCD (BiCMOS/DMOS) process, the proposed buck converter works with an input voltage ranging from 5.5 to 15 V, an output voltage ranging from 0.5 to 5 V, and an output load ranging up to 5 A. The experimental results show that based on the dual modulation adaptive on-time controlled mode, the transient recovery time from light to heavy load and from heavy load to light load is 13 μs and 15 μs, respectively. An overshot voltage of 57 mV and an undershot voltage of 53 mV are also achieved.

Keywords: DC-DC; buck converter; adaptive on-time; fast response; constant frequency; pulse skip modulation

Citation: Sun, M.; Chen, C.; Wang, L.; Xie, X.; Wang, Y.; Xu, M. A Fast Transient Adaptive On-Time Controlled BUCK Converter with Dual Modulation. *Micromachines* **2023**, *14*, 1868. https://doi.org/10.3390/mi14101868

Academic Editor: Kun Li

Received: 22 July 2023
Revised: 23 September 2023
Accepted: 26 September 2023
Published: 29 September 2023

Copyright: © 2023 by the authors. Licensee MDPI, Basel, Switzerland. This article is an open access article distributed under the terms and conditions of the Creative Commons Attribution (CC BY) license (https://creativecommons.org/licenses/by/4.0/).

1. Introduction

As communication and semiconductor technologies continue to advance, electronic equipment is also evolving rapidly and becoming more intelligent. As electronic systems grow more complex, the power supply modules have played an increasingly critical role in these systems [1,2]. The DC-DC power management chip is indispensable for electronic equipment to ensure the stability of the power supply and reduce system power consumption [3]. The DC-DC converter chip adjusts itself by monitoring the power output energy and response for the transformation, distribution, detection, and management of electrical energy in the system so that the battery power can be reasonably distributed to different modules in the system to save energy and ensure stable power supply [4,5].

The power management chips for microprocessors are fast-changing loads, so the power management chip needs to respond to load changes quickly [6,7]. With the increasing demand for intelligence and modularity, the load will change to be faster and more complex, which puts higher demands on the dynamic characteristics of the power management chip [8–10]. Feedback control is the main factor to ensure the converter's output is stable under different load conditions. The approaches for feedback control are generally voltage mode and current mode [11,12]. The buck converter with voltage mode control has a simpler structure but has more complex compensation circuits and slower transient response, and it is unable to monitor current [13]. The constant on-time (COT) controlled buck converters have recently been proposed due to their simple compensation structure and fast load transient response [14]. However, subharmonic oscillations may occur in some cases [15,16]. In this paper, a dual-loop feedback control is proposed to enhance the response speed while ensuring loop stability [17].

Since the on-time of power transistors per cycle is kept constant in COT control, the duty cycle is fixed for a defined input and output voltage. Thus, the operating frequency of the converter will vary with different input and output voltages [18]. However, once the switching frequency of the converter changes with the working conditions, it will bring serious electromagnetic interference (EMI) problems, and the converter will be restricted to applications that require fixed frequency. The buck converter proposed in this paper used an adaptive on-time generator to produce the on time that varies with the working conditions, which can keep the system operating at a constant frequency. In order to improve the system efficiency under a wide load range, pulse width modulation (PWM) combined with pulse skip modulation (PSM) has been used [19]. Under heavy load conditions, the main loss of the system is the conduction loss due to the load current going through the power transistors, and the PWM control can obtain high efficiency. The switching loss under light load is mainly caused by the charging and discharging of the gate capacitor of power transistors. The light load power consumption could be reduced by reducing the operating frequency to adopt PSM control [20]. The proposed zero-current detection (ZCD) circuit enables the converter to switch smoothly between two modulation modes.

This paper presents a fast adaptive on-time controlled buck converter with a dual feedback control loop compared with voltage-controlled and current-controlled modes. It can achieve a higher response speed, and compared with the traditional constant-on-time controlled mode, it can stabilize the output frequency. This converter could work under PWM control, which keeps the operating frequency stable and reduces the adverse effects of frequency changes. Furthermore, the PSM can be used under light loads to improve efficiency. The rest of this paper is organized as follows. The system of the proposed buck converter is described in Section 2. Section 3 illustrates the circuit-level implementations of the adaptive on-time generator and the zero-current detection circuit. The experimental results are presented in Section 4. Finally, Section 5 concludes this paper.

2. Proposed Fast Transient AOT-Controlled Buck Converter

Figure 1 illustrates the simplified system control architecture of the conventional COT control and the proposed dual-loop feedback AOT-controlled buck converter. The system is composed of a power stage and a feedback control stage. The power stage primarily includes a high-side and a low-side power transistor for switching and an LC filter that passes the energy from the input to the output.

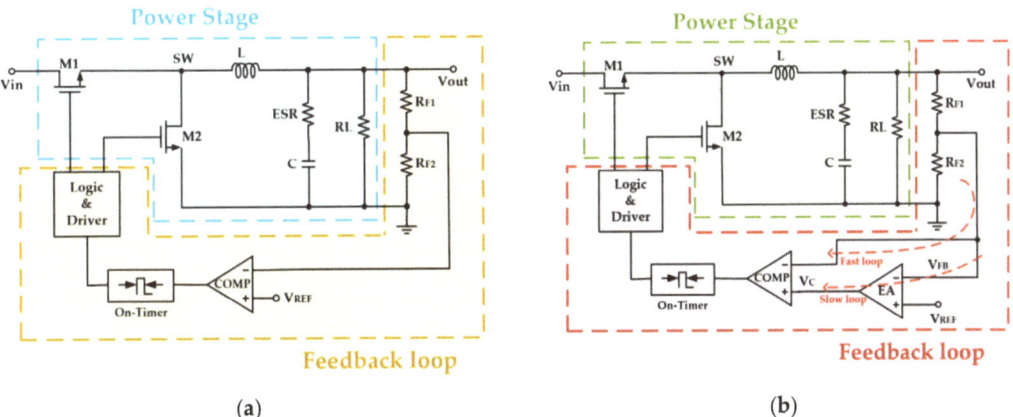

Figure 1. System control architecture of the AOT buck converter: (**a**) Conventional COT control; (**b**) the dual-loop feedback control applied in this paper.

In order to enhance the load transient response speed, a dual-loop feedback control stage is applied. A fast loop ensures a fast transient response speed, and a slow loop compensates for the output DC offset introduced by the output voltage step. The feedback loops are, as shown in Figure 1, the output voltage is divided by RF1 and RF2 to produce a feedback voltage VFB, which is the input of loop comparator COMP. The comparator controls the on-time generator to generate the Ton, which is the input of the logic and driver module to control the power transistors. This loop is the fast feedback loop, which, from the output voltage Vout directly to the comparator, without the need to go through the error amplifier (EA) and its compensation network, can achieve a fast response when the load transient step occurs. The presence of the fast feedback loop greatly improves the transient response speed of the system. However, the output voltage has a large DC detuning due to the limited gain of the loop. Therefore, another slow feedback loop is designed to amplify the feedback voltage VFB with a reference voltage VREF in the error amplifier first and then input to the loop comparator COMP to keep the DC value of the output voltage stable. Figure 2 shows the block diagram architecture of the proposed AOT buck converter.

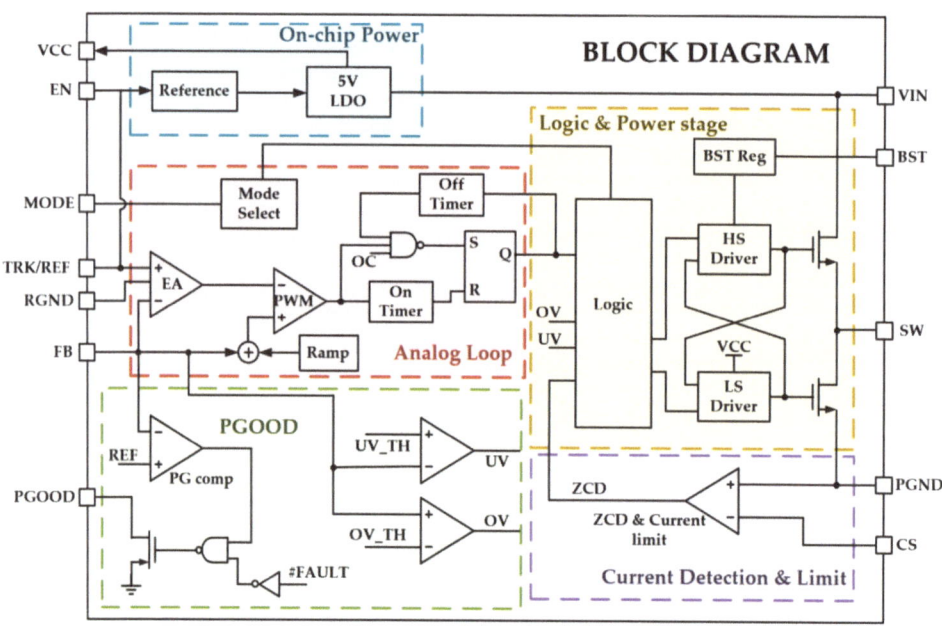

Figure 2. Block diagram of the proposed AOT buck converter.

Since the system sampled the output voltage in loop feedback control, the output voltage contains a capacitive voltage that is phase lagged compared to the inductor current. It may cause the output voltage ripple and inductor current ripple to be out of phase if this portion of the voltage is too large, which would make the system operating unstable [21,22]. The stability of the system was analyzed using the descriptive function method, with the addition of the perturbation $v_c(t) = r_0 + \hat{r} * sin(2\pi f_m * t + \theta)$ at the control voltage Vc. Considering the output voltage ripple, Equation (1) was derived according to the law of modulation.

$$v_c \cdot \left(t_{i-1} + T_{off(i-1)}\right) + S_n \cdot T_{on} - \int_{t_{i-1}+T_{off(i-1)}}^{t_i+T_{off(i)}} \frac{i_L(t) - v_{out}(t)/R_L}{C} dt \\ = v_c \cdot \left(t_i + T_{off(i)}\right) + S_f \cdot T_{off(i)} \quad (1)$$

where $T_{off(i)}$ is the turn-off time of the high-side transistor in the cycle i, and S_n and S_f are the rising and falling slopes of the voltage signal converted via inductor current.

$$S_n = \frac{R_i \cdot (V_{in} - V_{out})}{L} \quad (2)$$

$$S_f = \frac{R_i \cdot V_{out}}{L} \quad (3)$$

Suppose $T_{off(i)} = T_{off} + \Delta T_{off(i)}$, where T_{off} is the off time of the high-side transistor at steady state, and $\Delta T_{off(i)}$ is the perturbation of the high-side transistor off time at cycle i. Assuming that the steady state period is T_{sw}, the beginning time of the cycle i can be expressed as

$$t_i = (i-1) \cdot T_{sw} + \sum_{k=1}^{i-1} \Delta T_{off(k)} \quad (4)$$

Substitute Equation (4) into Equation (1) to obtain

$$\begin{aligned}
&S_f \cdot \left[\left(1 + \frac{T_{off}}{2 \cdot C \cdot R_{ESR}}\right) \cdot \Delta T_i - \left(1 - \frac{2T_{on} + T_{off}}{2 \cdot C \cdot R_{ESR}}\right) \cdot \Delta T_{i-1} \right] \\
&= \left[v_c \cdot \left(t_{i-1} + T_{off(i-1)}\right) - v_c \cdot \left(t_i + T_{off(i)}\right)\right] - \left[v_c \cdot \left(t_{i-2} + T_{off(i-2)}\right) - v_c \cdot \left(t_{i-1} + T_{off(i-1)}\right)\right] - \\
&\quad \frac{v_c \cdot \left(t_{i-1} + T_{off(i-1)} + \frac{\pi}{2\pi f_m \cdot 2}\right) - v_c \cdot \left(t_i + T_{off(i)} + \frac{\pi}{2\pi f_m \cdot 2}\right)}{2\pi f_m \cdot R_L C} + \\
&\quad \frac{v_c \cdot \left(t_{i-2} + T_{off(i-2)} + \frac{\pi}{2\pi f_m \cdot 2}\right) - v_c \cdot \left(t_{i-1} + T_{off(i-1)} + \frac{\pi}{2\pi f_m \cdot 2}\right)}{2\pi f_m \cdot R_L C}
\end{aligned} \quad (5)$$

Similarly, the perturbed duty cycle and the inductor current can be expressed as

$$d(t) = \sum_{i=1}^{M} \left[u \cdot \left(t - t_i - T_{off(i)}\right) - u \cdot \left(t - t_i - T_{off(i)} - T_{on}\right) \right] \quad (6)$$

$$i_L(t) = i_{L0} + \int_0^i \left[\frac{V_{in}}{L} \cdot d(t) - \frac{V_{out}}{L} \right] \cdot dt \quad (7)$$

Fourier analysis of the inductive current and substitution into Equation (1) gives

$$\begin{aligned}
c_m &= j \cdot \frac{2f_m}{N} \sum_{i=1}^{M} \int_{t_i + T_{off(i)}}^{t_i + T_{off(i)} + T_{on}} e^{-j \cdot 2\pi f_m \cdot t} \cdot dt \\
&\approx \frac{f_s}{S_f} \cdot \frac{\left(1 - e^{-j \cdot 2\pi f_m \cdot T_{on}}\right)\left(1 - e^{-j \cdot 2\pi f_m \cdot T_{sw}}\right)\left(1 - \frac{e^{j\pi/2}}{2\pi f_m \cdot R_L \cdot C}\right)}{\left(1 + \frac{T_{off}}{2 \cdot C \cdot R_{ESR}}\right) - \left(1 - \frac{2T_{on} + T_{off}}{2 \cdot C \cdot R_{ESR}}\right) e^{-j \cdot 2\pi f_m \cdot T_{sw}}} \cdot \frac{V_{in} \cdot e^{-j\theta}}{j \cdot 2\pi f_m \cdot L}
\end{aligned} \quad (8)$$

According to the equation of the control voltage, the transfer function from the control signal V_c to the inductor current in the s domain can be obtained as

$$\frac{i_L(s)}{v_c(s)} = \frac{f_s}{s_f} \cdot \frac{\left(1 - e^{-s \cdot T_{on}}\right)\left(1 - e^{-s \cdot T_{sw}}\right)}{\left(1 + \frac{T_{off}}{2 \cdot C \cdot R_{ESR}}\right) - \left(1 - \frac{2T_{on} + T_{off}}{2 \cdot C \cdot R_{ESR}}\right) e^{-s \cdot T_{sw}}} \cdot \frac{V_{in}}{s \cdot L} \quad (9)$$

The transfer function from the output voltage V_{out} to the control voltage V_c is obtained.

$$\frac{V_{out}(s)}{v_c(s)} = \frac{i_L(s)}{v_c(s)} \cdot \frac{R_L(R_{ESR} C \cdot s + 1)}{(R_L + R_{ESR}) C \cdot s + 1} \quad (10)$$

When $R_L \gg R_{ESR}$, the transfer function is

$$\frac{V_{out}(s)}{V_c(s)} \approx \frac{f_s}{s_f} \cdot \frac{V_{in}}{sL} \cdot \frac{1}{1+\frac{s}{Q_1\omega_1}+\frac{s^2}{\omega_1^2}} \cdot \frac{(R_{ESR} \cdot C \cdot s + 1)}{1+\frac{s}{Q_2\omega_2}+\frac{s^2}{\omega_2^2}} \quad (11)$$

where $Q_1 = 2/\pi$, $\omega_1 = \pi/T_{on}$ and $Q_2 = T_{sw}/(R_{ESR} \cdot C - T_{on}/2) \cdot \pi$, $\omega_2 = \pi/T_{sw}$.

As can be seen from the formula, Q_2 will vary with the output capacitance Co, equivalent series resistance (ESR), conduction time, and switching period, which have the possibility of splitting the poles of the right half plane. Therefore, the loop stability condition is

$$R_{ESR} \cdot C > \frac{T_{on}}{2} \quad (12)$$

In addition, Q_2 varies sensitivity with different applications and peripheral devices. To ensure loop stability, the TYPE-I compensation is selected, which introduces only one main pole, and the low-frequency real poles in extreme cases should be placed outside the gain-bandwidth product (GBW) of the feedback loop [23,24]. The loop model under this compensation mode in Simplis is established, and the actual values of external components are substituted. The Simplis simulation of the loop stability is shown in Figure 3. As can be seen in the figure, in this compensation mode, the feedback loop can remain stable.

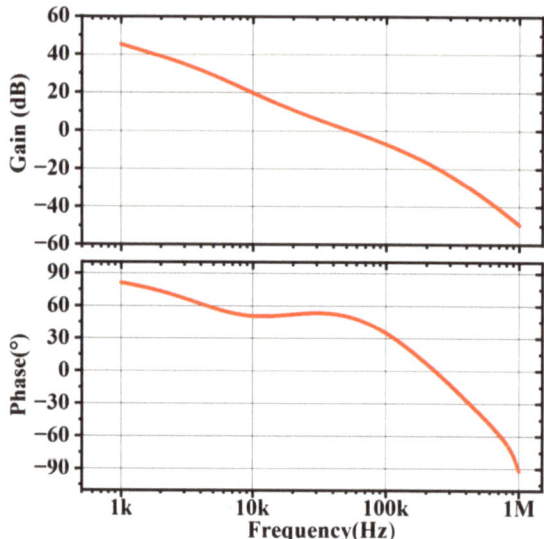

Figure 3. Bode plot of the system.

3. Circuit Implementations

3.1. AOT Generator

When the COT-controlled buck converter operates under the steady state of PWM mode without the consideration of the parasitic parameters or conversion efficiency, the on time of the high-side transistor can be formulated as

$$T_{on} = D \cdot T_{sw} = \frac{V_{out}}{V_{in}} \cdot \frac{1}{f_{sw}} \quad (13)$$

The above equation shows that the operating frequency of the converter is related to the input and output voltages with a constant on-time Ton of the transistor. In order to obtain a fixed operating frequency, it is necessary to dynamically adjust the on time at different input and output voltage [25,26].

The structure of the adaptive on-time generator used in this paper is shown in Figure 4, which consists of three main parts: V-I converter, a capacitor charging and discharging circuit, and a voltage comparator.

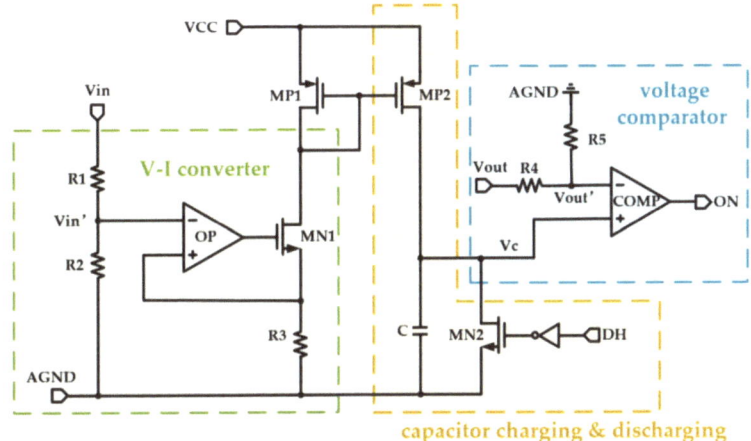

Figure 4. The structure of the adaptive on-time generator.

Resistors R1~R3, operational amplifier, and MN1 constitute the V-I converter. R1 and R2 are used to divide the voltage while isolating the high voltage at Vin. The operational amplifier makes the voltage on R3 equal to the divided voltage to determine the charge current, which is shown in Equation (14). This current is then mirrored to MP2 to finally charge the capacitor.

$$I_c = V_{in} \cdot \frac{R2}{R1+R2} \cdot \frac{1}{R3} \tag{14}$$

Since the range of Vout is close to the supply voltage, it is also necessary to divide Vout before comparison. The capacitor is charged until the voltage is equal to the division of Vout, the comparator flips, and the on time of the high-side transistor ends. At the same time, MN2 turns on and discharges the capacitor in preparation for the next charging cycle. Therefore, without consideration of the delay, the adaptive on-time of the generator can be expressed as

$$T_{on} = \frac{C}{I_c} \cdot \frac{R5}{R4+R5} \cdot V_{out} = \frac{C \cdot R3 \cdot R5 \cdot (R1+R2)}{R2 \cdot (R4+R5)} \cdot \frac{V_{out}}{V_{in}} \tag{15}$$

So, the on time would vary with changes in input voltage and output voltage at the stable state of this buck converter. Substituting Equation (14) into Equation (15), we can see the switching frequency is now constant:

$$f_{sw} = \frac{R2 \cdot (R4+R5)}{C \cdot R3 \cdot R5 \cdot (R1+R2)} \tag{16}$$

3.2. Zero Current Detection

When the buck converter works under a light load, the inductor current may drop to zero at each cycle, which is shown in Figure 5b. If the inductor current drops below 0 A, there will be a reverse current going through the body diode of the power transistor, causing a reduction in efficiency [27]. In order to improve efficiency under light loads, the buck converter proposed in this paper could operate in PSM under light load, and it turns off the low-side transistor once the inductor current drops to zero and keeps the inductor current at zero until the energy stored in the output capacitor is depleted [28,29].

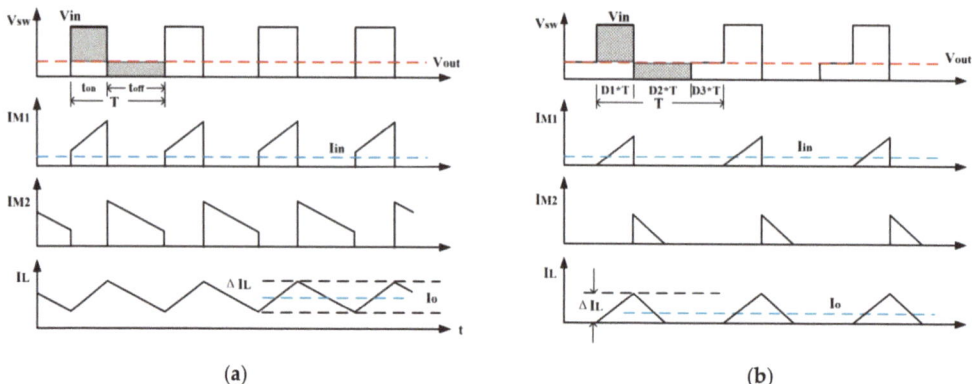

(a)   (b)

Figure 5. Operating waveforms of the BUCK converter: (**a**) heavy load operating and (**b**) light load operating.

To improve the response speed and reduce unnecessary losses, an accurate inductor zero-current detection circuit is applied [30]. Since the on resistance of the power transistor and the voltage change at node SW is relatively small, the comparator needs to work under negative input. Figure 6 shows the circuit of the inductor zero-current detection comparator designed in this paper. It comprises a three-stage differential amplifier. Since the SW voltage variation is small and may be negative, the common-mode input range needs to contain zero voltage, which also needs to reduce noise and improve sensitivity at the same time. So, the BJT emitter input and collector output are used to sensitively amplify the error between the ZCD_SW and ZCD_REF. The switching speed of these zero current detection comparators mainly depends on the delay of the amplifiers and the inverter [31].

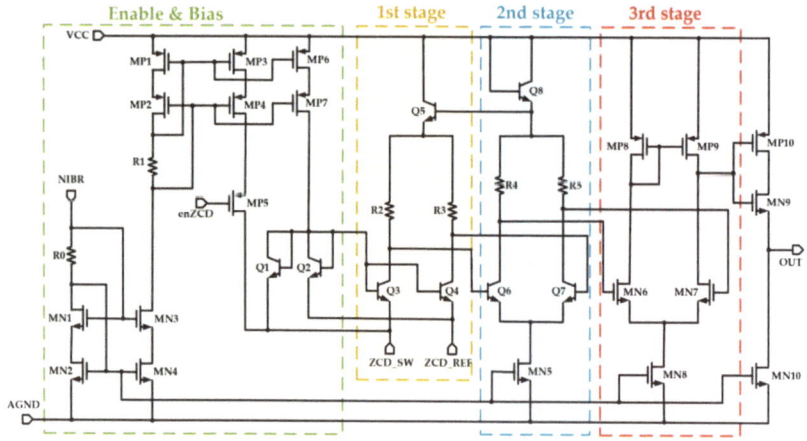

Figure 6. Inductor zero current detection comparator circuit.

4. Experimental Results

The proposed fast transient AOT-controlled buck converter was implemented and fabricated in the TSMC 0.18 μm BCD process. Figure 7 shows the chip microphotograph. The chip size is 3000 μm × 3000 μm, which includes the integrated power transistors, PAD, and electrostatic discharge (ESD) protection circuit. The input voltage range is 5.5 to 15 V, the output voltage range is 0.5 to 5 V, and the output load current range is up to 5 A.

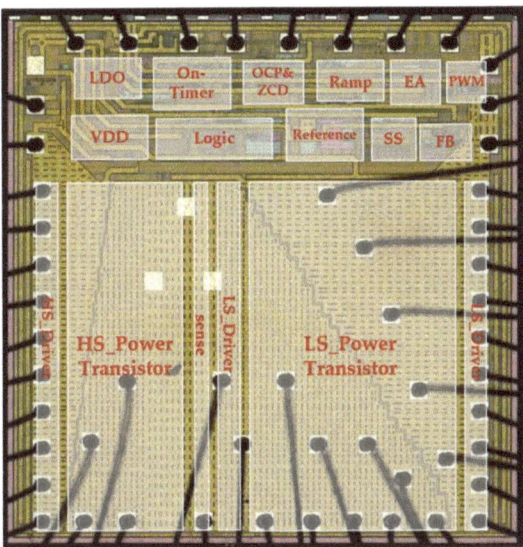

Figure 7. The chip microphotograph.

The steady-state PWM operating waveform of the system is illuminated in Figure 8. The input voltage is 15 V, the output voltage is 5 V, and the load current is 1 A. As shown in Figure 8, the converter works normally under PWM and is capable of driving the output load. As can be seen from Figure 8, the converter works smoothly in PWM mode, and there is no large overshoot and oscillation. The measured output voltage is about 4.89 V, and the output voltage ripple is about 11 mV. The overshoot and undershoot of the VSW and VBST is significantly less than 1 V. The steady operating frequency is about 2.17 MHz.

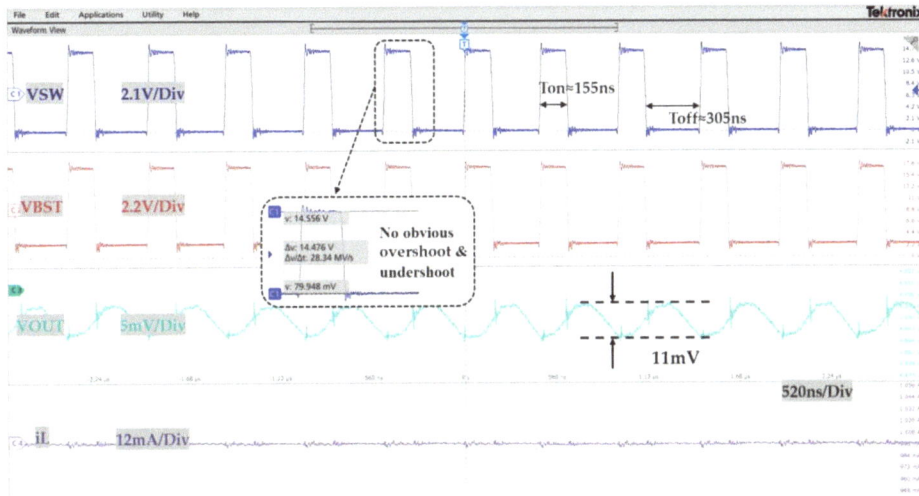

Figure 8. The steadystate PWM operating waveform.

As the load current decreases, the converter could work in the PSM mode, and Figure 9 shows the operating waveform under steady-state PSM, with an input voltage that is 15 V, an output voltage of 5 V, and a load current of 10 mA. As can be seen from the figure, under light loads, the converter reduced frequency and operated in PSM mode, the inductor current waveform was in the form of a single pulse, and there was no sub-harmonic oscillation, showing that the error amplifier and its compensation network, PWM comparators, etc., could operate stable. The measured output voltage is about 4.71 V, and the output voltage ripple is about 24 mV.

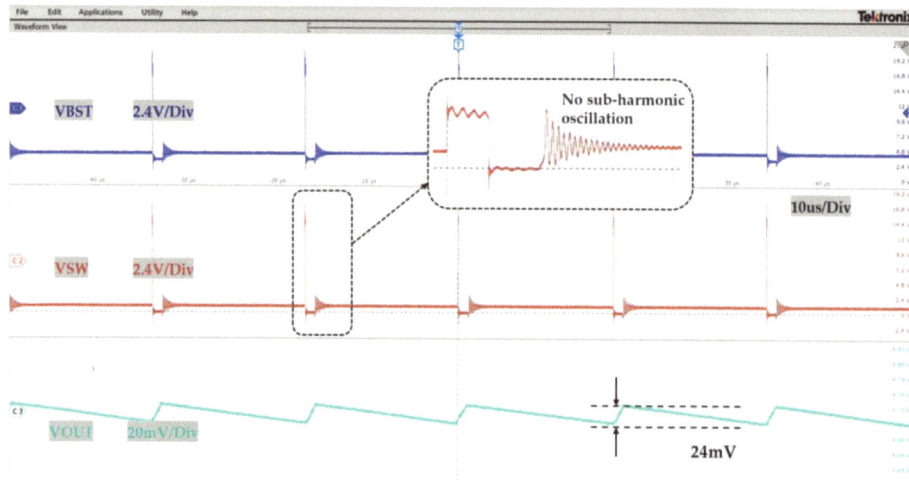

Figure 9. The steady-state PSM operating waveform.

Figure 10 shows the load transient response waveform when the load current was changed from 1 A to 200 mA and from 200 mA to 1 A. As can be seen from the figure, when the load was changed transiently, the VOUT returned quickly to the stable state without sub-harmonic oscillation and burrs. The stable value is around 4.9 V. When the load current was changed from 1 A to 200 mA, the overshoot was about 57 mV, and the recovery time was about 15 μs. When the load was changed from 200 mA to 1 A, the undershoot was about 53 mV, and the recovery time was about 13 μs.

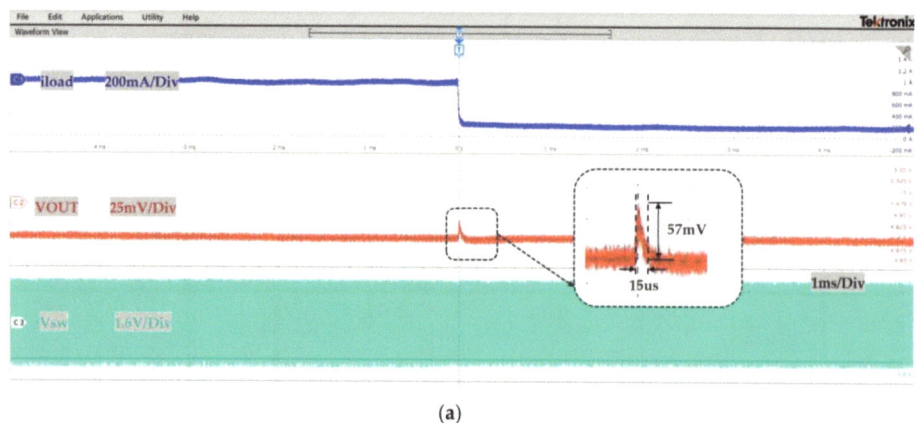

(**a**)

Figure 10. *Cont.*

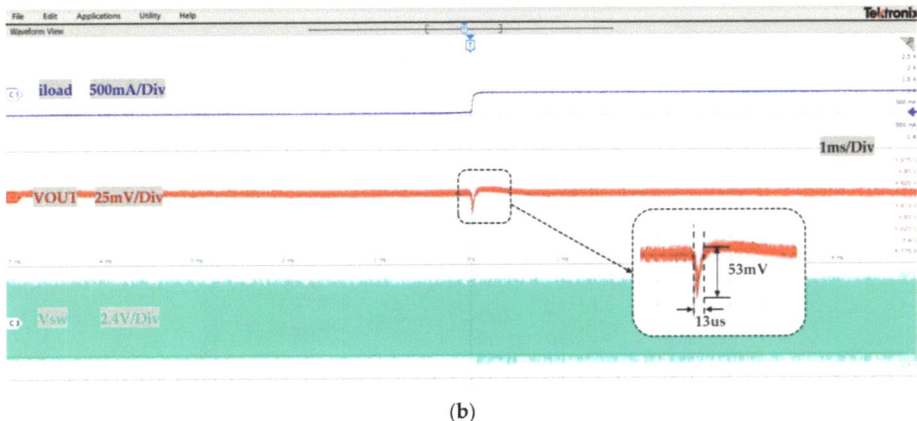

(b)

Figure 10. The load transient response waveform: (**a**) load current falling and (**b**) load current rising.

The converter frequency changes with input and output voltages are shown in Figure 11. Due to the adaptive on-time generator proposed in this paper, the switching frequency can be maintained at a quasi-fixed value over a large input and output range. The test results show that in the input voltage range from 5 V to 15 V, the frequency changes can be controlled within 9%. In the output voltage from 0.5 V to 4.5 V, the frequency changes can be controlled within 14%.

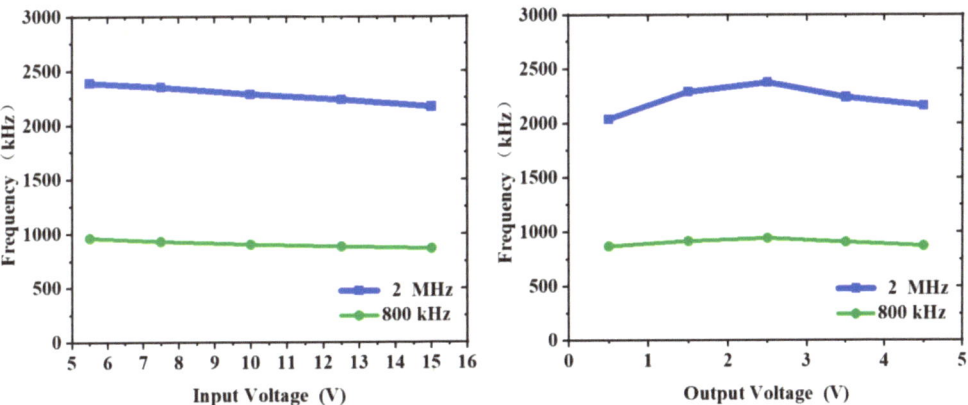

Figure 11. The frequency varied with input and output voltage.

The converter efficiency is shown in Figure 12. The input voltage and the output voltage are 7.5 V and 2.5 V, respectively, and the load current was set to 100 mA, 500 mA, 1 A, 2 A, and 5 A. The peak efficiency is 85.14% at 2 A load current. As the load current decreases, the system efficiency decreases because of the decay of the output power, but the system loss almost does not change. Due to the accurate zero current detection and PSM control used at light load, at a load current of 100 mA, the system efficiency only dropped to 71%. If the converter is forced to operate in PWM mode at light load, the efficiency will drop to 19% at the load current of 100 mA. And with the increasing of load current, the conduction loss of power transistors rises, which means the conversion efficiency gradually decreases. Under the maximum load of 5 A, the conversion efficiency dropped to about 83.15%.

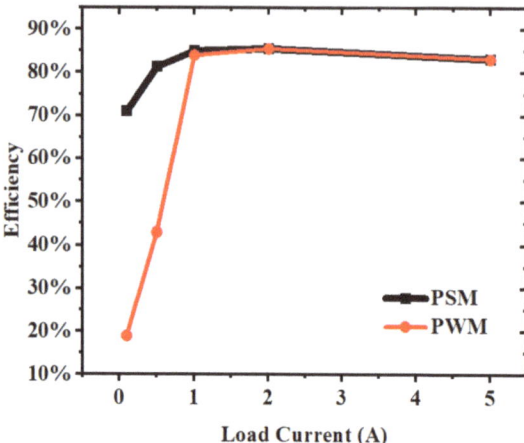

Figure 12. Efficiency at different load currents.

The main performance metrics of the proposed buck converter are summarized and compared with other recently published buck converters in Table 1. References [32,33] are conventional current mode control. References [34,35] are the COT control same as this paper. It can be seen that the dual modulation adaptive-on-time controlled mode buck converter designed in this paper achieves a better response speed than both the current mode controlled and the conventional constant-on-time controlled converters compared over a wide input and output range, significantly, which keeps the upper and lower overshoot voltage within 60 mV with the load transient response recovery time of about 13 μs and 15 μs.

Table 1. Main performance summary and comparison.

Reference	[32]	[33]	[34]	[35]	This Work
Process	0.5 μm	0.25 μm	0.18 μm	0.18 μm	0.18 μm
Operation Mode	Current mode	Current mode	ACOT	RBCOT	ACOT
Input Voltage Range [V]	2.5~5.5	3~4.5	2.1~5.5	3.3	5.5~15
Output Voltage Range [V]	0.9~5	1.8	1.8	1.03	0.5~5
Inductor [μH]	1.8	1	2.2	10	4.7
Output Capacitor [μF]	100	4.7	10	10	47
Load Range [A]	0~5	0~5	0~200 m	0~1	0~5
Switching Frequency	1 MHz	5 MHz	-	-	2 MHz
Output ripple [mV]	<10	50	19	18	11
Recovery time (rise) [μs]	20	14	30	18	13
Recovery time (fall) [μs]	50	15	24	30	15
Overshoot [mV]	60	60	48	330	57
Undershoot [mV]	90	60	43	200	53

5. Conclusions

A fully integrated AOT-controlled buck converter with fast load transient response speed is proposed in this paper. The converter was implemented in the TSMC 0.18 μm BCD process, and a fast and stable load transient response was achieved due to the dual-loop feedback control. The proposed circuit solved the problem caused by the frequency variations in the traditional COT method via an adaptive on-time generator. The converter could operate in PSM under light load to improve light load efficiency via the zero-current detection circuit. Efficiency can still be improved under heavy loads. The experimental results showed that based on the dual modulation adaptive-on-time controlled mode, the system could operate stably under different conduction modes and transform without inducing large undershoot voltage and response time. The rising and falling load transient response recovery time was approximately 13 μs and 15 μs, respectively, with the overshoot and undershoot voltages kept below 60 mV. The improved load transient response time and improved power efficiency under light load make the proposed buck converter a promising candidate for microprocessors.

Author Contributions: Methodology, M.S. and L.W.; Software, M.S. and Y.W.; Validation, M.S.; Formal analysis, M.S.; Investigation, M.S.; Data curation, C.C. and X.X.; Writing—original draft, M.S.; Writing—review & editing, M.X. All authors have read and agreed to the published version of the manuscript.

Funding: This research received no external funding.

Data Availability Statement: Not applicable.

Conflicts of Interest: The authors declare no conflict of interest.

References

1. Koudar, I. Robust DCDC Converter for Automotive Applications. In *Analog Circuit Design: High-Speed Clock and Data Recovery, High-Performance Amplifiers, Power Management*; Springer: Amsterdam, The Netherlands, 2006. [CrossRef]
2. Blaauw, D.; Sylvester, D.; Dutta, P.; Lee, Y.; Lee, I.; Bang, S.; Kim, Y.; Kim, G.; Pannuto, P.; Kuo, Y.-S.; et al. IoT design space challenges: Circuits and systems. In Proceedings of the 2014 IEEE Symposium on VLSI Technology, Honolulu, HI, USA, 9–12 June 2014; IEEE: Piscataway, NJ, USA, 2014. [CrossRef]
3. Nizami, T.K.; Mahanta, C. An intelligent adaptive control of DC–DC buck converters. *J. Frankl. Inst.* **2016**, *353*, 2588–2613. [CrossRef]
4. Tiwari, V.; Donnelly, R.; Malik, S.; Gonzalez, R. Dynamic power management for microprocessors: A case study. In Proceedings of the Tenth International Conference on VLSI Design, Bangalore, India, 26 February–2 March 2002; IEEE: Piscataway, NJ, USA, 2002. [CrossRef]
5. Ryu, Y.C.; Hwang, Y.W. A new soft-start method with abnormal over current protection function for switching power supplies. In Proceedings of the IEEE International Conference on Electric Machines & Drives, San Antonio, TX, USA, 15–18 May 2005; IEEE: Piscataway, NJ, USA, 2005. [CrossRef]
6. Izci, D.; Ekinci, S. A Novel Improved Version of Hunger Games Search Algorithm for Function Optimization and Efficient Controller Design of Buck Converter System. *E-Prime—Adv. Electr. Eng. Electron. Energy* **2022**, *2*, 100039. [CrossRef]
7. Sorouri, H.; Sedighizadeh, M.; Oshnoei, A.; Khezri, R. An intelligent adaptive control of DC–DC power buck converters. *Int. J. Electr. Power Energy Syst.* **2022**, *141*, 108099. [CrossRef]
8. Calderon, A.J.; Vinagre, B.M.; Feliu, V. Fractional order control strategies for power electronic buck converters. *Signal Process.-Amst.* **2006**, *86*, 2803–2819. [CrossRef]
9. Komurcugil, H. Adaptive terminal sliding-mode control strategy for DC-DC buck converters. *ISA Trans.* **2012**, *51*, 673–681. [CrossRef] [PubMed]
10. Yang, Y.; Wang, L.; Xu, Z.; Zhen, S.; Luo, P.; Zhang, B. A Hybrid DC-DC Converter for RF Power Supply Application. *Microelectronics* **2015**, *45*, 594–598.
11. Qin, M.; Xu, J.; Zhou, G.; Mu, Q. Analysis and comparison of voltage-mode and current-mode pulse train control buck converter. In Proceedings of the IEEE Conference on Industrial Electronics & Applications, Xian, China, 25–27 May 2009; IEEE: Piscataway, NJ, USA, 2009. [CrossRef]
12. Roy, P.; Banerjee, K.; Saha, S. Comparison study on the basis of transient response between Voltage Mode Control (VMC) & Current Mode Control (CMC) of buck converter. In Proceedings of the International Symposium on Devices, Circuits and Systems, Online, 14 June 2023. [CrossRef]
13. Siu, M.; Mok, P.; Leung, K.N.; Lam, Y.-H.; Ki, W.-H. A voltage-mode PWM buck regulator with end-point prediction. *IEEE Trans. Circuits Syst. II Express Briefs* **2006**, *53*, 294–298. [CrossRef]
14. Wang, P.Y.; Wu, L.T.; Kuo, T.H. A current-mode buck converter with bandwidth reconfigurable for enhanced efficiency and improved load transient response. In Proceedings of the 2014 IEEE Asian Solid-State Circuits Conference (A-SSCC), KaoHsiung, Taiwan, 10–12 November 2014. [CrossRef]
15. Jeon, I.; Min, K.; Park, J.; Roh, J.; Moon, D.-J.; Kim, H.-R. A Constant On-Time Buck Converter with Fully Integrated Average Current Sensing Scheme. In Proceedings of the 2021 18th International SoC Design Conference (ISOCC), Jeju Island, Republic of Korea, 6–9 October 2021. [CrossRef]
16. Ain, Q.U.; Khan, D.; Jang, B.G.; Basim, M.; Shehzad, K.; Asif, M.; Verma, D.; Ali, I.; Pu, Y.G.; Hwang, K.C.; et al. A High Efficiency Fast Transient COT Control DC-DC Buck Converter with Current Reused Current Sensor. *IEEE Trans. Power Electron.* **2021**, *36*, 9521–9535. [CrossRef]

17. Zhao, J.; Ye, Q.; Lai, X. A Frequency Stable On-Time Control Buck Converter with Reference and Frequency Compensation Technique Using Low ESR Output Capacitor. *IEEE Trans. Ind. Electron.* **2022**, *69*, 3536–3545. [CrossRef]
18. Chiu, M.L.; Yang, T.H.; Lin, T.H. A High Accuracy Constant-On-Time Buck Converter with Spur-Free On-Time Generator. In Proceedings of the 2019 IEEE International Symposium on Circuits and Systems (ISCAS), Sapporo, Japan, 26–29 May 2019; IEEE: Piscataway, NJ, USA, 2019. [CrossRef]
19. Zeng, W.L.; Ren, Y.; Lam, C.S.; Sin, S.W.; Che, W.K.; Ding, R.; Martins, R.P. A 470-nA Quiescent Current and 92.7%/94.7% Efficiency DCT/PWM Control Buck Converter with Seamless Mode Selection for IoT Application. *IEEE Trans. Circuits Syst. I. Regul. Pap. A Publ. IEEE Circuits Syst. Soc.* **2020**, *67*, 4085–4098. [CrossRef]
20. Park, Y.-J.; Park, J.-H.; Kim, H.-J.; Ryu, H.; Kim, S.; Pu, Y.; Hwang, K.C.; Yang, Y.; Lee, M.; Lee, K.-Y. A Design of a 92.4% Efficiency Triple Mode Control DC–DC Buck Converter with Low Power Retention Mode and Adaptive Zero Current Detector for IoT/Wearable Applications. *IEEE Trans. Power Electron.* **2017**, *32*, 6946–6960. [CrossRef]
21. Li, J.; Lee, F.C. New Modeling Approach for Current-Mode Control. In Proceedings of the IEEE Applied Power Electronics Conference & Exposition, Washington, DC, USA, 15–19 February 2009; IEEE: Piscataway, NJ, USA, 2009. [CrossRef]
22. Zhang, X.; Zhang, Z.; Bao, H.; Bao, B.; Qu, X. Stability Effect of Control Weight on Multi-Loop COT Controlled Buck Converter with PI Compensator and Small Output Capacitor ESR. *IEEE J. Emerg. Sel. Top. Power Electron.* **2020**, *9*, 4658–4667. [CrossRef]
23. Izci, D.; Hekimoğlu, B.; Ekinci, S. A new artificial ecosystem-based optimization integrated with Nelder-Mead method for PID controller design of buck converter. *Alex. Eng. J.* **2022**, *61*, 2030–2044. [CrossRef]
24. Yan, Y.; Liu, P.H.; Lee, F.; Li, Q.; Tian, S. V Control with Capacitor Current Ramp Compensation using Lossless Capacitor Current Sensing. In Proceedings of the 2013 IEEE Energy Conversion Congress and Exposition, Denver, CO, USA, 15–19 September 2013.
25. Xu, X.; Wu, X.; Yan, X. A quasi fixed frequency constant on time controlled boost converter. In Proceedings of the IEEE International Symposium on Circuits & Systems, Seattle, DC, USA, 18–21 May 2008; IEEE: Piscataway, NJ, USA, 2008. [CrossRef]
26. Jing, X.; Mok, P.K.T. Fixed-frequency adaptive-on-time boost converter with fast transient response and light load efficiency enhancement by auto-frequency-hopping. In *2011 Symposium on VLSI Circuits-Digest of Technical Papers*; IEEE: Piscataway, NJ, USA, 2011.
27. Zhu, L.F.; He, L.N.; Ye, Y.D. High voltage PWM/PSM dual-mode asynchronous BUCK converter. *J. Zhejiang Univ.* **2011**, *45*, 185–190. [CrossRef]
28. Hong, W.; Lee, M. A 7.4-MHz Tri-Mode DC-DC Buck Converter with Load Current Prediction Scheme and Seamless Mode Transition for IoT Applications. *Circuits Syst. I Regul. Pap. IEEE Trans.* **2020**, *67*, 4544–4555. [CrossRef]
29. Santoro, F.; Kuhn, R.; Gibson, N.; Rasera, N.; Tost, T.; Graeb, H.; Wicht, B.; Brederlow, R. A Hysteretic Buck Converter with 92.1% Maximum Efficiency Designed for Ultra-Low Power and Fast Wake-Up SoC Applications. *IEEE J. Solid-State Circuits* **2018**, *53*, 1856–1868. [CrossRef]
30. Liu, W.-C.; Chen, C.-J.; Cheng, C.-H.; Chen, H.-J. A Novel Accurate Adaptive Constant On-Time Buck Converter for Wide-Range Operation. *IEEE Trans. Power Electron.* **2019**, *35*, 3729–3739. [CrossRef]
31. Jiang, C.; Chai, C.; Han, C.; Yang, Y. A high performance adaptive on-time controlled valley-current-mode DC–DC buck converter. *J. Semicond.* **2020**, *41*, 53–60. [CrossRef]
32. Yuan, B.; Lai, X.-Q.; Wang, H.-Y.; Shi, L.-F. High-Efficient Hybrid Buck Converter with Switch-on-Demand Modulation and Switch Size Control for Wide-Load Low-Ripple Applications. *IEEE Trans. Microw. Theory Tech.* **2013**, *61*, 3329–3338. [CrossRef]
33. Tsai, J.-C.; Huang, T.-Y.; Lai, W.-W.; Chen, K.-H. Dual Modulation Technique for High Efficiency in High-Switching Buck Converters over a Wide Load Range. *IEEE Trans. Circuits Syst. I Regul. Pap.* **2011**, *58*, 1671–1680. [CrossRef]
34. Huang, W.; Liu, L.; Liao, X.; Xu, C.; Li, Y. A 240-nA Quiescent Current, 95.8% Efficiency AOT-Controlled Buck Converter with A2-Comparator and Sleep-Time Detector for IoT Application. *IEEE Trans. Power Electron.* **2021**, *36*, 12898–12909. [CrossRef]
35. Chang, C.H.; Chen, W.C.; Huang, K.S. A 93.4% Efficiency 8-mV Offset Voltage Constant On-Time Buck Converter with an Offset Cancellation Technique. *IEEE Trans. Circuits Syst. II Express Briefs* **2020**, *67*, 2069–2073. [CrossRef]

Disclaimer/Publisher's Note: The statements, opinions and data contained in all publications are solely those of the individual author(s) and contributor(s) and not of MDPI and/or the editor(s). MDPI and/or the editor(s) disclaim responsibility for any injury to people or property resulting from any ideas, methods, instructions or products referred to in the content.

Article

Z-Increments Online Supervisory System Based on Machine Vision for Laser Solid Forming

Junhua Wang [1,2,3,*], Junfei Xu [1], Yan Lu [4], Tancheng Xie [1,2,3], Jianjun Peng [1] and Junliang Chen [5,*]

1. School of Mechanical and Electrical Engineering, Henan University of Science and Technology, Luoyang 471003, China; mecha_xjf@163.com (J.X.); xietc@haust.edu.cn (T.X.); pjjsdu@163.com (J.P.)
2. Henan Intelligent Manufacturing Equipment Engineering Technology Research Center, Luoyang 471003, China
3. Henan Engineering Laboratory of Intelligent Numerical Control Equipment, Luoyang 471003, China
4. School of Materials Science and Engineering, Henan University of Science and Technology, Luoyang 471023, China; luyan@haust.edu.cn
5. College of Food and Bioengineering, Henan University of Science and Technology, Luoyang 471023, China
* Correspondence: wangjh@haust.edu.cn (J.W.); junliangchen@126.com (J.C.)

Abstract: An improper Z-increment in laser solid forming can result in fluctuations in the off-focus amount during the manufacturing procedure, thereby exerting an influence on the precision and quality of the fabricated component. To solve this problem, this study proposes a closed-loop control system for a Z-increment based on machine vision monitoring. Real-time monitoring of the precise cladding height is accomplished by constructing a paraxial monitoring system, utilizing edge detection technology and an inverse perspective transformation model. This system enables the continuous assessment of the cladding height, which serves as a control signal for the regulation of the Z-increments in real-time. This ensures the maintenance of a constant off-focus amount throughout the manufacturing process. The experimental findings indicate that the proposed approach yields a maximum relative error of 1.664% in determining the cladding layer height, thereby enabling accurate detection of this parameter. Moreover, the real-time adjustment of the Z-increment quantities results in reduced standard deviations of individual cladding layer heights, and the height of the cladding layer increases. This proactive adjustment significantly enhances the stability of the manufacturing process and improves the utilization of powder material. This study can, therefore, provide effective guidance for process control and product optimization in laser solid forming.

Keywords: machine vision; Z-increments; off-focus amount; closed-loop control system

Citation: Wang, J.; Xu, J.; Lu, Y.; Xie, T.; Peng, J.; Chen, J. Z-Increments Online Supervisory System Based on Machine Vision for Laser Solid Forming. *Micromachines* **2023**, *14*, 1558. https://doi.org/10.3390/mi14081558

Academic Editor: Ion Stiharu

Received: 29 June 2023
Revised: 1 August 2023
Accepted: 3 August 2023
Published: 4 August 2023

Copyright: © 2023 by the authors. Licensee MDPI, Basel, Switzerland. This article is an open access article distributed under the terms and conditions of the Creative Commons Attribution (CC BY) license (https://creativecommons.org/licenses/by/4.0/).

1. Introduction

Laser solid forming (LSF) is a promising advanced digital additive manufacturing methodology. It seamlessly integrates the advantages of unrestricted solid shaping from rapid prototyping alongside high-performance cladding deposition facilitated by synchronous powder feeding laser cladding [1]. Due to its inherent benefits such as cost-effectiveness, reduced cycle time, exceptional performance, and rapid response capability, LSF has gained substantial traction in various industries, including the aerospace, marine, automotive, and defense sectors, in recent years [2,3]. The effect of LSF forming during the manufacturing process is influenced by a number of factors. Fluctuating parameters and environmental changes during the forming process can cause the cladding height to fluctuate away from the set value. The conventional approach of employing fixed Z-increments during the laser solid forming process has been shown through extensive research to result in substantial variations in the off-focus amount due to fluctuations in the cladding height. These variations, in turn, have been empirically established to exert a direct influence on the dimensional accuracy and mechanical characteristics of the fabricated component [4–7]. Consequently, the preservation of a consistent off-focus level throughout the forming procedure assumes critical importance in safeguarding the dimensional accuracy and quality of the fabricated part. By implementing real-time monitoring of cladding height variations and establishing a closed-loop Z-increments control system, significant enhancements can be achieved in both the accuracy and quality of the forming process for the part.

A great deal of research is currently being carried out on real-time monitoring and process control of additive manufacturing forming processes. Chen B et al. [8] used a CCD camera to capture images of the melt pool and explored the effect of different process parameters on the melt pool area. It was demonstrated that different types of defects could be accurately detected by analyzing the melt pool area. Binega et al. [9] used a line laser scanner to scan the deposition layer profile in real time to extract the deposition layer geometry. Continuous monitoring of the deposition layer profile of the DED process is achieved by comparing the real-time data with the ideal profile. Takushima et al. [10] proposed an online monitoring system for the deposition layer height of laser-fused wire with a wire feed rate feedback control system. The method achieves high accuracy measurement of the deposition layer height by means of an imaging system and an oblique illumination system for the projected beam. The wire feeding rate is controlled according to the measured deposition layer height to maintain the gap between the wire feeding head and the feeding wire in the optimum zone. Zhang Bi et al. [11] used a coaxial high-speed camera to capture melt pool images and designed a Convolutional neural network model to learn melt pool features with a classification accuracy of 91.2% for porosity detection. Farshdianfar et al. [12] developed an infrared imaging system to monitor the melt pool temperature and cooling rate during the cladding process. Using the surface temperature as the feedback signal, a novel feedback PID controller was developed to control the cooling rate using the correlation between the cooling rate, travel speed, and the microstructure of the clad layer. Fleming et al. [13] monitored the morphology of each layer of the SLM process before and after processing by an inline coherent imaging system, identifying bumps and depressions, remelting the raised areas, and filling the depressed areas, enabling artificial closed-loop control of the surface quality of the solidified layer. Huang et al. [14] used a thermal image to monitor the temperature distribution of the solidification layer of the SLM process, established the relationship between scanning speed and temperature distribution, and maintained a stable solidification layer temperature by adjusting the scanning rate in real time to achieve closed-loop control of the SLM process. Numerous researchers have conducted extensive research to improve the quality of forming products for additive manufacturing [15]. The advancement of monitoring tools and control methodologies has played a pivotal role in the notable progress achieved in the realm of additive manufacturing process control. Nevertheless, a majority of the aforementioned studies primarily concentrated on optimizing the forming quality by controlling process parameters like laser power and scanning speed. In the present investigation of laser solid forming for thin-walled components, a predominant approach involves employing a constant Z-increments and often incorporating a negative off-focus amount to induce a self healing effect on the morphological attributes. However, the adjustment of the off-focus amount quantity in this manner proves to be inefficient and fails to effectively address the challenge of achieving optimal alignment between the layer height of thin-walled parts and the Z-increments during the cladding process.

To address the challenge of achieving appropriate alignment between the layer height and the Z-increments in laser solid forming fabrication of thin-walled parts, and ensure a constant off-focus amount throughout the manufacturing process, in the present study, we developed an off-axis camera monitoring system that leveraged edge detection techniques and an inverse perspective transformation model to facilitate real-time detection of the cladding height. By utilizing the cladding height as a feedback signal, the detected cladding height served as a control parameter for the robot to dynamically adjust the Z-increments. This control mechanism ensures the maintenance of a consistent off-focus amount throughout the manufacturing process, effectively mitigating potential deviations. The present research study presents a valuable contribution by offering effective guidance in achieving a precise alignment between the layer height and Z-increments, which plays a role in regulating the quality and accuracy of the forming process.

2. Design of an Off-Axis Camera Monitoring System

2.1. Hardware System Construction

The off-axis camera monitoring system consisted of a Basler industrial camera, filter components, and a workstation. The industrial camera was fixed horizontally to the side axis bracket, thereby ensuring that the optical axis of the camera remained parallel to the substrate. The filter component consisted of a filter lens, a neutral attenuator, and a protective lens. The light emitted from the laser solid forming processing site consisted of multiple wavelengths, intertwining cladding layer information with metal vapor and splashes. Filter lenses facilitated the transmission of specific wavelengths of light while eliminating interference from other wavelength radiations. This capability reduced stray light's impact on cladding layer images, ensuring a clear imaging of the cladding layer. Simultaneously, the filter lenses could reduce the entry of high-energy laser beams (at 1070 nm) into the CCD camera's photosensitive sensor, thus mitigating the risk of damaging the sensor. Through literature analysis and comparative testing, this study opted for infrared filter lenses with a range of 800–2500 nanometers. The function of the neutral attenuator laid in its ability to effectively diminish the intensity of light passing through it, thereby mitigating the risk of saturation in the photosensitive element caused by high levels of radiation during the processing stage. For this study, two central attenuation lenses with 10% light transmission were chosen. The laser solid forming process entailed the utilization of a significant quantity of high-temperature metal particles. To safeguard critical components such as cameras and lenses from potential harm or impairment, protective lenses were purposefully engineered, and the adverse consequences stemming from the presence of high-temperature metal particles could be effectively mitigated. In this study, a 2 mm thick quartz glass plate was chosen as the protective lens. The structure and mounting sequence of the filter assembly is shown in Figure 1.

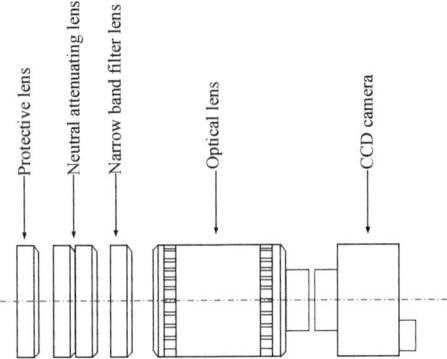

Figure 1. Diagram of the filter components.

The industrial camera was connected to the workstation and transmitted the captured images in real time to the workstation for image processing. The schematic diagram of the side axis camera monitoring system is shown in Figure 2a, and the site installation diagram is shown in Figure 2b.

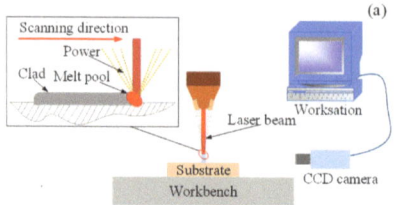
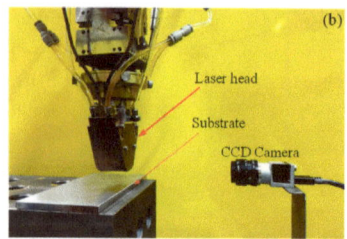

Figure 2. Side axis camera monitoring system. (**a**) Diagram of the off-axis camera monitoring system; (**b**) site installation drawings.

2.2. Image Coordinate System Transformation

In order to monitor the cladding layer height in real time through the off-axis camera, it was necessary to transform the coordinates of the images collected by the camera to obtain the real height value of the cladding layer.

2.2.1. Perspective Projection Model

The camera projected a three-dimensional scene onto the camera's two-dimensional plane through an imaging lens. The basic principle of the perspective projection was by converting the coordinates of objects in a three-dimensional scene to coordinates on a two-dimensional plane. The camera perspective projection model consisted of four coordinate systems: the world coordinate system ($O_W - X_W Y_W Z_W$), which served as a reference in the environment to describe the position of any object; the camera coordinate system ($O_C - X_C Y_C Z_C$), with the camera optical center O_C as the origin and the camera optical axis Z_C defining its direction; the image coordinate system ($O_{xy} - xy$), with the intersection of the camera optical axis and the image plane as the origin; and the pixel coordinate system ($O_{uv} - uv$), with the top-left corner of the image as the origin and pixels as the units. The transformation relationships between the coordinate points are illustrated in Figure 3, where Point P represents a point on the cladding layer, and the distance between the camera coordinate system and the image coordinate system origin $O_C O_{xy}$ is denoted as f, which represents the camera focal length. $P(X_C, Y_C, Z_C)$ denotes the coordinates of point P in the world coordinate system, $p(x, y)$ represents the projected coordinates of point P in the image coordinate system, and $p(u, v)$ represents the pixel coordinates of point P in the pixel coordinate system.

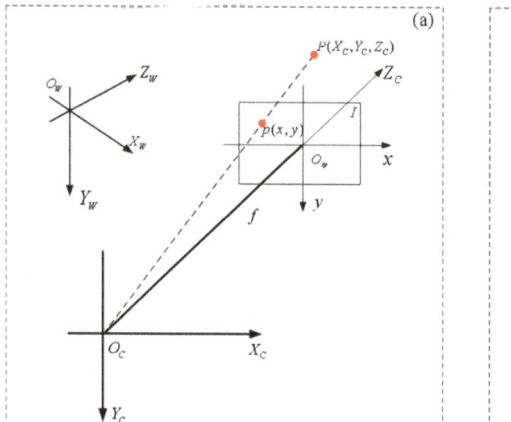

Figure 3. Diagram of the camera perspective projection model. (**a**) World coordinate system and camera coordinate system; (**b**) image coordinate system and pixel coordinate system.

In the transformation between the world coordinate system and the camera coordinate system, the distances, angles, and parallelism of points remained invariant. The transformation from the world coordinate system to the camera coordinate system comprised translation and rotation transformations, involving solely displacement and rotation without any scaling or non-rigid deformations. Hence, the transformation between the world coordinate system and the camera coordinate system was considered a rigid transformation. Utilizing the rotation matrix R and translation vector T, the transformation of coordinate point P between coordinate systems was expressed as follows:

$$\begin{bmatrix} X_C \\ Y_C \\ Z_C \\ 1 \end{bmatrix} = \begin{bmatrix} R & T \\ 0 & 1 \end{bmatrix} \begin{bmatrix} X_W \\ Y_W \\ Z_W \\ 1 \end{bmatrix} \quad (1)$$

where the rotation matrix R is an 3×3 orthogonal unit matrix, dimensionless; T is a three-dimensional translation matrix, which has units in mm.

The transformation of the camera coordinate system to the image coordinate system was a perspective projection, converting the coordinate point P from three-dimensional to two-dimensional. From the proportional relationship, it could be obtained:

$$Z_C \begin{bmatrix} x \\ y \\ 1 \end{bmatrix} = \begin{bmatrix} f & 0 & 0 & 0 \\ 0 & f & 0 & 0 \\ 0 & 0 & 1 & 0 \end{bmatrix} \begin{bmatrix} X_C \\ Y_C \\ Z_C \\ 1 \end{bmatrix} \quad (2)$$

where f indicates camera focal length; x and y represent the horizontal coordinate and vertical coordinates of the projection of coordinate point P onto the image coordinate system; the units of f, x, and y are mm.

As can be seen from Figure 3b, the image coordinate system and pixel coordinate system are both in the imaging plane but have different origins and units of measure. The conversion relation of coordinate points is:

$$\begin{bmatrix} u \\ v \\ 1 \end{bmatrix} = \begin{bmatrix} \frac{1}{dx} & 0 & u_0 \\ 0 & \frac{1}{dy} & v_0 \\ 0 & 0 & 1 \end{bmatrix} \begin{bmatrix} x \\ y \\ 1 \end{bmatrix} \quad (3)$$

where u and v denote the horizontal coordinate and vertical coordinates in the pixel coordinate system, respectively; the unit is px. dx and dy represent the physical size of each pixel in mm/px; u_0 and v_0 are the origin positions; the unit is px.

From Equations (1)–(3), the transformation relationship between the world coordinate system and the pixel coordinate system can be modelled as:

$$Z_C \begin{bmatrix} u \\ v \\ 1 \end{bmatrix} = \begin{bmatrix} f_x & 0 & u_0 & 0 \\ 0 & f_y & v_0 & 0 \\ 0 & 0 & 1 & 0 \end{bmatrix} \begin{bmatrix} R & T \\ 0 & 1 \end{bmatrix} \begin{bmatrix} X_W \\ Y_W \\ Z_W \\ 1 \end{bmatrix} \quad (4)$$

where $f_x = f/dx$ and $f_y = f/dy$ are the scale factors for the u axis and v axis, respectively. f_x, f_y, u_0, v_0 are internal camera parameters; R and T are external camera parameters. Simplifying the model representation:

$$M_1 = \begin{bmatrix} f_x & 0 & u_0 & 0 \\ 0 & f_y & v_0 & 0 \\ 0 & 0 & 1 & 0 \end{bmatrix}, M_2 = \begin{bmatrix} R & T \\ 0 & 1 \end{bmatrix} \quad (5)$$

M_1 is the camera internal parameter matrix, and M_2 is the camera external parameter matrix.

2.2.2. Camera Calibration

To effectively establish the mapping relationship between two-dimensional and three-dimensional images, it was imperative to incorporate the projection characteristics inherent in the transformation process from the camera to the image. This entailed solving for the pertinent parameters of this model by utilizing the corresponding relationship between the mathematical model of camera imaging and the underlying coordinate system. This procedure is commonly referred to as camera calibration [16]. In the present study, the calibration process of the CCD camera involved the utilization of a circular point calibration plate. The calibration plate consisted of 7×7 circular points, each possessing a diameter of 3.5 mm. These circular points were positioned at a uniform center distance of 7 mm from one another. Additionally, the circular point calibration plate featured a square inner frame measuring 56 mm in dimension.

Formula (4) represents the transformation relationship between the world coordinate system and the pixel coordinate system. M_1 denotes the intrinsic matrix, which is solely dependent on the camera's intrinsic properties and internal structure. M_2 is determined by the mapping relationship between the world coordinate system and the camera coordinate system. The camera calibration process involves the estimation of M_1 and M_2.

Due to lens imperfections, it was impossible for the camera's imaging model to achieve an ideal state, leading to distortions in the captured images. Nonlinear distortions mainly consisted of radial distortion and tangential distortion. To enhance the precision of camera calibration, this study not only obtained the radial distortion coefficient k_1, k_2, k_3 but also derived two tangential distortion coefficients p_1, p_2 during the calibration process. The nonlinear distortion model is represented as follows:

$$\begin{cases} x_u = x_d + \delta_x(x_d, y_d) \\ y_u = y_d + \delta_y(x_d, y_d) \end{cases} \tag{6}$$

where (x_u, y_u) represents the ideal coordinate values of the image point, (x_d, y_d) denotes the actual coordinate values of the image point, and δ_x, δ_y represents the nonlinear distortion values. The nonlinear distortion expression employed in this study is as follows:

$$\begin{cases} \delta_x(x_d, y_d) = x_d(k_1 r_d^2 + k_2 r_d^2 + k_3 r_d^2) + [p_1(3x_d^2 + y_d^2) + 2p_2 x_d y_d] \\ \delta_y(x_d, y_d) = y_d(k_1 r_d^2 + k_2 r_d^2 + k_3 r_d^2) + [p_2(3x_d^2 + y_d^2) + 2p_1 x_d y_d] \end{cases} \tag{7}$$

where $r_d^2 = x_d^2 + y_d^2$.

After calibration and calculation, the camera internal parameter matrix M_1 is obtained as:

$$M_1 = \begin{bmatrix} 5598.16 & 0 & 530.38 & 0 \\ 0 & 5597.33 & 404.37 & 0 \\ 0 & 0 & 1 & 0 \end{bmatrix}$$

The rotation matrix R and translation vector T in the camera external parameter matrix are:

$$R = \begin{bmatrix} 1 & -0.008 & -0.016 \\ 0.007 & 0.98 & -0.029 \\ 0.017 & 0.029 & 0.98 \end{bmatrix}, T = \begin{bmatrix} -1.52 \\ -6.055 \\ 392.78 \end{bmatrix}$$

2.2.3. Inverse Perspective Transformation

The establishment of transformation relationships among different coordinate systems was accomplished through camera calibration, which involved determining the internal and external matrix parameters of the camera. Visual measurement entailed an inverse perspective transformation process, distinct from the perspective transformation process described earlier. In this context, visual measurement involved the conversion of the image pixel dimensions of the target object from the image coordinate system to the world coordinate system, so as to obtain the actual size of the measured object. The inverse perspective transformation entailed the utilization of known camera internal parameter matrix M_1, camera external parameter matrix M_2, and image pixel coordinate points; Equation (4) is transformed into a linear equation about the three unknowns of X_W, Y_W, Z_W; and there is no unique solution to the system of equations. To ensure the existence of a unique solution for the aforementioned equation, the imposition of a constraint became necessary. This constraint facilitated the achievement of the inverse perspective transformation, enabling the conversion of two dimensions image pixel coordinate points to three-dimensional world coordinate points.

In this study, the relative positional relationship between the camera and the cladding layer did not change as the experiment proceeded. During the camera calibration process, the calibration plate and the cladding layer were on the same plane, and the plane $X_W Y_W$ in the world coordinate system coincided with the plane of the cladding layer, i.e., $Z_W = 0$. Therefore, by adding this constraint condition, the inverse perspective transformation of the camera was created.

Combining the camera perspective transformation model, let the rotation matrix R and translation matrix T be:

$$R = \begin{bmatrix} r_{11} & r_{12} & r_{13} \\ r_{21} & r_{22} & r_{23} \\ r_{31} & r_{32} & r_{33} \end{bmatrix}, T = \begin{bmatrix} t_1 \\ t_2 \\ t_3 \end{bmatrix} \tag{8}$$

Then, Equation (4) is converted to:

$$Z_C \begin{bmatrix} u \\ v \\ 1 \end{bmatrix} = \begin{bmatrix} f_x & 0 & u_0 & 0 \\ 0 & f_y & v_0 & 0 \\ 0 & 0 & 1 & 0 \end{bmatrix} \begin{bmatrix} r_{11} & r_{12} & r_{13} & t_1 \\ r_{21} & r_{22} & r_{23} & t_2 \\ r_{31} & r_{32} & r_{33} & t_3 \\ 0 & 0 & 0 & 1 \end{bmatrix} \begin{bmatrix} X_W \\ Y_W \\ Z_W \\ 1 \end{bmatrix} \tag{9}$$

As the plane of the world coordinate system $X_W Y_W$ coincides with the plane in which the fused layer is located during camera calibration, such that $Z_W = 0$, Equation (7) is simplified by matrix transformation as:

$$Z_C \begin{bmatrix} u \\ v \\ 1 \end{bmatrix} = \begin{bmatrix} f_x & 0 & u_0 \\ 0 & f_y & v_0 \\ 0 & 0 & 1 \end{bmatrix} \begin{bmatrix} r_{11} & r_{12} & t_1 \\ r_{21} & r_{22} & t_2 \\ r_{31} & r_{32} & t_3 \end{bmatrix} \begin{bmatrix} X_W \\ Y_W \\ 1 \end{bmatrix} \tag{10}$$

Then, it follows that:

$$Z_C \begin{bmatrix} u \\ v \\ 1 \end{bmatrix} = \begin{bmatrix} f_x r_{11} + u_0 r_{31} & f_x r_{12} + u_0 r_{32} & f_x t_1 + u_0 t_3 \\ f_y r_{21} + v_0 r_{31} & f_y r_{22} + v_0 r_{32} & f_y t_2 + v_0 t_3 \\ r_{31} & r_{32} & t_3 \end{bmatrix} \begin{bmatrix} X_W \\ Y_W \\ 1 \end{bmatrix} \tag{11}$$

Simplification of Equation (9)

$$H = \begin{bmatrix} f_x r_{11} + u_0 r_{31} & f_x r_{12} + u_0 r_{32} & f_x t_1 + u_0 t_3 \\ f_y r_{21} + v_0 r_{31} & f_y r_{22} + v_0 r_{32} & f_y t_2 + v_0 t_3 \\ r_{31} & r_{32} & t_3 \end{bmatrix} \tag{12}$$

Substituting Equation (10) into (9), the inverse perspective transformation model is obtained as follows:

$$\begin{bmatrix} X_W \\ Y_W \\ 1 \end{bmatrix} = Z_C H^{-1} \begin{bmatrix} u \\ v \\ 1 \end{bmatrix} \tag{13}$$

Utilizing the established inverse perspective transformation model, the coordinates of a point within the pixel coordinate system are utilized to derive its corresponding coordinates in the world coordinate system. This process relies on the camera's calibrated internal and external parameters, ultimately yielding the accurate size of the cladding layer within the world coordinate system.

2.3. Image Region of Interest Extraction

The initial image of the cladding layer possessed dimensions of 1280 × 1024 pixels, encompassing various elements such as the table, substrate, thin-walled components, and residual unmelted powder. Given the presence of numerous pixel points in the original image that were irrelevant to the study and potentially impeded the extraction of valuable information, it became necessary to isolate the region of interest (ROI) within the original image. In this study, the ROI was extracted from the original image, resulting in a rectangular area defined as [350:650, 70:1020]. The dimensions of the extracted image measured 950 × 300 pixels, as depicted in Figure 4. The extracted image showed mainly the cladding and the substrate, eliminating the interference of redundant pixel points and facilitating further processing of the image.

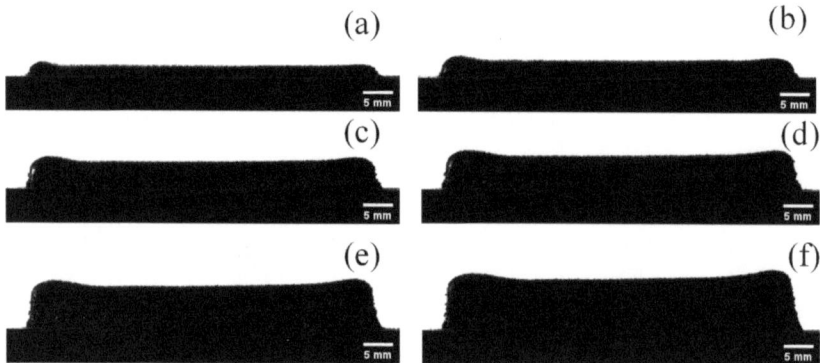

Figure 4. Images after ROI extraction. (**a**) Image after ROI extraction of 5 layers; (**b**) image after ROI extraction of 10 layers; (**c**) image after ROI extraction of 15 layers; (**d**) image after ROI extraction of 20 layers; (**e**) image after ROI extraction of 25 layers; and (**f**) image after ROI extraction of 30 layers.

2.4. Cladding Layer Contour Extraction

In this study, we addressed the issue of edge blurring encountered in the conventional Canny algorithm, which employed Gaussian filtering for noise reduction. To preserve the integrity of edges, we opted for bilateral filtering as an alternative to Gaussian filtering. Furthermore, we tackled the problem of artificially defined thresholds by employing an enhanced Otsu algorithm to derive image segmentation thresholds.

2.4.1. Image Filtering

The bilateral filter incorporated both spatial domain information of pixel points and value domain information based on a Gaussian filter framework. Traditional filtering methods tended to introduce edge blurring during gradual image transformations [17]. In contrast, bilateral filtering considered both the Euclidean distance between pixels and the gray value information of the image, enabling preservation of edge details while accomplishing denoising. Specifically, when the pixel values on either side of an edge differed, weights were diminished to give greater influence to neighboring pixels on the similar side, effectively preventing edge blurring. The bilateral filter pattern is shown below:

$$\hat{f}(i,j) = \frac{\sum_{(m,n) \in \Omega_{r,i,j}} \omega_d(m,n)\omega_r(m,n)f(m,n)}{\sum_{(m,n) \in \Omega_{i,j}} \omega_d(m,n)\omega_r(m,n)} \quad (14)$$

$$\omega_d(m,n) = \exp(-\frac{(i-m)^2 + (j-n)^2}{2\sigma_d^2}) \quad (15)$$

$$\omega_r(m,n) = \exp(-\frac{f(i,j) + f(m,n)^2}{2\sigma_r^2}) \quad (16)$$

where $f(m,n)$ is the gray value of the input image at coordinate (m,n); $\hat{f}(i,j)$ is the gray value of the filtered image at coordinate (i,j); r is the filter window radius; $\Omega_{r,i,j}$ is the set of coordinates of pixels in a square region with (i,j) as the center and sides of $(2r+1)$; $\omega_d(m,n)$ and $\omega_r(m,n)$ are the spatial weights and gray similarity weights at coordinates (m,n), respectively; σ_d and σ_r are spatial standard deviation and gray standard deviation, respectively.

When bilaterally filtering, as the strong noise differed significantly from the grey value of the central pixel, the obtained grey similarity was weighted more heavily, so this strong noise was retained as edge [18]. To address this issue, this study used a combination of adaptive median filtering, which had a high ability to remove strong noise and retain image information. The adaptive median filtering technique enabled dynamic adjustment of the filter window size based on varying noise densities encountered.

The steps of adaptive median filtering were as follows:

1. If $f_{min} < f_{med} < f_{max}$, go to step 2, otherwise increase window size S_{xy}. If $S_{xy} \leq S_{max}$, repeat step 1, otherwise output f_{med}.
2. If $f_{min} < f_{ij} < f_{max}$, output f_{ij}, otherwise output f_{med}.

Where S_{xy} is a window centered on the coordinates (x,y); S_{max} is the maximum size allowed for the window; f_{ij} is the gray value of the coordinate (x,y); f_{min}, f_{med}, and f_{max} are the minimum grey value, the median grey value, and the maximum grey value of S_{xy}, respectively.

2.4.2. Gradient Amplitude Calculation

Calculation of gradient amplitude after image filtering. The 45° and 135° directional gradient amplitudes are introduced in the 3 × 3 neighborhood, and the four directional gradient amplitudes are first obtained by means of the Sobel gradient template. The direction gradient templates for each direction are as follows:

Vertical (x) direction template:

$$\begin{bmatrix} -1 & -2 & -1 \\ 0 & 0 & 0 \\ 1 & 2 & 1 \end{bmatrix}$$

Horizontal (y) direction template

$$\begin{bmatrix} -1 & 0 & 1 \\ -2 & 0 & 2 \\ -1 & 0 & 1 \end{bmatrix}$$

45° orientation template

$$\begin{bmatrix} -2 & -1 & 0 \\ -1 & 0 & 1 \\ 0 & 1 & 2 \end{bmatrix}$$

135° orientation template

$$\begin{bmatrix} 0 & 1 & 2 \\ -1 & 0 & 1 \\ -2 & -1 & 0 \end{bmatrix}$$

The formula for calculating the gradient amplitude is as follows:

$$M(x,y) = \sqrt{G_{xy}^2(x,y) + G_{bevel}^2(x,y)} \tag{17}$$

$$G_{xy}^2 = G_x^2(x,y) + G_y^2(x,y) \tag{18}$$

$$G_{bevel}^2 = G_{45°}^2(x,y) + G_{135°}^2(x,y) \tag{19}$$

where G_x represents the gradient magnitude in the x-direction, G_y represents the gradient magnitude in the y-direction, $G_{45°}$ represents the gradient magnitude at a 45° angle, and $G_{135°}$ represents the gradient magnitude at a 135° angle.

2.4.3. Improved Otsu Algorithm for Threshold Segmentation

The Otsu algorithm constitutes a technique employed to determine the optimal threshold value by performing calculations on the image histogram as a fundamental basis. By leveraging the statistical characteristics of the histogram, this algorithm aims to identify the threshold value that maximizes the inter-class variance, thereby effectively segmenting the image into distinct regions. The Otsu algorithm initiates by determining an optimal threshold value, denoted as k, to facilitate the segmentation of the image into foreground and background regions. Subsequently, the algorithm computes the inter-class variance between these segmented regions. Notably, a higher disparity between the foreground and background intensities results in an augmented inter-class variance, indicative of an improved thresholding outcome. The classes square error σ_B^2 is defined as shown in Equation (18):

$$\sigma_B^2 = P_1(m_1 - m_G)^2 + P_2(m_2 - m_G)^2 \tag{20}$$

where P_1, P_2 are the probability that a pixel will be assigned to the foreground and background regions, respectively; m_1, m_2, m_G are the foreground, background, and global average greyscale values, respectively.

To evaluate the quality of the processed image at threshold k.

$$\eta = \frac{\sigma_B^2}{\sigma_G^2} \tag{21}$$

where σ_G^2 is the global variance. From Equation (19), as σ_G^2 is a constant, and σ_B^2 is a divisibility measure between classes, η is also a divisibility measure, and the two are maximally equivalent.

The Otsu algorithm demonstrated optimal performance when employed for segmenting images exhibiting prominent bimodal peaks within the histogram. However, its efficacy diminished when applied to labeled images containing a sparse distribution of target pixel points. In such cases, the segmentation threshold tended to exhibit a strong bias towards background regions characterized by a substantial proportion of pixels and significant intra-class variance [19]. The Otsu algorithm primarily determined thresholds by maximizing the interclass variance. Leveraging this characteristic, an enhanced segmentation outcome could be achieved by incorporating additional considerations, such as the grey value height within the threshold neighborhood and the average grey difference within the region. By incorporating these factors, the algorithm accentuated the discernibility of low grey value troughs, leading to improved differentiation between the foreground and background regions. The improved formula for the classes square error is:

$$\sigma_B^2 = (1 - \sum_{i=k-n}^{k+n} P_i)^a (P_1(m_1 - m_G)^2 + P_2(m_2 - m_G)^2 + |m_1 - m_2|) \tag{22}$$

where $\sum_{i=k-n}^{k+n} p_i$ is the probability of distribution of all pixels in the gray value interval $[k - n, k + n]$; a is the setting parameter; and a tends to take a larger value when the trough is not evident, making the threshold region trough.

The flowchart for image edge extraction is illustrated in Figure 5.

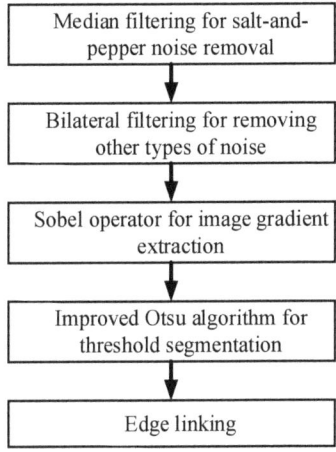

Figure 5. Flowchart of edge extraction.

To extract the pertinent information regarding the cladding layer height, the thresholded image underwent edge detection, specifically targeting the top edge profile of the cladding layer. This extracted profile encapsulated the essential details required for accurately calculating the cladding layer's height. By employing the edge detection technique, the prominent edges of the cladding layer were detected and extracted, thereby facilitating precise determination of its height through subsequent analysis. The image after the canny operator edge detection process is shown in Figure 6.

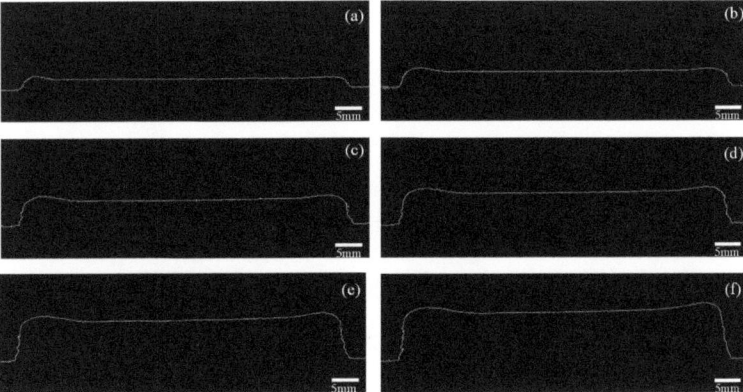

Figure 6. Images after Canny edge detection. (a) Image of edge detection of 5 layers; (b) image of edge detection of 10 layers; (c) image of edge detection of 15 layers; (d) image of edge detection of 20 layers; (e) image of edge detection of 25 layers; (f) image of edge detection of 30 layers.

2.5. Height Calculation and Analysis of Results

2.5.1. Height Calculation

The image after Canny operator edge detection was pixel traversed, the grey value of each pixel was traversed in turn, and the horizontal and vertical coordinates of the pixel with a grey value of 255 were recorded to obtain the horizontal and vertical coordinates of each pixel in the extracted edge profile of the thin-walled part. The vertical coordinate of the topmost contour of the substrate was subtracted from the vertical coordinate of each pixel point at the edge of the contour to obtain the pixel height of the thin-walled part. In this way, the overall pixel heights of the thin-walled parts in layers 5, 10, 15, 20, 25, and 30 were obtained, and the results are shown in Figure 7.

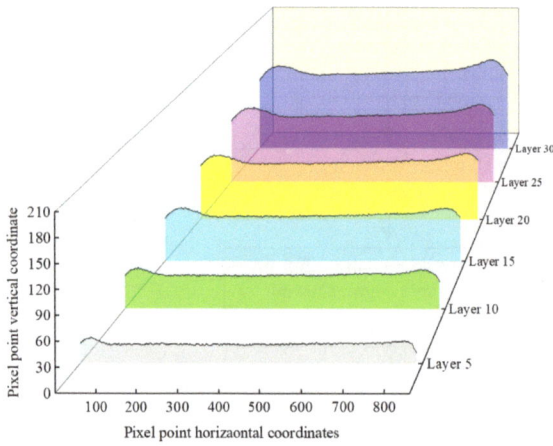

Figure 7. Pixel heights of thin-walled parts.

According to the camera calibration results and the inverse perspective transformation model, the coordinate points were transformed with inverse perspective to obtain the actual height of the cladding layer. The heights of the 5th, 10th, 15th, 20th, 25th, and 30th cladding layers are shown in Figure 8.

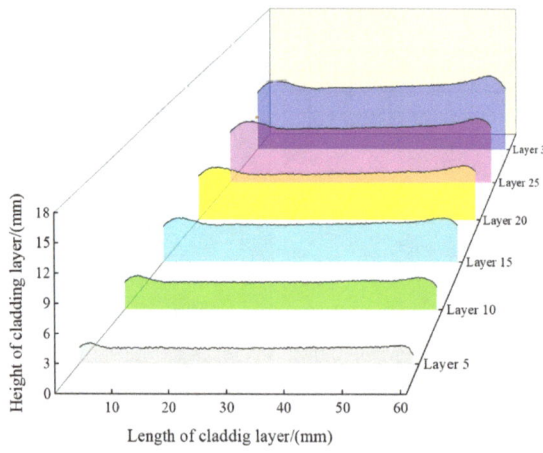

Figure 8. Calculated heights of thin-walled parts.

2.5.2. Height Error Analysis

The cladding height calculation error was obtained by comparing the cladding height calculation results with the height measurement values. Five positions were selected on the uppermost contour of the finished 30-ply thin-walled part, at approximately 4 mm, 20 mm, 35 mm, 48 mm, and 59 mm from the leftmost end in the horizontal direction, and the five selected measurement positions are shown in Figure 9.

Figure 9. Measurement sites of thin-walled parts. Point 1 is located 4 mm from the leftmost end; Point 2 is located 20 mm from the leftmost end; Point 3 is located 35 mm from the leftmost end; Point 4 is located 48 mm from the leftmost end; Point 5 is located 59 mm from the leftmost end.

Measurements were made using a height micrometer, with the measurement positions sorted from left to right, and the image calculation of the height against the actual measured data is shown in Figure 10.

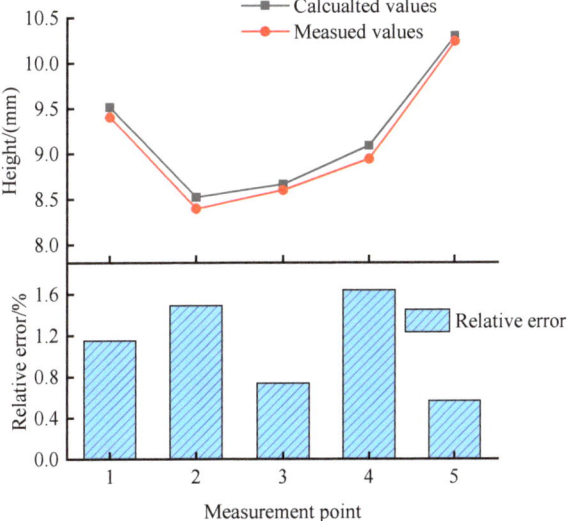

Figure 10. Comparison between calculated and measured values of height.

As illustrated in Figure 9, the analysis reveals a maximum relative error of 1.644% and a minimum relative error of 0.567% between the computed image-based cladding height values and the corresponding actual measured values. This demonstrates the high accuracy achieved in accurately quantifying the cladding height through the proposed methodology. The minimal relative error suggests precise measurement capabilities, thus enabling reliable assessment of the cladding height based on the image analysis. The observed errors can be attributed to several underlying factors. Firstly, inherent characteristics of the CCD camera chip itself may contribute to the issue. When capturing certain objects, inadequate contrast of the edge contour can arise, thus adversely affecting the efficacy of image processing. Secondly, errors within the imaging system can significantly impact the detection accuracy. The resolution of the industrial camera, in particular, plays a crucial role in achieving precise measurements. Additionally, geometric distortion constitutes another influential factor that impairs detection accuracy. Lastly, the presence of vibrations emerges as a prominent contributor to variations in the visual inspection results. Even slight vibrations can result in blurred and distorted images, thereby exerting a detrimental influence on the accuracy of detection.

3. Design of Closed-Loop Control System for Z-Increments

When thin-walled parts are formed under constant Z-increment conditions, the off-focus amount will accumulate layer by layer and gradually increase, resulting in a gradual reduction in the height of the thin-walled part, which ultimately leads to a lower total height of the thin-walled part at the corresponding position, an uneven surface of the thin-walled part, and poor-forming dimensional accuracy. To facilitate the attainment of precise and high-quality manufacturing of thin-walled components through LSF technology, this study introduced a Z-increment regulation method. This method is based on monitoring the forming process of thin-walled parts from the side-axis, enabling the determination of the total height of the fabricated thin-walled parts. By leveraging this information, the proposed method regulates the Z-increments, ensuring accurate control of the additive manufacturing process.

In this study, the Z-increments were used to monitor the layer height of thin-walled parts to ensure that the off-focus amount was kept within a constant range during the manufacturing process and to reduce the impact of the off-focus amount on the LSF forming quality, thereby achieving the goal of regulating the shape size and forming quality of thin-walled parts.

To accurately determine the incremental increase in height for each layer of cladding during the forming process of a thin-walled part, a calculation approach was employed. Specifically, the height value at the conclusion of the current layer of cladding was subtracted from the height value prior to the current layer of cladding. By performing this subtraction, the actual layer height for each layer of cladding within the thin-walled part could be ascertained. The calculation formula is shown below:

$$h_n = H_n - H_{n-1} \tag{23}$$

where h_n is the height of the nth cladding layer, and H_n is the total height of the thin-walled part at the nth layer.

The control system flow is shown in Figure 11.

1. The image of the cladding layer was captured in real time using a side axis camera and transmitted to the host computer workstation;
2. Image processing of the acquired images, including ROI extraction, image filtering, noise reduction, image thresholding, and edge detection;
3. The camera was calibrated, and the results of the camera calibration were used to perform an inverse perspective transformation of the image pixel points to calculate the cladding height;
4. The layer height of the cladding layer was calculated by Equation (21) and transmitted to the PLC from the host computer as the Z-increments;
5. The PLC sent a command to KUKA Robotics with the Z-increments acquired in step 4, which, in turn, ensured that the off-focus amount remained constant during the manufacturing process;
6. Detected whether the number of layers processed had reached the preset value. If the preset value was reached, the process ended; if the preset value was not reached, the process re-entered step 1.

This study used the monitored clad height as a Z-increment to effectively ensure a constant off-focus amount during the manufacturing process, which, in turn, provided guidance for clad quality regulation.

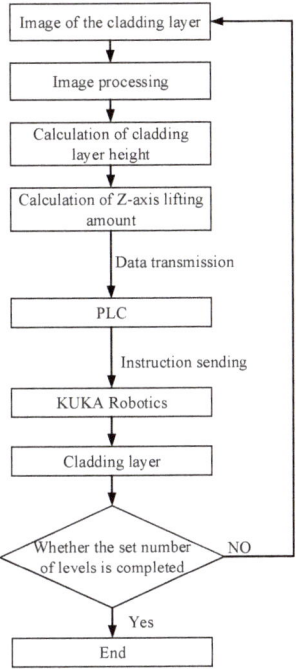

Figure 11. Control flow system diagram.

4. Experiments and Analysis of Results

4.1. Materials and Setup

All experiments in this study were carried out on a laser solid forming equipment, which consisted of a KUKA robot, a co-flying water cooler, a carrier air powder feeder, and a 3 Kw power All-Light laser and laser head. The laser head was connected to a water cooler to avoid damage to the equipment due to high temperatures during the manufacturing process. The powder feed gas and protective gas were both 99.99% argon with a gas flow rate of 12 L/min. The laser solid forming system is shown in Figure 12.

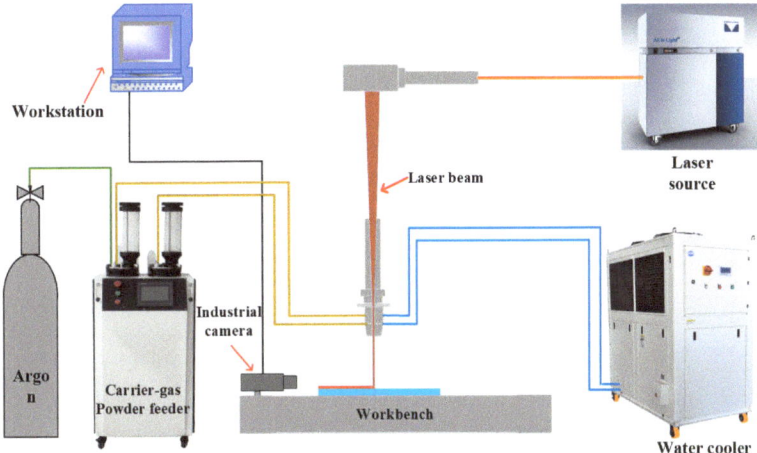

Figure 12. Diagram of the laser stereo forming system.

In the present experiment, a substrate composed of 45 steel, with dimensions measuring 20 mm by 10 mm by 8 mm, was employed. To mitigate any potential temperature-related interferences resulting from multiple cladding, only a single cladding experiment was conducted for each individual substrate. In this study, 17-4PH powder was used as the cladding material, and the chemical composition is shown in Table 1.

Table 1. 17-4PH chemical composition.

Element	C	Mn	Si	S	P	Cr	Ni	Cu	Nb
Wt%	0.07	1.0	1.0	0.025	0.035	15.0	3.0	3.0	0.15

4.2. Design of Experiments

Comparative experiments using the laser solid forming system were carried out with constant Z-increment cladding and real-time, regulated Z-increment cladding. The constant Z-increments were set at 0.25 mm for the cladding experiments, and the other experimental conditions are shown in Table 2.

Table 2. Design of experiments.

Process Parameters (Symbol, Unit)	Value
Laser power (P, W)	1800
Scan speed (v, mm/s)	12
Argon gas flux (Q, L/min)	8
Laser spot diameter (d, mm)	6×2
Number of layers of cladding (n, /)	30
Powder feed rate (f, g/min)	15

4.3. Analysis of Experimental Results

The evaluation of the modulation effect in this study focused on assessing the smoothness of each cladding layer's height. To achieve a more objective and accurate quantification of height smoothness, the standard deviation of the height values associated with each layer was employed as an evaluation metric. The standard deviation formula is shown below:

$$\sigma = \sqrt{\frac{1}{N}\sum_{i=1}^{N}(x_i - \mu)^2} \tag{24}$$

where N is the number of data points; i is the i-th data; and μ is the overall mean.

Each layer of cladding in the experiment was monitored paraxially, the standard deviation of its height value per layer was calculated, and the results of the experiment are shown in Figure 13. The blue line in Figure 13 represents the standard deviation of the height per level for the constant z-increments experiment, and the red line represents the standard deviation of the height per level for the real-time, regulated z-increments experiment.

As can be seen from Figure 13, the standard deviation of the height per level for the real-time regulated Z-increments experiment is significantly smaller than the standard deviation of the height for a constant Z-increments experiment, and this difference becomes larger as the number of levels increases. The analysis suggests that the change in off-focus amount produces cumulative errors as the number of layers increases under constant Z-increments conditions, making the instability of the cladding process worse. Real-time regulation of the Z-increments ensures that the off-focus amount remains constant during the manufacturing process, avoiding the accumulation of errors.

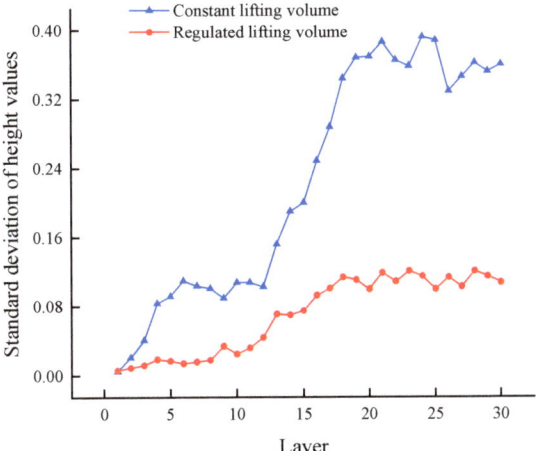

Figure 13. Height standard deviation for different experimental conditions.

Due to the heat accumulation caused by the continuous cladding process, fluctuations in height standard deviation could occur. The height standard deviation stabilized after the experiment reached 20 layers. The analysis suggested that this phenomenon was due to the "dynamic equilibrium" between the melt pool, the cladding layer, and the substrate at this point, where the heat input and heat transfer became balanced, and the whole process became relatively stable, with height fluctuations stabilizing.

The values of the height of each cladding layer monitored during the experiment are shown in Figure 14.

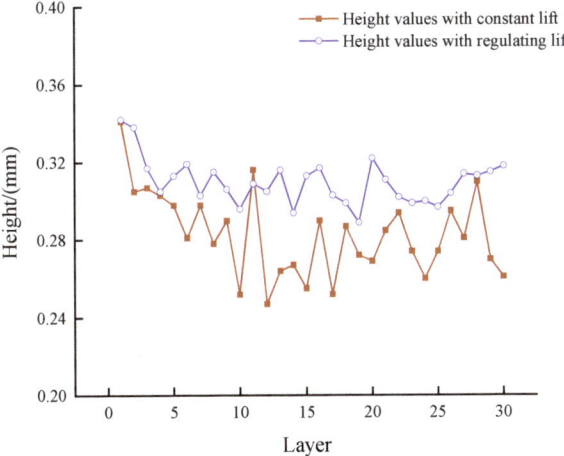

Figure 14. Height values for different experimental conditions.

As can be seen from Figure 14, the height per layer for the real-time, regulated Z-increments experiment is significantly greater than the height of the cladding layer for the constant Z-increments experiment. The analysis concluded that the effect of the off-focus amount on the cladding height was reduced by regulating the Z-increments, increasing the amount of powder entering the melt pool and improving powder utilization. At the same time, the standard deviation of the layer height was calculated. The layer height of the cladding layer in the controlled Z-increments experiment fluctuated less, and the standard deviation of the layer height was 0.015 mm; the layer height of the constant Z-increments experiment fluctuated more, and the standard deviation of the layer height was 0.027 mm. By controlling the z-increments, the layer height of the thin-walled part was relatively stable.

5. Conclusions

Inadequate z-increments during laser solid forming could lead to variations in the off-focus amount and thus affect the quality of the formed part. To solve this problem, this study proposed an LSF regulation method based on machine vision technology using a side axis camera to monitor the cladding height in real time and use the real time cladding height as a control signal to regulate the Z-increments. Through experimental verification, the following conclusions were obtained:

1. In this study, an off-axis camera was used to capture the cladding height image in real time, and, after ROI region extraction, edge detection, camera calibration, and inverse perspective transformation, the actual cladding height was obtained. Through experimental verification, the maximum measurement error was 1.664%. This method could measure the cladding layer height more accurately and in real time;
2. This study was based on machine vision using an off-axis camera to measure the cladding height in real time and use the cladding height as a control signal to regulate the z-increments. The results of the comparative experiments showed that the height of the cladding layer was more stable, the forming accuracy was improved, and the powder utilization rate was increased by real-time adjustment of the Z-increments. The results proved that the study could effectively improve the stability of the forming process and provide effective guidance for practical production.

Author Contributions: Conceptualization, J.W. and J.X.; methodology, J.W. and J.X.; validation, J.W., J.X. and Y.L.; formal analysis, J.X.; investigation, Y.L. and J.C.; resources, J.W.; data curation, Y.L. and J.P.; writing—original draft preparation, J.X.; writing—review and editing, J.W. and J.X.; visualization, J.X. and T.X.; supervision, J.W. and J.C.; project administration, J.W.; funding acquisition, J.W., J.C. and J.P. All authors have read and agreed to the published version of the manuscript.

Funding: This research was funded by the Joint Funds of Science Research and Development Program in Henan Province (222103810039, 222103810030), Science Fund of State Key Laboratory of Tribology in Advanced Equipment (SKLTKF22B12), Henan Province Science and Technology Key Issues (232102111064), Key Scientific Research Project of Colleges and Universities in Henan Province (22A460014, 20A460012), and Special Program for the Introduction of Foreign Intelligence (HNGD2023011).

Data Availability Statement: Not applicable.

Conflicts of Interest: The authors declare no conflict of interest.

References

1. Huang, C.; Liang, R.; Liu, F.; Yang, H.; Lin, X. Effect of dimensionless heat input during laser solid forming of high-strength steel. *J. Mater. Sci. Technol.* **2022**, *99*, 127–137. [CrossRef]
2. Xiao, L.; Peng, Z.; Zhao, X.; Tu, X.; Cai, Z.; Zhong, Q.; Wang, S.; Yu, H. Microstructure and mechanical properties of crack-free Ni-based GH3536 superalloy fabricated by laser solid forming. *J. Alloys Compd.* **2022**, *921*, 165950. [CrossRef]
3. Yang, H.-O.; Zhang, S.-Y.; Lin, X.; Hu, Y.-L.; Huang, W.-D. Influence of processing parameters on deposition characteristics of Inconel 625 superalloy fabricated by laser solid forming. *J. Cent. South Univ.* **2021**, *28*, 1003–1014. [CrossRef]
4. Zhao, P.; Zhang, Y.; Liu, W.; Zheng, K.; Luo, Y. Influence mechanism of laser defocusing amount on surface texture in direct metal deposition. *J. Mater. Process. Technol.* **2023**, *312*, 117822. [CrossRef]
5. Yao, X.; Li, J.; Wang, Y.; Gao, X.; Li, T.; Zhang, Z. Experimental and numerical studies of nozzle effect on powder flow behaviors in directed energy deposition additive manufacturing. *Int. J. Mech. Sci.* **2021**, *210*, 106740. [CrossRef]
6. Metelkova, J.; Kinds, Y.; Kempen, K.; de Formanoir, C.; Witvrouw, A.; Van Hooreweder, B. On the influence of laser defocusing in Selective Laser Melting of 316L. *Addit. Manuf.* **2018**, *23*, 161–169. [CrossRef]
7. Paraschiv, A.; Matache, G.; Condruz, M.R.; Frigioescu, T.F.; Ionică, I. The influence of laser defocusing in selective laser melted in 625. *Materials* **2021**, *14*, 3447. [CrossRef] [PubMed]

8. Chen, B.; Yao, Y.; Huang, Y.; Wang, W.; Tan, C.; Feng, J. Quality detection of laser additive manufacturing process based on coaxial vision monitoring. *Sens. Rev.* **2019**, *39*, 512–521. [CrossRef]
9. Binega, E.; Yang, L.; Sohn, H.; Cheng, J.C. Online geometry monitoring during directed energy deposition additive manufacturing using laser line scanning. *Precis. Eng.* **2022**, *73*, 104–114. [CrossRef]
10. Takushima, S.; Morita, D.; Shinohara, N.; Kawano, H.; Mizutani, Y.; Takaya, Y. Optical in-process height measurement system for process control of laser metal-wire deposition. *Precis. Eng.* **2020**, *62*, 23–29. [CrossRef]
11. Zhang, B.; Liu, S.; Shin, Y.C. In-Process monitoring of porosity during laser additive manufacturing process. *Addit. Manuf.* **2019**, *28*, 497–505. [CrossRef]
12. Farshidianfar, M.H.; Khajepour, A.; Gerlich, A. Real-time control of microstructure in laser additive manufacturing. *Int. J. Adv. Manuf. Technol.* **2016**, *82*, 1173–1186. [CrossRef]
13. Fleming, T.G.; Nestor, S.G.; Allen, T.R.; Boukhaled, M.A.; Smith, N.J.; Fraser, J.M. Tracking and controlling the morphology evolution of 3D powder-bed fusion in situ using inline coherent imaging. *Addit. Manuf.* **2020**, *32*, 100978. [CrossRef]
14. Huang, X.-K.; Tian, X.-Y.; Zhong, Q.; He, S.-W.; Huo, C.-B.; Cao, Y.; Tong, Z.-Q.; Li, D.-C. Real-time process control of powder bed fusion by monitoring dynamic temperature field. *Adv. Manuf.* **2020**, *8*, 380–391. [CrossRef]
15. Li, K.; Ma, R.; Qin, Y.; Gong, N.; Wu, J.; Wen, P.; Tan, S.; Zhang, D.Z.; Murr, L.E.; Luo, J. A review of the multi-dimensional application of machine learning to improve the integrated intelligence of laser powder bed fusion. *J. Mater. Process. Technol.* **2023**, *318*, 118032. [CrossRef]
16. Wang, X.; Chen, H.; Li, Y.; Huang, H. Online extrinsic parameter calibration for robotic camera–encoder system. *IEEE Trans. Ind. Inform.* **2019**, *15*, 4646–4655. [CrossRef]
17. Sajjad, M.; Haq, I.U.; Lloret, J.; Ding, W.; Muhammad, K. Robust image hashing based efficient authentication for smart industrial environment. *IEEE Trans. Ind. Inform.* **2019**, *15*, 6541–6550. [CrossRef]
18. Gavaskar, R.G.; Chaudhury, K.N. Fast adaptive bilateral filtering. *IEEE Trans. Image Process.* **2018**, *28*, 779–790. [CrossRef] [PubMed]
19. Tan, J.; Tang, Y.; Liu, B.; Zhao, G.; Mu, Y.; Sun, M.; Wang, B. A Self-Adaptive Thresholding Approach for Automatic Water Extraction Using Sentinel-1 SAR Imagery Based on OTSU Algorithm and Distance Block. *Remote. Sens.* **2023**, *15*, 2690. [CrossRef]

Disclaimer/Publisher's Note: The statements, opinions and data contained in all publications are solely those of the individual author(s) and contributor(s) and not of MDPI and/or the editor(s). MDPI and/or the editor(s) disclaim responsibility for any injury to people or property resulting from any ideas, methods, instructions or products referred to in the content.

Article

Non-Buffer Epi-AlGaN/GaN on SiC for High-Performance Depletion-Mode MIS-HEMTs Fabrication

Penghao Zhang [1,†], Luyu Wang [1,†], Kaiyue Zhu [2], Qiang Wang [1], Maolin Pan [1], Ziqiang Huang [1], Yannan Yang [1], Xinling Xie [1], Hai Huang [1], Xin Hu [1], Saisheng Xu [1], Min Xu [1,*], Chen Wang [1,*], Chunlei Wu [1] and David Wei Zhang [1,*]

[1] State Key Laboratory of ASIC and System, School of Microelectronics, Fudan University, Shanghai 200433, China; phzhang19@fudan.edu.cn (P.Z.); wangly20@fudan.edu.cn (L.W.); 21112020111@m.fudan.edu.cn (Q.W.); mlpan21@m.fudan.edu.cn (M.P.); zqhuang20@fudan.edu.cn (Z.H.); yangyn20@fudan.edu.cn (Y.Y.); 21212020013@m.fudan.edu.cn (X.X.); 22212020008@m.fudan.edu.cn (H.H.); 22212020078@m.fudan.edu.cn (X.H.); ssxu@fudan.edu.cn (S.X.); wuchunlei@fudan.edu.cn (C.W.)

[2] Department of Materials, Imperial College London, London SW7 2AZ, UK; fz322@ic.ac.uk

[*] Correspondence: xu_min@fudan.edu.cn (M.X.); chen_w@fudan.edu.cn (C.W.); dwzhang@fudan.edu.cn (D.W.Z.)

[†] These authors contributed equally to this work and should be considered co-first authors.

Citation: Zhang, P.; Wang, L.; Zhu, K.; Wang, Q.; Pan, M.; Huang, Z.; Yang, Y.; Xie, X.; Huang, H.; Hu, X.; et al. Non-Buffer Epi-AlGaN/GaN on SiC for High-Performance Depletion-Mode MIS-HEMTs Fabrication. *Micromachines* 2023, *14*, 1523. https://doi.org/10.3390/mi14081523

Academic Editor: Kun Li

Received: 30 June 2023
Revised: 22 July 2023
Accepted: 25 July 2023
Published: 29 July 2023

Copyright: © 2023 by the authors. Licensee MDPI, Basel, Switzerland. This article is an open access article distributed under the terms and conditions of the Creative Commons Attribution (CC BY) license (https:// creativecommons.org/licenses/by/ 4.0/).

Abstract: A systematic study of epi-AlGaN/GaN on a SiC substrate was conducted through a comprehensive analysis of material properties and device performance. In this novel epitaxial design, an AlGaN/GaN channel layer was grown directly on the AlN nucleation layer, without the conventional doped thick buffer layer. Compared to the conventional epi-structures on the SiC and Si substrates, the non-buffer epi-AlGaN/GaN structure had a better crystalline quality and surface morphology, with reliable control of growth stress. Hall measurements showed that the novel structure exhibited comparable transport properties to the conventional epi-structure on the SiC substrate, regardless of the buffer layer. Furthermore, almost unchanged carrier distribution from room temperature to 150 °C indicated excellent two-dimensional electron gas (2DEG) confinement due to the pulling effect of the conduction band from the nucleation layer as a back-barrier. High-performance depletion-mode MIS-HEMTs were demonstrated with on-resistance of 5.84 Ω·mm and an output current of 1002 mA/mm. The dynamic characteristics showed a much smaller decrease in the saturation current (only ~7%), with a quiescent drain bias of 40 V, which was strong evidence of less electron trapping owing to the high-quality non-buffer AlGaN/GaN epitaxial growth.

Keywords: GaN; MIS-HEMTs; buffer layer; SiC substrate; current collapse

1. Introduction

AlGaN/GaN high electron mobility transistors (HEMTs), as a representative of GaN-based wide-band gap semiconductor devices, are becoming one of the most promising candidates for next-generation high-density power conversion systems due to high breakdown voltage, low on-resistance, a wide range of operating temperatures, and high frequency switching [1,2]. Considering the advantages of substrate size and cost, GaN on Si technology has been widely used in AlGaN/GaN HEMTs fabrication [3]. However, large mismatches of lattice dimension and thermal expansion between Si and (Al)GaN will lead to high-density dislocation, poor crystalline quality, and even wafer cracks in the (Al)GaN epilayer due to the failure of stress control. To solve these problems, some strain management procedures have been developed such as AlN/GaN superlattice [4], gradient Al component from AlN to GaN [5,6], and an AlN insert layer [7].

Due to the presence of impurities as Si and O in the metal-organic chemical vapor deposition (MOCVD) chamber, n-type (Al)GaN has often been obtained for the buffer layer, even in the unintentional doping (UID) epitaxial process. Therefore, in order to reduce the vertical leakage and increase the breakdown voltage of devices [8], several methods have been proposed, such as a thicker buffer layer [9], optimized concentration of C-doping or Fe-doping, and so on [10–12]. It should be pointed out that a large number of deep-level traps will be introduced during compensation doping in the thick buffer layer, which is one of the main causes of the current collapse phenomenon of AlGaN/GaN HEMTs, leading to increasing dynamic on-resistance and a decreasing output current. In other words, the buffer layer on the Si substrate has both positive as well as negative effects on device performance.

Improving the epi-structure and substrate design is a promising method to enhance the dynamic performance of devices. For the 4H-SiC substrate, its lattice mismatch with (Al)GaN is much lower [13,14]; hence, there is no urgent need for the thick buffer layer to reduce dislocation. Compared with p-type doping Si, the 4H-SiC substrate is semi-insulating, which could suppress the vertical leakage current without the help of a high-resistance buffer. In addition, the thermal conductivity of SiC (substrate thickness of 0.5 mm) is much higher than that of Si (substrate thickness of 1 mm), indicating that there is a significant advantage in its heat dissipation capability [13]. Recently, the growth technique of SiC has been rapidly developed due to the huge demand for electric vehicles. Thus, the cost of SiC substrates will come down gradually, causing GaN on SiC to become a potential solution for high-end power-switching applications, instead of GaN on Si.

Chen [15] and Hult [16] et al. proposed an epi-structure named QuanFINE based on a SiC substrate that skipped the thick buffer layer between the AlN nucleation layer and GaN channel layer, showing some superiority in their reports, respectively. However, the properties of the epitaxial wafer and the device performance based on this novel epi-structure have not been studied in detail yet. In this work, a variety of characterization methods were employed to systematically evaluate the non-buffer epi-structure without a buffer layer compared with conventional epitaxial wafers based on SiC and Si substrates. Temperature-changing tests were performed for the Schottky diodes on the three epi-structures to further investigate the surface and transport properties of two-dimensional electron gas (2DEG). Normally-on metal insulator semiconductor-HEMTs (MIS-HEMTs) were also fabricated. The DC and dynamic performances of the devices were studied and compared, and the advantages of the non-buffer structure were revealed.

2. Experimental Section

Three AlGaN/GaN hetero-structures epitaxially grown by MOCVD on 4-inch semi-insulating SiC substrates and 6-inch p-doping Si substrates were used in this paper. The novel epi-structure without a buffer layer was GaN (2 nm)/$Al_{0.26}Ga_{0.74}N$ (17 nm)/AlN (1 nm)/UID GaN (150 nm)/AlN (200 nm)/SiC (0.5 mm). It should be pointed out that a unique epitaxy process of the AlN nucleating layer and interface treatment was developed to play an important role in the high crystal quality of the AlN growth. The other two conventional epitaxial wafers on SiC and Si substrates were grown simultaneously with the same structure: GaN (2 nm)/$Al_{0.25}Ga_{0.75}N$ (18 nm)/AlN (1 nm)/UID GaN (150 nm)/Buffer (Fe-doping, 2.2 μm)/AlN (200 nm)/SiC (0.5 mm) or Si (1 mm). They are referred to as Epi A, Epi B, and Epi C in the rest of this article. The schematic structures of the above epitaxial wafers are shown in Figure 1a–c.

Four characterization methods were performed to investigate the effects of the non-buffer design on the heterojunction and surface properties. A WITec alpha300R Raman Imaging Microscope (WITec, Ulm, Germany) was used to study the growth stress with a 405 nm solid-state laser. Rigaku SmartLabII X-ray Diffraction (XRD) (Rigaku, Tokyo, Japan) was employed to characterize the crystalline quality of the epilayer through X-ray rocking curves of (002) and (102). The surface morphology was evaluated using a Park NX10 Atomic Force Microscope (AFM) (Park Systems Corp., Suwon, Republic of Korea) in no-contact mode (NCM). The carrier concentration and mobility in the heterojunction channel were measured by the Ecopia AHT55T3 (Ecopia, Anyang, Republic of Korea) Hall Effect Measurement System.

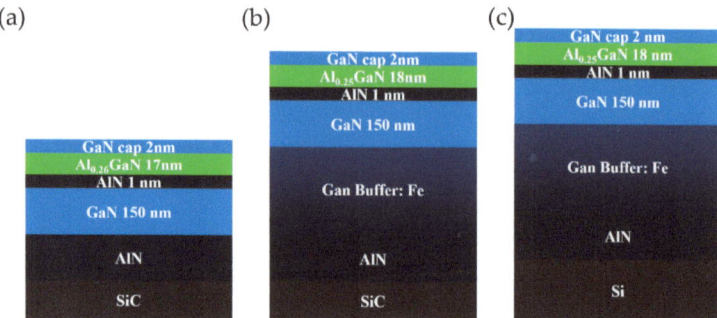

Figure 1. Schematic structures of (**a**) non-buffer epi-structure on SiC substrate, i.e., Epi A, (**b**) conventional epi-structure on SiC substrate, i.e., Epi B, and (**c**) conventional epi-structure on Si substrate, i.e., Epi C.

The fabrication of Schottky diodes on the three epi-structures started with mesa etch using an inductively coupled plasma (ICP) etcher with Cl-based gas. Ti/Al/Ni/Au (20/120/50/50 nm) metal stacks were deposited as the source/drain electrodes by electron beam evaporation (EBE). The ohmic contact was formed by rapid thermal annealing (RTA) at the temperature range of 790 °C to 860 °C for 30 s in N_2 ambient, according to the different crystalline quality in the AlGaN barrier layer on different substrates. The I-V results of the transmission line model (TLM) suggested that the contact resistance (R_c) of each epi-structure was 0.82 Ω·mm, 0.73 Ω·mm, and 0.77 Ω·mm, sequentially. Finally, the gate electrodes of Ni/Au (40/60 nm) were deposited by EBE.

The process flow for MIS-HEMTs was similar with Schottky diodes, with the addition of a gate dielectric process between the ohmic contact formation and the gate metal deposition, as shown in Figure 2a. Before 20 nm Al_2O_3 deposition by thermal atomic layer deposition (ALD), in-situ remote NH_3/N_2 plasma treatment was carried out in a Sentech Si-500 PEALD system (Sentech, Berlin, Germany), which was applied to the gate region in sequence at an RF plasma power of 100/50 W, gas flow of 100/100 sccm, process time of 4/10 min, and substrate temperature of 300 °C. NH_3 plasma acted as a deoxidation step to remove the native oxide, while the subsequent N_2 plasma compensated the N vacancy beneath the surface and formed the nitridated interface. After the following ALD process, post-deposition annealing (PDA) was implemented at 450 °C in O_2 ambient for 3 min to reduce the defects in the film. Transmission electron microscopy (TEM) was performed on the Al_2O_3/GaN/AlGaN interface to examine the effect of in-situ remote plasma treatment, as shown in Figure 2b. The boundary of the interface without any treatment was rough, which tended to be sharp after the low-damage in-situ NH_3/N_2 plasma treatment.

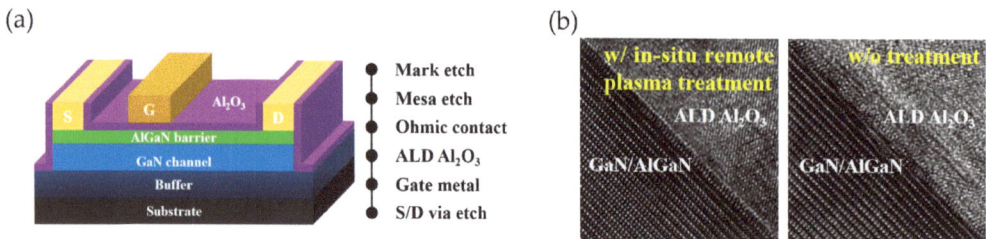

Figure 2. (**a**) Schematic process flow of MIS-HEMTs fabrication. (**b**) Cross-sectional TEM micrographs of the Al_2O_3/GaN/AlGaN interface with and without in-situ plasma treatment.

3. Results and Discussion
3.1. Epitaxial Wafer Quality Characterization

Figure 3a,b display the cross-section views of the non-buffer epi-structure, i.e., Epi A, and conventional epi-structure on the SiC substrate, i.e., Epi B, by scanning electron microscopy (SEM). The thickness of the total epilayer on Epi A was approximately 380 nm, which was only 1/7 compared with the conventional epilayer on Epi B, indicating a clear advantage of heat dissipation of the devices. However, the surface stress on Epi A should be investigated due to the absence of a buffer layer, which is considered to play a crucial role in stress control on conventional epi-structures. The scattering peaks of GaN in E2(TO) mode could be measured by Raman spectroscopy; Raman shift is sensitive enough to characterize the growth stress for a GaN epilayer. Figure 3c shows the Raman spectrum of the above three epi-structures. The peaks of SiC, Si, and GaN exhibited different strengths, consistent with the thickness of the respective epilayers and substrates. Figure 3d shows the Lorentz fitting results of the GaN E2(TO) peak for the above samples. The Raman shifts were 567.37 cm^{-1}, 567.44 cm^{-1}, and 567.29 cm^{-1}, respectively. Compared with the theoretical position of 567.6 cm^{-1} [17,18], slight red shifts occurred in all samples, which suggests that there was tensile stress on the surface with intensities of 0.07 GPa, 0.10 GPa, and 0.13 GPa by calculation [19]. The peak positions of GaN E2(TO) for the three epi-structures were very close to the theoretical value, indicating that the epitaxial process on the SiC substrate without a buffer layer could still achieve effective stress control.

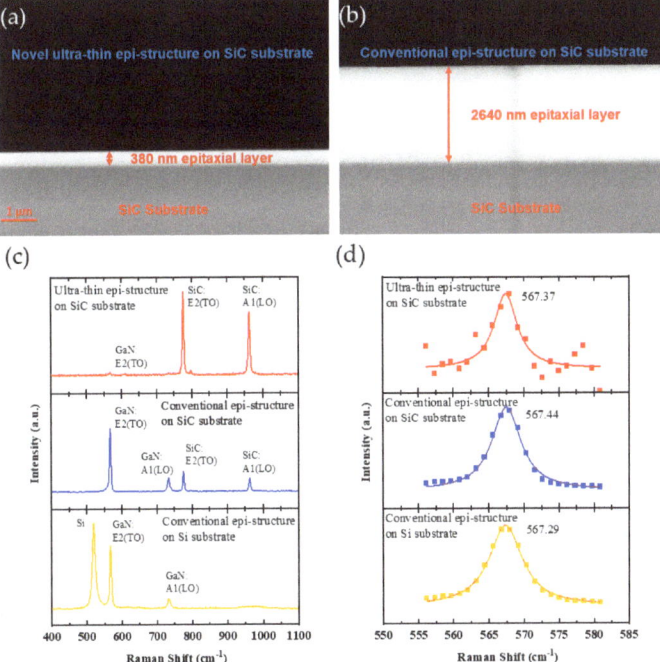

Figure 3. Cross-section views of (**a**) Epi A and (**b**) Epi B by SEM. (**c**) Room temperature Raman spectra of the three epi-structures on SiC and Si substrates under 405 nm laser line excitation. (**d**) Lorentz fitting curves of GaN E2(TO) for above samples.

The crystalline quality of the (Al)GaN epilayer could be characterized by the full width at half maximum (FWHM) in the XRD rocking curves of (002) and (102) planes, respectively. The FWHM of (002) and (102) represents the densities of helical and mixed dislocation [20], as shown in Figure 4a. The non-buffer epi-structure on the SiC substrate had the lowest dislocation density, with FWHMs of (002) and (102) of only 145 arcsec and 267 arcsec, confirming that the growing doping buffer layer was not essential on the SiC substrate. For the conventional epi-structure on SiC, the FWHMs were 196 arcsec and 395 arcsec. The FWHMs of the conventional epi-structure on Si were measured as 433 arcsec and 681 arcsec. This result fully indicates that the epi-structures on the 4H-SiC substrate had a lower dislocation density and higher crystalline quality than that on the Si substrate.

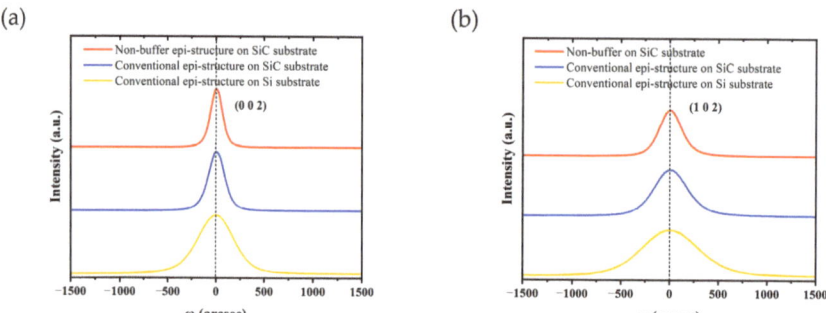

Figure 4. X-ray rocking curves of (**a**) (002) and (**b**) (102) for the three epi-structures by XRD.

Figure 5 shows the surface morphology of the above three epi-structures measured by AFM (5 × 5 µm^2). Atomic step-flow patterns were observed in all of them. The mean square root (RMS) roughness of two epi-structures on the SiC substrate was much less than those on the Si substrate. There were plenty of pits due to dislocation at both ends of the step flow curve on the surface of the epi-structure on the Si substrate. For comparison, no clear dislocation points could be observed on the surface of the other two epi-structures on the SiC substrate. Among these epi-structures, the non-buffer one had the smallest RMS roughness of 0.231 nm, indicating the lowest density of dislocation results and best surface morphology.

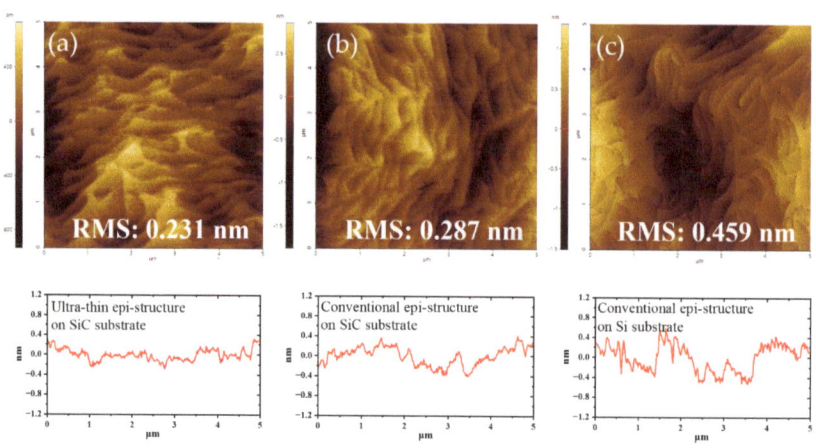

Figure 5. Surface morphology of (**a**) Epi A, (**b**) Epi B, and (**c**) Epi C by AFM.

Hall measurements were used to study the electrical performance of these heterojunction channels. The results are recorded in Table 1, where R_{sheet} is the sheet resistance of the heterojunction, μ_{2DEG} is 2DEG mobility, and n_s is electron density. The smallest R_{sheet} of 284 Ω/sq and largest n_s up to 9.8 × 10^{13} cm^{-2} were measured on the conventional epi-structure on the SiC substrate. Although owning the same epilayer, R_{sheet} of the conventional epi-structure on Si was nearly 30% larger. The non-buffer epi-structure on the SiC substrate had the highest μ_{2DEG} of 1835 cm^2/V·s and n_s of 9.3 × 10^{12} cm^{-2}, and its R_{sheet} was slightly higher than that for the conventional epi-structure. These results indicate that the 2DEG properties were almost independent of the buffer design but strongly related to the crystalline quality of the (Al)GaN epilayer.

Table 1. Hall test results of the three epitaxial structures.

Epi-Structures	R_{sheet} (Ω/sq)	μ_{2DEG} (cm^2/V·s)	n_s (10^{12} cm^{-2})
Non-buffer epi-structure on SiC	292	2287	9.4
Conventional epi-structure on SiC	284	2245	9.8
Conventional epi-structure on Si	365	1835	9.3

3.2. Temperature Changing Tests for Schottky Diodes

The I-V and C-V properties of the Schottky diodes on the three epi-structures were measured by an MPITS2000-SE probe platform at changing temperatures in the range of room temperature (RT) to 150 °C. The voltage scanning range on the anode was −8 V to 2 V, while the cathode was connected to the ground. As shown in Figure 6, while the temperature rose, the forward current and reverse leakage both increased on each diode. The forward current increased mainly because the high temperature improved the ability concerning 2DEG spilling over the AlGaN barrier. At 150 °C, the reverse leakage of the two epi-structures on the SiC substrate increased by approximately one order of magnitude, from ~10^{-4} mA/mm to ~10^{-3} mA/mm. The reverse current of the epi-structure on the Si substrate increased by more than 2 orders of magnitude, from ~10^{-3} mA/mm to ~10^{-1} mA/mm. The significant increase in reverse leakage at high temperatures was mainly due to the trap-assisted tunneling mechanism caused by helical dislocation [21]. The non-buffer epi-structure on the SiC substrate had the minimum reverse leakage increment at high temperatures, indicating the best quality of Schottky contact. These results prove the reliability of XRD whereby the novel epitaxial design had the smallest dislocation density.

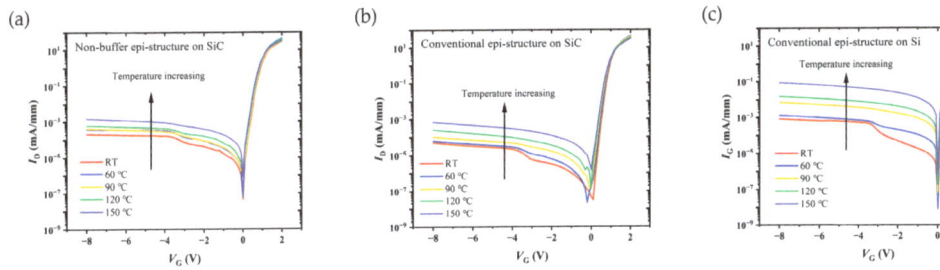

Figure 6. Measured I-V characteristics for Schottky diodes at different temperatures of RT ~ 150 °C on (a) Epi A, (b) Epi B, and (c) Epi C.

C-V measurements of these Schottky diodes were also carried out. Carrier distribution at different temperatures could be obtained through further calculation, as shown in Figure 7. The highest carrier concentration was found at 20~22 nm beneath the surface of the three epi-structures, which was the location of the AlGaN/GaN heterojunction channel. For the conventional epi-structures on the SiC and Si substrates, the distribution of the carrier in the direction to the channel and buffer layers was clearly widened as the temperature increased, which indicated that the high temperature led to the degradation of 2DEG confinement, and a few electrons spilled out from the channel.

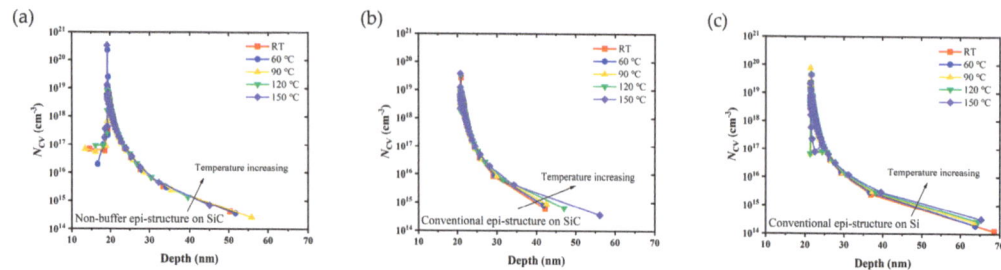

Figure 7. Calculated carrier distribution at different temperatures of RT ~150 °C on (**a**) Epi A, (**b**) Epi B, and (**c**) Epi C.

The following theory explains the effect of the non-buffer design on 2DEG confinement from the perspective of the energy band [15]. As shown in Figure 8, the GaN channel layer was located upon the AlN nuclear layer in the epi-structure without a buffer layer; thus, the energy band of the GaN channel layer was pulled up by AlN due to its larger band gap. AlN could serve as the back barrier of the AlGaN/GaN interface, making the back conduction band of the AlGaN/GaN heterojunction potential well very steep. However, for the conventional epi-structures, the pulling effect from the AlN nucleation layer was not clear due to the existence of a GaN buffer layer with the thickness of a micron order.

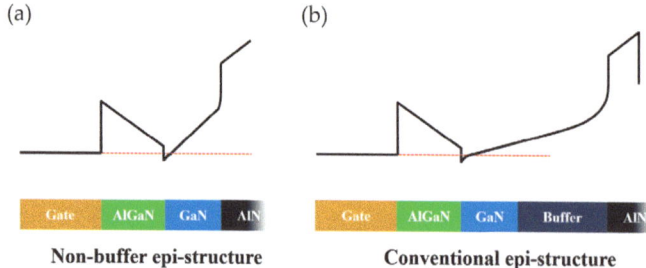

Figure 8. Conduction band (E_c) (black line) of a (**a**) novel non-buffer epi-structure and (**b**) conventional epi-structure.

3.3. Device Performance of the MIS-HEMTs

The fabricated MIS-HEMTs had a gate length (L_g) of 4 µm, gate-source distance (L_{gs}) of 4 µm, gate-drain distance (L_{gd}) of 6 µm, and gate width (W_g) of 50 µm. Transfer and output characteristics were measured with a parameter analyzer of Keithley 4200A (Keithley, Solon, OH, USA), as shown in Figure 9. The linear transfer characteristics were obtained for V_{DS} = 10 V in both forward and reverse sweep directions with V_{GS} steps of 0.1 V. After in-situ NH_3/N_2 remote plasma interface treatment, the Al_2O_3/GaN/AlGaN interfaces on the three epi-structures showed low interfacial state density with small hysteresis voltage ΔV_{th}. The ΔV_{th} of the MIS-HEMTs on the SiC substrate were both only 10 mV, which was only 1/10 of those on the Si substrate. The transconductance (G_m) of the two types of devices on the SiC substrates were ~120 mS/mm and 121 mS/mm, respectively, which was 25% higher than that on the Si substrate. MIS-HEMTs on the SiC substrate also had almost the same subthreshold swings (SSs), which were 97 mV/dec and 96 mV/dec, respectively, much less than those on the Si substrate of ~130 mV/dec.

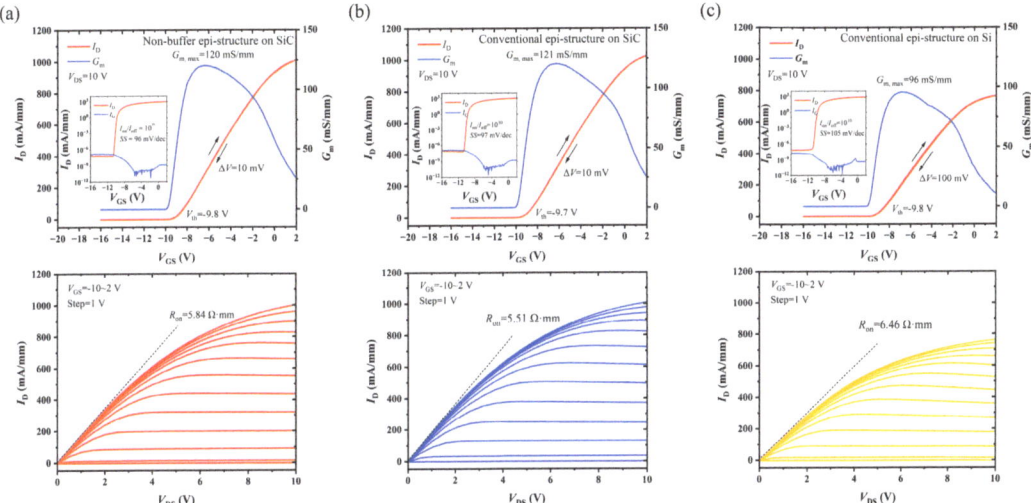

Figure 9. DC transfer and output characteristics for MIS-HEMTs on (**a**) Epi A, (**b**) Epi B, and (**c**) Epi C.

The output characteristics were measured at V_{DS} up to 10 V, with V_{GS} in the range of −10 V to 2 V in steps of +1 V. The MIS-HEMTs on all the epi-structures exhibited kink-free characteristics. MIS-HEMTs on the SiC substrate with a thick, doped buffer showed the highest maximum current $I_{D,max}$ of 1007 mA/mm and lowest on-resistance (R_{on}) of 5.51 Ω·mm. The output characteristic of the devices with a novel non-buffer epi-structure was almost the same as that with the conventional structure, while the $I_{D,max}$ and R_{on} of the MIS-HEMTs on the Si substrate were only 760 mA/mm and 6.46 Ω·mm, respectively.

Pulsed-I-V measurements under slow switching were performed to characterize the degree of current collapse of the fabricated MIS-HEMTs. The dynamic characteristics were investigated using the Keithley 4200A PMU module with the drain quiescent bias voltage V_{DSQ}. The period of the square wave pulse signal was 1 ms, with a width of 10 µs, duty cycle of 1%, and time of pulse rising and falling of 500 ns. The device was synchronously switched from a quiescent bias of V_{GSQ} = 0 V, V_{DSQ} = 0/10/20/30/40 V to a measurement state of V_{GS} = 2 V and V_{DS} from 0 V to 10 V.

Figure 10 illustrates the current collapse phenomenon of the MIS-HEMTs on the different epi-structures under increasing V_{DSQ}. When V_{DSQ} = 40 V, the current collapse of the conventional epitaxial structure on the Si substrate was 29%, which was the most serious among all the structures. For the two types of devices on the SiC substrate, no significant current collapse was observed when V_{DSQ} was within 20 V, indicating that the "virtual gate" effect did not occur at a weak field. When the V_{DSQ} was up to 40 V, the degree of current collapse on the novel non-buffer epi-structure was only 7%, while approximately 15% of current decrement occurred on the conventional structure. The non-buffer epi-structure on the SiC substrate had the best dynamic performance, which was mainly due to the following reasons: First, the novel epi-structure had a better crystalline quality and surface morphology, lower dislocation density, and resulted in fewer surface and body defects. Second, the novel structure offered better 2DEG confinement, which effectively reduced the probability of hot electron tunneling. Third, there were no deep-level traps caused by the doped thick buffer; thus, there was no contribution to current collapse.

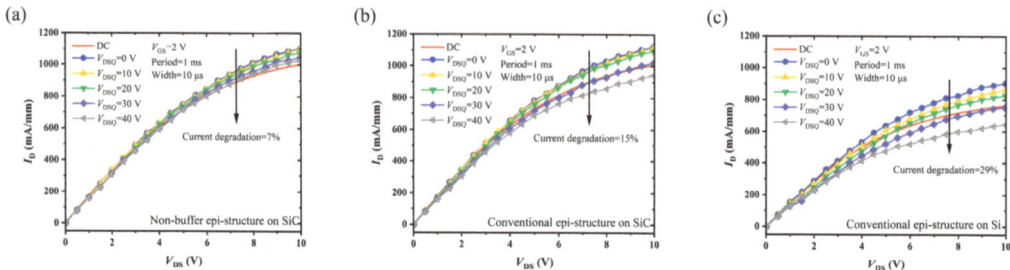

Figure 10. Pulsed-*I-V* measurements performed at different V_{DSQ} MIS-HEMTs on (**a**) Epi A, (**b**) Epi B, and (**c**) Epi C.

4. Conclusions

In this study, a novel AlGaN/GaN epitaxial growth without a conventional buffer layer on a SiC substrate was systematically investigated. The ultra-thin 380 nm AlGaN/GaN/AlN had better crystalline quality, as characterized by AFM, XRD, and Hall measurements. Carrier distribution at high temperatures extracted from the *C-V* curves of Schottky diodes indicated that the non-buffer epi-structure could be favorable for improving 2DEG confinement. MIS-HEMTs on the non-buffer epi-structure almost exhibited the same performance in DC characteristics as the conventional epi-structure on the SiC substrates, which were much better than the current commercial structure on the Si substrate. The pulsed-*I-V* measurements demonstrated the advantage of the non-buffer design, which showed the lowest current collapse compared to the conventional structures on both the SiC and Si substrates. This was attributed to the improved crystalline quality, enhanced electron confinement, and significantly reduced deep-level traps induced by the thick buffer doping. These characteristics made the non-buffer epi-structure on the SiC substrate an excellent candidate in AlGaN/GaN power HEMTs applications.

Author Contributions: Conceptualization, P.Z. and L.W.; methodology, P.Z. and L.W.; validation, Z.H., K.Z., Q.W. and M.P.; formal analysis, L.W. and K.Z.; investigation, P.Z. and L.W.; resources, S.X. and C.W. (Chunlei Wu); data curation, P.Z. and L.W.; writing—original draft preparation, P.Z.; writing—review and editing, M.X. and C.W. (Chen Wang); visualization, Y.Y., X.X., X.H. and H.H.; supervision, M.X., C.W. (Chen Wang) and D.W.Z.; project administration, M.X. and C.W. (Chen Wang); funding acquisition, M.X., C.W. (Chen Wang) and D.W.Z. All authors have read and agreed to the published version of the manuscript.

Funding: This research received no external funding.

Data Availability Statement: Not applicable.

Conflicts of Interest: The authors declare no conflict of interest.

References

1. Ishida, M.; Ueda, T.; Tanaka, T.; Ueda, D. GaN on Si Technologies for Power Switching Devices. *IEEE Trans. Electron Devices* **2013**, *60*, 3053–3059. [CrossRef]
2. Marcon, D.; Saripalli, Y.N.; Decoutere, S. 200 mm GaN-on-Si epitaxy and e-mode device technology. In Proceedings of the Electron Devices Meeting, San Francisco, CA, USA, 3–7 December 2016.
3. Chen, K.J.; Hberlen, O.; Lidow, A.; Tsai, C.L.; Ueda, T.; Uemoto, Y.; Wu, Y. GaN-on-Si Power Technology: Devices and Applications. *IEEE Trans. Electron Devices* **2017**, *64*, 779–795. [CrossRef]
4. Feltin, E.; Beaumont, B.; Laügt, M.; Mierry, P.D.; Gibart, P. Stress control in GaN grown on silicon (111) by metalorganic vapor phase epitaxy. *Appl. Phys. Lett.* **2001**, *79*, 3230–3232. [CrossRef]
5. Marchand, H.; Zhao, L.; Zhang, N.; Moran, B.; Freitas, J.A. Metalorganic chemical vapor deposition of GaN on Si(111): Stress control and application to field-effect transistors. *J. Appl. Phys.* **2001**, *89*, 7846–7851. [CrossRef]
6. Raghavan, S.; Redwing, J. Growth stresses and cracking in GaN films on (111) Si grown by metal-organic chemical-vapor deposition. I. AlN buffer layers. *J. Appl. Phys.* **2005**, *98*, 2653–2672. [CrossRef]

7. Reiher, A.; Bläsing, J.; Dadgar, A.; Diez, A.; Krost, A. Efficient stress relief in GaN heteroepitaxy on Si(1 1 1) using low-temperature AlN interlayers. *J. Cryst. Growth* **2003**, *248*, 563–567. [CrossRef]
8. Rowena, I.B.; Selvaraj, S.L.; Egawa, T. Buffer Thickness Contribution to Suppress Vertical Leakage Current with High Breakdown Field (2.3 MV/cm) for GaN on Si. *IEEE Electron Device Lett.* **2011**, *32*, 1534–1536. [CrossRef]
9. Drechsel, P.; Stauss, P.; Bergbauer, W.; Rode, P.; Steegmüller, U. Impact of buffer growth on crystalline quality of GaN grown on Si(111) substrates. *Phys. Status Solidi* **2012**, *209*, 427–430. [CrossRef]
10. Aggerstam, T.; Sjödin, M.; Lourdudoss, S. AlGaN/GaN high-electron-mobility transistors on sapphire with Fe-doped GaN buffer layer by MOVPE. *Phys. Status Solidi* **2010**, *3*, 2373–2376. [CrossRef]
11. Haffouz, S.; Tang, H.; Bardwell, J.A.; Hsu, E.M.; Webb, J.B.; Rolfe, S. AlGaN/GaN field effect transistors with C-doped GaN buffer layer as an electrical isolation template grown by molecular beam epitaxy. *Solid State Electron.* **2005**, *49*, 802–807. [CrossRef]
12. Meneghini, M.; Rossetto, I.; Bisi, D.; Stocco, A.; Chini, A.; Pantellini, A.; Lanzieri, C.; Nanni, A.; Meneghesso, G.; Zanoni, E. Buffer Traps in Fe-Doped AlGaN/GaN HEMTs: Investigation of the Physical Properties Based on Pulsed and Transient Measurements. *IEEE Trans. Electron Devices* **2014**, *61*, 4070–4077. [CrossRef]
13. Kaminski, N.; Hilt, O. SiC and GaN devices–wide bandgap is not all the same. *IET Circuits Devices Syst.* **2014**, *8*, 227–236. [CrossRef]
14. Levinshtein, M.E.; Rumyantsev, S.L.; Shur, M.S. *Properties of Advanced Semiconductor Materials: GaN, AlN, InN, BN, SiC, SiGe*; John and Wiley and Sons: Hoboken, NJ, USA, 2001.
15. Chen, D.Y.; Malmros, A.; Thorsell, M.; Hjelmgren, H.; Rorsman, N. Microwave Performance of 'Buffer-Free' GaN-on-SiC High Electron Mobility Transistors. *IEEE Electron Device Lett.* **2020**, *41*, 828–831. [CrossRef]
16. Hult, B.; Thorsell, M.; Chen, J.-T.; Rorsman, N. High Voltage and Low Leakage GaN-on-SiC MISHEMTs on a "Buffer-Free" Heterostructure. *IEEE Electron Device Lett.* **2022**, *43*, 781–784. [CrossRef]
17. Davydov, V.Y.; Kitaev, Y.E.; Goncharuk, I.; Smirnov, A.; Graul, J.; Semchinova, O.; Uffmann, D.; Smirnov, M.; Mirgorodsky, A.; Evarestov, R. Phonon dispersion and Raman scattering in hexagonal GaN and AlN. *Phys. Rev. B* **1998**, *58*, 12899. [CrossRef]
18. Perlin, P.; Jauberthie-Carillon, C.; Itie, J.P.; San Miguel, A.; Grzegory, I.; Polian, A. Raman scattering and x-ray-absorption spectroscopy in gallium nitride under high pressure. *Phys. Rev. B Condens. Matter* **1992**, *45*, 83–89. [CrossRef] [PubMed]
19. Kuball, M. Raman spectroscopy of GaN, AlGaN and AlN for process and growth monitoring/control. *Surf. Interface Anal.* **2001**, *31*, 987–999. [CrossRef]
20. Heinke, H.; Kirchner, V.; Einfeldt, S.; Hommel, D. X-ray diffraction analysis of the defect structure in epitaxial GaN. *Appl. Phys. Lett.* **2000**, *77*, 2145–2147. [CrossRef]
21. Li, Y.; Ng, G.I.; Arulkumaran, S.; Ye, G.; Liu, Z.H.; Ranjan, K.; Ang, K.S. Investigation of gate leakage current mechanism in AlGaN/GaN high-electron-mobility transistors with sputtered TiN. *J. Appl. Phys.* **2017**, *121*, 044504. [CrossRef]

Disclaimer/Publisher's Note: The statements, opinions and data contained in all publications are solely those of the individual author(s) and contributor(s) and not of MDPI and/or the editor(s). MDPI and/or the editor(s) disclaim responsibility for any injury to people or property resulting from any ideas, methods, instructions or products referred to in the content.

Communication

Improving Performance of Al$_2$O$_3$/AlN/GaN MIS HEMTs via In Situ N$_2$ Plasma Annealing

Mengyuan Sun [1,†], Luyu Wang [1,†], Penghao Zhang [1] and Kun Chen [1,2,*]

[1] State Key Laboratory of ASIC and System, School of Microelectronics, Fudan University, Shanghai 200433, China; 20212020003@fudan.edu.cn (M.S.); wangly20@fudan.edu.cn (L.W.); phzhang19@fudan.edu.cn (P.Z.)
[2] Shanghai Integrated Circuit Manufacturing Innovation Center Co., Ltd., Shanghai 200433, China
* Correspondence: chenk18@fudan.edu.cn
[†] These authors contributed equally to this work.

Abstract: A novel monocrystalline AlN interfacial layer formation method is proposed to improve the device performance of the fully recessed-gate Al$_2$O$_3$/AlN/GaN Metal-Insulator-Semiconductor High Electron Mobility Transistors (MIS-HEMTs), which is achieved by plasma-enhanced atomic layer deposition (PEALD) and in situ N$_2$ plasma annealing (NPA). Compared with the traditional RTA method, the NPA process not only avoids the device damage caused by high temperatures but also obtains a high-quality AlN monocrystalline film that avoids natural oxidation by in situ growth. As a contrast with the conventional PELAD amorphous AlN, C-V results indicated a significantly lower interface density of states (D_it) in a MIS C-V characterization, which could be attributed to the polarization effect induced by the AlN crystal from the X-ray Diffraction (XRD) and Transmission Electron Microscope (TEM) characterizations. The proposed method could reduce the subthreshold swing, and the Al$_2$O$_3$/AlN/GaN MIS-HEMTs were significantly enhanced with ~38% lower on-resistance at $V_\text{g} = 10$ V. What is more, in situ NPA provides a more stable threshold voltage (V_th) after a long gate stress time, and ΔV_th is inhibited by about 40 mV under $V_\text{g,stress} = 10$ V for 1000 s, showing great potential for improving Al$_2$O$_3$/AlN/GaN MIS-HEMT gate reliability.

Keywords: fully recessed-gate MIS HEMTs; AlN film; plasma annealing; threshold voltage

Citation: Sun, M.; Wang, L.; Zhang, P.; Chen, K. Improving Performance of Al$_2$O$_3$/AlN/GaN MIS HEMTs via In Situ N$_2$ Plasma Annealing. *Micromachines* **2023**, *14*, 1100. https://doi.org/10.3390/mi14061100

Academic Editor: Kun Li

Received: 12 April 2023
Revised: 14 May 2023
Accepted: 17 May 2023
Published: 23 May 2023

Copyright: © 2023 by the authors. Licensee MDPI, Basel, Switzerland. This article is an open access article distributed under the terms and conditions of the Creative Commons Attribution (CC BY) license (https:// creativecommons.org/licenses/by/ 4.0/).

1. Introduction

Gallium nitride (GaN) and its related wide-band gap compound semiconductors have been considered candidates for the next generation of RF and power conversion applications [1–5]. Compared with Si, high electron mobility transistors (HEMTs) based on AlGaN/GaN heterostructures have excellent performance due to their inherent high breakdown strength, low on-resistance, and high temperature operating capability [6,7]. In addition to these inherent advantages, there is also great interest in the possibility of growing GaN-based semiconductors on large-area (up to 200 mm) and low-cost Si substrates due to the large market potential and the possibility of integrating GaN power switches with Si CMOS technology [2]. The conventional AlGaN/GaN HEMTs operate in depletion mode (D-mode) because of the high-density 2DEG induced by the polarization effect. However, in order to reduce power loss during switching and to simplify circuit configuration, normally-off devices are necessary for power applications. Many technologies that could enable devices to achieve enhancement mode (E-mode) have been proposed, including fluorine ion treatment [8,9], p-GaN gate [10], thin AlGaN barrier [11], recessed gate [12,13], and so on. While the F ion implantation method appeared earlier, the F ion is easy to diffuse in the barrier layer at high temperatures, resulting in an unstable device threshold voltage. The thin AlGaN barrier structure will reduce the polarization effect of the draft region, resulting in a decrease in two-dimensional electron gas (2DEG) concentration and device output characteristics. p-GaN gate structure requires very precise etching conditions [14] and etch-induced damage is inevitable to the draft region. Among them, recessed gate HEMTs are considered one of the most promising approaches.

In order to make the device normally off, the polarization-induced charges will be eliminated when the AlGaN barrier under the gate region is fully removed. The barrier-removing process can introduce extra etching damage and surface states. Therefore, gate dielectric is significant for the recessed-gate structure because it can decrease off-state gate leakage and driver losses. At the same time, with the merits of suppressed gate leakage and enlarged gate swing, AlGaN/GaN metal-insulator-semiconductor high electron mobility transistors (MIS-HEMTs) are highly preferred over the conventional Schottky-gate HEMTs for high-voltage power switches. However, the insertion of gate dielectric creates an additional dielectric/AlGaN/GaN interface, where a high density of interface traps usually exists [15]. The Al_2O_3 film grown by atomic layer deposition (ALD) is commonly used as the gate dielectric in the fabrication of AlGaN/GaN MIS-HEMTs [16]. However, under high fabrication process temperatures and high operation voltage stress conditions, the oxygen element in ALD Al_2O_3 may diffuse to the surface of the AlGaN barrier layer, causing reliability concerns [17].

In terms of interface quality, Hinkle et al. [18] found that natural oxides (Ga_2O_3) on the surface of (Al)GaN compound semiconductors are the main reason for Fermi level pinning and high interface trap density (D_{it}) in GaN-based transistors. Robertson [19] proposed that natural defects such as Ga-suspended bonds at the oxide/(Al)GaN interface also restrict Fermi level variation. In addition to traditional natural oxide removal methods, nitridation interfacial layer (NIL) ahead of the gate dielectric deposition has been adopted in AlGaN/GaN HEMTs, such as AlN deposition [20], thermal nitridation [21], and remote plasma nitridation [22–24]. Implemented in a plasma-enhanced atomic layer deposition (PEALD) system, NIL is an excellent choice for interface quality improvement. It is because PEALD not only can achieve precise deposition thickness and avoid surface damage caused by plasma but also facilitates implementing the function of in situ dielectric deposition.

Monocrystalline AlN film has been reported as an interfacial dielectric layer in MIS-HEMTs, which can improve interface quality and device reliability [25]. Generally, high-temperature (over 600 °C) processes such as molecular beam epitaxy (MBE) and metal-organic vapor deposition (MOCVD) are considered necessary for the formation of high-quality AlN crystal [26]. Nevertheless, such a high temperature is not compatible with the subsequent fabrication process [27,28]. Although the ALD technique could facilitate the growth of AlN at a lower temperature (about 300 °C), the amorphous film still needs high temperature rapid thermal annealing (RTA) to transform to the monocrystalline state.

In this work, a novel in situ AlN crystal interfacial layer formation process is proposed to obtain a high-quality Al_2O_3/AlN/GaN interface. Compared with the traditional RTA method, the NPA process not only avoids the device damage caused by high temperatures but also obtains a high-quality AlN monocrystalline film that avoids natural oxidation by in situ growth. Namely, the post PEALD in situ N_2 plasma annealing (NPA) process not only promotes the crystallization of the AlN interfacial layer but also significantly suppresses interface states and reduces V_{th} shift after long-term gate stress as compared to fully recessed-gate MIS-HEMTs fabricated with conventional amorphous AlN.

2. Device Structure and Fabrication

For the fabrication of fully recessed-gate MIS-HEMTs, the epitaxial structure was grown on a 6-in Si (111) wafer by MOCVD. The epitaxial III-Nitride layers were composed of a 5 μm C-doped buffer layer, a 180 nm GaN channel, and a 20 nm $Al_{0.22}Ga_{0.78}N$ barrier layer. The density and mobility of 2DEG were 8.5×10^{12} cm^{-2} and 1960 cm^2/Vs, respectively, by Hall measurement at room temperature.

Figure 1a shows the schematic of the fully recessed-gate MIS-HEMTs, in which gate lengths of 4 μm and gate widths of 50 μm are prepared for the following characterization. After device isolation by Cl-based inductively coupled plasma (ICP) deep etch, a 50-nm SiN_x passivation stack was deposited by ICP-CVD. After selectively removing the passivation layers in the gate window by F-based ICP dry etching, the gate recess process was performed by O_2 and BCl_3 atomic layer etch (ALE) technology at an etch rate of 0.75 nm/cyc [29,30]. As shown in Figure 1b, a total recess depth of ~20 nm was reached after 26 cycles of ALE, indicating complete removal of the barrier layer. Then, the AlN insertion layer was deposited by PEALD. Trimethylaluminum and NH_3 were used as the metal precursor and the N source, respectively. The purge gas was high-purity N_2. The plasma RF power and chamber temperature were set at 60 W and 250 °C, respectively. In order to avoid excessive polarization charge at the AlN/GaN interface, the thickness of the AlN film with 20 cycles of ALD was nominally 1.5 nm. An in situ NPA process was first applied to make the amorphous AlN transform to the crystalline form, as shown in Table 1. Considering that thin AlN was the interface layer, an RF of 200 W and a process time of 300 s were chosen for high quality and low damage requirements. Afterwards, a 20 nm Al_2O_3 layer was in situ deposited by ALD, followed by 500 °C RTA for 90 s to eliminate dangling bonds. Finally, the Ti/Al/Ni/Au ohmic contact and Ni/Au gate were fabricated, respectively. In addition, MOS diodes for capacitance–voltage (C-V) tests were also prepared on the same wafer, as shown in Figure 1c. For comparison, MIS HEMTs with two different properties of the AlN interfacial layer were fabricated, which are also distinguished as Scheme I and Scheme II.

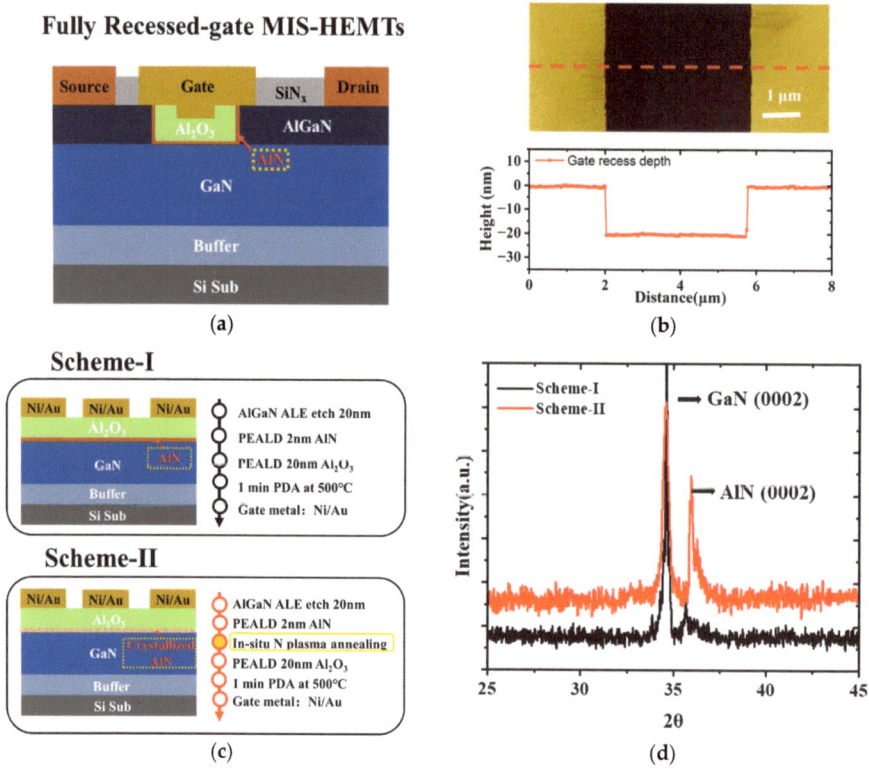

Figure 1. (a) Schematic cross-sectional view of normally-off Al_2O_3/AlN/GaN MIS-HEMT on silicon substrate; (b) AFM measurement of the trench profile along the recessed window; (c) The MOS diodes for capacitance–voltage (C-V) tests; (d) XRD spectrum of AlN (0002) and GaN (0002).

Table 1. Conditions of N₂ plasma annealing for AlN properties.

Power/W	Time/s	AlN Property
100	100	amorphous
100	300	amorphous
100	500	amorphous
200	100	Weak signal in AlN (0002)
200	**300**	**monocrystalline**
200	500	monocrystalline
300	100	monocrystalline
300	300	monocrystalline
300	500	Weak signal in AlN (0002)

Grazing incidence X-ray diffraction (GIXRD) was used to investigate the crystallization characteristics of the AlN, as shown in Figure 1d. It can be observed that the AlN peak appeared in the (0002) orientation. The diffraction peak intensity after NPA was significantly enhanced in Scheme II. This indicates that the proposed process can promote the crystallization of AlN films on the GaN substrate.

3. Results and Discussion

The multi-frequency capacitance–voltage characteristics of the MOS diodes are plotted in Figure 2a,b. With the frequency varying from 10 kHz to 10 MHz, the MOS diodes with an N₂ plasma-enhanced AlN crystal interfacial layer show much smaller frequency dispersion compared to Scheme II, indicating an improved interface with a lower trap density. The insets in Figure 2a,b show the C-V hysteresis characteristics at the frequency of 1 MHz. The shift of flat-band voltage is 53 mV and 129 mV for the MOS diodes in Schemes I and II, respectively. Trap density (D_{it}) can be obtained from multi-frequency C-V curve frequency dispersion [31]. For Scheme I, the measured ΔV_{FB} between 100 kHz and 1 MHz was 40 mV, indicating 9.7×10^{11} cm^{-2} eV^{-1} of trap states with a time constant in the range of 0.16~16 μs. The corresponding trap densities for Scheme II were 8×10^{12} cm^{-2} eV^{-1}. It can be seen that the MOS diodes in Scheme I have a lower V_{FB}. This is because the AlN crystal interfacial layer had an enhanced polarization effect, and more polarization charges were generated at the interface of AlN/GaN [17]. The cross-sectional TEM micrographs of Al₂O₃/AlN/GaN interfaces in the recessed region are shown in Figure 2c,d. A sharp monocrystal interfacial layer is formed through the NPA process. In contrast, Scheme II exhibits a rough interface.

Figure 3a shows the fully recessed-gate MIS-HEMTs transfer characteristics at $V_d = 10$ V of Scheme I and Scheme II, respectively. A normally-off operation with a V_{th} of 1.6 V is achieved. The subthreshold swing of Scheme I is much lower than Scheme II. What is more, the device of Scheme I exhibited well-suppressed off-state gate leakage compared with Scheme II at $V_d = 10$ V. The max saturated drain current (I_{sat}) is 371 mA/mm and 301 mA/mm at $V_g = 8$ V of Scheme I and Scheme II, respectively, as illustrated in Figure 3b. The extracted on-resistance (R_{ON}) of Scheme I and Scheme II are 10.1 Ω·mm and 15.93 Ω·mm at $V_g = 10$ V, respectively.

The threshold voltage (V_{th}) instability after a positive forward-reverse gate sweep or V_{th} shift during a positive gate bias stress, which is generally referred to as positive bias temperature instability (PBTI), has been reported for different gate dielectrics [32–34]. The PBTI represents serious reliability issues in fully recessed-gate MIS HEMTs for E-mode applications since a high-gate overdrive ($V_g - V_{th}$) is needed for fast switching [35]. Figure 4a shows the curves of threshold voltage shift (ΔV_{th}) versus stress period (t_{stress}) at $V_{g,stress} = 6$ V, 8 V, and 10 V, respectively. We observe that ΔV_{th} increases with $V_{g,stress}$, and t_{stress} increasing. After being stressed for 1000 s, ΔV_{th} exhibits 90 mV and 130 mV for $V_{g,stress} = 10$ V of Scheme I and Scheme II, respectively, which results from the high quality AlN crystal interfacial layer reducing the defect density and maintaining a more stable threshold voltage (V_{th}). After the stress phase, all devices are immediately biased at $V_{g,recovery} = 0$ V to record the V_{th} recovery at room temperature (Figure 4b). It is difficult to recover completely even when biased at $V_{g,recovery} = 0$ V, indicating that a higher density of trap states may be introduced by high overdrive voltage.

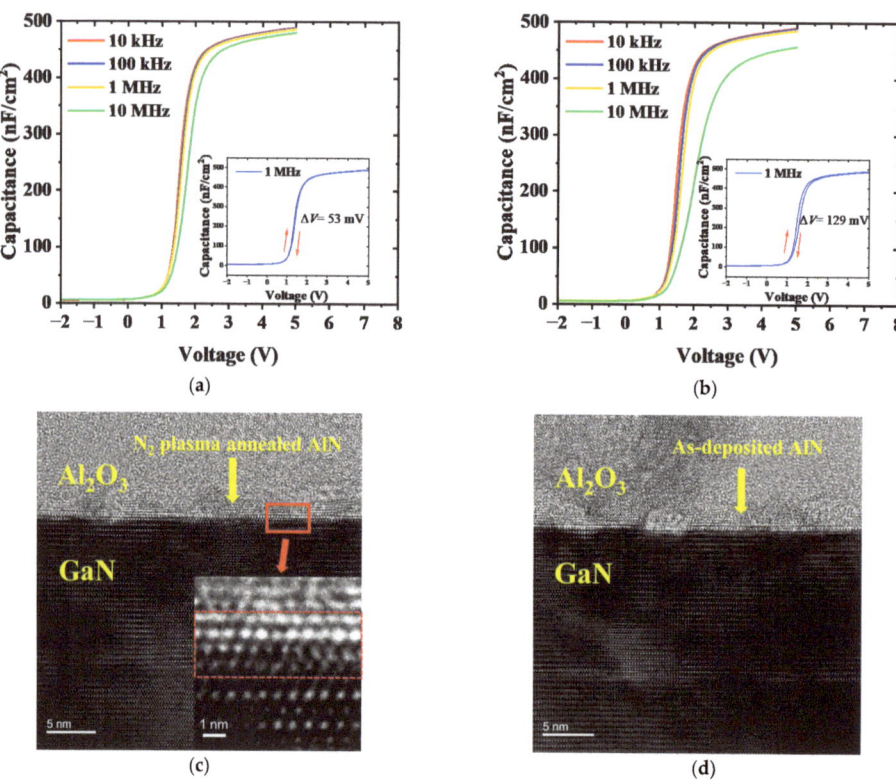

Figure 2. Multi-frequency C-V curves for diodes in Scheme I (**a**) and Scheme II (**b**). The insets were the hysteresis curves at 1 MHz and cross-sectional TEM images for the MOS structure in Scheme I (**c**) and Scheme II (**d**).

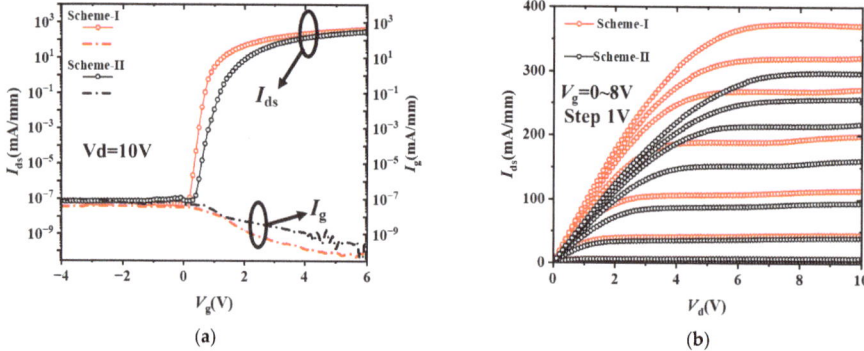

Figure 3. Device performance of the fabricated fully recessed-gate MIS HEMTs (**a**) transfer characteristics in semi-logarithm scale (insert graph linear scale) and (**b**) output characteristics.

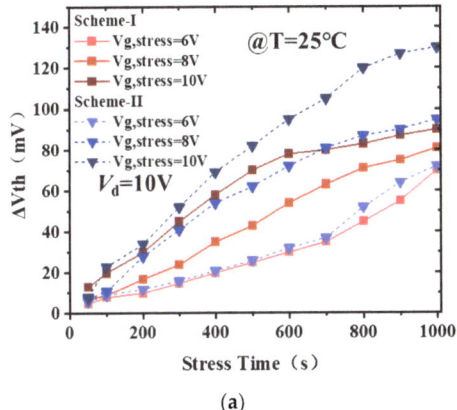 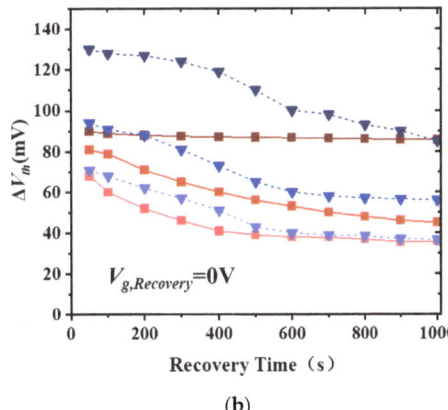

Figure 4. Changes of (a) ΔV_{th} versus t_{stress} under different gate voltage stresses at room temperature and (b) ΔV_{th} recovery at $V_{g,recovery} = 0$ V.

4. Conclusions

In summary, a novel post-AlN growth in situ NPA process is proposed to improve device performance. The newly proposed in situ NPA process could effectively promote the crystallization of the AlN interfacial layer. Compared with the traditional RTA method, the NPA process not only avoids high temperature-induced damage to the devices but also produces a high-quality AlN monocrystalline film that avoids natural oxidation by in situ growth. As a contrast with the conventional PELAD amorphous AlN, C-V results indicated a significantly lower interface density of states (D_{it}) in a MIS diode C-V characterization, which could be attributed to the polarization effect induced by the AlN crystal. The NPA process was helpful for the subthreshold swing reduction, and the Al_2O_3/AlN/GaN MIS-HEMTs were significantly enhanced with ~38% lower on-resistance at $V_g = 10$ V. What is more, in situ NPA provides a more stable V_{th} after a long gate stress time, and ΔV_{th} is inhibited by about 40 mV under $V_{g,stress} = 10$ V for 1000 s, showing great potential for improving Al_2O_3/AlN/GaN MIS-HEMT gate reliability.

Author Contributions: Conceptualization, M.S. and L.W.; methodology, M.S. and L.W.; validation, K.C. and P.Z.; formal analysis, M.S. and L.W.; investigation, P.Z. and K.C.; resources, P.Z. and K.C.; data curation, M.S. and L.W.; writing—original draft preparation, M.S.; writing—review and editing, P.Z. and K.C.; visualization, M.S.; supervision, L.W., P.Z. and K.C.; project administration, K.C.; funding acquisition, K.C. All authors have read and agreed to the published version of the manuscript.

Funding: This research received no external funding.

Data Availability Statement: Not applicable.

Conflicts of Interest: The authors declare no conflict of interest.

References

1. Saito, Y.; Tsurumaki, R.; Noda, N.; Horio, K. Analysis of Reduction in Lag Phenomena and Current Collapse in Field-Plate AlGaN/GaN HEMTs with High Acceptor Density in a Buffer Layer. *IEEE Trans. Device Mater. Reliab.* **2017**, *18*, 46–53. [CrossRef]
2. Sun, R.; Lai, J.; Chen, W.; Zhang, B. GaN Power Integration for High Frequency and High Efficiency Power Applications: A Review. *IEEE Access* **2020**, *8*, 15529–15542. [CrossRef]
3. Ajayan, J.; Nirmal, D.; Mohankumar, P.; Mounika, B.; Bhattacharya, S.; Tayal, S.; Fletcher, A.S. Challenges in material processing and reliability issues in AlGaN/GaN HEMTs on silicon wafers for future RF power electronics & switching applications: A critical review. *Mater. Sci. Semicond. Process.* **2022**, *151*, 106982. [CrossRef]
4. Li, W.; Romanczyk, B.; Guidry, M.; Akso, E.; Hatui, N.; Wurm, C.; Liu, W.; Shrestha, P.; Collins, H.; Clymore, C.; et al. Record RF Power Performance at 94 GHz From Millimeter-Wave N-Polar GaN-on-Sapphire Deep-Recess HEMTs. *IEEE Trans. Electron Devices* **2023**, *70*, 2075–2080. [CrossRef]

5. Mounika, B.; Ajayan, J.; Bhattacharya, S.; Nirmal, D. Recent developments in materials, architectures and processing of AlGaN/GaN HEMTs for future RF and power electronic applications: A critical review. *Micro Nanostruct.* **2022**, *168*, 207317. [CrossRef]
6. Pu, T.; Younis, U.; Chiu, H.C.; Xu, K.; Kuo, H.C.; Liu, X. Review of Recent Progress on Vertical GaN-Based PN Diodes. *Nanoscale Res. Lett.* **2021**, *16*, 101. [CrossRef]
7. Meneghini, M.; De Santi, C.; Abid, I.; Buffolo, M.; Cioni, M.; Khadar, R.A.; Nela, L.; Zagni, N.; Chini, A.; Medjdoub, F.; et al. GaN-based power devices: Physics, reliability, and perspectives. *J. Appl. Phys.* **2021**, *130*, 181101. [CrossRef]
8. Cai, Y.; Zhou, Y.; Lau, K.M.; Chen, K.J. Control of threshold voltage of AlGaN/GaN HEMTs by fluoride-based plasma treatment: From depletion mode to enhancement mode. *IEEE Trans. Electron Devices* **2006**, *53*, 2207–2215. [CrossRef]
9. Zhou, K.; Shan, L.; Zhang, Y.; Lu, D.; Ma, Y.; Chen, X.; Luo, L.; Wu, C. Fluorine Plasma Treatment for AlGaN/GaN HEMT-Based Ultraviolet Photodetector with High Responsivity and High Detectivity. *IEEE Electron Device Lett.* **2023**, *44*, 781–784. [CrossRef]
10. Hwang, I.; Kim, J.; Choi, H.S.; Choi, H.; Lee, J.; Kim, K.Y.; Park, J.B.; Lee, J.C.; Ha, J.; Oh, J.; et al. p-GaN Gate HEMTs With Tungsten Gate Metal for High Threshold Voltage and Low Gate Current. *IEEE Electron Device Lett.* **2013**, *34*, 202–204. [CrossRef]
11. Huang, S.; Liu, X.; Wang, X.; Kang, X.; Zhang, J.; Fan, J.; Shi, J.; Wei, K.; Zheng, Y.; Gao, H.; et al. Ultrathin-Barrier AlGaN/GaN Heterostructure: A Recess-Free Technology for Manufacturing High-Performance GaN-on-Si Power Devices. *IEEE Trans. Electron Devices* **2017**, *65*, 207–214. [CrossRef]
12. Hsieh, T.E.; Chang, E.Y.; Song, Y.Z.; Lin, Y.C.; Wang, H.C.; Liu, S.C.; Salahuddin, S.; Hu, C.C. Gate Recessed Quasi-Normally OFF Al_2O_3/AlGaN/GaN MIS-HEMT With Low Threshold Voltage Hysteresis Using PEALD AlN Interfacial Passivation Layer. *IEEE Electron Device Lett.* **2014**, *35*, 732–734. [CrossRef]
13. He, Y.; Gao, H.; Wang, C.; Zhao, Y.; Lu, X.; Zhang, C.; Zheng, X.; Guo, L.; Ma, X.; Hao, Y. Comparative Study Between Partially and Fully Recessed-Gate Enhancement-Mode AlGaN/GaN MIS HEMT on the Breakdown Mechanism. *Phys. Status Solidi (A)* **2019**, *216*, 1900115. [CrossRef]
14. Buttari, D.; Chini, A.; Chakraborty, A.; Mccarthy, L.; Xing, H.; Palacios, T.; Shen, L.; Keller, S.; Mishra, U.K. Selective dry etching of GaN over AlGaN in BCl3/SF6 mixtures. In Proceedings of the IEEE Lester Eastman Conference on High Performance Devices, Troy, NY, USA, 4–6 August 2004.
15. Chen, K.J.; Yang, S.; Tang, Z.; Huang, S.; Lu, Y.; Jiang, Q.; Liu, S.; Liu, C.; Li, B. Surface nitridation for improved dielectric/III-nitride interfaces in GaN MIS-HEMTs. *Phys. Status Solidi (A)* **2015**, *212*, 1059–1065. [CrossRef]
16. Kanamura, M.; Ohki, T.; Kikkawa, T.; Imanishi, K.; Hara, N. Enhancement-mode GaN MIS-HEMTs with n-GaN/i-AlN/n-GaN triple cap layer and high-κ gate dielectrics. *IEEE Electron Device Lett.* **2010**, *31*, 189–191. [CrossRef]
17. Liu, S.; Yang, S.; Tang, Z.; Jiang, Q.; Liu, C.; Wang, M.; Shen, B.; Chen, K.J. Interface/border trap characterization of Al_2O_3/AlN/GaN metal-oxide-semiconductor structures with an AlN interfacial layer. *Appl. Phys. Lett.* **2015**, *106*, 295–298. [CrossRef]
18. Hinkle, C.L.; Milojević, M.; Brennan, B.; Sonnet, A.M.; Aguirre-Tostado, F.S.; Hughes, G.; Vogel, E.M.; Wallace, R.M. Detection of Ga suboxides and their impact on III-V passivation and Fermi-level pinning. *Appl. Phys. Lett.* **2009**, *94*, 162101. [CrossRef]
19. Robertson, J. Model of interface states at III-V oxide interfaces. *Appl. Phys. Lett.* **2009**, *94*, 152104. [CrossRef]
20. Gao, F.; Lee, S.J.; Li, R.; Whang, S.J.; Balakumar, S.; Chi, D.Z.; Kean, C.C.; Vicknesh, S.; Tung, C.H.; Kwong, D.L. GaAs p- and n-MOS devices integrated with novel passivation (plasma nitridation and AlN-surface passivation) techniques and ALD-HfO2/TaN gate stack. In Proceedings of the 2006 International Electron Devices Meeting, San Francisco, CA, USA, 11–13 December 2006; pp. 1–4. [CrossRef]
21. Losurdo, M.; Capezzuto, P.; Bruno, G.; Perna, G.; Capozzi, V. N2–I2N2–H2 remote plasma nitridation for GaAs surface passivation. *Appl. Phys. Lett.* **2002**, *81*, 16–18. [CrossRef]
22. Romero, M.F.; JimÉnezJimenez, A.; Miguel-SÁnchezMiguel-Sanchez, J.; BraÑaBrana, A.F.; GonzÁlez-PosadaGonzalez-Posada, F.; Cuerdo, R.; Calle, F.; MuÑozMunoz, E. Effects of Plasma Pretreatment on the SiN Passivation of AlGaN/GaN HEMT. *Electron Device Lett.* **2008**, *29*, 209–211. [CrossRef]
23. Romero, A.; Jiménez, F.; González-Posada, S.; Martín-Horcajo, F.C.; Muñoz, E. Impact of N_2 Plasma Power Discharge on AlGaN/GaN HEMT Performance. *IEEE Trans. Electron Devices* **2012**, *59*, 374–379. [CrossRef]
24. Chen, K.J.; Huang, S. AlN passivation by plasma-enhanced atomic layer deposition for GaN-based power switches and power amplifiers. *Semicond. Sci. Technol.* **2013**, *28*, 074015. [CrossRef]
25. Liu, S.; Yang, S.; Tang, Z.; Jiang, Q.; Liu, C.; Wang, M.; Chen, K.J. Al_2O_3/AlN/GaN MOS-Channel-HEMTs With an AlN Interfacial Layer. *IEEE Electron Device Lett.* **2014**, *35*, 723–725. [CrossRef]
26. Koshelev, O.A.; Nechaev, D.V.; Brunkov, P.N.; Ivanov, S.V.; Jmerik, V.N. Stress control in thick $AlN/c-Al_2O_3$ templates grown by plasma-assisted molecular beam epitaxy. *Semicond. Sci. Technol.* **2021**, *36*, 035007. [CrossRef]
27. Kakanakova-Georgieva, A.; Ivanov, I.G.; Suwannaharn, N.; Hsu, C.W.; Cora, I.; Pécz, B.; Giannazzo, F.; Sangiovanni, D.G.; Gueorguiev, G.K. MOCVD of AlN on epitaxial graphene at extreme temperatures. *CrystEngComm* **2021**, *23*, 385–390. [CrossRef]
28. Xie, H.; Liu, Z.; Hu, W.; Zhong, Z.; Lee, K.; Guo, Y.X.; Ng, G.I. GaN-on-Si HEMTs Fabricated With Si CMOS-Compatible Metallization for Power Amplifiers in Low-Power Mobile SoCs. *IEEE Microw. Wirel. Compon. Lett. A Publ. IEEE Microw. Theory Tech. Soc.* **2021**, *31*, 141–144. [CrossRef]
29. Liu, S.; Peng, M.; Hou, C.; He, Y.; Li, M.; Zheng, X. PEALD-Grown Crystalline AlN Films on Si (100) with Sharp Interface and Good Uniformity. *Nanoscale Res. Lett.* **2017**, *12*, 279. [CrossRef]

30. Marcon, D.; Hove, M.V.; Jaeger, B.D.; Posthuma, N.; Decoutere, S. Direct comparison of GaN-based e-mode architectures (recessed MISHEMT and p-GaN HEMTs) processed on 200mm GaN-on-Si with Au-free technology. *Proc. SPIE-Int. Soc. Opt. Eng.* **2015**, *9363*, 117–128. [CrossRef]
31. Ramanan, N.; Lee, B.; Misra, V. Comparison of Methods for Accurate Characterization of Interface Traps in GaN MOS-HFET Devices. *IEEE Trans. Electron Devices* **2015**, *62*, 546–553. [CrossRef]
32. Lagger, P.; Reiner, M.; Pogany, D.; Ostermaier, C. Comprehensive Study of the Complex Dynamics of Forward Bias-Induced Threshold Voltage Drifts in GaN Based MIS-HEMTs by Stress/Recovery Experiments. *IEEE Trans. Electron Devices* **2014**, *61*, 1022–1030. [CrossRef]
33. Wu, T.L.; Marcon, D.; Bakeroot, B.; De Jaeger, B.; Lin, H.C.; Franco, J.; Stoffels, S.; Van Hove, M.; Roelofs, R.; Groeseneken, G.; et al. Correlation of interface states/border traps and threshold voltage shift on AlGaN/GaN metal-insulator-semiconductor high-electron-mobility transistors. *Appl. Phys. Lett.* **2015**, *107*, 93507. [CrossRef]
34. Kuo, H.M.; Chang, T.C.; Chang, K.C.; Lin, H.N.; Kuo, T.T.; Yeh, C.H.; Lee, Y.H.; Lin, J.H.; Tsai, X.Y.; Huang, J.W.; et al. Investigation of Threshold Voltage and Drain Current Degradations in Si_3N_4/AlGaN/GaN MIS-HEMTs Under X-Ray Irradiation. *IEEE Trans. Electron Devices* **2023**, *70*, 2216–2221. [CrossRef]
35. Wu, T.L.; Marcon, D.; Jaeger, B.D.; Hove, M.V.; Decoutere, S. The impact of the gate dielectric quality in developing Au-free D-mode and E-mode recessed gate AlGaN/GaN transistors on a 200mm Si substrate. In Proceedings of the 27th International Symposium on Power Semiconductor Devices and ICs, Hong Kong, China, 10–14 May 2015. [CrossRef]

Disclaimer/Publisher's Note: The statements, opinions and data contained in all publications are solely those of the individual author(s) and contributor(s) and not of MDPI and/or the editor(s). MDPI and/or the editor(s) disclaim responsibility for any injury to people or property resulting from any ideas, methods, instructions or products referred to in the content.

Article

Investigation into Photolithography Process of FPCB with 18 µm Line Pitch

Ke Sun [1], Gai Wu [1,2,*], Kang Liang [1,2,*], Bin Sun [3] and Jian Wang [3]

1. The Institute of Technological Sciences, Wuhan University, Wuhan 430072, China; sunke1720@whu.edu.cn
2. School of Power and Mechanical Engineering, Wuhan University, Wuhan 430072, China
3. Jiangsu Leader-Tech Semiconductor Co., Ltd., Pizhou 221300, China; sunbin@leader-techcn.com (B.S.); wangjian@leader-techcn.com (J.W.)
* Correspondence: wugai1988@whu.edu.cn (G.W.); liangkang@whu.edu.cn (K.L.)

Abstract: Due to the widespread application of flexible printed circuit boards (FPCBs), attention is increasing being paid to photolithography simulation with the continuous development of ultraviolet (UV) photolithography manufacturing. This study investigates the exposure process of an FPCB with an 18 µm line pitch. Using the finite difference time domain method, the light intensity distribution was calculated to predict the profiles of the developed photoresist. Moreover, the parameters of incident light intensity, air gap, and types of media that significantly influence the profile quality were studied. Using the process parameters obtained by photolithography simulation, FPCB samples with an 18 µm line pitch were successfully prepared. The results show that a higher incident light intensity and a smaller air gap result in a larger photoresisst profile. Better profile quality was obtained when water was used as the medium. The reliability of the simulation model was validated by comparing the profiles of the developed photoresist via four experimental samples.

Keywords: lithography simulation; FPCB; FDTD

Citation: Sun, K.; Wu, G.; Liang, K.; Sun, B.; Wang, J. Investigation into Photolithography Process of FPCB with 18 µm Line Pitch. *Micromachines* **2023**, *14*, 1020. https://doi.org/10.3390/mi14051020

Academic Editors: Niall Tait and Kun Li

Received: 10 April 2023
Revised: 27 April 2023
Accepted: 9 May 2023
Published: 10 May 2023

Copyright: © 2023 by the authors. Licensee MDPI, Basel, Switzerland. This article is an open access article distributed under the terms and conditions of the Creative Commons Attribution (CC BY) license (https://creativecommons.org/licenses/by/4.0/).

1. Introduction

Bearing the benefits of good portability [1], light weight, small size, and excellent bending performance, flexible printed circuit boards (FPCBs) are widely used in mobile communication equipments, flexible wearable devices [2], and automotive electronic products [3]. With increasing demand in the FPCB market, the photolithography process [4] has been a crucial driving force for achieving a narrower FPCB line pitch. In the process of ultraviolet (UV) lithography [5,6], patterns are constructed in a light-sensitive material [7] (photoresist) using ultraviolet light. During the exposure process [8], part of the light is absorbed by the photoresist [9], which generates photoacids [10] that change its solubility in the developer. Obtaining the desired size of the pattern requires controlling the photoacid distribution related to the light intensity distribution [11]. The expensive equipments and complicated process steps of the lithography process are time-consuming and costly. Lithography simulation technology can be used to study the propagation process of light in the photoresist. It has become an efficient means for analyzing and optimizing the manufacturing process and it efficiently reduces time and cost.

Currently, some investigations are being carried out into lithography simulation. Based on the scalar diffraction theory, Bourdillon et al. [12] used the Fresnel diffraction model to simulate 1 nm-proximity X-ray lithography. They demonstrated the influences of the gap width, spectral bandwidth, outriggers, T junctions, blur, etc. Tian et al. [13] simulated deep ultraviolet (DUV) photolithography with regard to the SU-8 photoresist using the modified Fresnel diffraction model, which can achieve a higher precision simulation than the Fresnel diffraction model. Different from the traditional Fresnel diffraction model, the modified Fresnel diffraction model divides the mask area into countless small squares to study the heterogeneity of diffraction in different directions. If the squares are small enough, the results can be very accurate. Zhou et al. [14] proposed a comprehensive aerial image model using Fresnel diffraction to calculate the three-dimensional (3D) inclined/vertical UV light intensity distribution in SU-8. Meanwhile, Geng et al. [15] constructed a high precision photolithography simulation of the thick SU-8 photoresist using the waveguide method (WG) in two dimensions and predicted the profiles of the photoresist based on the calculated light intensity distribution in SU-8. Koyama et al. [16] simulated the UV curing process of the photoresist via nanoimprint photolithography utilizing the molecular mechanics method. Majumder et al. [17] designed a full electromagnetic wave solution model using the finite element method to simulate the photochemical processes involved in absorbance modulation optical lithography. Kerim et al. [18] simulated electron beam lithography in relation to curved and inclined surfaces using a graphical-processing-unit-accelerated 3D Monte Carlo simulation based on first-principle scattering models. Liu et al. [19] developed a fast model for simulating the mask diffraction spectrum of extreme ultraviolet photolithography by combining an improved thin mask model and an equivalent layer method to simulate the mask diffraction for 22 nm space features. However, none of these studies have investigated 3D photolithography simulations based on the finite difference time domain (FDTD) method. Moreover, at present, many companies have achieved the manufacturing of FPCBs with a 20 μm line pitch, such as the LG company in South Korea, Chipbond Technology Co., Ltd., Taiwan, China and FLEXCEED Co., Ltd., Naka-shi, Japan [2]. The breakthrough in the production process of an FPCB with a 20 μm line pitch plays an extremely important driving role in the development of the flexible electronics field. The manufacturing of FPCBs with an 18 μm line pitch is the next important step in the development of the FPCBs. The production technology of an FPCB with an 18 μm line pitch or narrower line pitch is still not clear.

In this study, the exposure process of FPCBs with an 18 μm line pitch is studied. A 3D optical model of the FPCB is established and simulated using the FDTD method to obtained the profiles of the developed photoresist structures. The simulated profiles of the developed photoresist predicted via the light intensity distribution in the photoresist are compared with the experimental results for model validation purposes.

2. Materials and Methods

2.1. Governing Equations

Figure 1 shows the schematic of the exposure process. The incident light is assumed to be a beam of uniform and parallel light perpendicular to the mask surface. Diffraction occurs when the light passes through the mask. When the light passes through the photoresist surface, part of the light is refracted into the photoresist. The other part is reflected back into the air via the photoresist surface. The light entering the photoresist is partly absorbed by the photoresist to form photoacid. The light not absorbed by the photoresist is transmitted through the photoresist to the substrate, which is reflected back into the photoresist and superimposed with the incident light.

The main lithography simulation methods include the Fresnel diffraction model, FEM, WG, and FDTD. The Fresnel diffraction model considers the electric and magnetic field components in Maxwell's equations as scalars, without considering the actual coupling between the electric and magnetic vectors. It greatly simplifies the diffraction process, reduces computational complexity and improves computational speed, but the calculation is not accurate enough. Different from the the Fresnel diffraction model, FEM, WG and FDTD are based on the rigorous electromagnetic field theory, considering electromagnetic field components as vectors to achieve the accurate simulation of diffraction processes. The FEM has very high computational accuracy but requires a large amount of computing resources and time. The WG method has a shorter computing time but lower calculation accuracy than the FDTD method. The FDTD method has the advantage of high computational accuracy and wide applicability and the calculation time can be reduced via parallel computing. Therefore, in order to obtain appropriate simulation accuracy and computation time, we finally chose the FDTD method to calculate the light intensity distribution. Maxwell's equations are the core equations that explain the propagation process of the electromagnetic waves in nonmagnetic materials. In modern notation, Maxwell's equations are presented as follows:

$$\frac{\partial D}{\partial t} = \nabla \times H - J \tag{1}$$

$$-\frac{\partial B}{\partial t} = \nabla \times E \tag{2}$$

where D and H denote the electric displacement and the magnetic field vector, respectively; t is time; J is the electric charge current density; B is the magnetic flux density; E is the electric field intensity; and ∇ is the gradient differential operator.

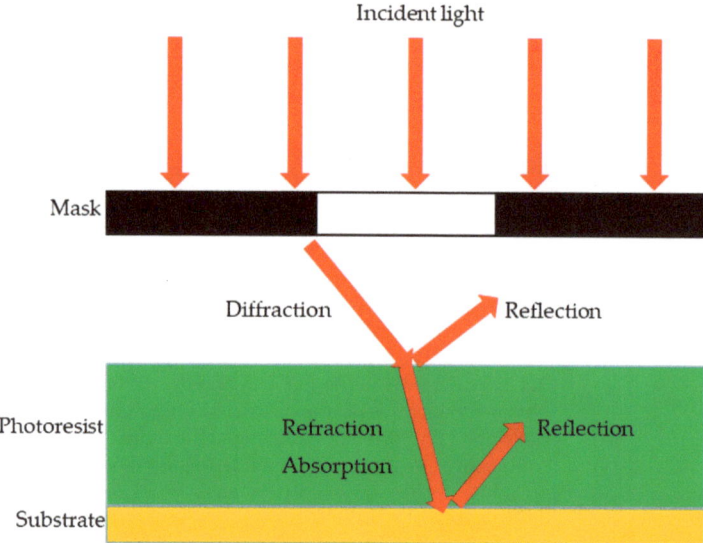

Figure 1. Schematic of the exposure process.

The constitutive relationship is a necessary condition for supplementing Maxwell's equations and characterizing the material parameters. The constitutive relationships beteeen the isotropic linear materials are as follows:

$$D = \varepsilon E \tag{3}$$

$$B = \mu H \tag{4}$$

$$J = \sigma E \tag{5}$$

where ε is the dielectric constant of the medium, μ is the magnetic permeability, and σ is the electrical conductivity.

In a rectangular coordinate system, Equations (1) and (2) can be transformed into the following form:

$$\frac{\partial H_z}{\partial y} - \frac{\partial H_y}{\partial z} = \varepsilon \frac{\partial E_x}{\partial t} + \sigma E_x \tag{6}$$

$$\frac{\partial H_x}{\partial z} - \frac{\partial H_z}{\partial x} = \varepsilon \frac{\partial E_y}{\partial t} + \sigma E_y \tag{7}$$

$$\frac{\partial H_y}{\partial x} - \frac{\partial H_x}{\partial y} = \varepsilon \frac{\partial E_z}{\partial t} + \sigma E_z \tag{8}$$

$$\frac{\partial E_z}{\partial y} - \frac{\partial E_y}{\partial z} = -\mu \frac{\partial H_x}{\partial t} \tag{9}$$

$$\frac{\partial E_x}{\partial z} - \frac{\partial E_z}{\partial x} = \mu \frac{\partial H_y}{\partial t} \tag{10}$$

$$\frac{\partial E_y}{\partial x} - \frac{\partial E_x}{\partial y} = \mu \frac{\partial H_z}{\partial t} \tag{11}$$

Using the central difference formula to replace the first order partial derivative, the three-dimensional electric and magnetic fields were sampled and calculated at discrete positions in time and space. $Fq\ (i, j, k)$ was assumed to be the discrete value of a component of the electric field E or the magnetic field H when the time was t and the coordinate was (i, j, k) in the rectangular coordinate system and $q = x, y, z$. In space, different components with the same index (i, j, k) can form a specific rectangular Yee cell. Figure 2 shows the actual position of each point in the Yee cell. The FDTD method was used to perform instantaneous sampling and to calculate of electric and magnetic field components at different discrete times. The electric field component corresponds to time $\Delta t, 2\ \Delta t, 3\ \Delta t, \ldots, n\ \Delta t$, while the magnetic field component corresponds to time $1.5\ \Delta t, 2.5\ \Delta t, 3.5\ \Delta t, \ldots, (n + 0.5)\ \Delta t$, with the offset always being $0.5\ \Delta t$. In the subsequent processing of the calculation results, it was necessary to unify the electric and magnetic field components to the same time.

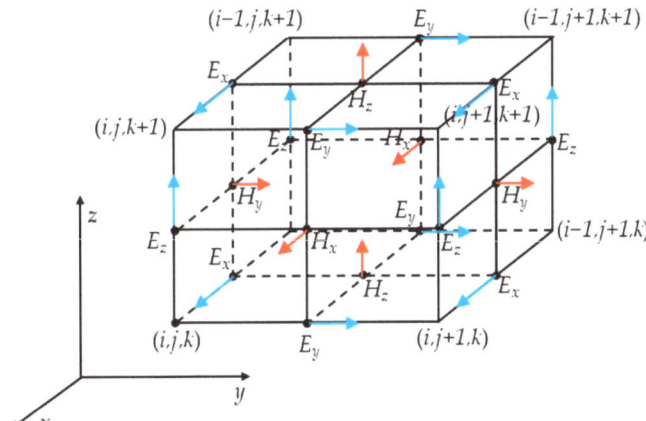

Figure 2. Positions and directions of various field components.

Figure 2 shows that the electric field component nodes are located at the center of the edge of the cell and the vector direction is parallel to their respective edges, while the magnetic field component nodes are located at the center of each surface and the direction is perpendicular to the corresponding surface. This also means that each electric field component node is surrounded by four magnetic field component nodes, simulating Ampere's law, while each magnetic field component node is surrounded by four electric field components, simulating Faraday's law.

The central difference equation is as follows:

$$\frac{\partial F_x^t(i,j,k)}{\partial x}\bigg|_{x=i\Delta x} = \frac{F_x^t(i+0.5,j,k) - F_x^t(i-0.5,j,k)}{\Delta x} \tag{12}$$

$$\frac{\partial F_q^t(i,j,k)}{\partial x}\bigg|_{t=n\Delta t} = \frac{F_q^{t+0.5}(i,j,k) - F_q^{t-0.5}(i,j,k)}{\Delta t} \tag{13}$$

For Equation (6), we have

$$\frac{E_x^{n+1}(i,j,k) - E_x^n(i,j,k)}{\Delta t} + \frac{\sigma_x(i,j,k)}{\varepsilon_x(i,j,k)} E_x^{n+0.5}(i,j,k) = \frac{1}{\varepsilon_x(i,j,k)} \left(\frac{H_z^{n+0.5}(i,j,k) - H_z^{n+0.5}(i,j-1,k)}{\Delta y} - \frac{H_y^{n+0.5}(i,j,k) - H_y^{n+0.5}(i,j,k-1)}{\Delta z} \right) \quad (14)$$

The equations corresponding to (7) and (8), respectively, can be similarly constructed.
For Equation (9), we have

$$H_x^{n+0.5}(i,j,k) = H_x^{n-0.5}(i,j,k) - \frac{\Delta t}{\mu_x(i,j,k)} \left(\frac{E_z^n(i,j+1,k) - E_z^n(i,j,k)}{\Delta y} - \frac{E_y^n(i,j,k+1) - E_y^n(i,j,k)}{\Delta z} \right) \quad (15)$$

The equations corresponding to (10) and (11), respectively, can be similarly constructed.
The instantaneous value at the intermediate time is assumed to be the average value:

$$F_q^{n+0.5}(i,j,k) = 0.5 \times (F_q^{n+1}(i,j,k) + F_q^n(i,j,k)) \quad (16)$$

For Equation (14), we have

$$E_x^{n+1}(i,j,k) = M_x(i,j,k) E_x^n(i,j,k) + N_x(i,j,k) \left(\frac{H_z^{n+0.5}(i,j,k) - H_z^{n+0.5}(i,j-1,k)}{\Delta y} - \frac{H_y^{n+0.5}(i,j,k) - H_y^{n+0.5}(i,j,k-1)}{\Delta z} \right) \quad (17)$$

$$M_x(i,j,k) = \frac{2\varepsilon_x(i,j,k) - \Delta t \sigma_x(i,j,k)}{2\varepsilon_x(i,j,k) - \Delta t \sigma_x(i,j,k)} \quad (18)$$

$$N_x(i,j,k) = \frac{2\Delta t}{2\varepsilon_x(i,j,k) + \Delta t \sigma_x(i,j,k)} \quad (19)$$

Equations (15) and (17) are time domain update equations for electric and magnetic field components, respectively.

According to the principle of field strength superposition, the relationship between power density and electric field strength during light propagation is as follows:

$$P = \frac{c n \varepsilon_0}{2} E^2 \quad (20)$$

where P is the power density, c is the light speed, and n is the refractive index.

2.2. Simulation Model

Figure 3 shows the geometric model of an FPCB circuit with an 18 μm line pitch, which selected a period region in the mask pattern as the simulation object. The spacing between the neighbor Cr layers was 8 μm. The Cr layer thickness was 0.1 μm. The thickness of the mask (SiO_2) was set to 2 μm. An air gap (t_a) is defined as the distance between the mask and the photoresist layer. The computer used in this simulation could complete the calculation at micron level but not the actual meter level calculation. Thus, the simulation simplified the actual production conditions. Five air gaps (i.e., 2, 4, 6, 8, and 10 μm) are studied herein. The incident light was parallel, uniform, and perpendicular to the mask surface along the negative Z direction, and its wavelength was 365 nm. The incident light was a plane wave in this simulation. The corresponding refractive index and permittivity of the materials with a 365 nm wavelength in the model are shown in Table 1. Among them, the material parameters of SiO_2, Cr, and copper can be found in Handbook of Optical Constants of Solids [20]. The material parameters of the photoresist were provided by Jiangsu Leader-Tech Semiconductor Co., Ltd., Pizhou, China. The data of SiO_2, Cr, and copper is derived from experiments described by sample data model. And the parameters of photoresist is set by (n, k) material model.

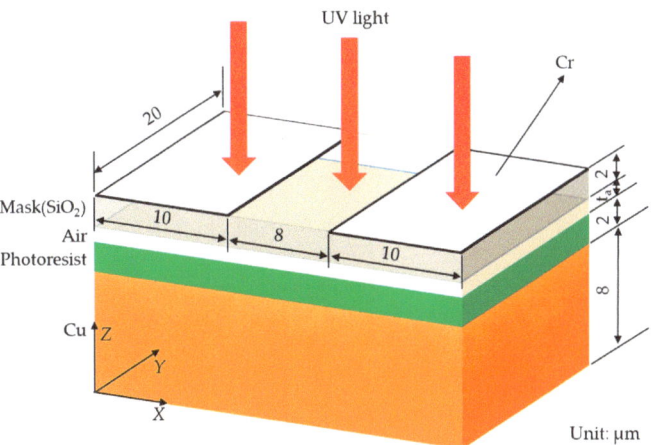

Figure 3. Geometric model of an FPCB circuit with an 18 µm line pitch.

Table 1. Material parameters in the model.

Materials	Cr	SiO$_2$	Photoresist	Cu
Refractive index	$1.4 + 3.26i$	1.47	$1.68 + 0.0058i$	$1.27 + 1.95i$
Permittivity	$-8.66 + 9.13i$	2.17	$2.82 + 0.019i$	$-2.21 + 4.96i$

In this work, all numerical simulations were performed using Lumerical 2020 R2 (ANSYS Inc., Pittsburgh, PA, USA), which is based on the FDTD method. Considering the periodicity and the symmetry of the simulation model in the X and Y directions, a plane wave source and periodic boundary conditions were added to save the computation resources and reduce the calculation time. Meanwhile, the symmetric boundary conditions in the X and Y directions were set to reduce the memory and time required for calculations. The mesh was auto non-uniform. The second level of mesh accuracy was used to complete the simulation calculations and the minimum mesh step was 0.00045 µm. The perfectly matched layer (PML) boundary condition with a steep angle in the Z direction is used to minimize the effects of reflections and improve the result accuracy. A refractive index monitor and field time monitor were set to check whether simulation calculations converge sufficiently to confirm the reliability of the simulation. Additionally, a 3D frequency domain-field and power monitor is added to collect electric field data from the photoresist.

3. Results and Discussion

3.1. Effects of Incident Light Intensity

Keeping the other parameters invariant, we focus herein on the effects of the incident light intensity. Figures 4–6 present the simulation results.

Figure 4 shows the intensity curves for the vertical exposure of different depths with varying incident light intensities. Via the screening effect of the mask on light, the light intensity reached its peak at the edge of the exposure area, and fluctuated continuously around a certain value within the exposure area, as shown in Figure 4. Due to the incident light diffraction and absorption in the photoresist, its intensity was simultaneously attenuated along the radiation direction with its increasing depths. According to the traditional optics theory, Fresnel diffraction exists near the photoresist surface and gradually changes to Fraunhofer diffraction with a depth increase. Due to the small size of the transmittance region (Figure 4), the Fresnel diffraction rapidly degenerated to Fraunhofer diffraction with increasing depths, where the intensity at the same depth sharply fluctuated in the x direction. Meanwhile, Figure 4 shows that the shape of the light intensity curve under different incident light intensities was basically the same, but the average value of light intensity increased with the increase of the incident light intensity. This indicates that a change in incident light intensity does not affect the propagation characteristics of light during the exposure process, but only affects the intensity of the light field in the photoresist.

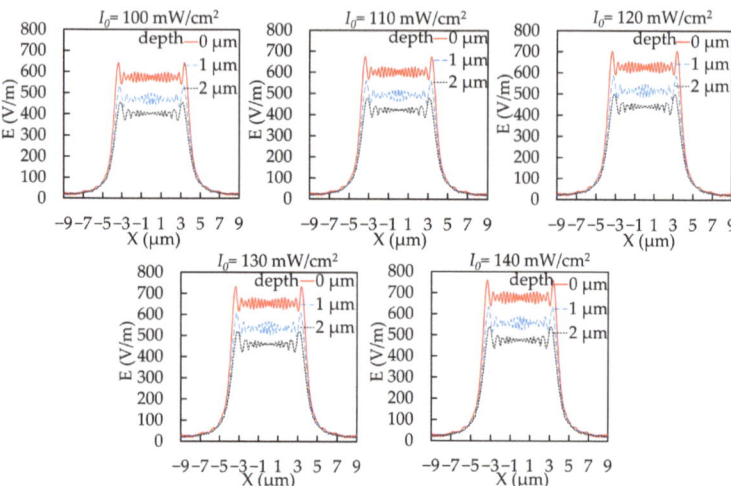

Figure 4. Intensity curves for the vertical exposure at different depths.

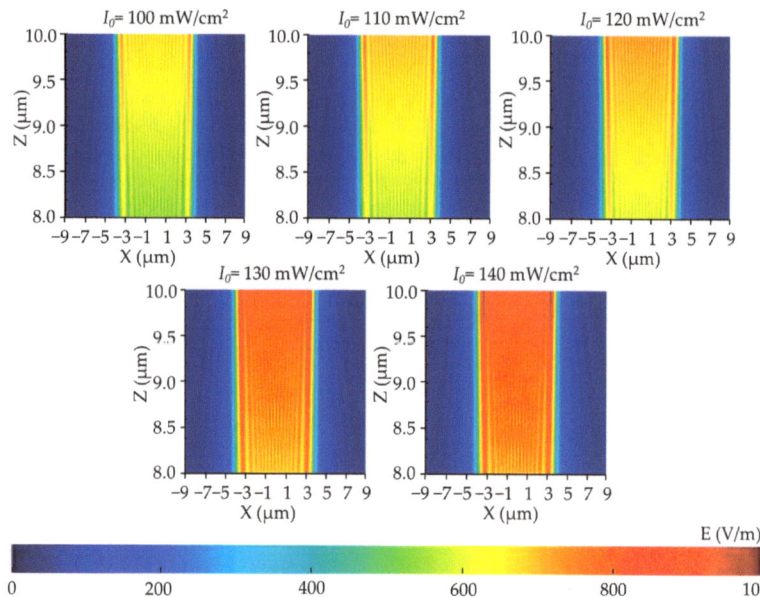

Figure 5. Vertical cross-section distribution of the electric field intensity in the photoresist with different incident light intensities.

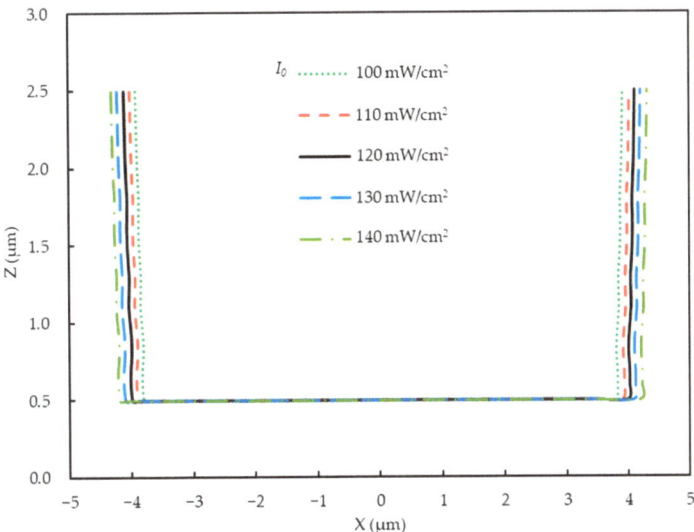

Figure 6. Simulated profiles of the developed photoresist with different incident light intensities.

Figure 5 illustrates the corresponding contour maps. We can see from the results that the light field was mainly distributed in the exposure area of the corresponding mask, while the light intensity in the non-exposure area was extremely low and can be ignored. Evidently, the effectiveness of the mask has been well demonstrated. In Figure 5, as the incident light intensity increased from 100 to 140 mW/cm^2, the light intensity in the exposed area increased, the exposure area shifted to the sides, the diffraction effect became more intense.

Figure 6 displays the profiles of the developed photoresist structures predicted via the light intensity distribution with different incident light intensities. The photoresist profile in Figure 6 was extracted from the light intensity distribution of the photoresist (Figure 5), which is the boundary between the exposed area and the unexposed area in the vertical section. At the same time, it was assumed that the development process was perfect, and so the profile of the exposure area extracted from the light intensity distribution can be seen as equivalent to the profile of the developed photoresist. The bottom of the profiles of the developed photoresist structures had widths of 7.78 μm, 7.88 μm, 8.01 μm, 8.13 μm, and 8.26 μm, respectively. In actual production, an 8 μm bottom width is ideal for obtaining an FPCB with an 18 μm line pitch.

With an incident light intensity of 120 mW/cm^2, the width at the bottom of the profiles was 8.01 μm, which was close to the ideal condition. Therefore, the incident light intensity during the production process should be controlled at approximately 120 mW/cm^2.

3.2. Effects of Air Gap and Types of Media

A parametric study on the air gap (t_a) was performed considering the influence of diffraction and absorption in air. Figure 7 shows that the light intensity distribution became non-uniform when the air gap increased. The light intensity of the photoresist reduced because when the air gap increased, more areas were generated. In these areas, Fraunhofer diffraction occured, which caused a severe intensity fluctuation. As can be seen in Figure 8, the simulated profile widths of the developed photoresist began to decrease with increasing air layer thickness because more diffraction and absorption resulted in inadequate exposure.

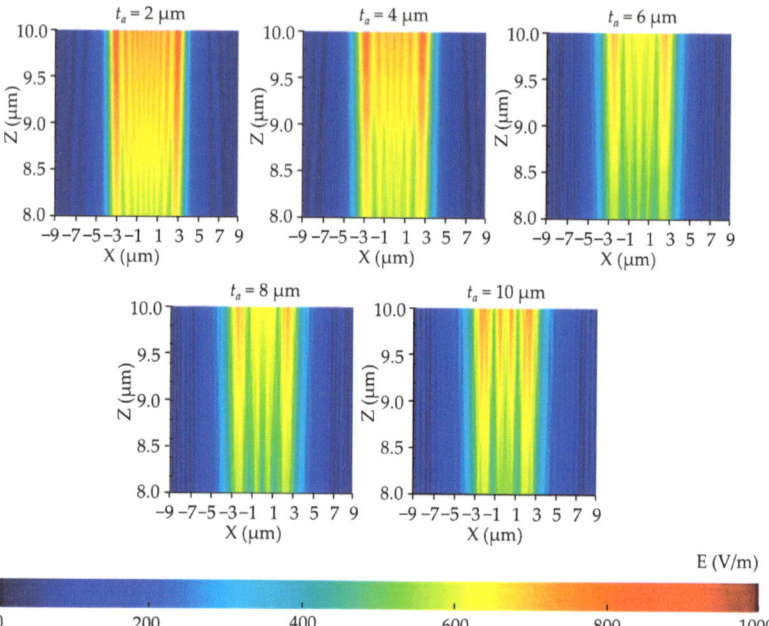

Figure 7. Vertical cross-section distribution of the electric field intensity in the photoresist with different air layer thicknesses.

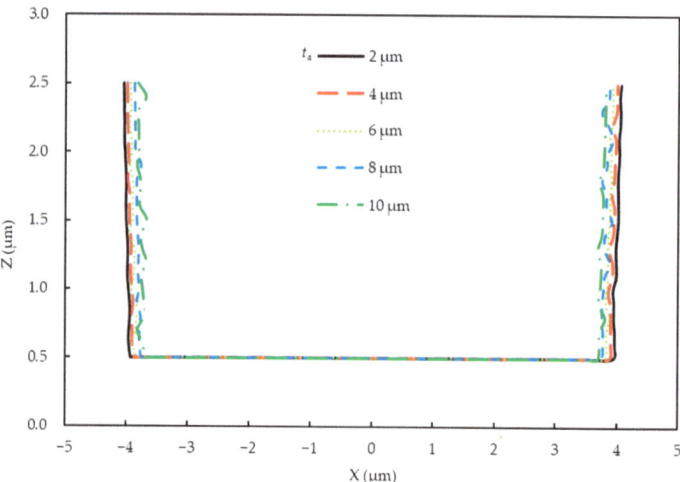

Figure 8. Simulated profiles of the developed photoresist with different air layer thicknesses.

The types of media between the mask and the photoresist greatly influenced the diffraction and absorption in the exposure process, thereby affecting the light intensity distribution of the photoresist. The exposure process of water being used as a medium in the field of FPCBs via simulation was explored to provide guidance for future breakthroughs in production technology. Figure 9 shows the vertical cross-section distribution of the light intensity in the photoresist with different water layer thicknesses (t_w). Figure 10 depicts the developing surface displacement of the photoresist. The results show that as the water layer thickness increased from 2 to 10 µm, the light intensity decreased, but almost no significant differences were found in the simulation profiles of the FPCB compared to the condition when air was the medium between the mask and the photoresist. Water had higher refractive index and transmissivity than air when light was rapidly attenuated in water. The water layer thickness only slightly influenced the developing surface displacement of the photoresist. In other words, water will be an excellent medium for replacing air if we can solve the other problems caused by water (e.g., photoresist stability and lens cleaning).

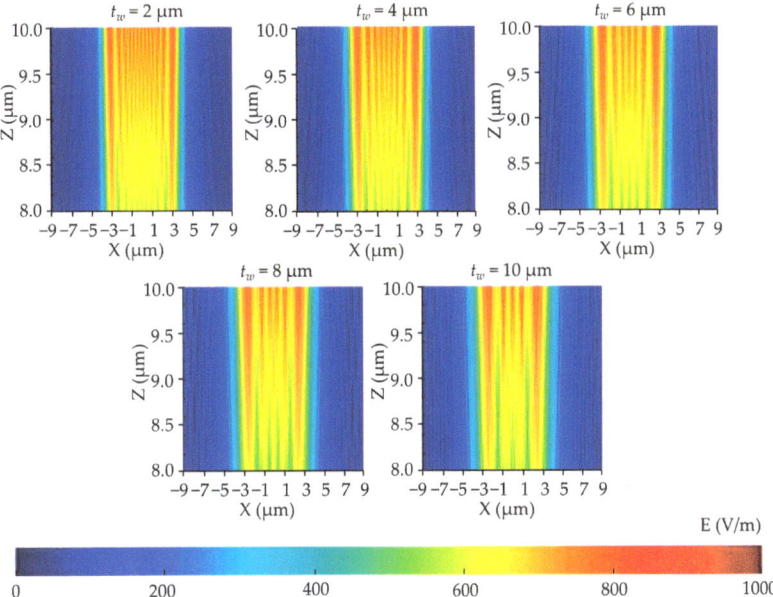

Figure 9. Vertical cross-section distribution of the electric field intensity in the photoresist with different water layer thicknesses.

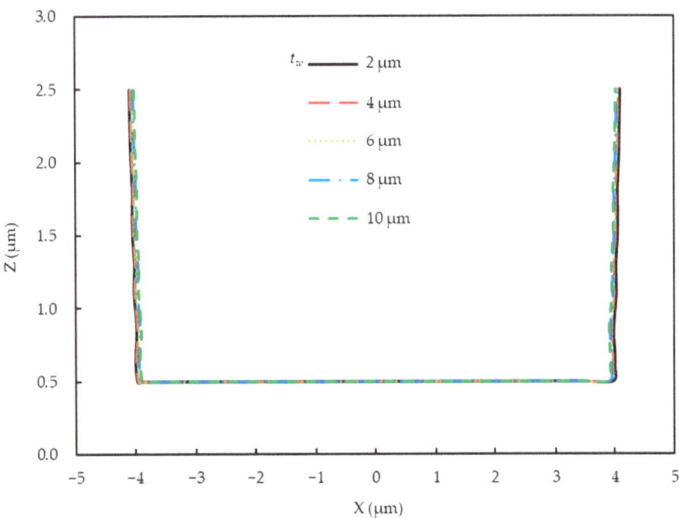

Figure 10. Simulated profiles of the developed photoresist with different water layer thicknesses.

3.3. Experimental Results and Analysis

After exposure, the photoresist was dissolved in the developer and a photoresist pattern is formed on the photoresist surface. Figure 11 shows the three-dimensional profile images of the FPCB sample (fabricated by Jiangsu Leader-Tech Semiconductor Co., Ltd.) obtained using the White Light Interferometer (Newview 9000, ZYGO, Middlefield, CT, USA) under the same exposure parameters and experimental conditions. In this experiment, the incident light intensity was 120 mW/cm^2, and its wavelength was 365 nm. The medium between the mask and the photoresist was air. The sample was exposed via an I-line exposure machine and developed using NaOH solution.

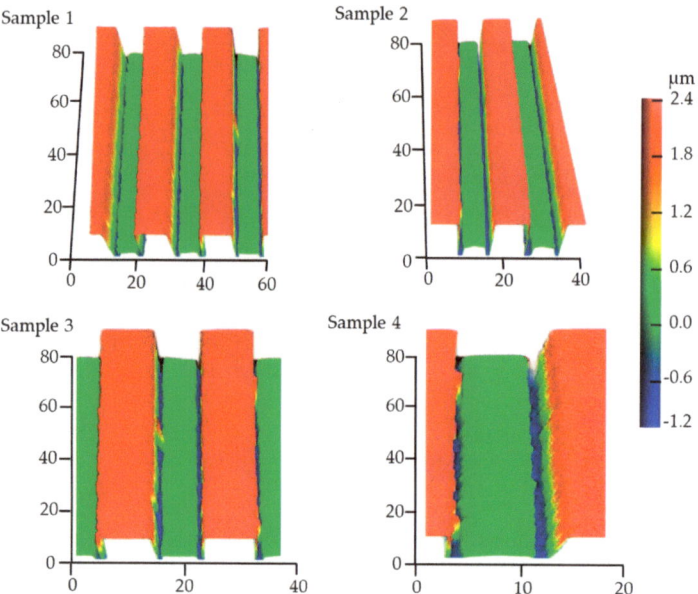

Figure 11. Three-dimensional profile images of four FPCB samples under the same exposure parmeters and experimental conditions.

Due to production and measurement errors, there are slight differences in the morphology of the samples. In order to reduce these errors, we produced four samples under the same parameters and production conditions. In general, the contour shapes of the samples are basically consistent and shows an isosceles trapezoid due to appropriate production parameters. Furthermore, the two-dimensional cross-section profiles of the developed photoresist were extracted from experimental results (Figure 11) and compared with the simulated profiles (Figure 12). The simulation profile (Figure 12) was obtained under the following parameters: the incident light intensity was 120 mW/cm^2 with a wavelength of 365 nm and the air gap was 2 μm. Good agreement was found, especially in the horizontal exposure width. This validated that the proposed method can achieve accurate simulation of FPCB exposure process, and the parameters obtained from the simulation are valuable.

No significant differences were found in the maximum lateral exposure of the sidewalls, because the photoresist thickness was only 2 μm, which was not thick enough for obvious diffraction. The width at the bottom of the profiles of the developed photoresist structures mainly depended on the diffraction and absorption in the mask, air, and photoresist.

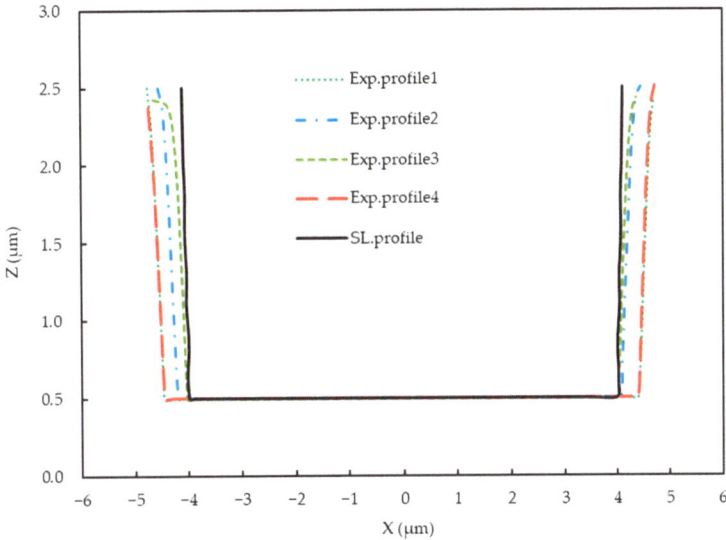

Figure 12. Profiles of the FPCB samples obtained via simulation and experiment.

4. Conclusions

This paper proposed a simulation method based on the rigorous electromagnetic field theory to predict the profiles of the developed photoresist structures. A 3D simulation model of the FPCB circuit was established to study the light intensity distribution in the photoresist based on the mask, light source, and photoresist information. According to the parametric study of the quality of the profiles of the developed photoresist structures, an incident light intensity of 120 mW/cm^2 was determined and used to expose the FPCBs. A smaller air gap can result in a larger photoresist profile. Better profile quality will be obtained when water is used as the medium. The simulation model was successfully verified by comparing the simulation and experiment results.

Author Contributions: Conceptualization, K.S. and G.W.; methodology, K.S. and K.L.; software, B.S.; validation, K.S. and J.W.; formal analysis, K.S.; investigation, K.S., G.W. and K.L.; resources, K.L. and G.W.; data curation, B.S.; visualization, K.S.; writing—original draft preparation, K.S.; writing—review and editing, G.W. and K.L. All authors have read and agreed to the published version of the manuscript.

Funding: This research was funded by Hubei Provincial Natural Science Foundation of China, grant number 2020CFA032.

Data Availability Statement: The data presented in this study are available from the corresponding author upon request.

Acknowledgments: The authors thank all reviewers for their great help in this article.

Conflicts of Interest: The authors declare no conflict of interest.

References

1. Meng, B.; Tang, W.; Zhang, X.S.; Han, M.D.; Liu, W.; Zhang, H.X. Self-powered flexible printed circuit board with integrated triboelectric generator. *Nano Energy* **2013**, *2*, 1101–1106. [CrossRef]
2. Sheng, J.Z.; Li, H.; Shen, S.N.; Ming, R.J.; Sun, B.; Wang, J.; Zhang, D.D.; Tang, Y.G. Investigation on chemical etching process of FPCB with 18 μm line pitch. *IEEE Access* **2021**, *9*, 50872–50879. [CrossRef]
3. Ming, R.J.; Li, H.; Chen, A.J.; Sheng, J.Z.; Sun, B.; Wang, J.; Huang, C.S.; Yang, J. Investigation on spraying uniformity in etching process of FPCB with 18 μm line pitch. *Int. J. Precis. Eng. Manuf.* **2022**, *23*, 479–488. [CrossRef]
4. Zhou, Z.F.; Huang, Q. Comprehensive simulations for ultraviolet lithography process of thick SU-8 photoresist. *Micromachines* **2018**, *9*, 341. [CrossRef] [PubMed]

5. Kim, S.K.; Lee, J.E.; Park, S.W.; Oh, H.K. Optical lithography simulation for the whole resist process. *Curr. Appl. Phys.* **2006**, *6*, 48–53. [CrossRef]
6. Koyama, M.; Shirai, M.; Kawata, H.; Hirai, Y.; Yasuda, M. Computational study on UV curing characteristics in nanoimprint lithography: Stochastic simulation. *Jpn. J. Appl. Phys.* **2017**, *56*, 06GL03. [CrossRef]
7. Tsvetkov, Y.B. Computer simulation of diffraction focusing in proximity lithography. *AIP Conf. Proc.* **2019**, *2195*, 020064.
8. Yasuda, M.; Koyama, M.; Imai, K.; Shirai, M.; Kawata, H.; Hirai, Y. Stochastic simulation of pattern formation for negative-type chemically amplified resists in extreme ultraviolet lithography. *J. Photopolym. Sci. Technol.* **2020**, *33*, 53–56. [CrossRef]
9. Rudolph, O.H.; Evanschitzky, P.; Erdmann, A.; Bar, E.; Lorenz, J. Rigorous electromagnetic field simulation of the impact of photomask line-edge and line-width roughness on lithographic processes. *J. Micro/Nanolith. MEMS MOEMS* **2012**, *11*, 013004. [CrossRef]
10. Ichikawa, T.; Yagisawa, T.; Furukawa, S.; Taguchi, T.; Nojima, S.; Murakami, S.; Tamaoki, N. Cooperative simulation of lithography and topography for three-dimensional high-aspect-ratio etching. *Jpn. J. Appl. Phys.* **2018**, *57*, 06JC01. [CrossRef]
11. Yasuda, M.; Tada, K.; Kotera, M. Multiphysics simulation of nanopatterning in electron beam lithography. *J. Photopolym. Sci. Technol.* **2016**, *29*, 725–730. [CrossRef]
12. Bourdillon, A.J.; Boothroyd, C.B.; Kong, J.R.; Vladimirsky, Y. A critical condition in Fresnel diffraction used for ultra-high resolution lithographic printing. *J. Phys. Appl. Phys.* **2000**, *33*, 2133–2141. [CrossRef]
13. Tian, X.; Liu, G.; Tian, Y.; Zhang, P.; Zhang, X. Simulation of deep UV lithography with SU-8 resist by using 365 nm light source. *Microsyst. Technol.* **2005**, *11*, 265–270. [CrossRef]
14. Zhou, Z.F.; Shi, L.L.; Zhang, H.; Huang, Q.A. Large scale three-dimensional simulations for thick SU-8 lithography process based on a full hash fast marching method. *Microelectron. Eng.* **2014**, *123*, 171–174. [CrossRef]
15. Geng, Z.C.; Zhou, Z.F.; Dai, H.; Huang, Q.A. A 2D waveguide method for lithography simulation of thick SU-8 photoresist. *Micromachines* **2020**, *11*, 972. [CrossRef] [PubMed]
16. Koyama, M.; Shirai, M.; Kawata, H.; Hirai, Y.; Yasuda, M. Stochastic simulation of the UV curing process in nanoimprint lithography: Pattern size and shape effects in sub-50 nm lithography. *J. Vac. Sci. Technol.* **2017**, *35*, 06G307. [CrossRef]
17. Majumder, A.; Helms, P.L.; Andrew, T.L.; Menon, R. A comprehensive simulation model of the performance of photochromic films in absorbance-modulation-optical-lithography. *AIP Adv.* **2016**, *6*, 035210. [CrossRef]
18. Arat, K.T.; Zonnevylle, A.C.; Ketelaars, W.S.M.M.; Belic, N.; Hofmann, U.; Hagen, C.W. Electron beam lithography on curved or tilted surfaces: Simulations and experiments. *J. Vac. Sci. Technol.* **2019**, *37*, 051604. [CrossRef]
19. Liu, X.L.; Wang, X.Z.; Li, S.K.; Yan, G.Y.; Erdmann, A. Fast model for mask spectrum simulation and analysis of mask shadowing effects in extreme ultraviolet lithography. *J. Micro/Nanolith. MEMS MOEMS* **2014**, *13*, 033007. [CrossRef]
20. Palik, E.D. *Handbook of Optical Constants of Solids*; Academic Press: Pittsburgh, PA, USA, 1998; Volume 3, pp. 749–763.

Disclaimer/Publisher's Note: The statements, opinions and data contained in all publications are solely those of the individual author(s) and contributor(s) and not of MDPI and/or the editor(s). MDPI and/or the editor(s) disclaim responsibility for any injury to people or property resulting from any ideas, methods, instructions or products referred to in the content.

Article

Simulation and Experimental Validation of a Pressurized Filling Method for Neutron Absorption Grating

Eryong Han, Kuanqiang Zhang, Lijuan Chen, Chenfei Guo, Ying Xiong, Yong Guan, Yangchao Tian and Gang Liu *

National Synchrotron Radiation Laboratory, University of Science and Technology of China, Hefei 230029, China; haneryong@mail.ustc.edu.cn (E.H.)
* Correspondence: liugang@ustc.edu.cn

Abstract: The absorption grating is a critical component of neutron phase contrast imaging technology, and its quality directly influences the sensitivity of the imaging system. Gadolinium (Gd) is a preferred neutron absorption material due to its high absorption coefficient, but its use in micro-nanofabrication poses significant challenges. In this study, we employed the particle filling method to fabricate neutron absorption gratings, and a pressurized filling method was introduced to enhance the filling rate. The filling rate was determined by the pressure on the surface of the particles, and the results demonstrate that the pressurized filling method can significantly increase the filling rate. Meanwhile, we investigated the effects of different pressures, groove widths, and Young's modulus of the material on the particle filling rate through simulations. The results indicate that higher pressure and wider grating grooves lead to a significant increase in particle filling rate, and the pressurized filling method can be utilized to fabricate large-size grating and produce uniformly filled absorption gratings. To further improve the efficiency of the pressurized filling method, we proposed a process optimization approach, resulting in a significant improvement in the fabrication efficiency.

Keywords: neutron absorption grating; particle filling method; pressurized filling; filling rate

Citation: Han, E.; Zhang, K.; Chen, L.; Guo, C.; Xiong, Y.; Guan, Y.; Tian, Y.; Liu, G. Simulation and Experimental Validation of a Pressurized Filling Method for Neutron Absorption Grating. *Micromachines* **2023**, *14*, 1016. https://doi.org/10.3390/mi14051016

Academic Editor: Kun Li

Received: 18 April 2023
Revised: 4 May 2023
Accepted: 6 May 2023
Published: 9 May 2023

Copyright: © 2023 by the authors. Licensee MDPI, Basel, Switzerland. This article is an open access article distributed under the terms and conditions of the Creative Commons Attribution (CC BY) license (https://creativecommons.org/licenses/by/4.0/).

1. Introduction

Neutron imaging has emerged as a powerful nondestructive inspection technique for detecting the interior of metallic materials owing to the high penetration ability of neutrons [1–3]. Neutron phase contrast imaging has been developed for materials that exhibit low attenuation to neutron beams, thereby allowing for higher resolution, higher image contrast, and higher measurement sensitivity [4,5]. The Talbot Lau interferometer, which employs three gratings, is the most common approach used for neutron phase contrast imaging. The absorption grating (G_0) generates spatially coherent neutron beamlets, which are phase shifted by the second grating (G_1) to produce transverse intensity modulation. However, this spatial modulation is too small to be resolved by the current neutron detector. Therefore, a third absorption grating (G_2) should be used to analyze the signal [6–8].

Despite its potential, the fabrication of neutron absorption gratings remains a key challenge in this technology. Gd is the preferred material for the absorber of neutron absorption gratings due to its high neutron absorption coefficient [9–11]. For strong absorption (97%), approximately 11 μm of Gd is required for a neutron wavelength of 4.1 Å or higher [12,13]. However, Gd is a very rare material in micro-nanofabrication and there is no universal deposition process [4]. Consequently, various fabrication processes have been developed to achieve this goal for different requirements, such as particle filling methods [9,13], the evaporation of pure gadolinium metal [10–12,14,15], and the embossing of gadolinium-alloyed metallic glasses [16–18], each of which has its own characteristics and applicability.

The particle filling method is advantageous for making large-size gratings, which can provide a large imaging field of view [4,19]. However, during particle deposition, the pores between the particles are large, resulting in a low particle packing density. This leads to a low particle filling rate and the required absorber height being greater than the design height [9,13]. Accordingly, an increase in the aspect ratio of the grating will increase the difficulty of fabrication. In the fabrication of the silicon grating structure, although current processes can achieve high aspect ratio structures (>50:1) [20], there are nonuniformity issues when fabricating large-size gratings. Moreover, when filling particles into the grating structure to fabricate an absorption grating, particle filling becomes difficult as the aspect ratio of the grating increases, which also poses the issue of nonuniformity. Therefore, it is necessary to increase the filling rate of the particles to reduce the aspect ratio of the grating. Ultrasound and high-speed centrifugation are commonly employed to address the low particle filling rate [21–23]. However, the grating structure may be damaged during ultrasonication, and the high-speed centrifuge is only suitable for making small-size gratings, which will lead to uneven particle filling for making large-size absorption gratings. In this paper, we utilize the pressurized filling method to enhance the particle filling rate and explore the factors influencing the rate during pressurized filling through simulation. Additionally, we propose a method for optimizing the particle filling process to improve the fabrication efficiency of absorption gratings.

2. Experimental Methods and Simulation

2.1. Methods

The filling rate is related to the particle size and particle arrangement. When the particle size is fixed, the particle arrangement can be changed to improve the particle filling rate. Particle rearrangement occurs when the particles are subjected to external forces during the stacking process. Ultrasound is used to uniformly distribute particles using vibration, while high-speed centrifugation is used to redistribute particles using centrifugal force. In this paper, external pressure was used to rearrange the particles as shown in Figure 1. The particles were deposited into the grating grooves, and then a soft material was placed on the surface of the grating structure, followed by uniform pressure to deform the material downward into the grating grooves. The particles in the grooves were compressed by the material and moved downward, causing their rearrangement and an increase in the particle stacking density. Then, we continued to deposit the particles into the grooves of the grating, applying pressure to the surface of the material so that the accumulation density of the particles in the grooves continued to increase. We repeated the pressure filling and gradually increased the filling rate.

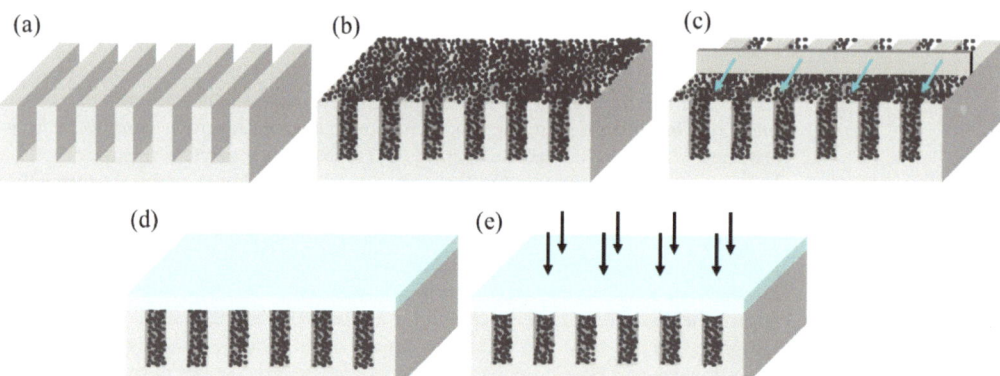

Figure 1. The fabrication process of neutron absorption grating. (**a**) Wet etching to obtain silicon gratings; (**b**) Particle deposition; (**c**) Scraping off excess particles; (**d**) Placement of material on the grating surface; (**e**) Pressure applied to the surface.

2.2. Experiment

A neutron absorption grating was fabricated using the pressurized filling method, as depicted in Figure 1. To create a neutron absorption grating, a silicon grating structure was initially fabricated using a wet etching method with a KOH solution. To prepare a particle suspension, gadolinium oxide (Gd_2O_3, Zhongnuo Advanced Materials (Beijing, China) Technology Co., Ltd.) particles with a size range of 1–3 μm were sonicated with ethanol at a mass ratio of 1:100, using an ultrasonic machine (XZ-3DTD, Ningbo Advanced Leaf Biological Technology Co., Ltd., Ningbo, China) at a power of 750 W for 20 min. This suspension was poured onto the surface of the silicon wafer and allowed to settle for 3 h. Once the ethanol had completely evaporated, the Gd_2O_3 particles were naturally deposited and filled the grooves of the grating. The remaining Gd_2O_3 particles on the surface of the silicon structure were then removed. Subsequently, a soft material such as PDMS (polydimethylsiloxane, Dow Corning, Midland, TX, USA) with Young's modulus of 7.5×10^5 Pa was placed on the surface of the grating, using an embossing device (ZKRYM-1, Institute of Optics and Electronics, CAS, Chengdu, China) with a pressure head size of 40 cm × 40 cm to ensure that the pressure head was much larger than the pressed material so that the pressure was applied uniformly onto the PDMS surface. After releasing the pressure, the suspension was used to continue depositing particles into the grating grooves. Pressure was reapplied repeatedly to gradually increase the filling rate. The grating structures and experimental parameters are shown in Table 1.

Table 1. Grating structures and experimental parameters.

	Period (μm)	Duty Cycle	Groove Depth (μm)	Grating Area (mm²)	Pressure (10^7 Pa)
Grating (a)	140	0.5	70	60 × 60	0.5
Grating (b)	40	0.5	70	60 × 60	0.7

2.3. Simulation

To investigate the pressurized filling method, COMSOL Multiphysics was used to simulate the material deformation during the particle filling process. As shown in Figure 2a, the model consisted of silicon, PDMS, and porous particle models. The deformation of the material in the grating groove was simulated when pressure was applied to the material on the grating surface. The mass of the particles in the grating groove remained unchanged after pressurization, and the volume of the particle model decreased. Using the particle mass divided by the total mass of fully filled gadolinium oxide material at the current volume, we obtained the particle packing density after each pressurization, which resulted in the corresponding particle filling rate. The filling rate of the particles in the grating groove in the simulation and experiment is calculated as

$$D = \frac{M}{(V_1 - V_2)\rho} \qquad (1)$$

where D is the filling rate, M is the mass of particles filled into the grating tank before pressurization, ρ is the density of Gd_2O_3, V_1 is the volume of the grating groove, and V_2 is the deformation volume of PDMS after pressurization. Based on this, the change in material deformation on the grating surface under repeated pressure filling was further simulated to obtain the final particle filling rate.

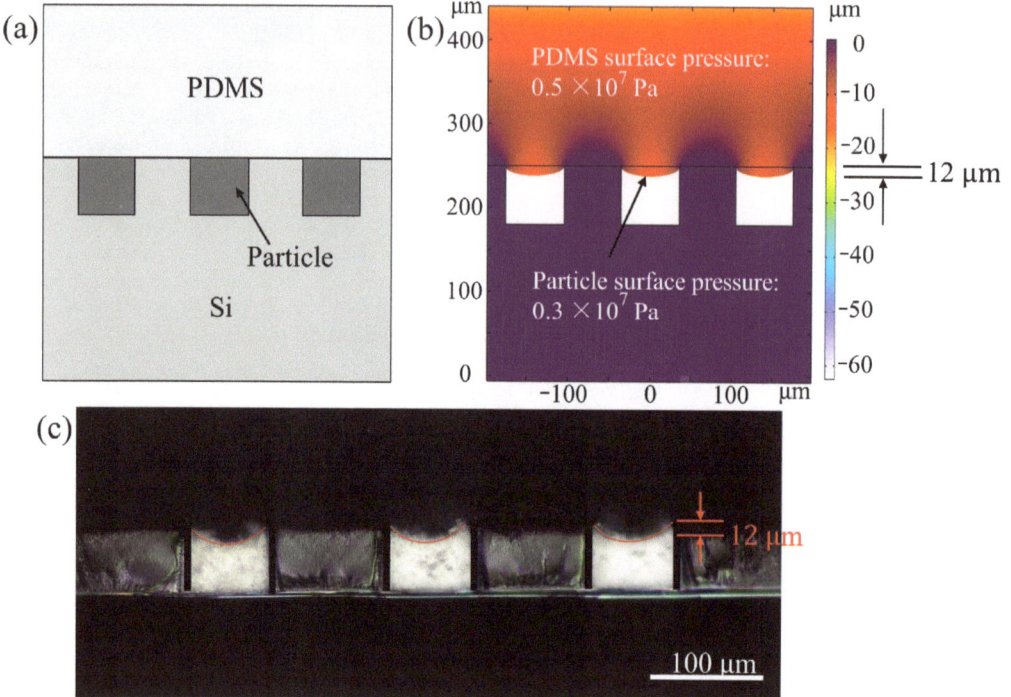

Figure 2. Pressurized filling method. (**a**) Diagram of the simulation model. (**b**) Deformation in simulation. (**c**) Actual experimental particle surface deformation diagram.

While silicon and PDMS can be found directly in the COMSOL Multiphysics materials library, the particle model is not available. Therefore, we used a porous model instead and obtained the relevant parameters of the porous model through experimentation.

In order to obtain the parameters of the particle model, we made a groove on the surface of a silicon wafer using SU8 2100 photoresist with an area of 10 mm × 20 mm. The particles were deposited in the groove and a hard material, such as a silicon wafer, was placed directly on the surface of the particles. Different pressures were applied to the material surface with an embossing machine in order to obtain the corresponding height of particle variation. The relationship between the variation of particle stacking density when different pressures were applied, i.e., particle filling rate, could be obtained.

3. Results and Discussion

3.1. Effect of Pressure

After deposition of the particles, the excess particles were removed from the grating surface and the mass of Gd_2O_3 particles filled into the grating grooves were obtained via the weighing method. Dividing the mass of the particles by the total mass of the fully filled Gd_2O_3 material at the current volume, we determined that the initial filling rate of the grating was 25%. When using Gd_2O_3 material as an absorber, an absorption grating with a depth of 70 µm needed to be fabricated in order to fully absorb the neutron at a wavelength of 4.1 Å. To facilitate this study, the grating groove depth was generally set to 70 µm for subsequent experiments and simulations.

The deformation of the PDMS in the grating groove, i.e., the height variation of the particles and the pressure on the particle surface, was obtained by inputting the porous model into the simulation and applying a boundary load to the PDMS surface, as shown in Figure 2b. From the simulation results, when applying a pressure of 0.5×10^7 Pa on the

PDMS material on a grating structure with a groove width of 70 µm, we were able to obtain a pressure of 0.3×10^7 Pa on the surface of the particles and a deformation variable of 12 µm for the PDMS. The plot in Figure 2c illustrates the variation in particle height resulting from a pressurized filling experiment conducted on a grating with a 70 µm groove width made of SU8 2100 photoresist. The experiment was performed using a PDMS with Young's modulus of 7.5×10^5 Pa and a pressure of 0.5×10^7 Pa applied to it. The deformation variable obtained from this experiment is in excellent accordance with Figure 2b. It could also further simulate the change in material deformation under the repeated pressurized filling to obtain the final particle filling rate.

Figure 3a shows the relationship between particle filling rate and pressure in experiments and simulations, where the simulation results were obtained by applying pressure onto the PDMS surface above a grating with a groove width of 20 µm, and the experimental results were obtained based on previous experiments involving porous particle models applying pressure onto the particle surface. It can be observed that, as the pressure applied to the surface of the PDMS increased, the filling rate increased correspondingly, which is consistent with the trend observed in the experimental results. However, it should be noted that the filling rate obtained by applying the same force on the surface of the PDMS was smaller than the experimental results. Simulations enabled us to obtain the pressure transmitted to the particle surface; it was smaller than that applied to the PDMS surface due to deformation. Comparing the relationship between the filling rate and the pressure on the particle surface, both the experimental and simulation results were consistent, indicating that the filling rate was determined by the pressure applied to the particle surface. As the pressure on the particle surface increased, the filling rate gradually increased. When the pressure on the surface of the particles reached 4×10^7 Pa, the filling rate increased significantly from 25% to 55%. Based on the pressurized filling method, it was possible to significantly reduce the aspect ratio of the grating, thus reducing the difficulty of producing the grating.

Figure 3b shows the actual experimental and simulated filling rate results for a silicon grating with a period of 40 µm, duty cycle of 0.5, and area of 60 mm × 60 mm, where the filling rate increased from 25% to 32%. Limited by the equipment and the area of grating, a pressure of 0.7×10^7 Pa was chosen for the pressurization experiments. From Figure 3b, it can be observed that the experimental results of repetitive filling were matched with the simulated results. It can also be seen that the particle filling rate increased fastest at the first pressurized filling and converged quickly, while the filling rate increased slowly with successive pressurized filling. The pressure on the particle surface after each pressurization were obtained from the simulation, as shown by the red dotted line in Figure 3b. Combining the results of points A and B in Figure 3a, it can be seen that when the pressure on the surface of PDMS material was 0.7×10^7 Pa, the particle surface pressure in the simulation was 0.35×10^7 Pa and the filling rate was 32%, which is consistent with the result in Figure 3b when the particle surface pressure was 0.35×10^7 Pa and the filling rate was about 32%. It can be observed that, by applying a fixed pressure to the surface of the grating material, the pressure on the particle surface gradually increased with repeated pressure filling, and the filling rate also increased accordingly. This result is consistent with the relationship between particle surface pressure and filling rate shown in Figure 3a, further indicating that the filling rate is determined by the pressure on the particle surface.

Uniformity is a critical issue to consider when fabricating gratings using the particle filling method, particularly for large-sized gratings where non-uniform filling is more prominent. The non-uniform filling rate during the pressurized filling process was analyzed via a simulation, as shown in Figure 3c, where the initial filling rates in the grating grooves were 20%, 25% and 29%, respectively. It can be observed that, when the initial particle filling rate in the grating groove was different, applying the same pressure produced an adaptive deformation, where the smaller the filling rate, the larger the deformation of PDMS. It can be seen from the simulations that each grating groove can be considered as individually filled and not affecting each other. Meanwhile, the filling rate in the grating grooves increased rapidly after a single pressurized filling, and then quickly approached the same filling rate. Therefore, the pressurized filling method was adapted to the fabrication of large-size absorption gratings to achieve uniform filling.

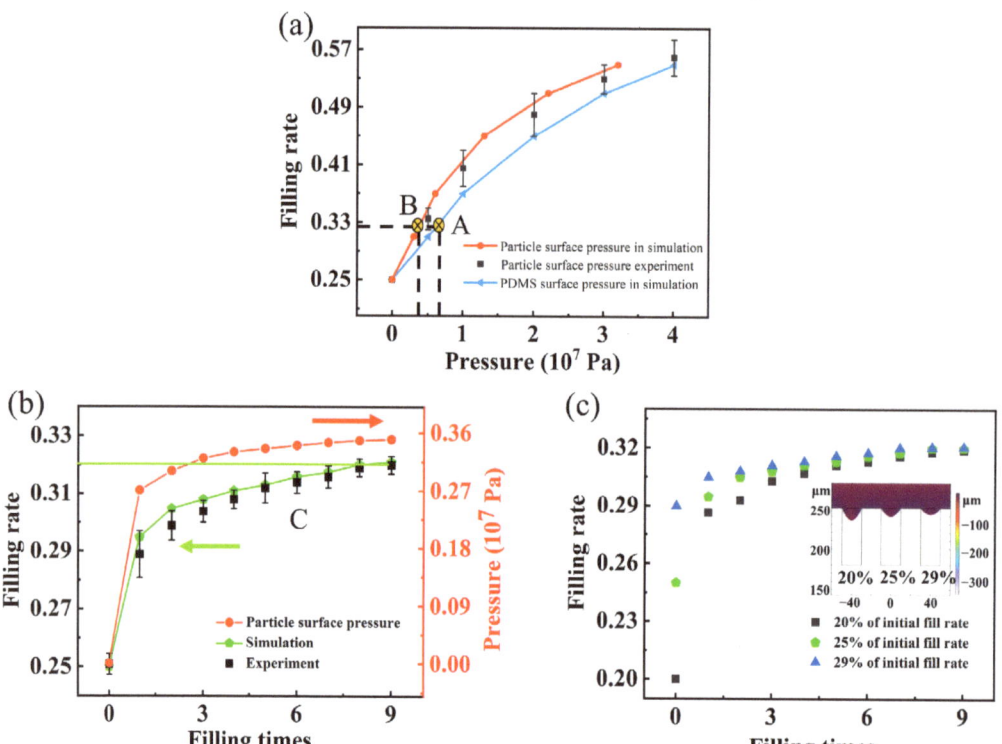

Figure 3. Experiment and simulation. (**a**) Relationship between pressure and particle filling rate, where points A and B indicate that the fill rate reaches 32% when the PDMS surface pressure is 0.7×10^7 Pa and the particle surface pressure is 0.35×10^7 Pa. (**b**) Comparison of experimental and simulated repetitive filling: The green solid line represents the maximum filling rate achieved during repetitive pressurization, while the arrows point to the y-axis corresponding to the different data. Point C denotes the filling rate achievable after six repetitions of filling. (**c**) Repetitive filling of gratings with different initial filling rates.

3.2. Effect of Structure and Material

The filling rate is related to the arrangement of the particles. From Figure 3a, it can be observed that the particle filling rate was determined by the pressure on the particle surface. The pressure applied on the surface of PDMS transferred to the particle surface was influenced by the grating groove width and the material of the grating surface. Therefore, the filling rate may also be related to the grating groove width and the surface material.

Figure 4 simulates the variation of the filling rate with the grating groove width obtained by means of repeated filling under a different Young's modulus of PDMS material. In order to study the effect of different materials on the filling rate and to simplify the simulation, the PDMS's Young's modulus was directly varied to perform the study, since the PDMS would have a different Young's modulus for the different rates of curing agents. The specific parameters were as follows: the pressure applied to the soft material PDMS on the grating surface was 4×10^7 Pa, the grating groove width increased from 3 μm to 20 μm, and the Young's modulus of PDMS were 5×10^5 Pa, 7.5×10^5 Pa, and 1.5×10^6 Pa. From Figure 4, it can be seen that as the grating groove width increased and the Young's modulus of PDMS decreased, the filling rate achieved by repeated filling gradually increased, and the effect of the Young's modulus on the filling was not significant. Moreover, it can be observed that the increase in filling rate using the pressurized filling method was not significant when the grating width was relatively small. Therefore, we have determined that, in order to improve the filling rate of the small-period grating, the surface pressure applied can be increased. However, obtaining high pressure for large-sized gratings is challenging. Thus, splicing and pressurizing small areas can be utilized to generate high pressure, enabling the fabrication of large-sized absorption gratings. With an overlap in the splicing and pressurization process, the pressurized filling method ensures a consistent filling rate without introducing uniformity problems because it is an adaptive filling process, thereby allowing the fabrication of large-size and uniform absorption gratings.

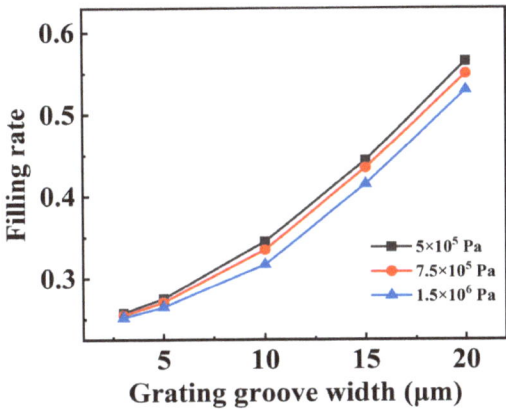

Figure 4. Variation of filling rate for different grating groove widths and Young's modulus of PDMS.

3.3. Process Optimization

As can be seen from Figure 3, repeated pressurized fillings were often required many times, resulting in a long production time and low efficiency. Therefore, as a solution, we propose depositing excessive particles on the grating surface, and then applying pressure on the deformed material surface. In order to facilitate the simulation, the model is simplified, as shown in Figure 5. When applying downward pressure to the material, the particles on the surface of the grating lines are compacted, which can be simplified to the grating lines, resulting in an increase in the depth of the grating grooves. Subsequently, the grating surface deformation material deforms between the grating grooves and presses the particles in the grooves. In order to obtain an absorption grating, it is finally necessary to remove the excess particles from the grating surface. The particle layer on the grating surface is immersed with ethanol, followed by the careful removal of particles using a scraper to ensure no residual particles remain on the grating surface.

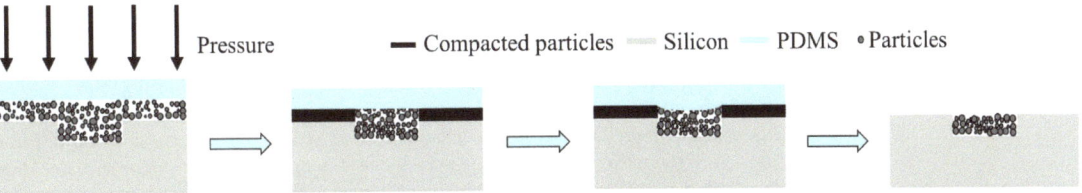

Figure 5. Process optimization diagram.

The effectiveness of the proposed optimized process was investigated through experiments and simulations by depositing excess particles on the grating surface and applying pressure onto the PDMS material surface. The grating structure with a 40 µm period and 20 µm groove width was simulated. The results of a single optimized filling are shown in Figure 6, demonstrating a significantly higher filling rate than that achieved by a single pressurized filling.

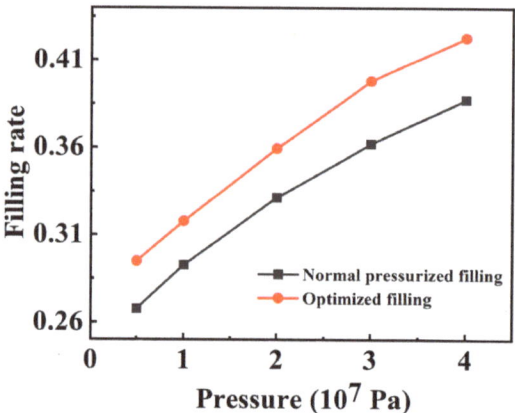

Figure 6. Comparison of filling rate after optimizing filling and normal pressurized filling.

After depositing particles on the surface of the grating structure, it was determined via simulation that the particle filling rate could reach about 31% after one filling by applying a pressure of 0.7×10^7 Pa on the material surface. Compared with point C in Figure 3b, this could be reached with the effect of six repeated fillings. The experimental results also demonstrated that a single filling of excess particles deposited on a grating could achieve a filling rate of 30.5%, which is consistent with the simulation results. During particle deposition, non-uniformity remains a concern, as can be seen in Figure 3c, where after a single pressurization, the filling rate was close to its maximum value. To be on the safe side, two optimization pressurizations can also be used to ensure a consistent filling rate.

The optimized process enables the production of a neutron-absorbing grating that can fully absorb neutron beams with a single pressurized filling, thereby greatly improving the filling efficiency. After pressure is applied, the excess layer of particles on the surface will be pressed down as a whole and the filling rate in the groove will increase, meaning that the optimized filling process is even more effective for the fabrication of small period gratings, making the pressurized filling process adaptable to the production of smaller-period gratings and increasing the scope of application.

4. Conclusions

In this study, we have introduced a pressurized filling method to enhance the filling rate of absorption gratings fabricated using the particle filling method. The results show that the filling rate is determined by the pressure, and the pressurized filling method can greatly improve the filling rate of the grating. Based on the simulation of the influence of pressure, grating groove width, and surface material on the filling rate, it can be concluded that the filling rate gradually increases with an increase in pressure and grating groove width. Moreover, the pressurized filling process is adaptable and allows for the fabrication of large-size gratings. To address the issue of low efficiency in fabricating the absorption grating using the pressurized filling method, we have proposed a process optimization approach that yields significantly higher efficiency. Therefore, our research provides a promising and practical solution for improving the filling rate and efficiency of the fabrication of absorption gratings.

Author Contributions: Conceptualization, E.H. and G.L.; Data curation, E.H. and K.Z; Formal analysis, E.H. and G.L.; Funding acquisition, G.L. and Y.G.; Investigation, E.H., K.Z., L.C. and C.G.; Methodology, E.H. and K.Z.; Software, E.H. and K.Z.; Project administration, G.L. and Y.T.; Supervision, Y.X., Y.T., Y.G. and G.L.; Validation, E.H.; Visualization, E.H.; Writing—original draft, E.H.; Writing—review and editing, G.L. All authors have read and agreed to the published version of the manuscript.

Funding: This work was supported by the National Key Basic Research Program of China (no. 2016YFA0400902) and the Youth Innovation Promotion Association, CAS (2020457).

Data Availability Statement: The data in the manuscript can be obtained from the corresponding author.

Conflicts of Interest: The authors declare no conflict of interest.

References

1. Artioli, G.; Hussey, D.S. Imaging with Neutrons. *Elements* **2021**, *17*, 189–194. [CrossRef]
2. Kardjilov, N.; Manke, I.; Woracek, R.; Hilger, A.; Banhart, J. Advances in neutron imaging. *Mater. Today* **2018**, *21*, 652–672. [CrossRef]
3. Strobl, M.; Manke, I.; Kardjilov, N.; Hilger, A.; Dawson, M.; Banhart, J. Advances in neutron radiography and tomography. *J. Phys. D-Appl. Phys.* **2009**, *42*, 21. [CrossRef]
4. Momose, A.; Takano, H.; Wu, Y.L.; Hashimoto, K.; Samoto, T.; Hoshino, M.; Seki, Y.; Shinohara, T. Recent progress in X-ray and neutron phase imaging with gratings. *Quantum Beam Sci.* **2020**, *4*, 9. [CrossRef]
5. Schillinger, B.; Calzada, E.; Lorenz, K. Modern neutron imaging: Radiography, tomography, dynamic and phase contrast imaging with neutrons. In *Materials in Transition, Proceedings*; Solid State Phenomena; Dobrzynski, L., Perzynska, K., Eds.; Trans Tech Publications Ltd.: Durnten, Switzerland, 2006; Volume 112, pp. 61–71.
6. Kim, J.; Lee, S.W.; Cho, G. Visibility evaluation of a neutron grating interferometer operated with a polychromatic thermal neutron beam. *Nucl. Instrum. Methods Phys. Res. Sect. A-Accel. Spectrometers Detect. Assoc. Equip.* **2014**, *746*, 26–32. [CrossRef]
7. Kim, Y.; Kim, J.; Kim, D.; Hussey, D.S.; Lee, S.W. Characterization of the phase sensitivity, visibility, and resolution in a symmetric neutron grating interferometer. *Rev. Sci. Instrum.* **2019**, *90*, 8. [CrossRef] [PubMed]
8. Reimann, T.; Muhlbauer, S.; Horisberger, M.; Betz, B.; Boni, P.; Schulz, M. The new neutron grating interferometer at the ANTARES beamline: Design, principles and applications. *J. Appl. Crystallogr.* **2016**, *49*, 1488–1500. [CrossRef]
9. Kim, J.; Lee, K.H.; Lim, C.H.; Kim, T.; Ahn, C.W.; Cho, G.; Lee, S.W. Fabrication and characterization of the source grating for visibility improvement of neutron phase imaging with gratings. *Rev. Sci. Instrum.* **2013**, *84*, 5. [CrossRef]
10. Samoto, T.; Takano, H.; Momose, A. Gadolinium oblique evaporation approach to make large scale neutron absorption gratings for phase imaging. *Jpn. J. Appl. Phys.* **2019**, *58*, 6. [CrossRef]
11. Samoto, T.; Takano, H.; Momose, A. Evaluation of obliquely evaporated gadolinium gratings for neutron interferometry by X-ray microtomography. *Mater. Sci. Semicond. Process.* **2019**, *92*, 91–95. [CrossRef]
12. Grunzweig, C.; Pfeiffer, F.; Bunk, O.; Donath, T.; Kuhne, G.; Frei, G.; Dierolf, M.; David, C. Design, fabrication, and characterization of diffraction gratings for neutron phase contrast imaging. *Rev. Sci. Instrum.* **2008**, *79*, 6. [CrossRef] [PubMed]
13. Gustschin, A.; Neuwirth, T.; Backs, A.; Schulz, M.; Pfeiffer, F. Fabrication of gadolinium particle-based absorption gratings for neutron grating interferometry. *Rev. Sci. Instrum.* **2018**, *89*, 7. [CrossRef] [PubMed]
14. Seki, Y.; Shinohara, T.; Ueno, W.; Parker, J.D.; Samoto, T.; Yashiro, W.; Momose, A. Experimental evaluation of neutron absorption grating fabricated by oblique evaporation of gadolinium for phase imaging. In Proceedings of the 8th International Topical Meeting on Neutron Radiography (ITMNR), Beijing, China, 4–8 September 2016; pp. 217–223. [CrossRef]
15. Harti, R.P.; Kottler, C.; Valsecchi, J.; Jefimovs, K.; Kagias, M.; Strobl, M.; Grunzweig, C. Visibility simulation of realistic grating interferometers including grating geometries and energy spectra. *Opt. Express* **2017**, *25*, 1019–1029. [CrossRef] [PubMed]
16. Chen, Y.C.; Tsai, T.R.; Chu, J.P.; Sung, H.; Jang, J.S.C.; Kato, H. Imprinting of metallic glasses: A simple approach to making durable terahertz high-pass filters. *Appl. Phys. Express* **2012**, *5*, 3. [CrossRef]
17. Sadeghilaridjani, M.; Kato, K.; Shinohara, T.; Yashiro, W.; Momose, A.; Kato, H. High aspect ratio grating by isochronal imprinting of less viscous workable Gd-based metallic glass for neutron phase imaging. *Intermetallics* **2016**, *78*, 55–63. [CrossRef]
18. Yashiro, W.; Noda, D.; Hattori, T.; Hayashi, K.; Momose, A.; Kato, H. A metallic glass grating for X-ray grating interferometers fabricated by imprinting. *Appl. Phys. Express* **2014**, *7*, 3. [CrossRef]
19. Kim, Y.; Kim, D.; Lee, S.; Kim, J.; Hussey, D.S.; Lee, S.W. Neutron grating interferometer with an analyzer grating based on a light blocker. *Opt. Express* **2020**, *28*, 23284–23293. [CrossRef]
20. Shi, Z.T.; Jefimovs, K.; Romano, L.; Stampanoni, M. Towards the Fabrication of High-Aspect-Ratio Silicon Gratings by Deep Reactive Ion Etching. *Micromachines* **2020**, *11*, 864. [CrossRef]
21. Lei, Y.H.; Li, Q.F.; Wali, A.; Liu, X.; Xu, G.W.; Li, J.; Huang, J.H. Tungsten nanoparticles-based x-ray absorption gratings for cascaded Talbot-Lau interferometers. *J. Micromech. Microeng.* **2019**, *29*, 6. [CrossRef]

22. Hojo, D.; Kamezawa, C.; Hyodo, K.; Yashiro, W. Fabrication of X-ray absorption grating using an ultracentrifuge machine. *Jpn. J. Appl. Phys.* **2019**, *58*, 3. [CrossRef]
23. Pinzek, S.; Gustschin, A.; Gustschin, N.; Viermetz, M.; Pfeiffer, F. Fabrication of X-ray absorption gratings by centrifugal deposition of bimodal tungsten particles in high aspect ratio silicon templates. *Sci. Rep.* **2022**, *12*, 5405. [CrossRef] [PubMed]

Disclaimer/Publisher's Note: The statements, opinions and data contained in all publications are solely those of the individual author(s) and contributor(s) and not of MDPI and/or the editor(s). MDPI and/or the editor(s) disclaim responsibility for any injury to people or property resulting from any ideas, methods, instructions or products referred to in the content.

Article

Plotter Cut Stencil Masks for the Deposition of Organic and Inorganic Materials and a New Rapid, Cost Effective Technique for Antimicrobial Evaluations

Andre Childs [1], Jorge Pereira [2], Charles M. Didier [3], Aliyah Baksh [3], Isaac Johnson [4], Jorge Manrique Castro [5], Edwin Davidson [2], Swadeshmukul Santra [2,3,6] and Swaminathan Rajaraman [1,3,5,6,*]

1. Department of Material Science and Engineering, University of Central Florida, Orlando, FL 32816, USA
2. Department of Chemistry, University of Central Florida, Orlando, FL 32816, USA
3. Burnett School of Biomedical Sciences, University of Central Florida, Orlando, FL 32827, USA
4. Department of Mechanical and Aerospace Engineering, University of Central Florida, Orlando, FL 32816, USA
5. Department of Electrical and Computer Engineering, University of Central Florida, Orlando, FL 32816, USA
6. NanoScience Technology Center, University of Central Florida, Orlando, FL 32826, USA
* Correspondence: swaminathan.rajaraman@ucf.edu; Tel.: +1-407-823-4339

Abstract: Plotter cutters in stencil mask prototyping are underutilized but have several advantages over traditional MEMS techniques. In this paper we investigate the use of a conventional plotter cutter as a highly effective benchtop tool for the rapid prototyping of stencil masks in the sub-250 μm range and characterize patterned layers of organic/inorganic materials. Furthermore, we show a new diagnostic monitoring application for use in healthcare, and a potential replacement of the Standard Kirby-Bauer Diffusion Antibiotic Resistance tests was developed and tested on both *Escherichia coli* and *Xanthomonas alfalfae* as pathogens with Oxytetracycline, Streptomycin and Kanamycin. We show that the reduction in area required for the minimum inhibitory concentration tests; allow for three times the number of tests to be performed within the same nutrient agar Petri dish, demonstrated both theoretically and experimentally resulting in correlations of R ≈ 0.96 and 0.985, respectively for both pathogens.

Keywords: plotter cutter; stencil mask; material patterning; Kirby-Bauer; antibiotics testing; minimal inhibitory concentration detection

1. Introduction

Traditionally MEMS stencils have been largely fabricated using laser micromachining, computer numerical control (CNC) milling, deep reactive-ion etching (DRIE), and other photolithographic approaches due to their great precision and ease of workflow integration achieving excellent feature resolutions down to 10's of nanometers [1–7]. However, with lengthier processing times, high cost for prototyping, and the requirements of a cleanroom facility, these technologies have become much less appealing lately particularly, due to accessibility concerns for low-resource settings and low-income countries [8–11]. With extreme process optimization, technologies such as plotter cutters and other electronic cutters could replace older stencil masking technologies [12]. As these tools have been shown to be more effective, especially in microfluidic and biosensor applications in recent years and are extremely appealing to the growing makerspace and makerspace microfabrication communities [10,13–16].

Previous work has demonstrated that these current benchtop plotter cutters can reach down to ~200 μm in minimum feature sizes and according to Bartholomeusz et al. most plotter cutters have a theoretical resolution of ~25 μm which is harder to demonstrate repeatably. The plotter cutter used by Bartholomeusz et al. was a FC5100A-75 from Graphtec (~$4000 in cost), demonstrated excellent features. With this tool microchannels of ~38 μm in width and serpentine patterns of 78 ± 23 μm in dimensions, with poor accuracies up to ~26% in discrepancy were defined [17]. The recent work by Islam et al., measured the accuracy of four different microfluidic patterns: straight lines, serpentines, zig zag patterns and square microchannels for electrophoresis, mixing, particles separation and bubble transport, respectively. They reported error values in the accuracy of the cuts were as high 26.5% and as low as 9.09% with the smallest cut being 208.07 ± 9.09 µm in a double-sided pressure sensitive adhesive. The plotter cutter used in this work was a Graphtec CE6000-40 (~$2000 in cost). Additionally work from Yuen and Goral showed a 200 µm limit for serpentine channels using a QuicKutz® Silhouette™ SD (~$300 in cost) [18]. All these results are summarized in Table S1.

In this work, a similar Silhouette™ Plotter Cutter Cameo 4 (SPC4) which is relatively inexpensive (~$300 in cost) is utilized. The SPC4 can cut up to 3 mm thick materials, has two carriages: one that can produce up to 210 gf (2.01 N) force and the other up to 5 kgf (200 N). The 5 kgf force is the highest reported force in this machine class and has a work envelope of approximately 30.48 cm by 30.48 cm cutting mat. The blade that is commonly used for the SPC4 Silhouette™ Plotter Cutter is the Autoblade, which can be fitted into one of the carriers of the system. Additionally, three more blades can be retrofitted in a second carrier in the system: a Kraft blade, a Rotary blade, and a Punch tool. Researchers have typically worked with the SPC4 using the kraft cutter with an engraving tip width of 100–300 µm, as this tool has produced channel resolutions widths of 45 ± 5 µm and 50–300 µm in depth [19,20]. However, to our knowledge, no one has done extensive research on the Autoblade SPC4.

Even though the utilization of plotter cutters to create custom stencils is not new, plenty of new research exploration in terms of tools and materials for cutting, as well as applications remain. In this paper, we design and microfabricate stencil masks with Kapton® 300 HN due to its low cost, favorable mechanical and electrical properties, flexibility, use as electrical insulation, chemical/radiation resistance, and ability to operate over wide temperature ranges (−269 to 400 °C). Other advantages of Kapton® for this application include a tensile strength of 24,000 MPa, shrinkage at 200 °C of just 0.35%, moisture absorption (4.0%), excellent dielectric strength (4500 kV/mm), volume resistivity (10^{12} ohm/cm), dielectric constant (3.5) and dissipation factor of just 0.0036 at 1 kHz [21].

Kapton® films have been used in cardiovascular applications [22], space station solar cell arrays [23], and flexible electronic sensors [24–26]. Kapton® films have also been used as stencil masks by our group and other researchers [15,26–28]. Specifically, the usage of these microfabricated stencil masks can be extrapolated to a myriad of applications such as Interdigitated Electrodes (IDEs) for impedance measurement, dielectric layer deposition, and microenvironment patterning for the manipulation and selective localization of bacterial and eukaryotic cells. Such localization is imperative to the study of cellular behavior such as morphology, and response to pharmaceutical drugs in critical disease states such as cancer [29,30]. With the rise of antibiotic resistant pathogens, it has become important to quickly assess the vulnerability of bacteria to different antibiotics to better determine medical treatment in humans and animals or crop disease management in plants. The Kirby-Bauer Disk Diffusion Test (DDT) allows for antimicrobial assessment of different compounds against bacterial population [31,32] that require large test area (10 s of mm^2) and sample volume (20 µL or more), limiting device use efficiency.

This work proposes a novel variation of the DDT that utilizes Kapton® stencils to minimize and confine the area required for testing. Due to the importance of this assessment for both the biomedical and agricultural fields the proposed test was assessed against a strain of Escherichia coli (*E. coli* K-12), a well known model human pathogen that can be transmitted from animal to human, and *Xanthomonas alfalfae* (*X. alfalfae*), a model pathogen of the Xanthomonas family, which is known to cause significant damage to crops, resulting in high yield loss [33–36].

The capabilities demonstrated in this paper showcase how the SPC4 tool provides precision with a multitude of swift cuts leading to the rapid prototyping of material layers for accurate diagnosis, analysis treatment or acute action. We analyze the best cutting parameters for 300 HN Kapton® and the drag method with force penetration using the tetragonally shaped Autoblade. We additionally utilized the stencil masks, for varied deposition of organic and inorganic materials such as Cr, Au, SiO2, and gelatin.

2. Materials and Methods

2.1. Methods of Imaging

The following tools were used for imaging all the reported data in the results and Supplementary Materials sections. The cutting surfaces and materials casting were analyzed using the Confocal Microscope (Keyence BZ-X800, Itasca, IL, USA), the Dektak Xt profilometer (Billerica, MA, USA), the Zeiss Ultra 55 Scanning Electron Microscope (SEM) (Oberkochen, Germany) and Apple iPhone 11 Pro (Cupertino, CA, USA).

2.2. Analysis of the Cutting Blade

The cutting blade from the Silhouette™ Plotter Cutter Cameo 4 (Autoblade) was removed from the cutter for analysis. The blade angle and the different portions of the cutting surface were analyzed thoroughly from the surface of the material, cutting blade edge dimensions, angles, etc., with the Confocal Microscope Keyence BZ-X800. Force measurements to determine the cutting forces were performed using a 400 series square Force Sensing Resistor (FSR) from Adafruit (New York, NY, USA). In order to elucidate the force required to cut through 2 layers Kapton® substrate, a Force Sensitive Resistor (FSR, Adafruit) was placed in between the substrate and the cutting mat. Straight line channels (Length: 12.7 mm × Width: 2 mm) were machined and the resultant forces measured using the FSR.

2.3. Stencil Mask Designs

Kapton® Films 300HN were cut using the Silhouette™ Cameo 4 plotter cutter (Figure 1A) with feature dimensions of widths 2 mm, 1 mm, 750 µm, 500 µm, 250 µm, 200 µm with lengths of 12.7 mm and 10 mm, for the "Straight Line Channels" and "Square Channels" designs, respectively. Both were designed using SOLIDWORKS (Dassault Systèmes, Paris, France). The "Serpentine Channels" were designed in SOLIDWORKS as well with feature dimensions: 4 mm, 2 mm, 1500 µm, 1000 µm and 500 µm with channel widths of 2 mm and length of 5 mm. The geometries were chosen due to the common use in many microfluidic devices and other microsystems. For instance, the straight channels are used in electrophoresis techniques, the serpentine channels for particles separation and the square channels for bubble transport [12]. A Metal Cr (Chromium, 50 nm) was deposited through these cut stencils to demonstrate conductive layer depositions.

2.4. Gelatin Casting

A 10 wt% gelatin solution was prepared using Gelatin from Bovine Skin (Powder) purchased from Sigma Aldrich (St. Louis, MO, USA). Kapton® stencil masks were attached to glass slides using Kapton® tape and then the gelatin was cast through the following stencil masks "Straight Line Channels", "Square Channels", "Serpentine Channels" on the glass slides using transfer pipettes. (Figure 1B).

2.5. Interdigitated Electrode Design (IDE) and Microfabrication

IDEs of design dimensions, 1 mm and 2 mm in width, 5 mm and 2.5 mm pitch were developed using SOLIDWORKS. These designs were additionally cut using the Silhouette Cameo 4 Plotter Cutter for the study of rapid microfabricated Interdigitated Electrodes. These Kapton® masks were attached to glass slides as described in Section 2.4, and Cr/Au (50/25 nm) was deposited using a Temescal E-beam evaporator (Ferrotec, Livermore, CA, USA) with a chamber pressure of 8×10^{-6} torr at a deposition rate of 1.5 Å/s (Figure 1C).

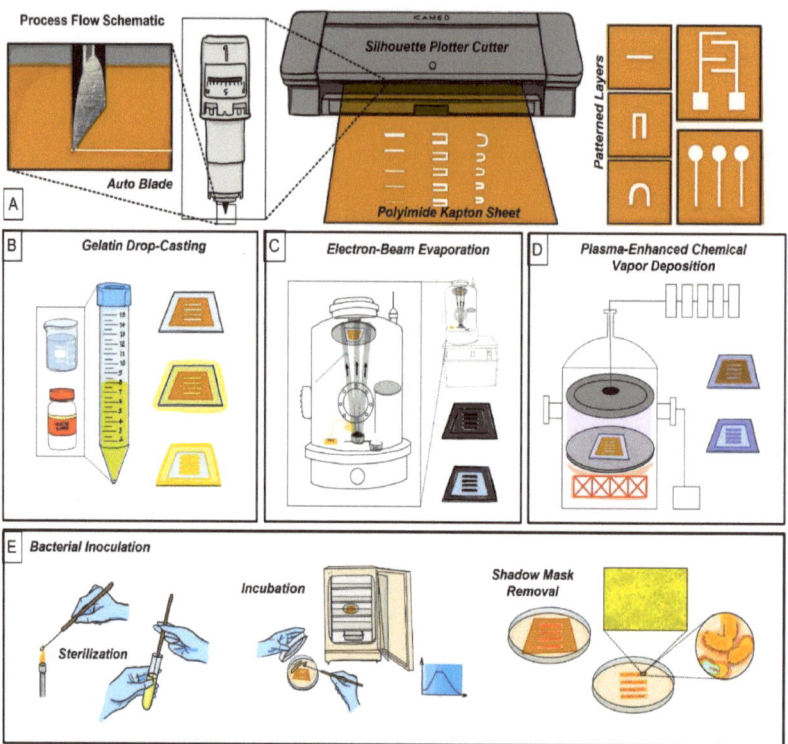

Figure 1. Process flow schematic: (**A**) Silhouette Plotter Cutter along with fabricated designs that were evaluated in this study (right). (**B**) Schematic of the casting process for 10 wt% gelatin. (**C**) E-beam evaporation process. (**D**) Plasma Enhanced Chemical Vapor Deposition (PECVD) process for the deposition of SiO_2. (**E**) Culturing of *E. coli* K-12 & *X. alfalfae*.

2.6. Silicon Dioxide Deposition

Kapton® stencil masks with "Straight Line Channels", "Square Channels", "Serpentine Channels" were developed as described in Section 2.3 and placed on glass slides such as in Section 2.4. Silicon dioxide deposition was performed with an Orion Trion Plasma Enhanced Chemical Vapor Deposition (PECVD) (Plasma Therm, Clearwater, FL, USA) at a rate of 50 nm/min for 10 min at 55 °C and 300 °C using TEOS and O_3 (Airgas, Radno, PA, USA). Analysis of the deposited SiO_2 layers was performed with the Dektak Xt profilometer and Zeiss SEM (Figure 1D).

2.7. Bacteria Handling

Bacteria stocks were stored at −80 °C in a Thermo Scientific REVCO Freezer ULT2186-5-AVA (Waltham, MA, USA). For the tests, bacteria were streaked onto Mueller Hinton II agar (MHA) (Becton Dickinson and Company, Franklin Lakes, NJ, USA) plates and incubated for two days. Afterward, individual colonies were utilized to inoculate 10 mL of nutrient broth and set on an incubator shaker at 150 rpm for 24 h before being utilized for the antimicrobial assessment. *E. coli* K-12 was incubated at 37 °C, while *X. alfalfae* was incubated at 28 °C.

2.8. Disk Diffusion Assay

The disk diffusion assay was performed according to a previously published protocol with some variations [31]. Briefly, 6.0 mm disks were imbibed in 20 µL of an antibiotic solution to deposit a defined mass of the compound. Separately, MHA plates were streaked with a sterile cotton applicator soaked in a bacterial solution set to a bacterial density of ~10^8 colony forming units per milliliter (CFU/mL). Once the agar plates were dry, the antibiotic discs were placed separately on the surface of the agar, the plate was then closed and sealed using parafilm. Before measuring the radius of inhibition *E. coli* K-12 and *X. alfalfae*, inoculated tests were incubated for 24 h at 37 °C and 48 h at 28 °C, respectively in a Fisher Scientific Isotemp Incubator (Model No: 637F). For the templated Stencil Disk Diffusion test (SDDT), MHA agar plates were covered with the Kapton® stencil masks (design dimensions: 6 mm diameter circle and a linear channel of 42 mm in length and 2 mm in width, Figure 1E) before streaking the bacteria only on the exposed agar surface. Antibiotics Streptomycin Sulfate (2.5–20 µg), Oxytetracycline Hydrochloride (7.5–60 µg), and Kanamycin Sulfate (7.5–60 µg) were utilized in the new diffusion test increasing doses linearly. All tests and characterizations were performed in triplicates (i.e., n = 3).

3. Results

3.1. Analysis of the Cutting Blade

The key cutting feature of the plotter cutter blade is shown in Figure 2. The Autoblade has an ellipsoidal curvature of length ~1400 µm and a tip that measures ~26 µm with a flat surface; rather than having a radius of curvature, creating a micron-scale flat features when cutting through materials such as Kapton®. Looking at the top left image we can see that the cutting tool's blade is a scalene tetragonal pyramidally shaped tool, with ellipsoidal cutting faces with a cutting angle of ~42° and 90° degree flat tip as depicted clearly in Figure 2. The cutting method utilized in this work is the drag knife method [17]. This involves dragging the blade without lifting it out of the material path. Further the path direction is depicted in Figure S1. The preset cutting forces as per the Silhoutte cutter ranges from F1–F33. These were measured using the FSR and are as follows (for the ones used in this work): F1 = 0.09415 N, F10 = 0.1285 N, F22 = 0.2962 N, F24 = 0.3236 N, F26 = 0.3433 N and F33 = 0.3629 N as shown in Table 1. Experiments with all of these forces were carried to determine the best cutting conditions for Kapton (Figures S2–S4). Increasing the force increased the chances of plastic deformation as shown in Figure S5. Forces of F1 and F10 did not allow for the blade to penetrate all the way through the substrate as shown in Figure S3, however cutting at forces F22 and above ensured through-substrate cutting as shown in Figure 3. Using the FSR method described in Materials and Methods led to the determination of a force value of 0.2962 N for cutting all the way through the substrate. Increasing the speed was shown to have a marginal increase on edge roughness (value: 5.7 µm) and thus a speed of one (corresponding to 1 mm/s) was used for all cuts [20].

Table 1. Characterization of Forces for Cutting Kapton.

Force	FSR (N)	Corners H Angles (°)	Corners V Angles (°)
F1	0.09415	~	~
F10	0.1285	~	~
F22	0.2962	92.48 ± 2.86	93.72 ± 1.02
F24	0.3236	93.32 ± 1.98	95.44 ± 0.97
F26	0.3433	91.18 ± 1.31	94.18 ± 1.31
F33	0.3629	~	~

Blade Depth of 10, Speed 1 and Pass 1 was used for all forces.

3.2. Design to Device Measurements

After the cutting tool, the force, and velocity of cutting were analyzed for a detailed look at the cutting parameters for stencil masks. In Figure 4, a "Straight Channels" was used as the starting reference for imaging and design to device calculations. Additionally, the error for SiO_2 stenciled patterns are mentioned with all the different designs as it was the highest of percent errors for all materials deposited. It was found that the smallest (left) pattern machined by the plotter cutter reached an average of 257.1 μm design, however we were able to define smaller than 100 μm traces (Figure S6) just not repeatably. For "single line slit cut" designs of dimensions, 2000 μm, 1000 μm, 750 μm, 500 μm, and 250 μm, the average micromachined stencil mask (n = 3) measured: 2051.2 μm, 1128.9 μm, 875.6 μm, 617 μm and 287.3 μm, respectively. These stencils were utilized in chromium deposition as described in Section 2.3. The measurement of the metal features using the stencils (Figure 5) above were:

2038.7 μm, 1104.7 μm, 858.3 μm, 617.4 μm, and 257.1 μm, respectively. As described in Section 2.4, to demonstrate inorganic material deposition through the stencils, gelatin was cast through the "single line slit cut" designs and they measured: 1989.6 μm, 930.1 μm, 738.5 μm, 549.9 μm and 314.9 μm, respectively. Lastly, for the *E. coli* inoculation stencils the measurements were: 2209.1 μm, 1185.7 μm, 946.5 μm, 614.8 μm and 303 μm, respectively.

For the dielectric SiO_2 deposition these values measured 2166.92 μm, 1160.68 μm, 947.06 μm, 657.07 μm, 314.11 μm, 338.99 μm with 70% error measured when depositing at 200 μm single line channels design. These values are depicted in Figure 4.

For the serpentine design in Figure 6, it was found that the smallest pattern cut by the plotter cutter reached an average of 331.7 μm, when cutting the 500 μm design. The designed values for the various serpentines were 4000 μm, 2000 μm, 1500 μm, 1000 μm and 500 μm when compared to actual stencil values of 3903.6 μm, 1896.9 μm, 1434.6 μm, 896.2 μm, and 548.5 μm, respectively. The average values for the measured serpentine patterns during chromium deposition were: 3953.1 μm, 1976.5 μm, 1498 μm, 1043.6 μm and 589.3 μm. For *E. coli* inoculation the measured numbers were: 3671.6 μm, 1742.7 μm, 1310.2 μm, 946.8 μm and 331.7 μm, respectively. For Gelatin the results were 4023.3 μm, 2107.9 μm, 1687.9 μm, 1177.9 μm, and 596.5 μm, respectively. The dielectric SiO_2 transitioned to 3761.2 μm, 1731.8 μm, 1260.9 μm, 870.6 μm, and 447.6 μm with the greatest percent error for SiO_2 being 15.9%.

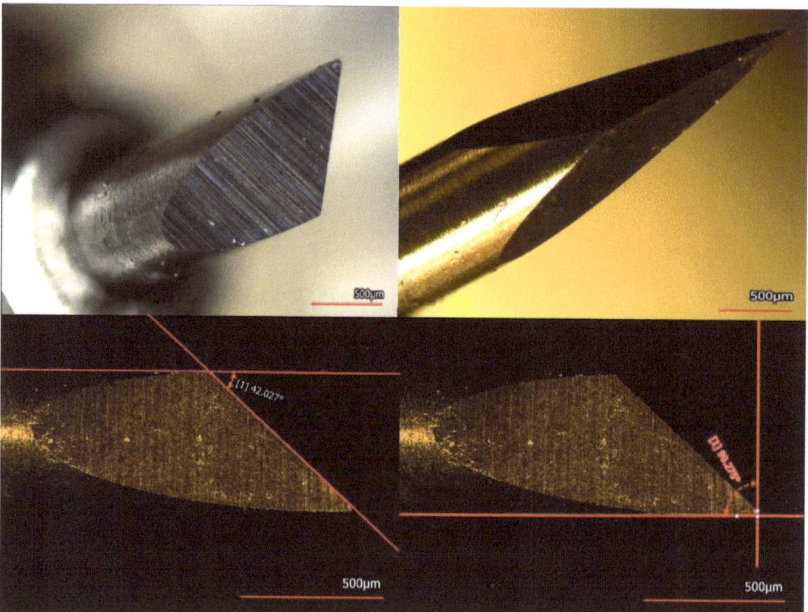

Figure 2. Laser confocal scanning images of the AutoBlade from the plotter cutter showcasing ellipsoidal curvature on the left and tip dimensions on the right. The blade characterization helps with assessment of cutting features on specific materials.

Figure 3. 3D confocal images images of "Good Cuts" depicting F22, F24, F26 Horizontally. The Cuts are 2 mm wide and 12.7 mm in length.

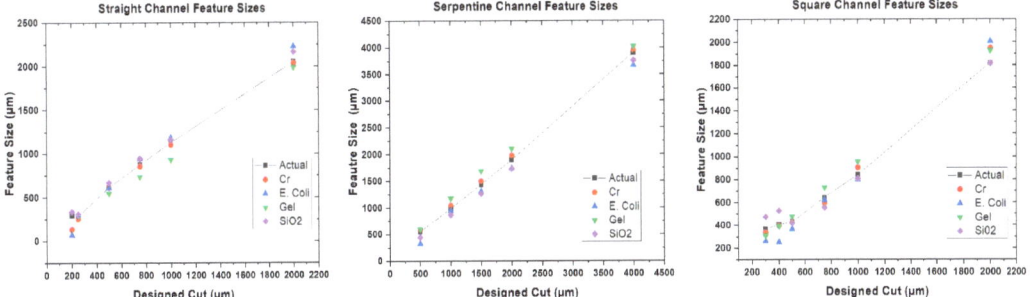

Figure 4. Design to device for: (**Left**) Straight Channels, (**Center**) Serpentine Channels, (**Right**) Square Channel Feature Sizes.

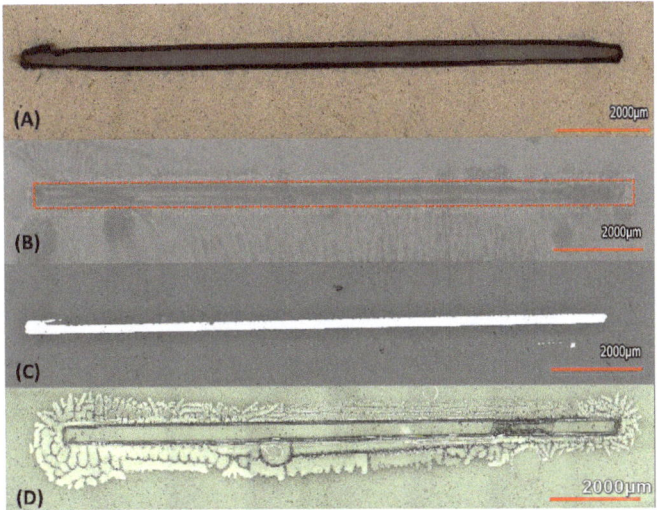

Figure 5. Straight Channel Cuts depicting the smallest cuts for (**A**) Stencil Mask, (**B**) *E. coli* (**C**) Cr, (**D**) Gelatin. Inconsistent Sub 100 μm gelatin structures shown in Figure S6.

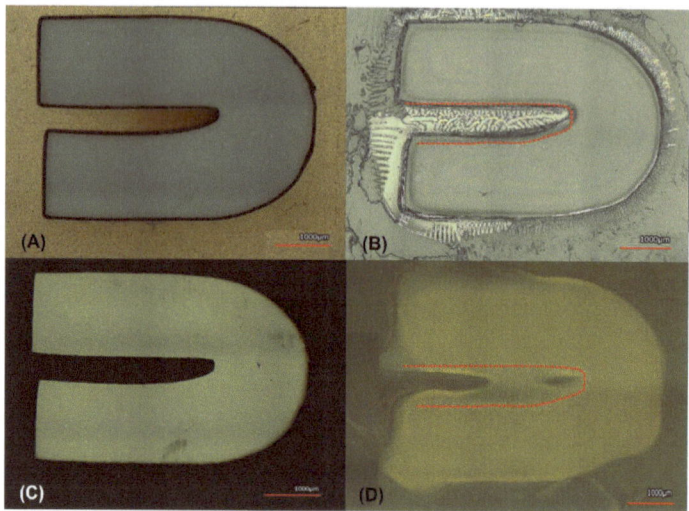

Figure 6. Serpentine Channel Depicting the Smallest feature sizes for (**A**) Stencil Mask, (**B**) Gelatin, (**C**) Cr, (**D**) *E. coli*.

For the Square Channel pattern in Figure 7, it was found that the smallest pattern cut by the plotter cutter reached an average of 368.9 µm, when cutting the 500 µm design. The designed value of 250 µm was unable to be resolved. The average actual measured values from designs were: (design) 2000 µm, 1000 µm, 750 µm, 500 µm and 250 µm to (measured value) 1817.2 µm, 841.9 µm, and 643.2 µm, respectively. For Chromium Deposition, the numbers were: 1948.4 µm, 904.9 µm, 596.1 µm, 368.9 µm and for *E. coli* inoculation the numbers were: 2006.9 µm, 800.4 µm, 622.6 µm and 368.9 µm, respectively. For Gelatin this translated to 1928.0 µm, 958.1 µm, 730.8 µm, 479.3 µm, 392.4 µm and 308.9 µm. For the dielectric SiO_2 this corresponded to 1815.8 µm, 807.5 µm, 556.1 µm, 419.6 µm, 530.5 µm and 476.0 µm with 58.7% being the largest error.

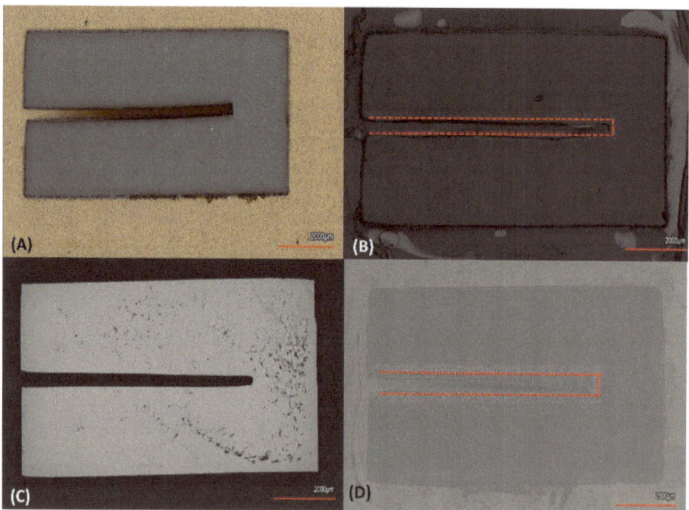

Figure 7. Square Channels depicting the smallest feature sizes for (**A**) Stencil Mask, (**B**) Gelatin, (**C**) Cr, (**D**) *E. coli*.

3.3. Interdigitated Electrodes Fabrication

The low-cost parallel planar interdigitated electrodes with Cr/Au layers were fabricated as discussed in Section 2.5. Figure 8 depicts the full spectrum impedance performance of the devices over the frequency range of 10 Hz–100 kHz. Such an impedance measure (real part of the impedance plotted in Figure 8) and its corresponding values of Cell Index can shine light on morphology and spread of living tissues such as bacteria over a wide range of physical conditions [37–39].

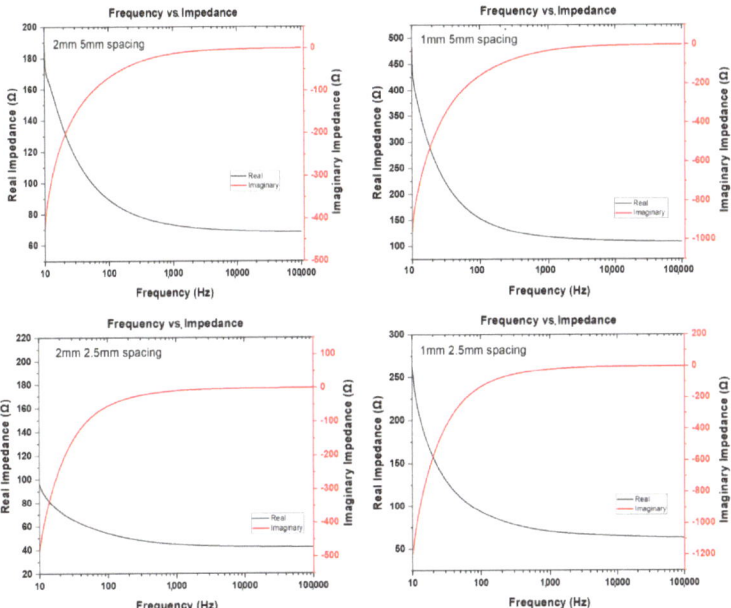

Figure 8. Full spectrum (Range from 1 Hz to 100 kHz) measurements for 1 mm and 2 mm pitch IDE device. (Showcased as metal patterned Cr/Au). Each device was recorded three times and a statistical average of the real and imaginary parts of the impedance are presented.

For both devices, finger width of 1 and 2 mm behave as expected showing interdigitated type signature and reduced impedance as the pitch (or spacing) is reduced and width is increased [37,40]. The ratio of the pitch difference in the range of values is a key indication of the proportion of the electrodes, number of fingers and based upon the electrolyte solution used [38]. At low frequencies the total impedance essentially becomes resistive. The sensitivity increases by increasing the area of the contact surface between the electrodes and the sample under test. A high impedance would result in a large, applied electrode voltage leading to undesirable electrochemical reactions that may be harmful to cellular cultures (not shown here).

3.4. Silicon Dioxide Deposition Results

In Figure 9 we see the amorphous and polycrystalline structures of SiO_2, and a dielectric IDE design showcased in the optical micrograph on the left. The different deposited architectures of SiO_2 occur based on the different temperatures of deposition, at higher temperature leads to greater absorbance due to much smother substrates [41,42]. We expect that at a lower temperature of 50° Celsius, the PECVD deposition structure to be similar to those deposited using ICP PECVD SiO_2 matrixes at these temperatures [43].

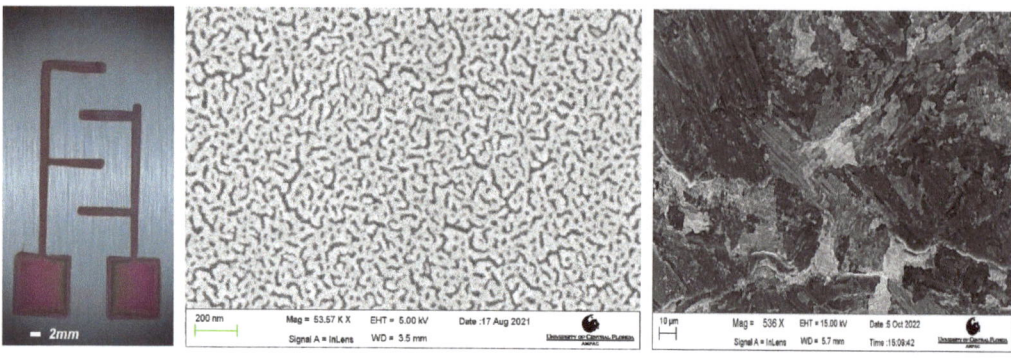

Figure 9. Deposition of a PECVD; (**Left**) Dielectric Interdigitated electrode, (**Center**) Deposition at 300 °C and (**Right**) Deposition at 50 °C.

3.5. Optimized Kirby Bauer Stencil Mask

On a disc diffusion assay, antibiotics slowly diffuse outwards from the impregnated discs toward the surface of the agar, creating a concentration gradient of the compound. Due to this effect, the zone of inhibition is the representation of the area containing antibiotics over the minimum concentration necessary to prevent bacterial growth. According to Bonev et al. the concentration of antibiotics decreases logarithmically from the discs [32]. Therefore, the similar inhibition length obtained from the DDT and SDDT demonstrates that the Kapton stencil does not change antibiotic diffusion as shown in Figures 10 and 11.

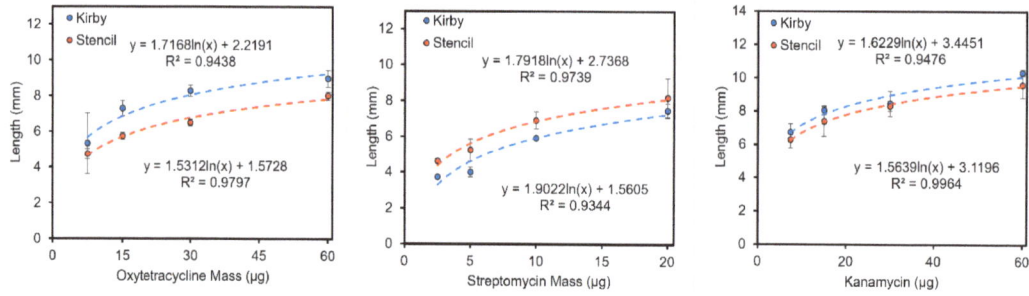

Figure 10. *E. coli* K-12 Antibiotic Resistance Test. (**Left**) Oxytetracycline, (**Center**) Streptomycin, (**Right**) Kanamycin.

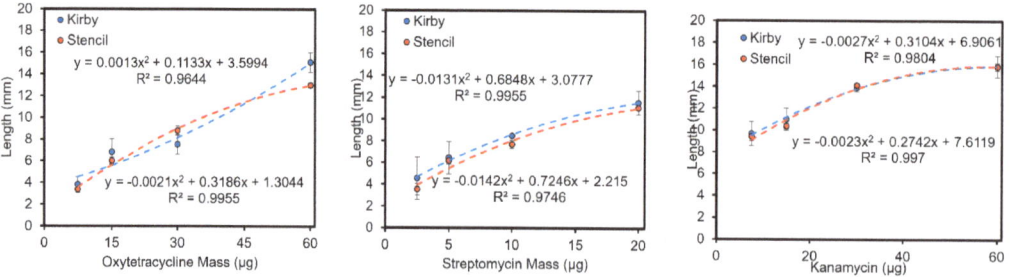

Figure 11. *X. alfalfae*. Plant Bacteria Antibiotic Resistance Test. (**Left**) Oxytetracycline, (**Center**) Streptomycin, (**Right**) Kanamycin.

3.5.1. Length of Inhibition

As customary for DDT, the diameter of the zone of inhibition was measured with an electronic caliper, but the length from the disc to the edge of the inhibition zone was utilized to compare results with SDDT (Figures S7 and S8). The results show similar inhibition lengths for both test variations (Figures 10 and 11). For *E. coli*, increasing the initial antibiotic mass results in a logarithmic increase of the inhibition length/radius. Interestingly, tests on *X. alfalfae* display a quadratic relation between antibiotic mass and inhibition length. This disparity might be due to the difference in bacteria, incubation, agar, and temperature between the tests, as it is expected these variables will have an impact on the diffusion of the compound [32].

3.5.2. Area of Inhibition

In Figure 12 we show the theoretical/qualitative framework behind the transition and comparison of the Kirby Bauer and Kapton Stencil Mask. Due to antibiotic diffusion from the disc, the Minimal Inhibitory Concentration is measured at the boundary of bacterial growth and antibiotic diffusion. A regular area of inhibition is key to interpreting the DDT correctly and must not have any interruptions so that its diameter is the same in all directions. The SDDT possesses 2 mm wide channels for bacterial growth which reduces ambiguity. Based on the measurements the inhibition zones were estimated. For the DDT the area of inhibition was calculated as a perfect circle: $A = \frac{1}{4}\pi d^2$, where d is the diameter of the zone of inhibition. For the SDDT the inhibition zone was determined as an exact rectangle combined with the area of the antibiotic disc: $A = 2 \cdot l + 28.27$ mm^2, where l is the length of inhibition, depictions of the Kirby Bauer DDT and SDDT are shown in Figures S7 and S8. The results show that due to the stencil the area of inhibition drastically decreases compared to the regular DDT.

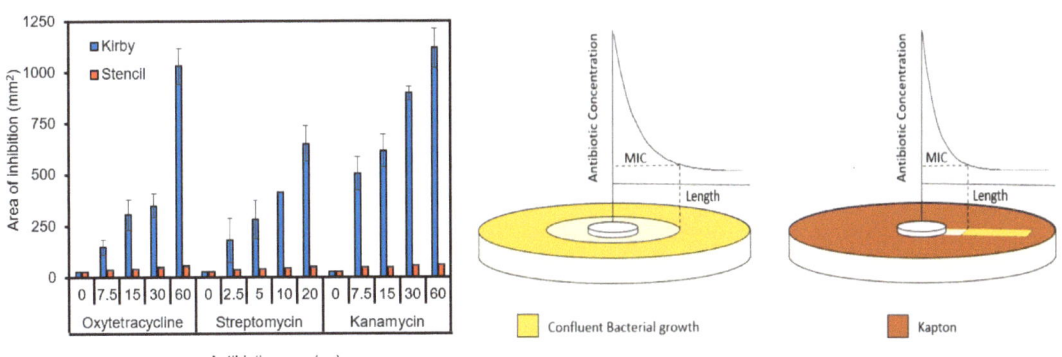

Figure 12. Theoretical MIC Justification. (**Left**) The change in size requirement for the area of inhibition, (**Right**) depiction of the required surface area of the bacteria growing in a Petri dish.

For the governing equations of the SDDT test compared to the DDT correlated with each other for *E. coli* K-12 and *X. alfalfae*.

E. coli K-12

$$y = C ln(x) + B \tag{1}$$

X. alfalfae.

$$y = C_1 X^2 + C_2 X + B \tag{2}$$

The equations governing the human *E. coli* K-12 followed the logarithmic equation predicted by Bonev [32]. The plant bacteria which *X. alfalfae* took on the form of a 2nd order polynomial, which to our knowledge has not been studied and requires further study.

A few circumstances of the traditional Kirby Bauer Test are non-favorable, with the DDT there is a limit of about 3–4 disks to be placed in a Petri dish. The circular diffusion pattern of the test can have a large overlap [44]. Our SDDT using Kapton® stencils, allows for at least 6 or more tests of the same potency or lower potency as shown in Bhargav et al. [45]. The stencil mask tests are also more favorable than another antimicrobial test known as the Epsilometer-test (E-test) being 5 mm wide and 60 mm long, which contains exponential gradients with a combination of dilution and diffusion, which claims to be able to test five drugs in the same diameter plate [46,47]. This test claims to have a symmetrical inhibition ellipse centered along the strip, with the end of the strip having the largest antibiotic concentration showing the maximum distance for MIC and lowest MIC on the other side. However, both claims do not occur frequently, and the ellipses are not always uniform requiring the test to be performed repeatedly [48–50]. The E-test is also extremely cost prohibitive when compared to the DDT and SDDT test at >$400 for 100 tests.

The reduction of the area needed to perform the test is beneficial because more assays can be performed on a single agar plate, decreasing the resources and bacteria needed for large susceptibility screenings. Based on these findings, it is envisioned that plotter cutter-generated stencils will be utilized in the future to produce economical standardized antibiotic testing kits.

4. Conclusions

We have successfully shown that an inexpensive plotter cutter can be used to design rapid microstructure prototypes to full-blown applications within seconds using 300HN Kapton® films. We found that these films were best suited for our application as they are thin enough to cut through in a single pass and prevent tearing while cutting. The effective parameters for cutting through the Kapton® films were determined as: F = 22–26 (which was determined to be 0.2962, 0.3236, 0.3433 N); Speed= 1 mm/s with fabrication time in seconds. Design to device measurements showcase an average error of 9.66% without SiO_2 results, with expected trends with smaller dimensions patterns showing larger errors. Both metal and silicon dioxide interdigitated electrodes were successfully fabricated and tested. Lastly, a novel growth restricted diffusion test for MIC detection of both human and plant pathogen bacteria were demonstrated. The data for the bacteria SDDT correlated with the DDT, inferring that using this SDDT will allow for a high throughput method allowing for 3 times the number of tests to be allowed in an agar Petri dish for antimicrobial characterization. Using this new protocol, we can additionally test smaller antibiotic concentrations and multiple antibiotics in the same Petri dish for assessment. Overall, the methodologies described in this work can be utilized in makerspaces, limited resource countries and laboratories without access to microfabrication facilities. The plotter cutter can rapidly design and microfabricate stencil mask prototypes for the deposition of various inorganic and organic materials and create reproducible bacterial susceptibility tests with correlation coefficients of R ≈ 0.96 and 0.985.

Supplementary Materials: The following supporting supplementary information can be downloaded at: https://www.mdpi.com/article/10.3390/mi14010014/s1, Figure S1: Drag Knife Method; Table S1: Plotter Cutter Applications; Figure S2: Bad Cuts Forces 1, 10, and 33; Figure S3: Confocal Images of Bad Cuts; Figure S4: Horizontal and Vertical cuts; Figure S5: Optical Images of Square Channels; Figure S6: Below 100 μm pattern gelatin, Figure S7: Kirby Bauer Disk Diffusion Test (DDT); Figure S8: Stencil Mask Disk Diffusion Test (SDDT).

Author Contributions: Conceptualization, S.R. and A.C.; methodology, A.C., J.P., I.J., E.D., J.M.C., A.C. and C.M.D.; validation, A.C., J.P., S.R. and S.S.; formal analysis, A.C., J.P. and J.M.C.; investigation, A.C.; resources, S.S. and S.R.; data curation, A.C. and J.P.; writing—original draft preparation, A.C.; writing—A.C. and J.P.; editing, A.C. and S.R.; visualization, A.C., A.B. and I.J.; supervision, S.R.; project administration, S.R.; funding acquisition, S.R. All authors have read and agreed to the published version of the manuscript.

Funding: This work is supported by industrial collaboration grants from Primordia BioSystems and industrial match grants from the Florida High Tech Corridor.

Data Availability Statement: The data of visualization and characterization and experimental tests can be sent on request. The files of Images and samples can be found in Dr. Swaminathan Rajaraman's Laboratory in the sample box that says Andre Childs.

Acknowledgments: This work is supported by grants from Primordia Biosystems and the Florida High Tech Corridor.

Conflicts of Interest: Prof. Rajaraman is a co-founder and major equity holder in Primordia BioSystems. The other authors declare no conflict of interest.

References

1. Yesilkoy, F.; Flauraud, V.; Rüegg, M.; Kim, B.J.; Brugger, J. 3D nanostructures fabricated by advanced stencil lithography. *Nanoscale* **2016**, *8*, 4945–4950. [CrossRef] [PubMed]
2. Whitesides, G.M. The origins and the future of microfluidics. *Nature* **2006**, *442*, 368–373. [CrossRef]
3. Brannon, J. *Excimer Laser Ablation and Etching*, 1st ed.; The Education Committee, American Vacuum Society: New York, NY, USA, 1993.
4. Fourkas, J.T. Nanoscale Photolithography with Visible Light. *J. Phys. Chem. Lett.* **2010**, *1*, 1221–1227. [CrossRef]
5. Liu, W.; Wang, J.; Xu, X.; Zhao, C.; Xu, X.; Weiss, P.S. Single-Step Dual-Layer Photolithography for Tunable and Scalable Nanopatterning. *ACS Nano* **2021**, *15*, 12180–12188. [CrossRef] [PubMed]
6. Shao, D.B.; Chen, S.C. Surface-plasmon-assisted nanoscale photolithography by polarized light. *Appl. Phys. Lett.* **2005**, *86*, 253107. [CrossRef]
7. Pavel, E.; Prodan, G.; Marinescu, V.; Trusca, R. Recent advances in 3- to 10-nm quantum optical lithography. *J. Micro Nanolithogr. MEMS MOEMS* **2019**, *18*, 020501. [CrossRef]
8. Hart, C.; Rajaraman, S. Low-Power, Multimodal Laser Micromachining of Materials for Applications in sub-5 µm Shadow Masks and sub-10 µm Interdigitated Electrodes (IDEs) Fabrication. *Micromachines* **2020**, *11*, 178. [CrossRef]
9. Tsao, C.-W. Polymer Microfluidics: Simple, Low-Cost Fabrication Process Bridging Academic Lab Research to Commercialized Production. *Micromachines* **2016**, *7*, 225. [CrossRef]
10. Walsh, D.I.; Kong, D.S.; Murthy, S.K.; Carr, P.A. Enabling Microfluidics: From Clean Rooms to Makerspaces. *Trends Biotechnol.* **2017**, *35*, 383–392. [CrossRef]
11. Nath, P.; Maity, T.S.; Pettersson, F.; Resnick, J.; Kunde, Y.; Kraus, N.; Castano, N. Polymerase chain reaction compatibility of adhesive transfer tape based microfluidic platforms. *Microsyst. Technol.* **2013**, *20*, 1187–1193. [CrossRef]
12. Islam, M.; Natu, R.; Martinez-Duarte, R. A study on the limits and advantages of using a desktop cutter plotter to fabricate microfluidic networks. *Microfluid. Nanofluid.* **2015**, *19*, 973–985. [CrossRef]
13. Qamar, A.Z.; Shamsi, M.H. Desktop Fabrication of Lab-On-Chip Devices on Flexible Substrates: A Brief Review. *Micromachines* **2020**, *11*, 126. [CrossRef] [PubMed]
14. Kundu, A.; Ausaf, T.; Rajasekaran, P.; Rajaraman, S. Multimodal Microfluidic Biosensor with Interdigitated Electrodes (IDE) and Microelectrode Array (MEA) for Bacterial Detection and Identification. In Proceedings of the 2019 20th International Conference on Solid-State Sensors, Actuators and Microsystems & Eurosensors XXXIII (TRANSDUCERS & EUROSENSORS XXXIII), Berlin, Germany, 23–27 June 2019. [CrossRef]
15. Johns, P.; Roseway, A.; Czerwinski, M. Tattio: Fabrication of Aesthetic and Functional Temporary Tattoos. In Proceedings of the 2016 CHI Conference Extended Abstracts on Human Factors in Computing Systems, San Jose, CA, USA, 7–12 May 2016; pp. 3699–3702.
16. Kundu, A.; Ausaf, T.; Rajaraman, S. 3D Printing, Ink Casting and Micromachined Lamination (3D PICLµM): A Makerspace Approach to the Fabrication of Biological Microdevices. *Micromachines* **2018**, *9*, 85. [CrossRef] [PubMed]
17. Bartholomeusz, D.A.; Boutte, R.W.; Andrade, J.D. Xurography: Rapid prototyping of microstructures using a cutting plotter. *J. Microelectromech. Syst.* **2005**, *14*, 1364–1374. [CrossRef]
18. Yuen, P.K.; Goral, V.N. Low-cost rapid prototyping of flexible microfluidic devices using a desktop digital craft cutter. *Lab Chip* **2009**, *10*, 384–387. [CrossRef]
19. Glavan, A.C.; Martinez, R.V.; Maxwell, E.J.; Subramaniam, A.B.; Nunes, R.M.D.; Soh, S.; Whitesides, G.M. Rapid fabrication of pressure-driven open-channel microfluidic devices in omniphobic RF paper. *Lab Chip* **2013**, *13*, 2922–2930. [CrossRef]
20. Samae, M.; Rodniam, C.; Chirasatitsin, S. Characterization of Microfluidic chips Fabricated by a low-cost technique using a vinyl cutter. In Proceedings of the 7th TSME Internationa Conference on Mechanical Engineering, Bali, Indonesia, 1–2 July 2015.
21. DuPont de Nemours, Inc. *DuPont™ Kapton®HN Polyimde Film*; Dupont: Wilmington, DE, USA, 2021.
22. Lindner, E.; Cosofret, V.V.; Ufer, S.; Buck, R.P.; Kusy, R.P.; Ash, R.B.; Nagle, H.T. Flexible (Kapton-based) microsensor arrays of high stability for cardiovascular applications. *J. Chem. Soc. Faraday Trans.* **1993**, *89*, 361–367. [CrossRef]

23. Zhang, Y.; Li, Q.; Yuan, H.; Yan, W.; Chen, S.; Qiu, M.; Liao, B.; Chen, L.; Ouyang, X.; Zhang, X.; et al. Mechanically Robust Irradiation, Atomic Oxygen, and Static-Durable CrO$_x$/CuNi Coatings on Kapton Serving as Space Station Solar Cell Arrays. *ACS Appl. Mater. Interfaces* **2022**, *14*, 21461–21473. [CrossRef]
24. Masihi, S.; Panahi, M.; Maddipatla, D.; Hajian, S.; Bose, A.K.; Palaniappan, V.; Narakathu, B.B.; Bazuin, B.J.; Atashbar, M.Z. Cohesion Failure Analysis in a Bi-layered Copper/Kapton Structure for Flexible Hybrid Electronic Sensing Applications. In Proceedings of the 2021 IEEE International Conference on Electro Information Technology (EIT), Pleasant, MI, USA, 14–15 May 2021; pp. 409–412. [CrossRef]
25. Yong, K.; Ashraf, A.; Kang, P.; Nam, S. Rapid Stencil Mask Fabrication Enabled One-Step Polymer-Free Graphene Patterning and Direct Transfer for Flexible Graphene Devices. *Sci. Rep.* **2016**, *6*, 24890. [CrossRef]
26. Kinnamon, D.S.; Krishnan, S.; Brosler, S.; Sun, E.; Prasad, S. Screen Printed Graphene Oxide Textile Biosensor for Applications in Inexpensive and Wearable Point-of-Exposure Detection of Influenza for At-Risk Populations. *J. Electrochem. Soc.* **2018**, *165*, B3084–B3090. [CrossRef]
27. Gnanasambanthan, H.; Nageswaran, S.; Maji, D. Fabrication of Flexible Thin Film Strain Sensors using Kapton Tape as Stencil Mask. In Proceedings of the 2019 International Conference on Vision towards Emerging Trends in Communication and Networking (ViTECoN), Vellore, India, 30–31 March 2019; pp. 1–4. [CrossRef]
28. Shim, H.; Jang, S.; Jang, J.G.; Rao, Z.; Hong, J.-I.; Sim, K.; Yu, C. Fully rubbery synaptic transistors made out of all-organic materials for elastic neurological electronic skin. *Nano Res.* **2021**, *15*, 758–764. [CrossRef]
29. Alexander, F., Jr.; Price, D.T.; Bhansali, S. Optimization of interdigitated electrode (IDE) arrays for impedance based evaluation of Hs 578T cancer cells. *J. Phys. Conf. Ser.* **2010**, *224*, 012134. [CrossRef]
30. Tandon, N.; Marolt, D.; Cimetta, E.; Vunjak-Novakovic, G. Bioreactor engineering of stem cell environments. *Biotechnol. Adv.* **2013**, *31*, 1020–1031. [CrossRef] [PubMed]
31. Hudzicki, J. Kirby-Bauer Disk Diffusion Susceptibility Test Protocol. *Am. Soc. Microbiol.* **2009**, *15*, 55–63.
32. Bonev, B.; Hooper, J.; Parisot, J. Principles of assessing bacterial susceptibility to antibiotics using the agar diffusion method. *J. Antimicrob. Chemother.* **2008**, *61*, 1295–1301. [CrossRef]
33. Armstrong, G.L.; Hollingsworth, J.; Morris, J.J.G. Emerging Foodborne Pathogens: Escherichia coil O157:H7 as a Model of Entry of a New Pathogen into the Food Supply of the Developed World. *Epidemiol. Rev.* **1996**, *18*, 29–51. [CrossRef]
34. Kaper, J.B. Pathogenic *Escherichia coli*. *Nat. Rev. Microbiol.* **2004**, *2*, 123–140. [CrossRef]
35. Mansfield, J.; Genin, S.; Magori, S.; Citovsky, V.; Sriariyanum, M.; Ronald, P.; Dow, M.; Verdier, V.; Beer, S.V.; Machado, M.A.; et al. Top 10 plant pathogenic bacteria in molecular plant pathology. *Mol. Plant Pathol.* **2012**, *13*, 614–629. [CrossRef]
36. Zarei, S.; Taghavi, S.M.; Hamzehzarghani, H.; Osdaghi, E.; Lamichhane, J. Epiphytic growth of *Xanthomonas arboricola* and *Xanthomonas citri* on non-host plants. *Plant Pathol.* **2017**, *67*, 660–670. [CrossRef]
37. Ibrahim, M.; Claudel, J.; Kourtiche, D.; Nadi, M. Geometric parameters optimization of planar interdigitated electrodes for bioimpedance spectroscopy. *J. Electr. Bioimpedance* **2013**, *4*, 13–22. [CrossRef]
38. Timms, S.; Colquhoun, K.; Fricker, C. Detection of *Escherichia coli* in potable water using indirect impedance technology. *J. Microbiol. Methods* **1996**, *26*, 125–132. [CrossRef]
39. Franks, W.; Schenker, I.; Schmutz, P.; Hierlemann, A. Impedance Characterization and Modeling of Electrodes for Biomedical Applications. *IEEE Trans. Biomed. Eng.* **2005**, *52*, 1295–1302. [CrossRef] [PubMed]
40. Hong, J.; Yoon, D.S.; Kim, S.K.; Kim, T.S.; Kim, S.; Pak, E.Y.; No, K. AC frequency characteristics of coplanar impedance sensors as design parameters. *Lab Chip* **2005**, *5*, 270–279. [CrossRef] [PubMed]
41. Juárez, H.; Pacio, M.; Diaz, T.; Rosendo, E.; Garcia, G.; Garcia, A.; Mora, F.; Escalante, G. Low temperature deposition: Properties of SiO$_2$ films from TEOS and ozone by APCVD system. *J. Phys. Conf. Ser.* **2009**, *167*, 012020. [CrossRef]
42. Fujino, K.; Nishimoto, Y.; Tokumasu, N.; Maeda, K. Silicon Dioxide Deposition by Atmospheric Pressure and Low-Temperature CVD Using TEOS and Ozone. *J. Electrochem. Soc.* **1990**, *137*, 2883. [CrossRef]
43. Mannino, G.; Ruggeri, R.; Alberti, A.; Privitera, V.; Fortunato, G.; Maiolo, L. Electrical Properties of Ultrathin SiO$_2$ Layer Deposited at 50 °C by Inductively Coupled Plasma-Enahnced Chemical Vapor Deposition. *Appl. Phys. Express* **2012**, *5*, 021103. [CrossRef]
44. Bauer, A.W.; Kirby, W.M.; Sherris, J.C.; Turck, M. Antibiotic susceptibility testing by a standardized single disk method. *Am. J. Clin. Pathol.* **1966**, *45*, 493–496. [CrossRef]
45. Bhargav, H.S.; Shastri, S.D.; Poornav, S.P.; Darshan, K.M.; Nayak, M.M. Measurement of the Zone of Inhibition of an Antibiotic. In Proceedings of the 2016 IEEE 6th International Conference on Advanced Computing (IACC), Bhimavaram, India, 27–28 February 2016; pp. 409–414. [CrossRef]
46. Wanger, A.; Mills, K. Etest for susceptibility testing of *Mycobacterium tuberculosis* and *Mycobacterium avium*-intracellulare. *Diagn. Microbiol. Infect. Dis.* **1994**, *19*, 179–181. [CrossRef]
47. Freixo, I.M.; Caldas, P.C.S.; Martins, F.; Brito, R.C.; Ferreira, R.M.C.; Fonseca, L.S.; Saad, M.H.F. Evaluation of Etest Strips for Rapid Susceptibility Testing of *Mycobacterium tuberculosis*. *J. Clin. Microbiol.* **2002**, *40*, 2282–2284. [CrossRef]
48. Hausdorfer, J.; Sompek, E.; Allerberger, F.; Dierich, M.P.; Rüsch-Gerdes, S. E-test for susceptibility testing of *Mycobacterium tuberculosis*. *Int. J. Tuberc. Lung Dis.* **1998**, *2*, 751–755.

49. Joyce, L.F.; Downes, J.; Stockman, K.; Andrew, J.H. Comparison of five methods, including the PDM Epsilometer test (E test), for antimicrobial susceptibility testing of *Pseudomonas aeruginosa*. *J. Clin. Microbiol.* **1992**, *30*, 2709–2713. [CrossRef] [PubMed]
50. Colombo, A.L.; Barchiesi, F.; A McGough, D.; Rinaldi, M.G. Comparison of Etest and National Committee for Clinical Laboratory Standards broth macrodilution method for azole antifungal susceptibility testing. *J. Clin. Microbiol.* **1995**, *33*, 535–540. [CrossRef] [PubMed]

Disclaimer/Publisher's Note: The statements, opinions and data contained in all publications are solely those of the individual author(s) and contributor(s) and not of MDPI and/or the editor(s). MDPI and/or the editor(s) disclaim responsibility for any injury to people or property resulting from any ideas, methods, instructions or products referred to in the content.

Article

Microextrusion Printing of Multilayer Hierarchically Organized Planar Nanostructures Based on NiO, $(CeO_2)_{0.8}(Sm_2O_3)_{0.2}$ and $La_{0.6}Sr_{0.4}Co_{0.2}Fe_{0.8}O_{3-\delta}$

Tatiana L. Simonenko *, Nikolay P. Simonenko *, Philipp Yu. Gorobtsov, Elizaveta P. Simonenko and Nikolay T. Kuznetsov

Kurnakov Institute of General and Inorganic Chemistry of the Russian Academy of Sciences, 31 Leninsky pr., Moscow 119991, Russia
* Correspondence: egorova.offver@gmail.com (T.L.S.); n_simonenko@mail.ru (N.P.S.)

Citation: Simonenko, T.L.; Simonenko, N.P.; Gorobtsov, P.Y.; Simonenko, E.P.; Kuznetsov, N.T. Microextrusion Printing of Multilayer Hierarchically Organized Planar Nanostructures Based on NiO, $(CeO_2)_{0.8}(Sm_2O_3)_{0.2}$ and $La_{0.6}Sr_{0.4}Co_{0.2}Fe_{0.8}O_{3-\delta}$. *Micromachines* 2023, 14, 3. https://doi.org/10.3390/mi14010003

Academic Editor: Kun Li

Received: 20 November 2022
Revised: 11 December 2022
Accepted: 18 December 2022
Published: 20 December 2022

Copyright: © 2022 by the authors. Licensee MDPI, Basel, Switzerland. This article is an open access article distributed under the terms and conditions of the Creative Commons Attribution (CC BY) license (https://creativecommons.org/licenses/by/4.0/).

Abstract: In this paper, NiO, $La_{0.6}Sr_{0.4}Co_{0.2}Fe_{0.8}O_{3-\delta}$ (LSCF) and $(CeO_2)_{0.8}(Sm_2O_3)_{0.2}$ (SDC) nanopowders with different microstructures were obtained using hydrothermal and glycol–citrate methods. The microstructural features of the powders were examined using scanning electron microscopy (SEM). The obtained oxide powders were used to form functional inks for the sequential microextrusion printing of NiO-SDC, SDC and LSCF-SDC coatings with resulting three-layer structures of (NiO-SDC)/SDC/(LSCF-SDC) composition. The crystal structures of these layers were studied using an X-ray diffraction analysis, and the microstructures were studied using atomic force microscopy. Scanning capacitance microscopy was employed to build maps of capacitance gradient distribution over the surface of the oxide layers, and Kelvin probe force microscopy was utilized to map surface potential distribution and to estimate the work function values of the studied oxide layers. Using SEM and an energy-dispersive X-ray microanalysis, the cross-sectional area of the formed three-layer structure was analyzed—the interfacial boundary and the chemical element distribution over the surface of the cross-section were investigated. Using impedance spectroscopy, the temperature dependence of the electrical conductivity was also determined for the printed three-layer nanostructure.

Keywords: hydrothermal synthesis; glycol–citrate synthesis; hierarchical structures; microextrusion printing; functional layer; LSCF; SDC; NiO

1. Introduction

The development of advanced "green" power generation technologies in the framework of the constantly increasing energy consumption level and anthropogenic impact on the environment is currently one of the most important tasks of materials science. The decarbonization of the economy has increased the interest of many countries in hydrogen energy devices, in particular, solid oxide fuel cells (SOFCs) [1–4]. This type of electrochemical generator allows for the direct electrocatalytic conversion of fuel into electricity and heat with high efficiency and extremely low emissions. The applications of such fuel cells are diverse: aerospace, the automotive industry, small and large power plants, portable power generators, combined heat power engineering and reserve power supplies [5,6]. Nevertheless, the widespread introduction and mass commercialization of classical SOFCs are largely limited by their high operating temperatures, which impose stringent requirements for the structural and functional materials used in the fuel cell, accelerating the degradation of the power properties of the device as a whole, thereby increasing the cost of the electricity generated [7,8]. There are two different ways of solving this problem: searching for new materials whose performance in the intermediate-temperature range is not inferior to that of high-temperature materials and reducing the thickness of the SOFC's main components (electrodes and electrolyte) in order to minimize the total resistance of the device [9,10]. One of the most frequently used compositions of intermediate-temperature electrolyte materials is solid solutions based on cerium dioxide doped with rare-earth oxides [11–13]. When considering electrode materials compatible with CeO_2-based electrolytes in terms of their mutual chemical inertness and thermophysical characteristics, composite cathode materials are usually chosen, representing a mixture of perovskite oxides based on lanthanum cobaltite containing strontium and iron ($La_{1-x}Sr_xCo_yFe_{1-y}O_{3-\delta}$), together with an electrolyte [14–16], as well as anode materials containing nickel or its oxide, combined with a solid electrolyte [17,18]. When searching for the optimal composition of cathode and anode materials, not only is the aim to reach a compromise between the linear thermal expansion coefficient and the value of the oxygen-ion conductivity of the electrodes, but it is also to reduce their polarization resistance and to promote an oxygen reduction reaction (ORR) on the cathode, as well as a hydrogen oxidation reaction (HOR) on the anode. It has been reported in a number of works that a cathode containing 50 wt% $La_{0.6}Sr_{0.4}Co_{0.2}Fe_{0.8}O_{3-\delta}$ and 50 wt% $(CeO_2)_{0.8}(Sm_2O_3)_{0.2}$ is characterized by restraining electrode grain growth and the formation of a microstructure that facilitates diffusion and surface exchange, as well as a lower total polarization resistance in comparison with similar cathodes [19,20]. According to the literature [21,22], the ratio of 50 wt% NiO to 50 wt% $(CeO_2)_{0.8}(Sm_2O_3)_{0.2}$ is one of the classic variants for anode or anode functional layer formation since this material can have a more developed triple-phase boundary surface (the area where NiO, SDC and hydrogen phases contact), which has a positive effect on HOR efficiency. In addition, this composition demonstrates sufficiently low resistance values and long-term stability.

It is known that the formation of planar devices for microelectronics and alternative energy is associated with high energy costs, when many stages of application and the processing of functional layers are carried out [23]. At the same time, the use of additive manufacturing methods for the fabrication of ceramic materials can significantly improve reproducibility, promote miniaturization (particularly in the context of μ-SOFC [24,25]) and reduce the number of production stages in the automated formation of functional coatings of the desired geometry. Currently, a number of printing technologies are used to produce planar components for alternative energy devices: inkjet printing [26–29], aerosol printing [30–32], screen printing [33–36], microplotter printing [37–39] and microextrusion printing [40–44]. It is worth noting the microextrusion printing method separately due to it having significantly less stringent requirements for the viscosity level of the ink used (compared, for example, with classical inkjet printing), as well as an advantage when forming thick films (the ability to quickly achieve the desired coating thickness when one or a few layers are applied) of the required geometry.

The microstructure of the functional layers is also an important factor that seriously affects the final characteristics of the entire fuel cell. When developing SOFC components with different geometries, researchers often aim to obtain powders with high dispersity, a narrow particle-size distribution and a hierarchical particle organization or ordered porous structure, which generally allows for the formation of materials with a developed surface, facilitates charge transfer and contributes to improved mass transport [45,46]. One of the most convenient approaches towards obtaining materials based on doped cerium dioxide and nickel oxide is the hydrothermal method, which allows for the fine tuning of the microstructural parameters (dispersity, particle shape, porosity level, the degree of crystallinity, etc.) of the resulting product (both in the form of powders and coatings) depending on the objectives to be achieved [47–54]. For oxides of complex composition based on lanthanum cobaltite doped with strontium and iron, combustion methods (glycol–citrate, citrate–nitrate, glycine–nitrate, etc.) are the most suitable way to synthesize them, corresponding to the specific chemical nature of their constituent oxides [16,55,56].

Thus, the purpose of this work was to develop an approach for the additive formation of multilayer hierarchically organized planar nanostructures based on oxides of the composition NiO, $(CeO_2)_{0.8}(Sm_2O_3)_{0.2}$ and $La_{0.6}Sr_{0.4}Co_{0.2}Fe_{0.8}O_{3-\delta}$ on various types of substrates.

2. Materials and Methods

2.1. Preparation of $La_{0.6}Sr_{0.4}Co_{0.2}Fe_{0.8}O_{3-\delta}$, $(CeO_2)_{0.8}(Sm_2O_3)_{0.2}$ and NiO Nanopowders

Nanopowders of $La_{0.6}Sr_{0.4}Co_{0.2}Fe_{0.8}O_{3-\delta}$ (LSCF), $(CeO_2)_{0.8}(Sm_2O_3)_{0.2}$ (SDC) and NiO composition were obtained using glycol–citrate (LSCF) and hydrothermal synthesis (SDC and NiO) similarly to the methods we described earlier [43,44,57]. In this case, more concentrated (0.3 mol/L) solutions of metal-containing reagents were used to synthesize SDC and NiO. In order to obtain the oxides of the indicated composition, the following reagents were used: $La(NO_3)_3$*$6H_2O$ (99.99%, LANHIT, Moscow, Russia), $Sr(NO_3)_2$ (>98%, Profsnab, St. Petersburg, Russia), $Co(NO_3)_2$*$6H_2O$ (>98%, Lenreactiv, St. Petersburg, Russia), $FeCl_3$*$6H_2O$ (>98%, Lenreactiv, St. Petersburg, Russia), $Ce(NO_3)_3·6H_2O$ (99.99%, LANHIT, Moscow, Russia), $Sm(NO_3)_3·6H_2O$ (99.99%, LANHIT, Moscow, Russia), $[Ni(C_5H_7O_2)_2]$ (>98%, RUSHIM, Moscow, Russia), $(NH_2)_2CO$ (99%, RUSHIM, Moscow, Russia) and n-butanol ($C_4H_{10}O$, 99%, Ekos-1, Moscow, Russia).

2.2. Microextrusion Printing of LSCF-SDC, SDC and NiO-SDC Functional Layers

The nanopowders obtained were further used to prepare functional inks for the microextrusion printing of the corresponding LSCF-SDC (ω(SDC) = 50%), SDC and NiO-SDC (ω(SDC) = 50%) coatings. The optimal rheological characteristics of the inks providing the necessary stability and wettability of the polycrystalline Al_2O_3 substrate and the Pt/Al_2O_3/Pt chip used, as well as maintaining the geometry of the formed oxide coatings, were achieved by using α-terpineol (>97%, Vekton, St. Petersburg, Russia) as a solvent and 20 wt% ethylcellulose (48.0–49.5% (w/w) ethoxyl base, Sigma Aldrich, St. Louis, MO, USA) as a binder. The optimum mass fraction of the solid-phase particles during the formation of disperse systems based on the $(CeO_2)_{0.80}(Sm_2O_3)_{0.20}$ powder was 25%, and in the case of the LSCF-SDC and NiO-SDC electrode materials, the total mass fraction of the oxide particles in the composite functional inks obtained was 15%. The features of the methodology used for the microextrusion printing of oxide coatings have been discussed in our previous papers [43,44]. In the present study, the microextrusion printing of each functional layer was performed sequentially on the surface of a specialized Pt/Al_2O_3/Pt chip (Ra = 100 nm, geometric dimensions 4.1 × 25.5 × 0.6 mm; Pt interdigitated electrodes are located on the front side of the Al_2O_3 substrate, and the Pt micro-heater is on the reverse side) to evaluate their local electrophysical characteristics with Kelvin probe force microscopy (KPFM) and scanning capacitance microscopy (SCM). A three-layer planar structure of (NiO-SDC)/SDC/(LSCF-SDC) composition was also formed via the layer-by-layer printing of a composite anode, a solid electrolyte and a composite cathode (with their certain displacement relative to each other in the lateral plane in order to avoid the direct contact of electrode materials) on the surface of a polycrystalline Al_2O_3 substrate (α- and γ-Al_2O_3 phase mixture). After the printing process was completed, each layer was dried (40 °C, 5 h), followed by additional heat treatment (600 °C, 1 h) to remove the residual solvent and to decompose the organic constituents of the functional inks.

2.3. Investigation of the Obtained Samples

An X-ray diffraction analysis (XRD) of the obtained planar materials was performed on a Bruker D8 Advance diffractometer (CuKα = 1.5418 Å, Ni-filter, E = 40 kV, I = 40 mA, 2θ range—20–80° and 0.02° resolution, and signal accumulation time per point was 0.3 s).

A cross-section of the (NiO-SDC)/SDC/(LSCF-SDC) 3-layer structure was made after its fixation in epoxy resin using a diamond-coated cutting wheel (followed by the polishing of the examined area). The surface microstructures of the oxide powders and coatings obtained, as well as the cross section of the 3-layer structure, were examined using scanning electron microscopy (Carl Zeiss NVision-40, Carl Zeiss, Inc., Oberkochen, Germany). The chemical compositions of the powders and coatings, as well as the mapping of the chemical element distribution across the cross-sectional area of the sample, were analyzed using an energy-dispersive X-ray microanalysis (energy-dispersive X-ray (EDX) spectrometer INCA X-MAX 80, Oxford Instruments, Abingone, UK).

The topography and local electrophysical properties of the oxide coatings applied sequentially to the surface of the Pt/Al$_2$O$_3$/Pt chip were additionally studied using atomic force microscopy (AFM), including KPFM and SCM. An NT-MDT Solver PRO-M microscope (NT-MDT, Zelenograd, Russia) and ETALON HA_HR probes (ScanSens, Bremen, Germany) with a W$_2$C conductive coating (rounding radius < 35 nm) were used for this purpose. In the course of the KPFM measurements, the work function of the coatings' surfaces (ϕ_{film}) was evaluated. For this purpose, their surface was scanned using a probe with a known output work (ϕ_{tip}), the average value of the contact potential (ϕ_{CPD}) was calculated for the area scanned, and then the value of ϕ_{film} was determined as the difference between ϕ_{tip} and ϕ_{CPD}.

The electrical conductivity of the (NiO-SDC)/SDC/(LSCF-SDC) three-layer planar structure printed on the surface of the Al$_2$O$_3$ substrate was measured using impedance spectroscopy utilizing an electrochemical test system based on potentiostat/halvanostat P-45X with the impedance measurement module FRA-24M (Electrochemical Instruments, Chernogolovka, Russia). The registration of the impedance hodographs was carried out in a frequency range of 1 MHz–0.1 Hz in the air in a temperature range of 150–600 °C. The sample was heated in a cell placed in a Miterm-12 tube furnace (Microinstrument, Moscow, Russia) with an accuracy of ±1° (the heating accuracy was ensured by the Microinstrument controller MicroCont 1010, Moscow, Russia). The electrical resistance values of the planar samples were calculated from the obtained electrochemical impedance spectroscopy data using ZView software (Scribner Associates Inc., Version3.3c, Southern Pines, NC, USA).

3. Results and Discussion

The microstructure of the obtained oxide powders was studied using scanning electron microscopy (Figure 1). As can be seen in the micrographs (Figure 1a,b), the resulting nickel(II) oxide powder consisted of aggregated spherical nanoparticles with a size of 25 ± 3 nm. No impurities differing from the basic substance in shape or particle size were detected in the powder composition.

In the case of the $(CeO_2)_{0.8}(Sm_2O_3)_{0.2}$ solid solution, the powder was characterized by a multilevel hierarchical organization of the particles (Figure 1c,d). Thus, the corresponding nanoscale isotropic particles were ordered into one-dimensional structures about 5 μm long and 200–300 nm wide, united into bundles, which, in turn, were coordinated at an angle of about 60° to each other, forming six-point structures. Besides the above formations, the structure of this powder also contained smaller nanosheets—about 500 nm long, 200–250 nm wide and about 50 nm thick. Thus, under hydrothermal conditions, samarium-doped cerium dioxide with a developed surface was obtained, which also frequently provides a change in the functional characteristics of the material. When synthesizing $La_{0.6}Sr_{0.4}Co_{0.2}Fe_{0.8}O_{3-\delta}$ oxide, due to large differences in the chemical properties of its constituent metals, the glycol–citrate method was used, which, in such cases, is the optimal approach to obtaining a single-phase nanomaterial of complex composition. The microstructure of the obtained oxide powder is representative of the products synthesized when using this method. As can be seen in the SEM images (Figure 1e,f), the material, due to the foaming of the reaction system, consisted of two-dimensional formations with lateral sizes up to tens of micrometers and thicknesses close to the average particle size. In our case, there was a bimodal distribution of particle sizes—in general, the films consisted of faceted grains sized 120 ± 15 nm, with smaller particles (sized about 20 ± 4 nm) on the surface. Thus, using hydrothermal and glycol–citrate methods, nanoscale oxides of the composition NiO, $(CeO_2)_{0.8}(Sm_2O_3)_{0.2}$ and $La_{0.6}Sr_{0.4}Co_{0.2}Fe_{0.8}O_{3-\delta}$ were synthesized with different microstructures—0D, 1D and 2D, respectively.

Using functional inks based on the obtained oxide nanopowders, NiO-SDC (the composite anode), SDC (the solid electrolyte) and LSCF-SDC (the composite cathode) coatings were sequentially applied to the surface of the specialized Pt/Al$_2$O$_3$/Pt chip by using microextrusion printing. Additionally, a (NiO-SDC)/SDC/(LSCF-SDC) three-layer structure was also fabricated on the Al$_2$O$_3$ substrate surface by using the same method according to the scheme of the functional layer arrangement (Figure 2 top). The appearance of the sample after applying each functional layer (Figure 2 bottom) confirms that the employed technique makes it possible to form both single- and multi-layer functional semiconductor oxide coatings with different chemical compositions, ensuring the targeting of the material and the specified geometry of the resulting planar nanostructures.

Figure 1. Microstructure of NiO (**a,b**), $(CeO_2)_{0.8}(Sm_2O_3)_{0.2}$ (**c,d**) and $La_{0.6}Sr_{0.4}Co_{0.2}Fe_{0.8}O_{3-\delta}$ (**e,f**) powders, used in the preparation of functional ink for microextrusion printing of oxide coatings.

The crystal structures of the coatings printed on the surfaces of substrates of different types were controlled by using the X-ray diffraction analysis (Figure 3). As can be seen in the X-ray diffraction patterns for the composite anodic coating (NiO-SDC) (Figure 3, sample 1), no chemical interaction between its components and the substrate material occurred during heat treatment—there were reflexes for all the corresponding constituents, while the nickel oxide retained the cubic crystal structure of the halite type (sp.gr. $Fm\text{-}3m$; the average size of the coherent scattering region (CSR) was 23 ± 2 nm), and for cerium dioxide, doped with samarium oxide, a cubic structure of the fluorite type was observed (sp.gr. $Fm\text{-}3m$; the average size of CSR is 11 ± 1 nm). The analysis of the two-layer coating of the composition (NiO-SDC)/SDC (Figure 3, sample 2) showed that, besides the reflexes of the $Pt/Al_2O_3/Pt$ substrate, only the signals of the $(CeO_2)_{0.80}(Sm_2O_3)_{0.20}$ electrolyte coating (average OCD size—11 ± 1 nm) were present, which confirms the complete overlapping of the composite electrode layer located below. The three-layer (NiO-SDC)/SDC/(LSCF-SDC) coatings applied to the surfaces of the specialized $Pt/Al_2O_3/Pt$ chip and the Al_2O_3 substrate (Figure 3, samples 3 and 4, respectively) also only showed a signal from the upper functional layer, the composite cathode, in the corresponding

XRD patterns. The type of crystal lattice and the value of the average CSR size for the components of this composite coating conform to the used powders of the corresponding composition ($La_{0.6}Sr_{0.4}Co_{0.2}Fe_{0.8}O_{3-\delta}$ has a perovskite-type cubic crystal structure, sp.gr. $Pm\text{-}3m$, and the average SCR size was 28 ± 3 nm; in the case of the $(CeO_2)_{0.80}(Sm_2O_3)_{0.20}$ oxide, the SCR size corresponded to the aforementioned value).

The microstructures of the oxide coatings applied sequentially to the surface of the Al_2O_3 substrate were studied using scanning electron microscopy (Figure 4). According to the SEM images, the NiO-SDC composite coating surface (Figure 4a) was porous and nanodispersed. The surface examination with the simultaneous use of secondary and backscattered electron detectors allowed for the presence of two phases in the composition of the studied material to be observed in the average atomic number distribution mode, and their target composition was confirmed using an energy-dispersive X-ray microanalysis. The analysis of the (NiO-SDC)/SDC two-layer coating (Figure 4b), where the upper layer was a $(CeO_2)_{0.80}(Sm_2O_3)_{0.20}$ solid electrolyte, showed that its microstructure was denser and that the material surface was characterized by the presence of anisotropic oxide particles, whose size and shape agree with the features of the powder used to prepare the corresponding functional inks. When studying the surface of the three-layer structure (Figure 4c), where the upper layer had the LSCF-SDC composition (the presence of the two corresponding phases was evident in the mode of the average atomic number distribution using the backscattered electron detector), an increased porosity, which is an important characteristic of solid oxide fuel cell electrodes to ensure their sufficient gas permeability, was noted.

Figure 2. Scheme of the (NiO-SDC)/SDC/(LSCF-SDC) three-layer structure formed (**top**) and the appearance of the functional layers (**bottom**) sequentially deposited on the Al_2O_3 substrate surface (left—composite anode layer, middle—electrolyte layer on the anode surface, right—composite cathode layer on the electrolyte surface).

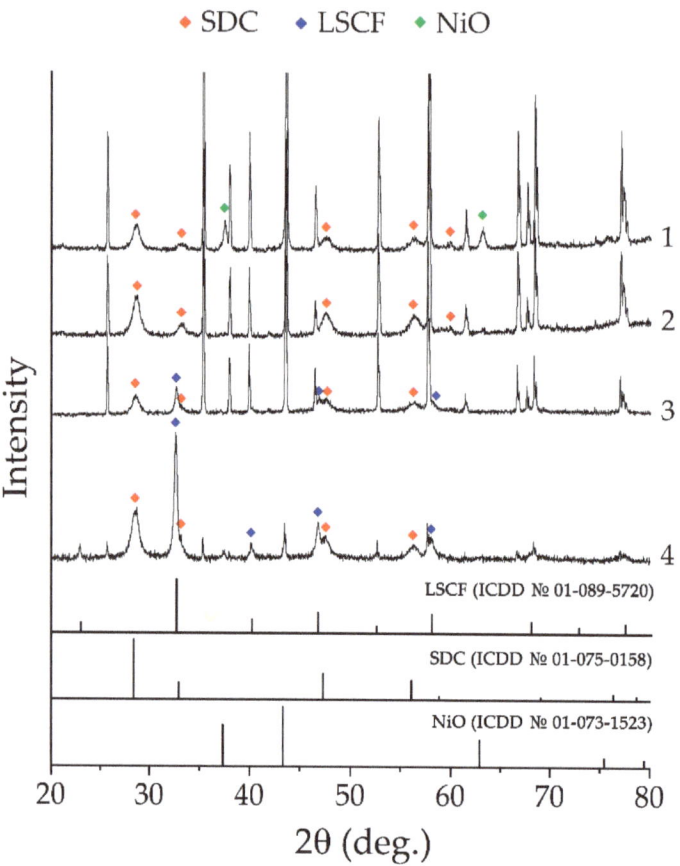

Figure 3. XRD patterns of single- ((NiO-SDC; 1), two- ((NiO-SDC)/SDC; 2), and three-layer ((NiO-SDC)/SDC/(LSCF-SDC); 3 and 4) coatings printed on the surface of Pt/Al$_2$O$_3$/Pt chip (1–3) and Al$_2$O$_3$ substrate (4). The unmarked reflections refer to the material of the corresponding substrate.

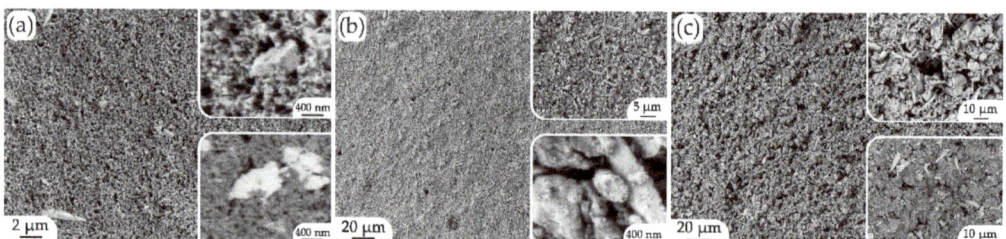

Figure 4. Surface microstructure of single- (**a**), two- (**b**) and three-layer (**c**) coatings deposited sequentially on an Al$_2$O$_3$ substrate: (**a**) NiO-SDC (top inset—secondary electron detector; bottom inset—reflected electron detector); (**b**) (NiO-SDC)/SDC; (**c**) (NiO-SDC)/SDC/(LSCF-SDC) (top inset—secondary electron detector; bottom inset—reflected electron detector).

Thus, it was shown that all oxide layers were nanoscale and did not contain defects in the form of delaminations, fractures or impurity inclusions that could negatively affect their functional properties.

To examine the microstructural features of the (NiO-SDC)/SDC/(LSCF-SDC) three-layer structure formed on the surface of the Al_2O_3 substrate, a corresponding cross-section was fabricated and further analyzed using scanning electron microscopy (Figure 5). As can be seen in the micrographs obtained with the backscattered electron detector, the differences in the chemical compositions of the functional layers appeared in phase contrast, which allowed us to observe clear interfacial boundaries. For the NiO-SDC and LSCF-SDC layers, their composite structures and the uniform distribution of $(CeO_2)_{0.80}(Sm_2O_3)_{0.20}$ particles in their volume were clearly visible. A color visualization (Cameo+ function) of the cross-section area of the three-layer oxide structure, which takes into account the results of the chemical element mapping on the studied surface, also clearly demonstrated the distinct boundaries between the separate layers and the absence of significant defects, and it also enabled a more accurate determination of the thickness of each layer, as well as confirming the compliance of light inclusions in the composition of the NiO-SDC and LSCF-SDC layers with $(CeO_2)_{0.80}(Sm_2O_3)_{0.20}$ solid solution particles. The element distribution maps of the cross-sectional area confirm that the lower functional layer mainly contained nickel oxide and samarium-doped ceria distributed in its volume. The second layer contained cerium and samarium, which is consistent with the expected structure of the material. The third layer contained lanthanum, iron and cobalt (the strontium signal was weaker), as well as cerium and samarium. Thus, the chemical element mapping of the surface of the studied cross-section confirms the formation of a three-layer structure of the composition (NiO-SDC)/SDC/(LSCF-SDC) with a thickness of about 36.5 μm. The two lower layers (NiO-SDC and SDC) were the same thickness (10.5 μm), and the upper composite layer (LSCF-SDC) was about 15.5 μm thick. Taking into account that each of these layers of different compositions was formed via microextrusion printing in a single pass, in general, such similar thickness parameters can be considered an important advantage of the employed approach.

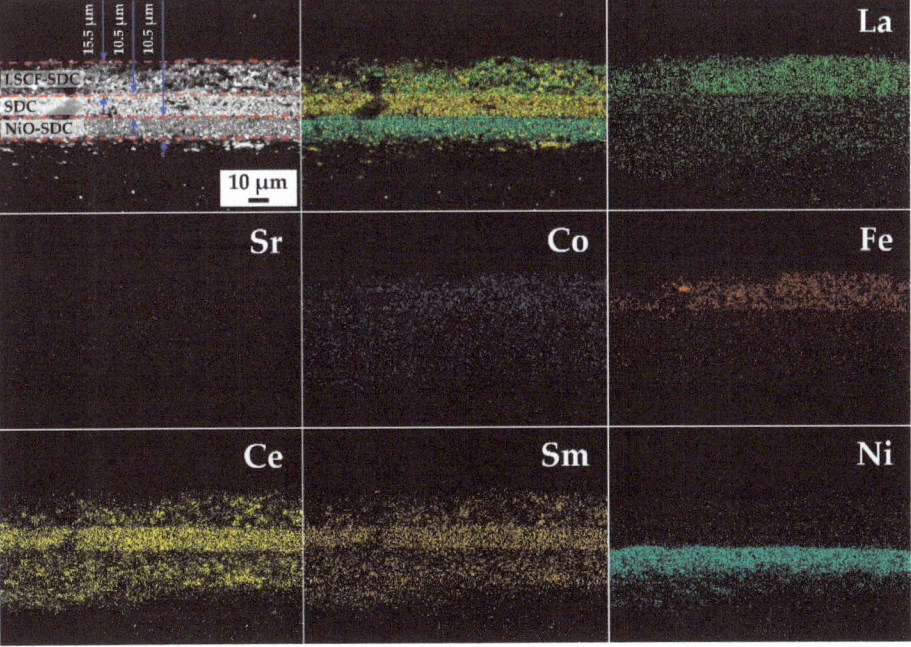

Figure 5. Microstructure of the cross-sectional area of the printed three-layer structure of the composition (NiO-SDC)/SDC/(LSCF-SDC), as well as chemical element distribution maps of its surface.

The morphology (the size and shape of particles and roughness) and local electrophysical characteristics (the work function of the material surface, distribution of charge carriers and capacitance values of the "probe tip—surface of the material" capacitor) of the oxide layers of different compositions formed successively on the Pt/Al$_2$O$_3$/Pt chip were studied by means of several AFM techniques. It was found that, generally, the microstructural characteristics of the particles in the oxide coatings are in good agreement with the corresponding parameters of the nanopowders used. Thus, in the case of the NiO-SDC coatings, particle agglomerates of about 200 nm in size were preserved in its composition (Figure 6a). According to the AFM results, when the second layer ((CeO$_2$)$_{0.80}$(Sm$_2$O$_3$)$_{0.20}$) was applied onto the NiO-SDC coating, the surface of the lower layer overlapped completely, which could be observed via a significant relief change: the surface was formed by elongated agglomerates from 200 nm to 2 µm long (Figure 6b). The oxide nanopowder particles used in the preparation of the corresponding functional inks had a similar microstructure. For the top layer of the (NiO-SDC)/SDC/(LSCF-SDC) structure, both the elongated agglomerates characteristic of samarium-doped ceria and the accumulation of less ordered nanoparticles characteristic of La$_{0.6}$Sr$_{0.4}$Co$_{0.2}$Fe$_{0.8}$O$_{3-\delta}$ oxide (Figure 6c) were observed. The roughness of the coatings varied in accordance with the microstructure features of a particular material. Thus, for the one-layer coating (NiO-SDC), the mean square roughness of an area of about 300 µm^2 was 58 nm; for the two-layer coating ((NiO-SDC)/SDC), it was 92 nm; and for the three-layer coating ((NiO-SDC)/SDC/(LSCF-SDC)), it was 70 nm. The increased surface roughness of the SDC layer is probably related to its composition of hierarchically organized agglomerates larger in size than the particles of the first layer. Surface potential distribution maps were built during the surface scanning of the oxide layers in the KPFM mode. One could see that, in the case of the NiO-SDC layer, there was a uniform potential distribution, which is usually due to the relatively high electronic conductivity of the material, probably ensured by the nickel oxide (Figure 6, middle column). For the second oxide layer surface, the existence of areas with an increased potential was noted, which indicates a reduced electronic conductivity, leading to the appearance of local accumulations of small charges. When studying the third layer surface, there was a noticeable contrast between the LSCF and SDC agglomerates in terms of the surface potential and a slight difference in the value of the work function (4.87 eV for the (CeO$_2$)$_{0.90}$(Sm$_2$O$_3$)$_{0.20}$ anisotropic agglomerates; 4.83 eV for the La$_{0.6}$Sr$_{0.4}$Co$_{0.2}$Fe$_{0.8}$O$_{3-\delta}$ particles). In general, the work function value of the studied material surface gradually decreased with each subsequent layer: for NiO-SDC, it was 5.02 eV; for the (NiO-SDC)/SDC two-layer coating, it was 4.93 eV; and for the three-layer structure of composition (NiO-SDC)/SDC/(LSCF-SDC), it was 4.83–4.87 eV (depending on the coating region). Concerning the SCM maps of the capacitance gradient distribution over the oxide layer surface (Figure 6, right column), one could observe lighter areas on the material grain boundaries in all cases. These are related to increased capacitance changes in the boundaries during the formation of depletion layers due to the higher concentration of charge carriers on the boundaries. This, in turn, indicates the grain-boundary mechanism of electronic conductivity in the oxide layers.

The electrical conductivity of the (NiO-SDC)/SDC/(LSCF-SDC) three-layer nanostructure was studied using electrochemical impedance spectroscopy. Figure 7a shows typical impedance hodographs of the sample under study—it can be seen that they have the form of semicircles slightly extended along the abscissa axis. The intercept of the impedance arc with the real axis at low frequencies represents the total resistance of the three-layer structure under consideration. It was found that the diameter of the semicircles decreased with an increase in the temperature, since the charge transfer process is more active, which leads to a decrease in the resistance of the sample. Using impedance spectroscopy data, the temperature dependence of the total conductivity of the material was estimated, and it was determined that the conductivity increased linearly by 1.5 orders of magnitude as the temperature was increased from 150 to 600 °C (Figure 7b). It is known that a high-density electrolyte layer is preferable when creating conventional SOFCs, since it helps to reduce the resistance of the fuel cell. In our case, the conductivity of the printed three-layer structure had a rather low value, but a significant increase in this parameter can be achieved by the additional sintering of the components at higher temperatures. One of the most important results of this study is the demonstration of the potential of the developed technology for the microextrusion printing of miniaturized multilayer hierarchically organized conducting nanostructures of a given geometry, which is promising for the evolution of SOFC miniaturization approaches.

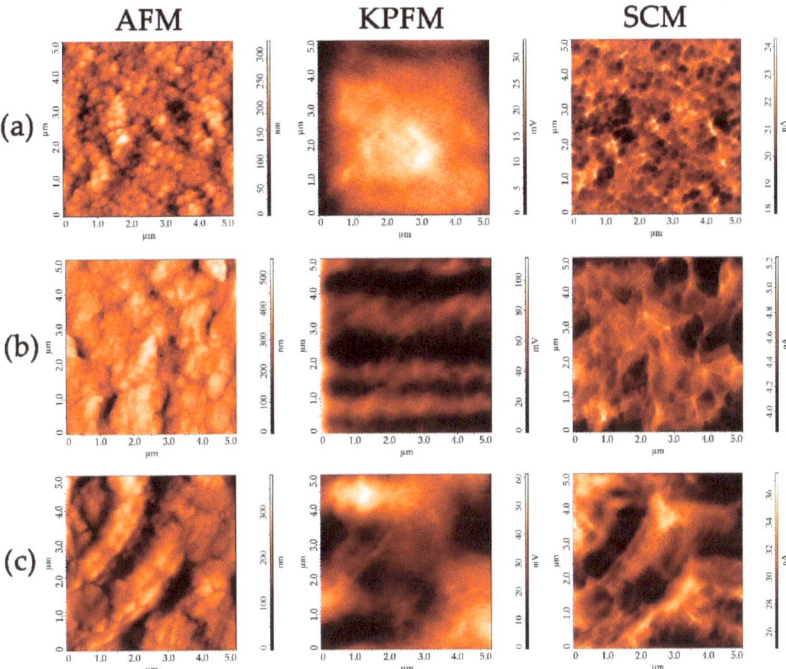

Figure 6. Topography of the coatings deposited sequentially on the Pt/Al$_2$O$_3$/Pt chip surface (left column), as well as surface potential (center) and capacity gradient (right) distribution maps for NiO-SDC (**a**), (NiO-SDC)/SDC (**b**) and (NiO-SDC)/SDC/(LSCF-SDC) layers (**c**).

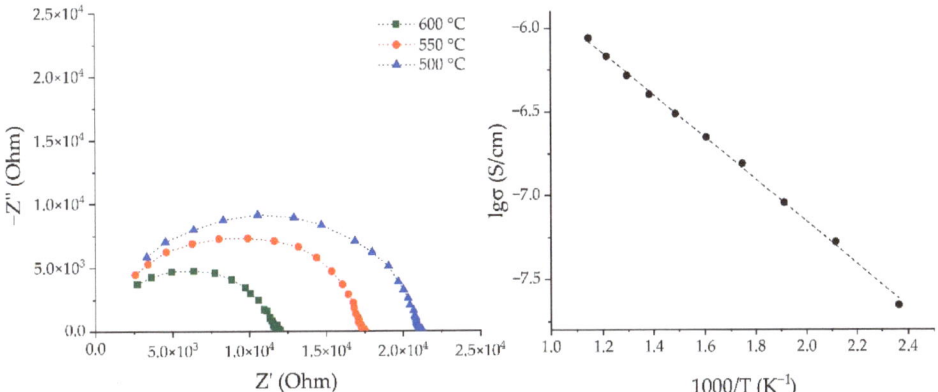

Figure 7. Frequency dependences of the impedance of the (NiO-SDC)/SDC/(LSCF-SDC) three-layer structure (**left**) and the temperature dependence of its specific conductivity (**right**).

4. Conclusions

In the course of this study, NiO, $La_{0.6}Sr_{0.4}Co_{0.2}Fe_{0.8}O_{3-\delta}$ (LSCF) and $(CeO_2)_{0.8}(Sm_2O_3)_{0.2}$ (SDC) nanopowders characterized by different microstructures—0D, 1D and 2D, respectively—were obtained using hydrothermal and glycol–citrate methods. It was shown that the nickel(II) oxide powder consisted of aggregated spherical nanoparticles sized 25 ± 3 nm. The LSCF was composed of two-dimensional formations with lateral sizes of up to tens of micrometers and a thickness of about 120 nm, with a bimodal particle size distribution—the films consisted mainly of faceted grains of 120 ± 15 nm with finer particles (about 20 ± 4 nm) on their surface. In the case of the SDC sample, the powder had a multilevel hierarchical self-organization of particles—nanoscale isotropic particles were ordered into one-dimensional structures about 5 μm long and 200–300 nm wide, combined into bundles, which, in turn, were coordinated at an angle of about 60° to each other to form six-point structures.

Using functional inks based on the obtained oxide nanopowders, the NiO-SDC, SDC and LSCF-SDC coatings were sequentially applied to the surfaces of various substrates using microextrusion printing, resulting in three-layer structures of the composition (NiO-SDC)/SDC/(LSCF-SDC).

According to the XRD data, a chemical interaction between the substrates and components of the formed functional layers of different compositions did not occur during additional heat treatment. NiO retained the crystal structure of the halite type (average CSR size was 23 ± 2 nm), SDC retained a structure of the fluorite type (11 ± 1 nm), and LSCF retained a structure of the perovskite type (28 ± 3 nm). Based on the SEM and AFM data, the microstructural features of the obtained oxide layers were inherited from the corresponding powders; the materials were nanoscale and contained no defects in the form of delaminations, fractures or impurity inclusions; and for the NiO-SDC and LSCF-SDC layers, the composite structures were confirmed. When studying the cross-sectional area of the three-layer structure of (NiO-SDC)/SDC/(LSCF-SDC) composition, element distribution maps were constructed, which confirmed the target compositions of all functional layers and distinct interfacial boundaries. The resulting thickness of the formed multilayer hierarchically organized structure was about 36.5 μm. The thicknesses of the two lower layers (NiO-SDC and SDC) were the same (10.5 μm), and the upper composite layer (LSCF-SDC) was about 15.5 μm thick. The resulting thickness of the formed multilayer hierarchically organized structure was about 36.5 μm. The thicknesses of the two lower layers (NiO-SDC and SDC) were the same (10.5 μm), and the upper composite layer (LSCF-SDC) was about 15.5 μm thick. Thus, microextrusion printing made it possible to achieve similar thickness values for the layers of different compositions in one pass in each case.

The SCM maps of the capacitance gradient distribution over the surface of the sequentially deposited oxide layers demonstrated that a shift in the charge carrier density to the boundaries between the grains occurred in all cases. Using KPFM, it was shown that the work function value of the studied material surface gradually decreased with the application of each successive layer: it was 5.02 eV for NiO-SDC, 4.93 eV for the two-layer coating (NiO-SDC)/SDC and 4.83–4.87 eV (depending on the coating region) for the (NiO-SDC)/SDC/(LSCF-SDC) three-layer structure.

For the printed three-layer nanostructure of composition (NiO-SDC)/SDC/(LSCF-SDC), the temperature dependence of conductivity was determined, and it was shown that the conductivity increased linearly by 1.5 orders of magnitude when the temperature was increased from 150 to 600 °C.

One of the key results of this study is the demonstration of the capability of the developed technology for the microextrusion printing of miniaturized multilayer hierarchically organized conducting nanostructures of a given geometry, which is promising for the development of SOFC miniaturization approaches.

Author Contributions: Conceptualization, N.P.S. and T.L.S.; investigation, T.L.S., N.P.S. and P.Y.G.; writing—original draft preparation, T.L.S., P.Y.G. and N.P.S.; writing—review and editing, T.L.S. and N.P.S.; visualization, T.L.S., N.P.S. and P.Y.G.; supervision, N.P.S., E.P.S. and N.T.K. All authors have read and agreed to the published version of the manuscript.

Funding: This work was supported by the Russian Science Foundation, project No. 21-73-00288, https://rscf.ru/en/project/21-73-00288/ (accessed on 19 November 2022).

Institutional Review Board Statement: Not applicable.

Informed Consent Statement: Not applicable.

Data Availability Statement: Not applicable.

Conflicts of Interest: The authors declare no conflict of interest.

References

1. Pirou, S.; Talic, B.; Brodersen, K.; Hauch, A.; Frandsen, H.L.; Skafte, T.L.; Persson, Å.H.; Høgh, J.V.T.; Henriksen, H.; Navasa, M.; et al. Production of a monolithic fuel cell stack with high power density. *Nat. Commun.* **2022**, *13*, 1263. [CrossRef]
2. Obara, S. Economic Performance of an SOFC Combined System with Green Hydrogen Methanation of Stored CO_2. *SSRN Electron. J.* **2022**, *262*, 125403. [CrossRef]
3. Alirahmi, M.; Behzadi, A.; Ahmadi, P.; Sadrizadeh, S. An innovative four-objective dragonfly-inspired optimization algorithm for an efficient, green, and cost-effective waste heat recovery from SOFC. *Energy* **2022**, *263*, 125607. [CrossRef]
4. Yu, D.; Hu, J.; Wang, W.; Gu, B. Comprehensive techno-economic investigation of biomass gasification and nanomaterial based SOFC/SOEC hydrogen production system. *Fuel* **2023**, *333*, 126442. [CrossRef]
5. Cammarata, A.; Díaz Lacharme, M.C.; Colbertaldo, P.; Donazzi, A.; Campanari, S. Numerical and experimental assessment of a novel SOFC-based system for micro-power generation. *J. Power Sources* **2022**, *551*, 232180. [CrossRef]
6. Chen, J.; Sun, S.; Chen, Y.; Zhang, H.; Lu, Z. Study on Model Evolution Method Based on the Hybrid Modeling Technology With Support Vector Machine for an SOFC-GT System. *J. Electrochem. Energy Convers. Storage* **2023**, *20*, 011015. [CrossRef]
7. Zhang, J.; Ricote, S.; Hendriksen, P.V.; Chen, Y. Advanced Materials for Thin-Film Solid Oxide Fuel Cells: Recent Progress and Challenges in Boosting the Device Performance at Low Temperatures. *Adv. Funct. Mater.* **2022**, *32*, 2111205. [CrossRef]
8. Alipour, S.; Sagir, E.; Sadeghi, A. Multi-criteria decision-making approach assisting to select materials for low-temperature solid oxide fuel cell: Electrolyte, cathode & anode. *Int. J. Hydrogen Energy* **2022**, *47*, 19810–19820. [CrossRef]
9. Chasta, G.; Himanshu; Dhaka, M.S. A review on materials, advantages, and challenges in thin film based solid oxide fuel cells. *Int. J. Energy Res.* **2022**, *46*, 14627–14658. [CrossRef]
10. Wang, Q.; Fan, H.; Xiao, Y.; Zhang, Y. Applications and recent advances of rare earth in solid oxide fuel cells. *J. Rare Earths* **2022**, *40*, 1668–1681. [CrossRef]
11. Maiti, T.K.; Majhi, J.; Maiti, S.K.; Singh, J.; Dixit, P.; Rohilla, T.; Ghosh, S.; Bhushan, S.; Chattopadhyay, S. Zirconia- and ceria-based electrolytes for fuel cell applications: Critical advancements toward sustainable and clean energy production. *Environ. Sci. Pollut. Res.* **2022**, *29*, 64489–64512. [CrossRef] [PubMed]
12. Simonenko, T.L.; Kalinina, M.V.; Simonenko, N.P.; Simonenko, E.P.; Khamova, T.V.; Shilova, O.A. Synthesis and Physicochemical Properties of Nanopowders and Ceramics in a CeO_2–Gd_2O_3 System. *Glas. Phys. Chem.* **2018**, *44*, 314–321. [CrossRef]
13. Egorova, T.L.; Kalinina, M.V.; Simonenko, E.P.; Simonenko, N.P.; Kopitsa, G.P.; Glumov, O.V.; Mel'nikova, N.A.; Murin, I.V.; Almásy, L.; Shilova, O.A. Study of the effect of methods for liquid-phase synthesis of nanopowders on the structure and physicochemical properties of ceramics in the CeO_2–Y_2O_3 system. *Russ. J. Inorg. Chem.* **2017**, *62*, 1275–1285. [CrossRef]
14. Bai, J.; Zhou, D.; Zhu, X.; Wang, N.; Chen, R.; Wang, B. New SOFC Cathode: 3D Core–Shell-Structured $La_{0.6}Sr_{0.4}Co_{0.2}Fe_{0.8}O_{3-\delta}$@$PrO_{2-\delta}$ Nanofibers Prepared by Coaxial Electrospinning. *ACS Appl. Energy Mater.* **2022**, *5*, 11178–11190. [CrossRef]
15. Ishfaq, H.A.; Khan, M.Z.; Mehran, M.T.; Raza, R.; Tanveer, W.H.; Bibi, S.; Hussain, A.; Muhammad, H.A.; Song, R. Boosting performance of the solid oxide fuel cell by facile nano-tailoring of $La_{0.6}Sr_{0.4}CoO_{3-\delta}$ cathode. *Int. J. Hydrogen Energy* **2022**, *47*, 37587–37598. [CrossRef]
16. Wang, P.; Cheng, J. Preparation and Performance of a $La_{0.6}Sr_{0.4}Co_xFe_{1-x}O_3$ Cathode for Solid Oxide Fuel Cells. *J. Electron. Mater.* **2022**, *51*, 6410–6415. [CrossRef]
17. Patil, S.P.; Jadhav, L.D.; Chourashiya, M. Investigation of quality and performance of Cu impregnated NiO-GDC as anode for IT-SOFCs. *Open Ceram.* **2022**, *9*, 3–9. [CrossRef]
18. Timurkutluk, C.; Bilgil, K.; Celen, A.; Onbilgin, S.; Altan, T.; Aydin, U. Experimental investigation on the effect of anode functional layer on the performance of anode supported micro-tubular SOFCs. *Int. J. Hydrogen Energy* **2022**, *47*, 19741–19751. [CrossRef]
19. Loureiro, F.J.A.; Macedo, D.A.; Nascimento, R.M.; Cesário, M.R.; Grilo, J.P.F.; Yaremchenko, A.A.; Fagg, D.P. Cathodic polarisation of composite LSCF-SDC IT-SOFC electrode synthesised by one-step microwave self-assisted combustion. *J. Eur. Ceram. Soc.* **2019**, *39*, 1846–1853. [CrossRef]
20. Lee, S.; Song, H.S.; Hyun, S.H.; Kim, J.; Moon, J. LSCF–SDC core–shell high-performance durable composite cathode. *J. Power Sources* **2010**, *195*, 118–123. [CrossRef]

21. Oveisi, S.; Khakpour, Z.; Faghihi-sani, M.A.; Kazemzad, M. Processing study of gel-cast tubular porous NiO/SDC composite materials from gel-combustion synthesized nanopowder. *J. Sol-Gel Sci. Technol.* **2021**, *97*, 581–592. [CrossRef]
22. Morales, M.; Laguna-Bercero, M.Á. Influence of Anode Functional Layers on Electrochemical Performance and Mechanical Strength in Microtubular Solid Oxide Fuel Cells Fabricated by Gel-Casting. *ACS Appl. Energy Mater.* **2018**, *1*, 2024–2031. [CrossRef]
23. Deepi, A.S.; Dharani Priya, S.; Samson Nesaraj, A.; Selvakumar, A.I. Component fabrication techniques for solid oxide fuel cell (SOFC)—A comprehensive review and future prospects. *Int. J. Green Energy* **2022**, *19*, 1600–1612. [CrossRef]
24. Evans, A.; Bieberle-Hütter, A.; Rupp, J.L.M.; Gauckler, L.J. Review on microfabricated micro-solid oxide fuel cell membranes. *J. Power Sources* **2009**, *194*, 119–129. [CrossRef]
25. Joo, J.H.; Choi, G.M. Simple fabrication of micro-solid oxide fuel cell supported on metal substrate. *J. Power Sources* **2008**, *182*, 589–593. [CrossRef]
26. Han, G.D.; Choi, H.J.; Bae, K.; Choi, H.R.; Jang, D.Y.; Shim, J.H. Fabrication of Lanthanum Strontium Cobalt Ferrite-Gadolinium-Doped Ceria Composite Cathodes Using a Low-Price Inkjet Printer. *ACS Appl. Mater. Interfaces* **2017**, *9*, 39347–39356. [CrossRef]
27. Bagishev, A.S.; Mal'bakhova, I.M.; Vorob'ev, A.M.; Borisenko, T.A.; Asmed'yanova, A.D.; Titkov, A.I.; Nemudryi, A.P. Layer-by-Layer Formation of the NiO/CGO Composite Anode for SOFC by 3D Inkjet Printing Combined with Laser Treatment. *Russ. J. Electrochem.* **2022**, *58*, 600–605. [CrossRef]
28. Rahumi, O.; Sobolev, A.; Rath, M.K.; Borodianskiy, K. Nanostructured engineering of nickel cermet anode for solid oxide fuel cell using inkjet printing. *J. Eur. Ceram. Soc.* **2021**, *41*, 4528–4536. [CrossRef]
29. Qu, P.; Xiong, D.; Zhu, Z.; Gong, Z.; Li, Y.; Li, Y.; Fan, L.; Liu, Z.; Wang, P.; Liu, C.; et al. Inkjet printing additively manufactured multilayer SOFCs using high quality ceramic inks for performance enhancement. *Addit. Manuf.* **2021**, *48*, 102394. [CrossRef]
30. Schnell, J.; Tietz, F.; Singer, C.; Hofer, A.; Billot, N.; Reinhart, G. Prospects of production technologies and manufacturing costs of oxide-based all-solid-state lithium batteries. *Energy Environ. Sci.* **2019**, *12*, 1818–1833. [CrossRef]
31. Erilin, I.S.; Agarkov, D.A.; Burmistrov, I.N.; Pukha, V.E.; Yalovenko, D.V.; Lyskov, N.V.; Levin, M.N.; Bredikhin, S.I. Aerosol deposition of thin-film solid electrolyte membranes for anode-supported solid oxide fuel cells. *Mater. Lett.* **2020**, *266*, 127439. [CrossRef]
32. Baek, S.W.; Jeong, J.; Schlegl, H.; Azad, A.K.; Park, D.S.; Baek, U.B.; Kim, J.H. Metal-supported SOFC with an aerosol deposited in-situ LSM and 8YSZ composite cathode. *Ceram. Int.* **2016**, *42*, 2402–2409. [CrossRef]
33. Li, D.; Zhang, X.; Liang, C.; Jin, Y.; Fu, M.; Yuan, J.; Xiong, Y. Study on durability of novel core-shell-structured $La_{0.8}Sr_{0.2}Co_{0.2}Fe_{0.8}O_{3-\delta}@Gd_{0.2}Ce_{0.8}O_{1.9}$ composite materials for solid oxide fuel cell cathodes. *Int. J. Hydrogen Energy* **2021**, *46*, 28221–28231. [CrossRef]
34. Dittrich, S.; Reitz, E.; Schell, K.G.; Bucharsky, E.C.; Hoffmann, M.J. Development and characterization of inks for screen printing of glass solders for SOFCs. *Int. J. Appl. Ceram. Technol.* **2020**, *17*, 1304–1313. [CrossRef]
35. Chen, J.; Yang, X.; Wan, D.; Li, B.; Lei, L.; Tian, T.; Chi, B.; Chen, F. Novel structured $Sm_{0.5}Sr_{0.5}CoO_{3-\delta}$ cathode for intermediate and low temperature solid oxide fuel cells. *Electrochim. Acta* **2020**, *341*, 136031. [CrossRef]
36. Baharuddin, N.A.; Abdul Rahman, N.F.; Abd. Rahman, H.; Somalu, M.R.; Azmi, M.A.; Raharjo, J. Fabrication of high-quality electrode films for solid oxide fuel cell by screen printing: A review on important processing parameters. *Int. J. Energy Res.* **2020**, *44*, 8296–8313. [CrossRef]
37. Fedorov, F.S.; Simonenko, N.P.; Trouillet, V.; Volkov, I.A.; Plugin, I.A.; Rupasov, D.P.; Mokrushin, A.S.; Nagornov, I.A.; Simonenko, T.L.; Vlasov, I.S.; et al. Microplotter-Printed On-Chip Combinatorial Library of Ink-Derived Multiple Metal Oxides as an "Electronic Olfaction" Unit. *ACS Appl. Mater. Interfaces* **2020**, *12*, 56135–56150. [CrossRef]
38. Simonenko, T.L.; Simonenko, N.P.; Gorobtsov, P.Y.; Vlasov, I.S.; Solovey, V.R.; Shelaev, A.V.; Simonenko, E.P.; Glumov, O.V.; Melnikova, N.A.; Kozodaev, M.G.; et al. Microplotter printing of planar solid electrolytes in the CeO_2–Y_2O_3 system. *J. Colloid Interface Sci.* **2021**, *588*, 209–220. [CrossRef]
39. Salmon, A.; Lavancier, M.; Brulon, C.; Coudrat, L.; Fix, B.; Ducournau, G.; Peretti, R.; Bouchon, P. Rapid prototyping of flexible terahertz metasurfaces using a microplotter. *Opt. Express* **2021**, *29*, 8617. [CrossRef]
40. Seo, H.; Iwai, H.; Kishimoto, M.; Ding, C.; Saito, M.; Yoshida, H. Microextrusion printing for increasing electrode–electrolyte interface in anode-supported solid oxide fuel cells. *J. Power Sources* **2020**, *450*, 227682. [CrossRef]
41. Seo, H.; Nishi, T.; Kishimoto, M.; Ding, C.; Iwai, H.; Saito, M.; Yoshida, H. Study of Microextrusion Printing for Enlarging Electrode–Electrolyte Interfacial Area in Anode-Supported SOFCs. *ECS Trans.* **2019**, *91*, 1923–1931. [CrossRef]
42. Kim, D.W.; Yun, U.J.; Lee, J.W.; Lim, T.H.; Lee, S.B.; Park, S.J.; Song, R.H.; Kim, G. Fabrication and operating characteristics of a flat tubular segmented-in-series solid oxide fuel cell unit bundle. *Energy* **2014**, *72*, 215–221. [CrossRef]
43. Simonenko, T.L.; Simonenko, N.P.; Gorobtsov, P.Y.; Klyuev, A.L.; Grafov, O.Y.; Ivanova, T.M.; Simonenko, E.P.; Sevastyanov, V.G.; Kuznetsov, N.T. Hydrothermally synthesized hierarchical $Ce_{1-x}Sm_xO_{2-\delta}$ oxides for additive manufacturing of planar solid electrolytes. *Ceram. Int.* **2022**, *48*, 22401–22410. [CrossRef]
44. Mokrushin, A.S.; Simonenko, T.L.; Simonenko, N.P.; Gorobtsov, P.Y.; Bocharova, V.A.; Kozodaev, M.G.; Markeev, A.M.; Lizunova, A.A.; Volkov, I.A.; Simonenko, E.P.; et al. Microextrusion printing of gas-sensitive planar anisotropic NiO nanostructures and their surface modification in an H_2S atmosphere. *Appl. Surf. Sci.* **2022**, *578*, 151984. [CrossRef]
45. Timurkutluk, B.; Ciflik, Y.; Altan, T.; Genc, O. Synthetical designing of solid oxide fuel cell electrodes: Effect of particle size and volume fraction. *Int. J. Hydrogen Energy* **2022**, *47*, 31446–31458. [CrossRef]

46. Solodkyi, I.; Borodianska, H.; Sakka, Y.; Vasylkiv, O. Effect of Grain Size on the Electrical Properties of Samaria-Doped Ceria Solid Electrolyte. *J. Nanosci. Nanotechnol.* **2012**, *12*, 1871–1879. [CrossRef]
47. Yao, X.; Li, P.; Yu, B.; Yang, F.; Li, J.; Zhao, Y.; Li, Y. Hydrothermally synthesized NiO-samarium doped ceria nano-composite as an anode material for intermediate-temperature solid oxide fuel cells. *Int. J. Hydrogen Energy* **2017**, *42*, 22192–22200. [CrossRef]
48. Jomjaree, T.; Sintuya, P.; Srifa, A.; Koo-amornpattana, W.; Kiatphuengporn, S.; Assabumrungrat, S.; Sudoh, M.; Watanabe, R.; Fukuhara, C.; Ratchahat, S. Catalytic performance of Ni catalysts supported on CeO_2 with different morphologies for low-temperature CO_2 methanation. *Catal. Today* **2020**, *375*, 234–244. [CrossRef]
49. Lee, M.J.; Hong, S.K.; Choi, B.H.; Hwang, H.J. Fabrication and performance of solid oxide fuel cell anodes from core-shell structured Ni/yttria-stabilized zirconia (YSZ) powders. *Ceram. Int.* **2016**, *42*, 10110–10115. [CrossRef]
50. Kim, J.; Choi, B.H.; Kang, M. Physicochemical properties of cubic Ni complex powders synthesized using urotropine chelating ligand for solid oxide fuel cells. *Adv. Powder Technol.* **2014**, *25*, 609–614. [CrossRef]
51. Yamamoto, K.; Hashishin, T.; Matsuda, M.; Qiu, N.; Tan, Z.; Ohara, S. High-performance Ni nanocomposite anode fabricated from Gd-doped ceria nanocubes for low-temperature solid-oxide fuel cells. *Nano Energy* **2014**, *6*, 103–108. [CrossRef]
52. Majumder, D.; Chakraborty, I.; Mandal, K. Room temperature blooming of CeO_2 3D nanoflowers under sonication and catalytic efficacy towards CO conversion. *RSC Adv.* **2020**, *10*, 22204–22215. [CrossRef] [PubMed]
53. Pan, C.; Zhang, D.; Shi, L.; Fang, J. Template-Free Synthesis, Controlled Conversion, and CO Oxidation Properties of CeO_2 Nanorods, Nanotubes, Nanowires, and Nanocubes. *Eur. J. Inorg. Chem.* **2008**, *2008*, 2429–2436. [CrossRef]
54. Gong, J.; Meng, F.; Yang, X.; Fan, Z.; Li, H. Controlled hydrothermal synthesis of triangular CeO_2 nanosheets and their formation mechanism and optical properties. *J. Alloys Compd.* **2016**, *689*, 606–616. [CrossRef]
55. Ahuja, A.; Gautam, M.; Sinha, A.; Sharma, P.K.; Patro, P.K.; Venkatasubramanian, A. Effect of processing route on the properties of LSCF-based composite cathode for IT-SOFC. *Bull. Mater. Sci.* **2020**, *43*, 129. [CrossRef]
56. Kumar, S.A.; Kuppusami, P.; Vengatesh, P. Auto-combustion synthesis and electrochemical studies of $La_{0.6}Sr_{0.4}Co_{0.2}Fe_{0.8}O_{3-\delta}$—$Ce_{0.8}Sm_{0.1}Gd_{0.1}O_{1.90}$ nanocomposite cathode for intermediate temperature solid oxide fuel cells. *Ceram. Int.* **2018**, *44*, 21188–21196. [CrossRef]
57. Simonenko, T.L.; Simonenko, N.P.; Simonenko, E.P.; Sevastyanov, V.G.; Kuznetsov, N.T. Obtaining of $La_{0.6}Sr_{0.4}Co_{0.2}Fe_{0.8}O_{3-\delta}$ Nanopowder Using the Glycol–Citrate Method. *Russ. J. Inorg. Chem.* **2021**, *66*, 477–481. [CrossRef]

Disclaimer/Publisher's Note: The statements, opinions and data contained in all publications are solely those of the individual author(s) and contributor(s) and not of MDPI and/or the editor(s). MDPI and/or the editor(s) disclaim responsibility for any injury to people or property resulting from any ideas, methods, instructions or products referred to in the content.

Article

Polishing Performance and Removal Mechanism of Core-Shell Structured Diamond/SiO₂ Abrasives on Sapphire Wafer

Guangen Zhao [1], Yongchao Xu [1,*], Qianting Wang [1], Jun Liu [1], Youji Zhan [2] and Bingsan Chen [2]

[1] School of Materials Science and Engineering, Fujian University of Technology, Fuzhou 350118, China
[2] Fujian Key Laboratory of Intelligent Machining Technology and Equipment, Fujian University of Technology, Fuzhou 350118, China
* Correspondence: 19862091@fjut.edu.cn

Abstract: Corrosive and toxic solutions are normally employed to polish sapphire wafers, which easily cause environmental pollution. Applying green polishing techniques to obtain an ultrasmooth sapphire surface that is scratch-free and has low damage at high polishing efficiency is a great challenge. In this paper, novel diamond/SiO₂ composite abrasives were successfully synthesized by a simplified sol-gel strategy. The prepared composite abrasives were used in the semi-fixed polishing technology of sapphire wafers, where the polishing slurry contains only deionized water and no other chemicals during the whole polishing process, effectively avoiding environmental pollution. The experimental results showed that diamond/SiO₂ composite abrasives exhibited excellent polishing performance, along with a 27.2% decrease in surface roughness, and the material removal rate was increased by more than 8.8% compared with pure diamond. Furthermore, through characterizations of polished sapphire surfaces and wear debris, the chemical action mechanism of composite abrasives was investigated, which confirmed the solid-state reaction between the SiO₂ shell and the sapphire surface. Finally, applying the elastic-plastic contact model revealed that the reduction of indentation depth and the synergistic effect of chemical corrosion and mechanical removal are the keys to improving polishing performance.

Citation: Zhao, G.; Xu, Y.; Wang, Q.; Liu, J.; Zhan, Y.; Chen, B. Polishing Performance and Removal Mechanism of Core-Shell Structured Diamond/SiO₂ Abrasives on Sapphire Wafer. *Micromachines* **2022**, *13*, 2160. https://doi.org/10.3390/mi13122160

Academic Editor: Kun Li

Received: 13 November 2022
Accepted: 22 November 2022
Published: 7 December 2022

Publisher's Note: MDPI stays neutral with regard to jurisdictional claims in published maps and institutional affiliations.

Copyright: © 2022 by the authors. Licensee MDPI, Basel, Switzerland. This article is an open access article distributed under the terms and conditions of the Creative Commons Attribution (CC BY) license (https://creativecommons.org/licenses/by/4.0/).

Keywords: polishing; sapphire; core shell; composite abrasives; surface roughness; material removal rate

1. Introduction

Sapphire, composed of single crystal alumina oxide (α-Al₂O₃), is an ideal material for infrared windows and aerospace [1,2] and is the main substrate material for optoelectronic devices, large-scale integrated circuits [3–6], and superconducting films due to its excellent mechanical and optical properties, such as high hardness, strong light transmittance, and stable chemical inertness. In particular, as the substrate material of GaN-based light-emitting diodes (LEDs), sapphire wafers have strict requirements for processing accuracy and surface quality, including nanoscale surface roughness, damage-free, and scratch-free [7]. However, given the high hardness and chemical inertia of sapphire [8,9], it's a great challenge to achieve satisfactory processing results.

The widely used free abrasive polishing is a traditional material removal strategy, which can provide a smooth surface in the field of electronic device substrate manufacturing [10]. However, because of the high hardness and brittleness of sapphire, the free abrasive process has the disadvantages of uncontrollable trajectory [11], low removal efficiency, and easy agglomeration, which will undoubtedly affect the polishing effect, resulting in high roughness and heavy damage on sapphire surfaces [12]. In addition, the free abrasive polishing of sapphire wafers usually applies strong acids, alkalis and toxic chemicals, leading to environmental pollution [13]. Among the reported nontraditional polishing technologies, the semi-fixed polishing pad using diamond abrasive has attracted great attention, which effectively avoids the problems of uncontrollable trajectory and the agglomeration of free abrasive [14]. The abrasives in the semi-fixed pad exhibited "yielding effects", for which surface damage and scratches induced by the larger abrasives can be reduced or even eliminated. Furthermore, in the whole polishing process, the slurry contains only deionized water, without any other chemicals, effectively avoiding environmental pollution. Hence, a semi-fixed abrasive polishing pad is one of the most promising polishing tools for processing hard and brittle materials such as sapphire, SiC, GaN, etc. [15]. Nevertheless, the relatively soft polishing pad inevitably reduces the polishing efficiency of the inner abrasive, making it difficult to achieve perfect surface quality with a high material removal rate (MRR) even when using hard abrasives. In order to overcome these problems, scholars have conducted in-depth research on polishing abrasives and made lots of significant progress.

Recent advances in composite abrasives with core-shell structures have provided a new direction for obtaining a supersmooth surface and a high MRR [16–19]. The core-shell composite abrasives overcome the limitations of single hard abrasives with many deep scratches and single soft abrasives with a low material removal rate, and they give full play to the excellent characteristics of different abrasives. Lu et al. [20] developed novel diamond/akageneite composite abrasives with a core-shell structure, which possessed stronger adhesion to the semi-fixed polishing pad, and the surface quality of sapphire has also been improved. Specifically, the surface roughness (Ra) of sapphire polished by diamond/akageneite composite abrasives was reduced from 1.70 nm to 1.39 nm, which is about 12.6% lower than that of pure diamond. The MRR of the diamond/akageneite composite abrasives is similar to that of diamond, which is about 0.28 nm/min. In our previous work [21], the prepared Al_2O_3/SiO_2 composite abrasives achieved excellent polishing performance, along with a 20.2% decrease in surface roughness, and the MRR was increased by more than 5.1% compared with pure Al_2O_3. The improvement in the MRR may be attributed to the solid-state reaction of the SiO_2 shell with the sapphire surface during polishing, resulting in a softened layer that can be easily removed by the mechanical action of the hard core. With a Mohs hardness of 10, diamond has a higher removal effect than Al_2O_3 (Mohs hardness of 9), which means that using diamond as the core material is expected to further improve MRR. However, the polishing performance of diamond/SiO_2 composite abrasives on the sapphire wafer has not been reported. In addition, research on the polishing behavior and material removal mechanism of core-shell composite abrasives is still insufficient.

In this study, the diamond/SiO_2 composite abrasives were successfully synthesized via a simplified sol-gel method and then characterized by field emission scanning electron microscopy (FESEM), transmission electron microscope (TEM) and EDS energy spectrum, X-ray diffraction (XRD), and Fourier transform infrared spectra (FT-IR), respectively. Subsequently, the polishing performance of pure diamond and diamond/SiO_2 composite abrasives on sapphire wafers was explored by using semi-fixed abrasive polishing pad under the same polishing parameters. The polishing results were investigated from the aspects of surface morphology, surface roughness, the material removal rate (MRR), and residual stress. Finally, combined with TEM and X-ray photoelectron spectroscopy (XPS), the polishing behavior and material removal mechanism of composite abrasives on sapphire wafers were discussed in terms of mechanical action and chemical corrosion.

2. Materials and Experimental Methods

2.1. Chemicals and Materials

Commercial diamond particles as the core material of composite abrasives, with a nominal particle size of 3 μm, were supplied by Yvxing Micro diamond Co. Ltd. (Zhengzhou, China). Tetraethyl orthosilicate (TEOS, AR), offered by Shanghai Yien Chemical Technology Co., Ltd. (Shanghai, China), was used as raw material to provide silicon shell through hydrolysis polycondensation and other reactions. Other chemicals, including ammonia solution ($NH_3·H_2O$, 25–28%) and absolute ethanol (C_2H_5OH, AR), were purchased from Shanghai Chemical Reagent Co., Ltd. (Shanghai, China).

2.2. Synthesis of Diamond/SiO$_2$ Composite Abrasives

Diamond/SiO$_2$ core-shell composite abrasives were synthesized via a facile sol-gel strategy on the basis of the hydrolysis and polycondensation reaction of TEOS. Firstly, a certain amount of diamond particles was added to the beaker containing absolute ethanol and was dispersed under ultrasound for 20 min until a uniform suspension was formed. Afterward, the diamond suspension was transferred to a thermostatic water bath; under continuous stirring, ammonia solution and deionized water were slowly dropped into the diamond suspension. After magnetic stirring at 30 °C for 15 min, TEOS was drop by drop added into the above mixed solution. The reaction was carried out at 30 °C for 12 h, during which magnetic stirring at low speed was maintained. Subsequently, the resultant precipitates were collected by centrifugation and washed three times with deionized water and anhydrous ethanol, separately. Finally, they were dried at 60 °C for 12 h, with which diamond/SiO$_2$ core-shell composite abrasives were obtained.

By means of a simplified sol-gel method, tetraethyl orthosilicate (TEOS, Si (OC$_2$H$_5$)$_4$) was catalyzed by ammonia to form a SiO$_2$ shell on the surface of diamond particles through hydrolysis and polycondensation. The formation of the SiO$_2$ shell can be summarized by the following chemical equations [22]:

$$Si(OC_2H_5)_4 + n\, H_2O \longrightarrow Si(OC_2H_5)_{4-n}(OH)_n + n\, C_2H_5OH \tag{1}$$

$$\underset{\underset{OH}{|}}{\overset{\overset{OH}{|}}{HO-Si-OH}} + \underset{\underset{OH}{|}}{\overset{\overset{OH}{|}}{HO-Si-OH}} \longrightarrow \underset{\underset{OH}{|}}{\overset{\overset{OH}{|}}{HO-Si}}-O-\underset{\underset{OH}{|}}{\overset{\overset{OH}{|}}{Si-OH}} + H_2O \tag{2}$$

$$\underset{\underset{OH}{|}}{\overset{\overset{OH}{|}}{HO-Si-OH}} + \underset{\underset{OC_2H_5}{|}}{\overset{\overset{OC_2H_5}{|}}{C_2H_5O-Si-OC_2H_5}} \longrightarrow \underset{\underset{OH}{|}}{\overset{\overset{OH}{|}}{HO-Si}}-O-\underset{\underset{OH}{|}}{\overset{\overset{OH}{|}}{Si-OH}} + C_2H_5OH \tag{3}$$

$$\text{Diamond} + n\,(Si-O-Si) \longrightarrow \text{Diamond} - (-Si-O-Si-)_n$$

Combined with the above chemical formulas, TEOS hydrolyzes to generate silanol, which can also be regarded as the release of active monomers, leading to a nucleation phenomenon [23]. On account of the existence of multiple O-H functional groups on the surface of diamond particles, coupled with its high specific surface area and high specific surface energy, the O-H functional groups are preferentially combined with silanol groups through a condensation reaction, which is the key for SiO$_2$ to nucleate on the surface of diamond. Subsequently, as-nucleated SiO$_2$ particles slowly gathered on the surface of diamond particles to form a discontinuous coating. Given the colloidal stability, there is a balance between the hydrolysis of TEOS and the primary particles. In other words, with the progress of reaction, the active monomer increases continuously, and the SiO$_2$ seeds on the diamond surface grow further to form a uniform and dense SiO$_2$ shell.

2.3. Characterizations

FESEM (NovaNanoSEM450, FEI Ltd., Natural Bridge Station, VA, USA) was employed to observe the external morphology of pure diamond and composite abrasives. The microstructure and elemental composition of abrasives were characterized by using the TEM (JEM-2100, JEOL Ltd., Tokyo, Japan) equipped with EDS (Oxford X-MaxN, Oxford Instruments Ltd., Abingdon, UK). XRD (D8ADVANCE, Bruker Ltd., Ettlingen, Germany) was adopted to analyze the phase components of pristine and composited abrasives by using Cu Kα radiation. To interpret the surface functional groups of abrasives, the spectra were measured within the wavenumber range of 400 cm^{-1} to 4000 cm^{-1} on FT-IR (Nicolet 6700, Thermo Ltd., Waltham, MA, USA).

2.4. Polishing Tests

Commercially available single crystal sapphire wafers (C-M plane; two-inch in diameter; Mohs hardness of 9) with original surface roughness values (Ra) of 10 ± 1 nm were bought from Wuxi Jingdian Semiconductor Material Co., Ltd. (Wuxi, China). Polishing texts of sapphire wafers were conducted on a rotary-type polishing tester (UNIPOL-1200S, Shenyang Kejing Co., Ltd., Shenyang, China) with a semi-fixed abrasive polishing pad. The abrasives (5 wt%) were evenly dispersed in a flexible matrix with unsaturated resin as the main component through full mixing and stirring, and then the semi-fixed polishing pad with a diameter of 300 mm was prepared through processes such as screeding and curing. Pure diamond and diamond/SiO_2 core-shell composite abrasives were separately used as abrasives for the semi-fixed flexible polishing pad. Before polishing, sapphire wafers were several times ultrasonically cleaned in deionized water and absolute ethanol to remove natural oxides and pollutants from their surfaces. The polishing parameters were described as follows: the polishing pressure was set as 5 kg, the polishing time was 3 h, and the rotation speed of the workpiece and that of the polishing pad were 60 rpm and 120 rpm, respectively. No chemicals were used in the polishing process, and only deionized water was used as coolant. After polishing, the sapphire wafers were cleaned repeatedly with deionized water and ethanol under sonication, and then they were dried in a drying oven. The schematics of the polishing process are shown in Figure 1.

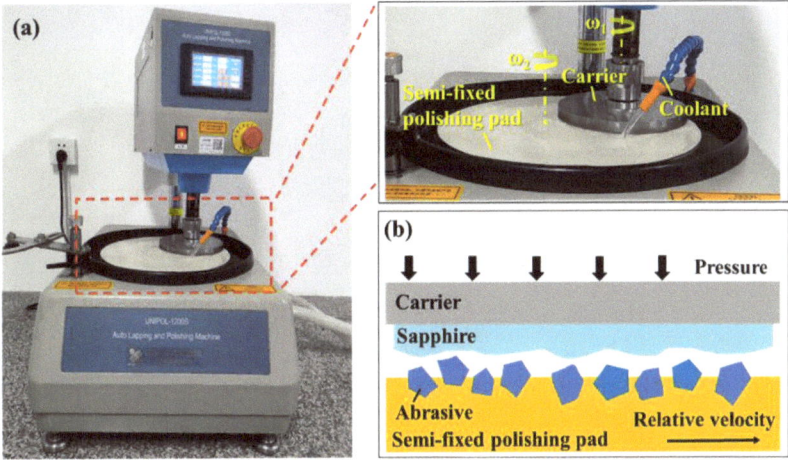

Figure 1. (a) Device diagram and (b) schematic diagram of the polishing process.

The atomic force microscope (AFM, Dimension Icon, Bruker Ltd., Germany) was utilized to investigate the surface topography and profile curve of the sapphire wafers before and after polishing, with an accuracy of 0.01 nm. During the polishing process, the contact surface roughness meter (MarSurf GD25, Mahr Ltd., Germany), with 0.1 nm accuracy, was used to measure the surface roughness (Ra) of workpieces every half an hour. For each machined workpiece, surface roughness (Ra) is the average value of 10 areas evenly distributed on the sapphire wafer surface.

To further analyze the material removal mechanism of sapphire wafers, the surface elements and existing forms of polished sapphire wafers were characterized by XPS (ESCALAB 250XI, Thermo Ltd., USA). TEM was used to analyze the wear debris removed from the surface of sapphire wafers during polishing. The material removal rate (nm/min) was calculated by Equation (4), and the masses (the average value of three measurements) of sapphire wafers before and after polishing were tested by an electron balance with 0.01 mg precision (GE0505, Shanghai YoKe, Ltd., Shanghai, China).

$$MRR = \frac{10^7 \times \Delta m}{\rho \times 2.54^2 \times \pi \times t} \quad (4)$$

Here, Δm (mg) is the mass loss of sapphire wafers before and after polishing, ρ (3.98 g/cm^3) is the sapphire density, and t (min) is the polishing time.

3. Results and Discussion

3.1. Characterizations of the Diamond/SiO$_2$ Composite Abrasives

FESEM images of pure diamond and diamond/SiO$_2$ composite abrasives are illustrated in Figure 2. The morphology of pure diamond shown in Figure 2a demonstrates that the pristine diamond particles have a uniform shape and smooth surface. As can be seen in Figure 2b, the prepared diamond/SiO$_2$ composite abrasives have good dispersion and no agglomeration. It is also noticed that there are no SiO$_2$ microspheres nucleated separately in Figure 2b, suggesting that the deposition of SiO$_2$ onto the diamond occurred in the form of a SiO$_2$ network rather than SiO$_2$ particles.

Figure 2. FESEM images of (**a**) pure diamond and (**b**) diamond/SiO$_2$ abrasive particles.

To give a better understanding the microstructure and elemental composition of diamond/SiO$_2$ composite abrasives, TEM images and an EDS spectrum of pure diamond and diamond/SiO$_2$ composite abrasives are demonstrated in Figure 3. The TEM images and EDS spectrum exhibited in Figure 3a clearly show that the surface of pure diamond is smooth and not covered with other impurities. As presented in the HRTEM image, the lattice fringes of the detected area are clearly visible, which reveals that the pure diamond is crystalline and that the lattice fringe spacing is 0.206 nm, corresponding to the (111) crystal plane. From the EDS elemental map of pure diamond, C and Cu elements could be found in the detection area. Among them, part of C element comes from diamond, the other part stems from carbon film, and Cu element is attributed to copper mesh. From the above results, the pure diamond has high purity and no impurities on the surface, which is conducive to the coating process of diamond abrasive.

Figure 3b exhibits the microstructure and elemental composition of diamond/SiO$_2$ composite abrasives. It is noticed that there is an obvious boundary between the core and shell [24]; under the HRTEM, the lattice fringes can be observed in the core; and the spacing of lattice fringes is 0.206 nm, which corresponds to the pure diamond in Figure 3a, whereas the fact that the shell has no lattice fringes confirms the amorphous structure of coating layer. Furthermore, the coating layer is uniform and dense, with a thickness of 10 nm. The EDS spectrum of diamond/SiO$_2$ composite abrasives is illustrated in Figure 3b. Compared with pure diamond, besides C and Cu elements, Si and O elements could be found in the composite abrasives. Combined with the TEM images, it can be explained that the coating layer is amorphous SiO$_2$.

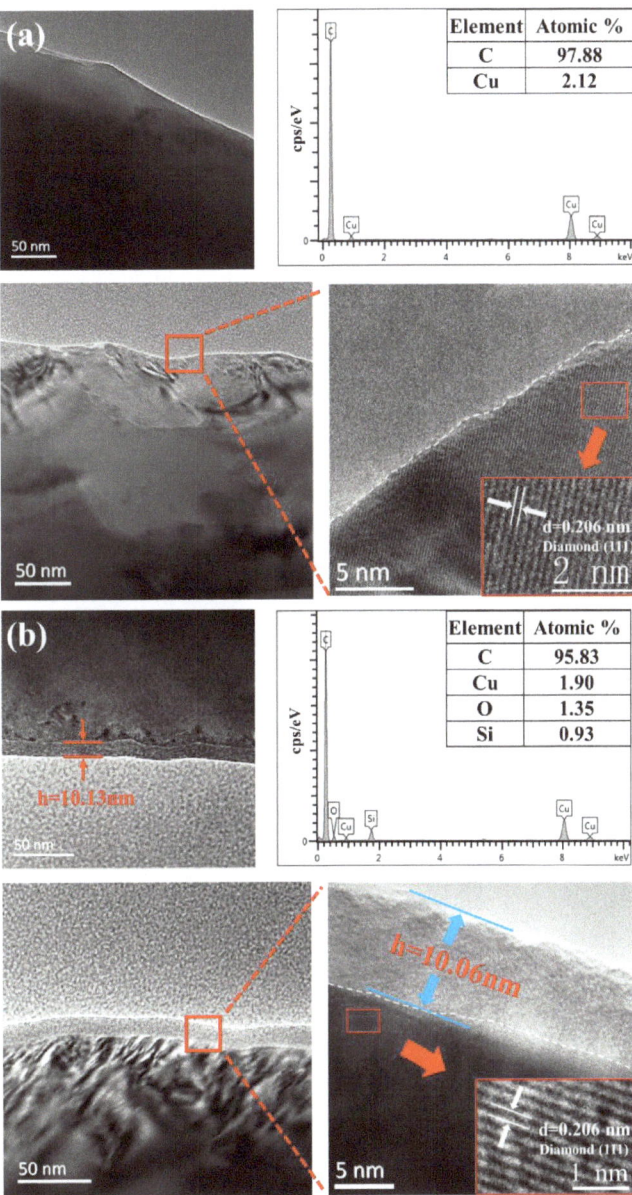

Figure 3. TEM images and EDS spectrum of (**a**) pure diamond and (**b**) diamond/SiO$_2$ abrasive particles.

An attempt can be made to research the surface functional groups of abrasives by using FTIR spectroscopy. Figure 4 presents the FT-IR spectra of pure diamond and diamond/SiO$_2$ composite abrasives. For pure diamond, the absorption peaks around 3461 cm^{-1} and 1632 cm^{-1} are attributed to O-H stretching and bending vibrations of adsorbed water [25], respectively. In contrast, the peak at 1076 cm^{-1} appears in the spectra of diamond/SiO$_2$ composite abrasives, which is ascribed to the asymmetric stretch vibration of Si-O-Si [26], indicating that SiO$_2$ is grafted on the surface of the diamond.

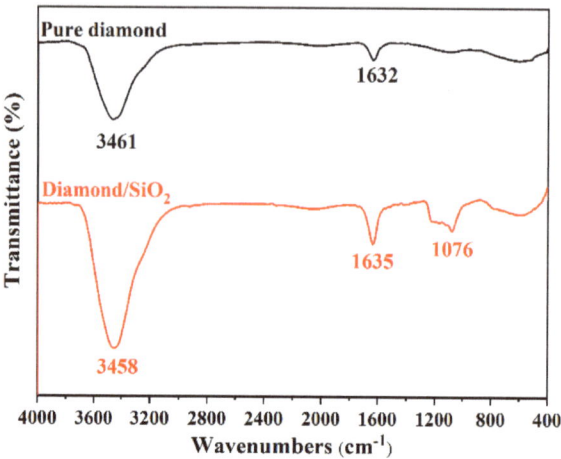

Figure 4. FT-IR spectra of pure diamond and diamond/SiO$_2$ composite abrasives.

The XRD patterns of pure diamond, amorphous SiO$_2$, and diamond/SiO$_2$ composite abrasives are demonstrated in Figure 5. As we can see, the characteristic diffraction peaks at 2-theta = 43.9° and 75.3° correspond to the (111) and (220) lattice planes [20], respectively, which could be indexed as the diamond standard card (PDF#06-0675). The diffraction peak of amorphous SiO$_2$ is a wide and low diffusion peak near 24.3°. The feature peaks of diamond/SiO$_2$ composite abrasives match well with all the diffraction peaks, suggesting that the composite abrasives both contain crystalline diamond and amorphous SiO$_2$, and they have a stable interfacial bonding in core-shell structure, which corresponds to the TEM characterization results. From the above characterization results, it can be seen that diamond/SiO$_2$ core-shell structure composite abrasives were successfully synthesized and that amorphous SiO$_2$ with a thickness of about 10 nm was closely coated on the diamond surface.

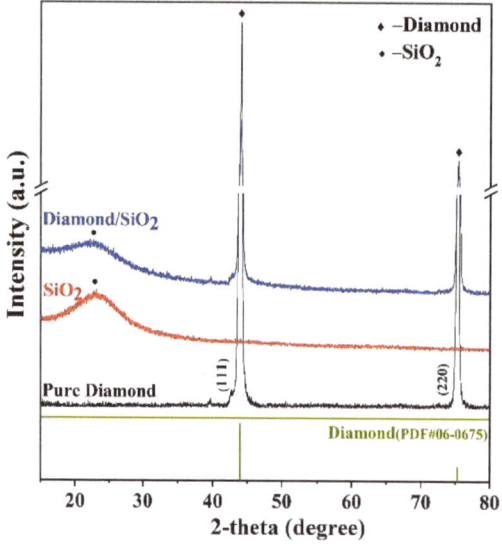

Figure 5. XRD patterns of pure diamond, pure SiO$_2$, and diamond/SiO$_2$ composite abrasives.

3.2. Polishing Test

Under the same testing conditions, sapphire wafers were polished with pure diamond and diamond/SiO$_2$ composite abrasives to investigate the polishing performance of as-prepared diamond/SiO$_2$ composite abrasives.

Figure 6 illustrates surface morphology and corresponding profile curves of the sapphire wafers before and after polishing with pure diamond and diamond/SiO$_2$ composite abrasives, respectively. As shown in Figure 6a, the AFM images of pristine sapphire wafers demonstrate poor flatness, rough surface, and numerous scratches, with a maximum scratch depth of more than 22.3 nm, as well as the PV (peak-to-valley) value of 44.13 nm. Figure 6b presents the morphology of the sapphire wafer after polishing with pure diamond. On the whole, the surface flatness has been improved. However, two obvious scratches pass through the sapphire surface, with a depth of 13.1 nm, and the value of PV is 23.48 nm. It could be found that pure diamond abrasives will inevitably bring pits and scratches, which offer limited improvement to the surface quality of sapphire. The morphology of a sapphire wafer after polishing it with diamond/SiO$_2$ composite abrasives is exhibited in Figure 6c, in which all the obvious defects, such as bumps and scratches on the pristine surface, are eliminated. By contrast, the micro profile curve is smooth, and the PV value is the lowest, at 2.52 nm. It can be concluded that the surface morphology of sapphire wafers polished by diamond/SiO$_2$ composite abrasives is superior to that treated by pure diamond, resulting in a smoother and scratch-free surface.

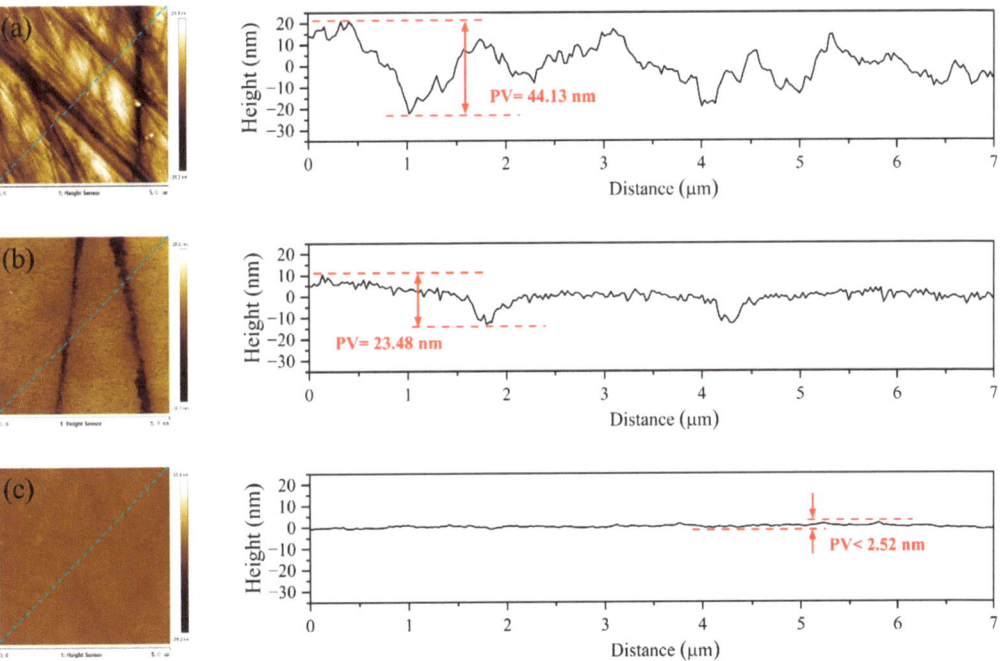

Figure 6. Morphologies and the corresponding profile curves of (**a**) the pristine sapphire wafer and sapphire wafers polished by (**b**) pure diamond and (**c**) diamond/SiO$_2$ composite abrasives.

The roughness at different processing time periods and the MRR of a sapphire wafer polished by pure diamond and diamond/SiO$_2$ composite abrasives are exhibited in Figure 7, in which the roughness shows a downward trend in 3 h of processing with both abrasives under the same parameters. Because of the limitations of pure diamond abrasives, the roughness finally tends to be flat, whereas the roughness of composite abrasives always keeps significantly dropping. Finally, the Ra value of pure diamond decreased to 7.16 nm, and the composite abrasives reached 5.21 nm, which decreased by 27.2%. Meanwhile, after machining a sapphire wafer with diamond/SiO_2 composite abrasives, the MRR is 1.47 nm/min, which is 8.8% larger than the 1.35 nm/min of pure diamond. Along with the morphology of sapphire wafers shown in Figure 6, it can be concluded that compared with pure diamond, the diamond/SiO_2 composite abrasives have achieved more-significant polishing performance on sapphire wafers with lower surface roughness and a higher MRR.

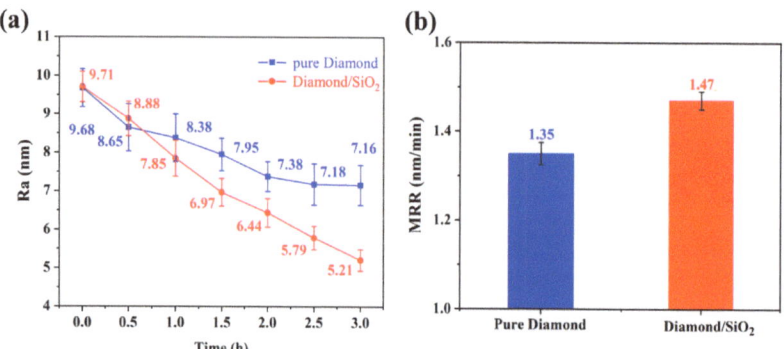

Figure 7. (**a**) Surface roughness (Ra) and (**b**) MRR of the sapphire wafer polished with pure diamond and diamond/SiO_2 composite abrasives.

3.3. Polishing Mechanism

The mechanical-chemical polishing of sapphire wafers includes chemical corrosion and mechanical action, which complement and promote each other. On the basis of these two aspects, the mechanism of improving the polishing quality of diamond/SiO_2 composite abrasives was investigated.

The wear debris, which was removed from the surface of sapphire wafers during polishing, can be used to analyze the polishing mechanism. TEM images, HRTEM images, EDS spectra, and SAED patterns of the wear debris generated by pure diamond and composite abrasives during polishing are presented in Figure 8. The C and Cu elements in the EDS spectra come from the copper mesh plated with carbon support film. As shown in Figure 8a, the wear debris produced by pure diamond is in the form of a block with a size of about 200 nm [27]. Under HRTEM, lattice stripes can be observed with a spacing of 0.24 nm, which matches the (11$\bar{2}$0) crystal plane of sapphire. Combined with the EDS energy spectrum, it can be preliminarily determined as sapphire wear debris. The corresponding SAED pattern shows regular diffraction points, but polycrystalline concentric rings can also be vaguely observed, demonstrating that the wear debris was most likely both crystalline and amorphous, indicating that the pure diamond abrasives are removed mainly by a single mechanical method in the polishing process. Figure 8b illustrates the wear debris generated by diamond/SiO_2 composite abrasives, and that debris is in the shape of fragments, showing the size of dozens of nanometers. Under the detection area shown in the figure, the HRTEM image has no lattice fringes, and the SEAD pattern presents a large diffuse halo, indicating that the wear debris is amorphous. Furthermore, the EDS spectrum contains not only Al and O elements but also an Si element. It can be inferred that the SiO_2 shell reacts with sapphire in the process of polishing to form a new amorphous Al_2O_3-SiO_2 compound, which means that the material removal process combines chemical corrosion and mechanical action [28].

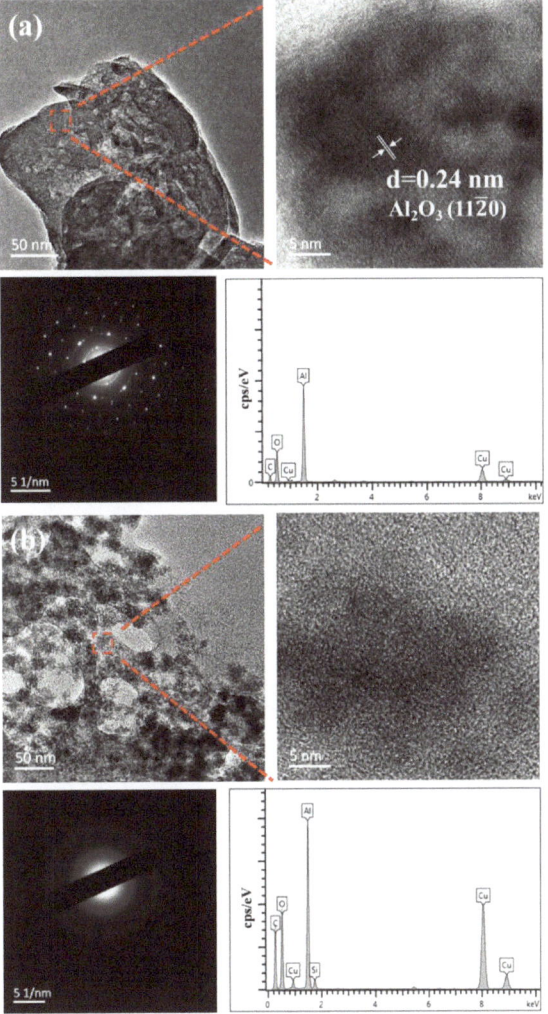

Figure 8. TEM images, HRTEM images, SAED patterns, and EDS spectrum of wear debris produced by (**a**) pure diamond, (**b**) diamond/SiO$_2$ composite abrasives.

XPS was used to characterize the composition and existing form of elements in the sample, which can confirm the chemical corrosion mechanism in the polishing process. Figure 9 demonstrates the XPS spectra of the Al and Si elements on the sapphire wafer surface polished by diamond/SiO$_2$ composite abrasives. According to the narrow scanning spectrum of Al 2p shown in Figure 9a, there are three main chemical states of Al on the surface of sapphire. The peak, centered at 73.57 eV, corresponds to Al$_2$O$_3$, which is the main component of sapphire wafer, and peaks at 74.10 eV and 74.60 eV can be assigned to AlOOH and Al$_2$Si$_2$O$_7$·H$_2$O [29], respectively. From Figure 9b, the peak of Si 2p with the binding energy of 100.43 eV could be attributed to Al$_2$Si$_2$O$_7$·H$_2$O [30]. This observation indicates that the SiO$_2$ shell of composite abrasives underwent a solid-state chemical reaction with the surface of sapphire, which yields AlOOH with a Mohs hardness of 3–3.5

and $Al_2Si_2O_7 \cdot H_2O$ with a Mohs hardness of 4. Possible reactions can be summarized as follows:

$$Al_2O_3 + H_2O \rightarrow 2AlOOH \qquad (5)$$

$$2AlOOH + 2SiO_2 \rightarrow Al_2Si_2O_7 \cdot H_2O \qquad (6)$$

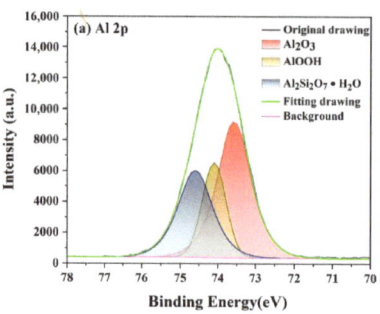 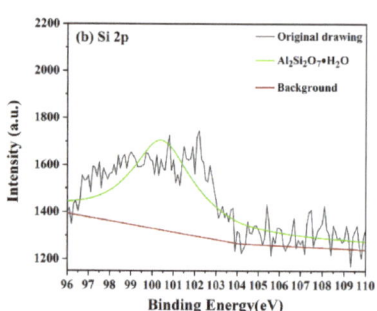

Figure 9. (**a**) Al 2p and (**b**) Si 2p narrow scan spectra on sapphire surface polished by diamond/SiO$_2$ composite abrasives.

In addition, the types and properties of the abrasive particles are the keys to mechanical action, which affects the whole mechanical-chemical polishing process. Compared with pure diamond, the composite abrasives coated with amorphous SiO$_2$ shells have a lower elastic modulus. On the basis of the microcontact mechanics model proposed by Chen [31], and given the deformation of abrasive particles in the machining process, the microcontact model between sapphire wafer, abrasive particles, and polishing pads was established, as shown in Figure 10. Equations (7)–(9) demonstrated the calculation formulas for the indentation depth into the sapphire wafer and the deformation of abrasive particles. Under the action of polishing pressure F, the indentation depth δ_w pressed into the sapphire wafer is reduced because of the lower Young's modulus E_s of composite abrasives. As a result, the diamond/SiO$_2$ composite abrasives became an ellipsoid during the polishing process, which makes the contact stress lower and more uniform. The reduction of the indentation depth can effectively avoid deep scratches and serious damage, which has a decisive impact on improving the surface roughness. In this case, the roughness of a processed sapphire wafer will be improved when diamond/SiO$_2$ composite abrasives are used in polishing, but the MRR will decrease because of the reduction in the indentation depth, which results in a weaker plow effect.

$$\delta = \left(\frac{9F^2}{8DE_{sw}^2}\right)^{\frac{1}{3}} \qquad (7)$$

$$\frac{1}{E_{sw}} = \frac{1-v_s^2}{E_s} + \frac{1-v_w^2}{E_w} \qquad (8)$$

$$\delta_w = D - \delta - \delta_p = D - \delta\left[1 + \left(\frac{E_{sw}}{E_{sp}}\right)^{\frac{3}{2}}\right] \qquad (9)$$

where F is the polishing pressure, δ is the deformation of the particle; D is the diameter of the particle; E_{sw} is the Young's modulus of the particle and wafer pair; E_s and v_s are the Young's modulus and the Poisson's ratio of the abrasive particle, respectively; E_w and v_w are the Young's modulus and the Poisson's ratio of the wafer, respectively; E_{sp} is the Young's modulus of the particle and pad pair; δ_p is the indentation depth of the particle into the polishing pad; and δ_w is the indentation depth of the particle into the sapphire wafer [31].

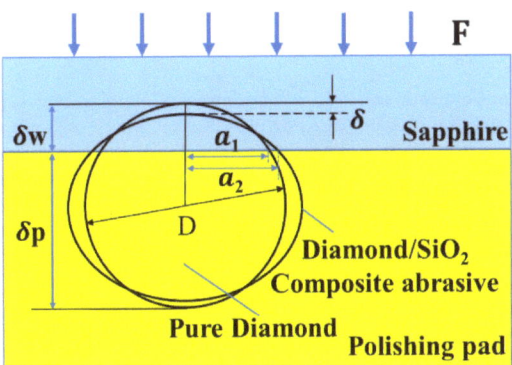

Figure 10. Microcontact diagram of pure diamond and diamond/SiO₂ composite abrasive during the polishing process.

In fact, when the diamond/SiO$_2$ composite abrasives are used in the polishing process of sapphire, the larger elastic deformation also obtains a larger contact area with the sapphire wafer, such that the solid-state reaction can proceed more continuously and fully [32]. Subsequently, the softening reaction products can be easily removed by the mechanical action of abrasives, which is more significant than the plow effect, thus further improving the MRR. Therefore, only by relying on the chemical corrosion and mechanical action of composite abrasives to balance and promote each other is it possible to at the same time improve the surface roughness and the MRR. Figure 11 shows the material removal model of diamond/SiO$_2$ composite abrasives for polishing sapphire wafers.

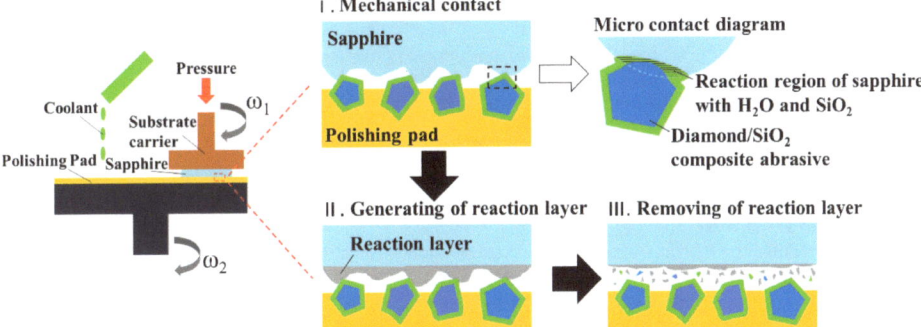

Figure 11. Material removal model of diamond/SiO$_2$ composite abrasives for polishing sapphire wafers.

4. Conclusions

By depositing SiO$_2$ on diamond surface, core-shell structured composite abrasives can be synthesized using a simplified sol-gel method. The diamond/SiO$_2$ core-shell composite abrasives showed excellent polishing performance in green polishing technology without corrosive slurry. Compared with pure diamond, composite abrasives obtained ultrasmooth and low-damage sapphire surfaces, which effectively reduced the surface roughness by 27.2%, accompanied by an 8.8% improvement in the MRR. The reduction in surface roughness may have been caused by the lower Young's modulus of diamond/SiO$_2$ composite abrasives, which resulted in the decrease in indentation depth pressed into the sapphire wafer. In addition, chemical reactions occurred between the sapphire and SiO$_2$ shell during the polishing process, which yielded AlOOH and Al$_2$Si$_2$O$_7$ with low hardness. Meanwhile, the sufficiently continuous solid-state reaction brought by soft SiO$_2$ shells enhanced the MRR in cooperation with the mechanical removal action of composite abrasives. Thus, diamond/SiO$_2$ composite abrasives are well-defined abrasives that meet the practical requirements of the high surface quality and high MRR of sapphire wafers.

Author Contributions: G.Z.: investigation, validation, writing—original draft preparation. Y.X.: conceptualization, methodology, formal analysis, writing—original draft. Q.W.: supervision, resources, funding acquisition, writing—reviewing and editing. J.L.: data curation. Y.Z.: visualization, data curation. B.C.: project administration. All authors have read and agreed to the published version of the manuscript.

Funding: This research was funded by the National Natural Science Foundation of China (Grant No. 51775113, 52275413), the General Science and Technology Projects of Fujian Province (Grant No. 2021I0022), the General Science and Technology Projects from the Fujian Provincial Department of Finance (Grant No. GY-Z21006), and the Program for Innovative Research Team in Science and Technology at Fujian Province University (IRTSTFJ). The APC was funded by the General Science and Technology Projects from the Fujian Provincial Department of Finance (Grant No. GY-Z21006).

Conflicts of Interest: The authors declare no conflict of interest.

References

1. Zhai, Q.; Zhai, W.; Gao, B. Investigation on the relationship between apparent viscosity of Fe$_3$O$_4$@SiO$_2$ abrasive-based magneto-rheological fluid and material removal rate of sapphire in magneto-rheological polishing. *Colloids Surf. A* **2022**, *640*, 20. [CrossRef]
2. Pan, J.; Chen, Z.; Yan, Q. Study on the rheological properties and polishing properties of SiO$_2$@CI composite particle for sapphire wafer. *Smart Mater. Struct.* **2020**, *29*, 114003. [CrossRef]
3. Zhai, Q.; Zhai, W.; Gao, B. Effects of quantities and pole-arrangements of magnets on the magneto-rheological polishing (MRP) performance of sapphire hemisphere. *Appl. Surf. Sci.* **2022**, *584*, 152589. [CrossRef]
4. Zhou, C.; Xu, X.; Dai, L.; Gong, H.; Lin, S. Chemical-mechanical polishing performance of core-shell structured polystyrene@ceria/nanodiamond ternary abrasives on sapphire wafer. *Ceram. Int.* **2021**, *47*, 31691–31701. [CrossRef]
5. Zhai, Q.; Zhai, W.; Gao, B. Modeling of forces and material removal rate in ultrasound assisted magnetorheological polishing (UAMP) of sapphire. *Colloids Surf. A* **2021**, *628*, 127272. [CrossRef]
6. Xie, W.; Zhang, Z.; Liao, L.; Liu, J.; Su, H.; Wang, S.; Guo, D. Green chemical mechanical polishing of sapphire wafers using a novel slurry. *Nanoscale* **2020**, *12*, 22518–22526. [CrossRef]
7. Zhang, Z.; Liu, J.; Hu, W.; Zhang, L.; Liao, L. Chemical mechanical polishing for sapphire wafers using a developed slurry. *J. Manuf. Process.* **2021**, *62*, 762–771. [CrossRef]
8. Li, Z.; Deng, Z.; Hu, Y. Effects of polishing parameters on surface quality in sapphire double-sided CMP. *Ceram. Int.* **2020**, *46*, 13356–13364. [CrossRef]
9. Lei, H.; Liu, T.; Xu, L. Synthesis of Sm-doped colloidal SiO$_2$ composite abrasives and their chemical mechanical polishing performances on sapphire substrates. *Mater. Chem. Phys.* **2019**, *237*, 121819. [CrossRef]
10. Lee, H.; Kim, H.; Jeong, H. Approaches to Sustainability in Chemical Mechanical Polishing (CMP): A Review. *Int. J. Pract. Eng. Man-Gt.* **2021**, *9*, 349–367. [CrossRef]
11. Hu, G.; Lu, J.; Xu, X. Polishing Silicon Wafers with the Nanodiamond Abrasive Tools Prepared by Sol-Gel Technique. *Key. Eng. Mater.* **2012**, *496*, 1–6. [CrossRef]
12. Lu, J.; Li, Y.; Xu, X. The effects of abrasive yielding on the polishing of SiC wafers using a semi-fixed flexible pad. *Proc. Inst. Mech. Eng. Part B J. Process Mech. Eng.* **2015**, *229*, 170–177. [CrossRef]
13. Liu, L.; Zhang, Z.; Wu, B.; Hu, W.; Meng, F.; Li, Y. A review: Green chemical mechanical polishing for metals and brittle wafer. *J. Phys. D Appl. Phys.* **2021**, *54*, 373001. [CrossRef]
14. Xu, Y.; Lu, J.; Xu, X. Study on planarization machining of sapphire wafer with soft-hard mixed abrasive through mechanical chemical polishing. *Appl. Surf. Sci.* **2016**, *389*, 713–720. [CrossRef]
15. Xu, Y.; Lu, J.; Xu, X.; Chen, C.; Lin, Y. Study on high efficient sapphire wafer processing by coupling SG-mechanical polishing and GLA-CMP. *Int. J. Mach. Tool. Manu.* **2018**, *130*, 12–19. [CrossRef]
16. Gao, B.; Zhai, W.; Zhai, Q.; Wang, C. Novel photoelectrochemically combined mechanical polishing technology for scratch-free 4H-SiC surface by using CeO$_2$-TiO$_2$ composite photocatalysts and PS/CeO$_2$ core/shell abrasives. *Appl. Surf. Sci.* **2021**, *570*, 151141. [CrossRef]
17. Zhai, Q.; Zhai, W.; Gao, B.; Shi, Y.; Cheng, X. Effect of core-diameters and shell-thicknesses of Fe$_3$O$_4$/SiO$_2$ composite abrasives on the performance of ultrasound-assisted magnetorheological polishing for sapphire. *Colloids Surf. A* **2021**, *625*, 126871. [CrossRef]
18. Wang, X.; Lei, H.; Chen, R. CMP behavior of alumina/metatitanic acid core–shell abrasives on sapphire substrates. *Precis. Eng.* **2017**, *50*, 263–268. [CrossRef]
19. Chen, Y.; Zuo, C.; Chen, A. Core/shell structured sSiO$_2$/mSiO$_2$ composite particles: The effect of the core size on oxide chemical mechanical polishing. *Adv. Powder Technol.* **2018**, *29*, 18–26. [CrossRef]
20. Lu, J.; Xu, Y.; Zhang, D.; Xu, X. The Synthesis of the Core/Shell Structured Diamond/Akageneite Hybrid Particles with Enhanced Polishing Performance. *Materials* **2017**, *10*, 673. [CrossRef]

21. Xu, Y.; Zhao, G.; Wang, Q.; Zhan, Y.; Chen, B. Synthesis of Al_2O_3/SiO_2 core–shell composite abrasives toward ultrasmooth and high-efficiency polishing for sapphire wafers. *Proc. Inst. Mech. Eng. Part E J. Process Mech. Eng.* **2022**. [CrossRef]
22. Jeffrey, C. *Sol-Gel Science: The Physics and Chemistry of Sol-Gel Processing*; Academic Press: Cambridge, MA, USA, 1990; pp. 2–8.
23. Lismont, M.; Páez, C.; Dreesen, L. A one-step short-time synthesis of $Ag@SiO_2$ core–shell nanoparticles. *J. Colloid Interf. Sci.* **2015**, *447*, 40–49. [CrossRef] [PubMed]
24. Yang, J.; Song, G.; Zhou, L.; Wang, X.; Li, J. Highly sensitively detecting tetramethylthiuram disulfide based on synergistic contribution of metal and semiconductor in stable Ag/TiO_2 core-shell SERS substrates. *Appl. Surf. Sci.* **2021**, *539*, 147744. [CrossRef]
25. Zheltova, V.; Vlasova, A.; Bobrysheva, N.; Abdullin, I.; Osmolovskaya, O. Fe_3O_4@HAp core–shell nanoparticles as MRI contrast agent: Synthesis, characterization and theoretical and experimental study of shell impact on magnetic properties. *Appl. Surf. Sci.* **2020**, *531*, 147352. [CrossRef]
26. Hsiang, H.; Wang, S.; Chen, C. Electromagnetic properties of FeSiCr alloy powders modified with amorphous SiO_2. *J. Magn. Magn. Mater.* **2020**, *514*, 167151. [CrossRef]
27. Luo, Q.; Lu, J.; Tian, Z.; Jiang, F. Controllable material removal behavior of 6H-SiC wafer in nanoscale polishing. *Appl. Surf. Sci.* **2021**, *562*, 150219. [CrossRef]
28. Luo, Q.; Lu, J.; Xu, X. A comparative study on the material removal mechanisms of 6H-SiC polished by semi-fixed and fixed diamond abrasive tools. *Wear* **2016**, *350*, 99–106. [CrossRef]
29. Remy, M.; Genet, M.; Poncelet, G.; Lardinois, P.; Notte, P. Investigation of dealuminated mordenites by X-ray photoelectron-spectroscopy. *Cheminform* **1992**, *96*, 2614–2617. [CrossRef]
30. Pitts, J.; Thomas, T.; Czanderna, A.; Passler, M. XPS and ISS of submonolayer coverage of Ag on SiO_2. *Appl. Surf. Sci.* **1986**, *26*, 107–120. [CrossRef]
31. Chen, X.; Zhao, Y.; Wang, Y. Modeling the effects of particle deformation in chemical mechanical polishing. *Appl. Surf. Sci.* **2012**, *258*, 8469–8474. [CrossRef]
32. Zhai, Q.; Zhai, W.; Gao, B.; Shi, Y.; Chen, X. Synthesis and characterization of nanocomposite Fe_3O_4/SiO_2 core–shell abrasives for high-efficiency ultrasound-assisted magneto-rheological polishing of sapphire. *Ceram. Int.* **2021**, *47*, 31681–31690. [CrossRef]

Article

Source/Drain Trimming Process to Improve Gate-All-Around Nanosheet Transistors Switching Performance and Enable More Stacks of Nanosheets

Kun Chen [1,2,3], Jingwen Yang [1], Tao Liu [1], Dawei Wang [1], Min Xu [1,2,3,*], Chunlei Wu [1,2,3,*], Chen Wang [1,2,3], Saisheng Xu [1], David Wei Zhang [1,2,3,*] and Wenchao Liu [4]

[1] State Key Laboratory of ASIC and System, School of Microelectronics, Fudan University, Shanghai 200433, China; chenk18@fudan.edu.cn (K.C.); 18112020011@fudan.edu.cn (J.Y.); tliu14@fudan.edu.cn (T.L.); 20112020120@fudan.edu.cn (D.W.); chen_w@fudan.edu.cn (C.W.); ssxu@fudan.edu.cn (S.X.)
[2] Shanghai Integrated Circuit Manufacturing Innovation Center Co., Ltd., Shanghai 200433, China
[3] Zhangjiang Fudan International Innovation Center, Shanghai 200433, China
[4] Primarius Technologies Co., Ltd., Shanghai 201306, China; liuwc@khai-long.com
* Correspondence: xu_min@fudan.edu.cn (M.X.); wuchunlei@fudan.edu.cn (C.W.); dwzhang@fudan.edu.cn (D.W.Z.)

Citation: Chen, K.; Yang, J.; Liu, T.; Wang, D.; Xu, M.; Wu, C.; Wang, C.; Xu, S.; Zhang, D.W.; Liu, W. Source/Drain Trimming Process to Improve Gate-All-Around Nanosheet Transistors Switching Performance and Enable More Stacks of Nanosheets. *Micromachines* 2022, 13, 1080. https://doi.org/10.3390/mi13071080

Academic Editor: Kun Li

Received: 5 June 2022
Accepted: 2 July 2022
Published: 8 July 2022

Publisher's Note: MDPI stays neutral with regard to jurisdictional claims in published maps and institutional affiliations.

Copyright: © 2022 by the authors. Licensee MDPI, Basel, Switzerland. This article is an open access article distributed under the terms and conditions of the Creative Commons Attribution (CC BY) license (https://creativecommons.org/licenses/by/4.0/).

Abstract: A new S/D trimming process was proposed to significantly reduce the parasitic RC of gate-all-around (GAA) nanosheet transistors (NS-FETs) while retaining the channel stress from epitaxy S/D stressors at most. With optimized S/D trimming, the 7-stage ring oscillator (RO) gained up to 27.8% improvement of delay with the same power consumption, for a 3-layer stacked GAA NS-FETs. Furthermore, the proposed S/D trimming technology could enable more than 4-layer vertical stacking of nanosheets for GAA technology extension beyond 3 nm CMOS technology.

Keywords: gate-all-around (GAA); nanosheet (NS); S/D stressor; channel stress enhancement

1. Introduction

Gate-All-Around (GAA) Nanosheet (NS) transistor is the most promising candidate for 3 nm node and beyond, owing to its superior electrostatics compared to FinFET [1]. For GAA NS-FETs technology, vertical stacking architecture with multiple parallel channels is the key to boost drive current capability at a given footprint [2]. However, GAA technology faces some critical fabrication challenges such as channel release, formation of inner-spacer, and epitaxy growth of source/drain (S/D) SiGe stressor. The channel release and inner-spacer require ultra-high selective SiGe etching to retain stacked Si channels integrity [3], and advanced ALD low-k dielectric deposition and precision etch to control inner-spacer thickness and uniformity [4]. In addition, for GAA, the main transport surface orientation changes from (110) to (100), which has higher electron mobility but lower hole mobility. Therefore, the epitaxy growth of S/D SiGe stressor becomes extremely important and challenging to achieve N/P current matching. Furthermore, the S/D parasitic RC would be a bottleneck for drive current boost due to normally large epitaxy S/D volume required for channel stress engineering [5]. However, how to balance the channel stress and S/D RC optimization, and its impacts on GAA NS-FETs have not been systematically investigation.

In this work, we present a new integration scheme of a self-align S/D trimming process to solve the trade-off between channel stress engineering and S/D RC optimization. Based on device and system TCAD studies, the proposed S/D trimming scheme offers superior switching performance with almost no sacrificing of transistor DC performance. Furthermore, this new integration scheme may provide a potential path for continuing track-height scaling and enable more vertical stacking of nanosheets.

2. S/D Epi Growth and Impact on GAA NS-FETS

2.1. Simulation Methodology

To evaluate realistic S/D process impact on GAA NS-FETs, a three-layer vertical stacking Si GAA nanosheets structure is selected for TCAD simulation. Key parameters of the assumed structure, including gate length of 12 nm, NS width of 20 nm, gate pitch of 44 nm, nanosheet thickness of 5 nm, spacing between nanosheets of 12 nm, and contact poly pitch of 44 nm, were adopted referring to the IRDS roadmap for 3 nm node.

3D full flow process simulation, from SiGe/Si super-lattice growth to M0 contact formation, is carried out by the Sentaurus Process simulator. The compressive $Si_{0.7}Ge_{0.3}$ sacrificial layers with initial in-plane biaxial stress of −2 GPa were assumed. Lattice Kinetic Monte Carlo (LKMC) model is employed to accurately simulate the epitaxial $Si_{0.6}Ge_{0.4}$ (p-FETs) S/D stressor growth.

Both drift–diffusion transport model and quantum potential model are employed in Sentaurus SDevice for DC performance simulation after structure generation. Low field ballistic mobility, auto-orientation inversion, accumulation layer mobility, and high field saturation velocity were also included to account for the electrical characteristics of the nanoscale device. The multi-valley electron and hole mobility model was enabled to calculate strain effects. The physical parameters of the baseline GAA Si NS-FET simulation models were carefully calibrated using experimental data [1]. The electrical behavior of simulated NS-FETs is modeled with the Primarius BSIMplus module. Inverter and Ring Oscillator (RO) are constructed for AC performance evaluation.

2.2. Stress Requirements and S/D RC Concern for NS-FETs

The majority of the surface area in FinFET is lateral (110)/<110>, while for NS-FET, the main transport surface becomes (100)/<110>. Because (100)/<110> orientation has higher electron mobility but lower hole mobility, it is much more difficult to achieve N/P current matching for 3 nm GAA NS-FETs technology. One of the most effective methods would be channel stress engineering by an epitaxial S/D stressor to boost carrier mobility. In order to investigate the stress engineering requirement for NS-FETs N/P current matching, transistor drive current Ion versus channel stress was simulated and plotted in Figure 1a for both n-FETs and p-FETs.

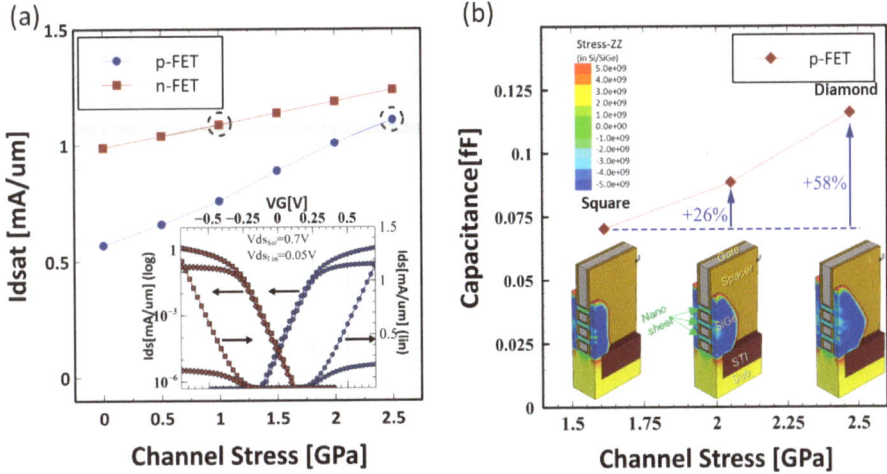

Figure 1. (a) Idsat response curve versus Channel stress for NS-FETs, with n-FET @ 1.0 GPa and p-FET @ 2.5 GPa, good N/P current matching is achieved; (b) capacitance versus stress for different S/D EPI sizes. The high stress EPI volume comes with capacitance penalty.

As clearly seen in Figure 1a, p-FET Idsat is only around half of n-FET Idsat without channel stress, but p-FET has much higher stress sensitivity than n-FET. As reported in Ref. [6], the channel would inherit 450–950 MPa tensile stress from the release of the sacrificial SiGe layer due to stress transfer. This favors n-FET and makes it very crucial to have higher channel stress in p-FET. As shown in the inset of Figure 1a, good N/P current matching could be realized with n-FET @ 1 GPa tensile stress and p-FET @ 2.5 GPa compressive stress, indicating it is extremely important to engineer the channel stress well, especially for p-FET.

As well known, S/D SiGe EPI volume strongly influences channel stress; therefore, LKMC EPI process simulation is performed to investigate the S/D SiGe EPI evolution and its effect on p-FET channel stress. As shown in Figure 1b, three selected points during S/D SiGe epitaxy growth are presented with a cut-away structure diagram to illustrate the S/D shape evolution, with corresponding stress and Cgg capacitance. Initially, the EPI growth was started on three small regions and evolve to a merged 'Square' shaped S/D, where the channel stress reaches around 1.6 GPa. As the epitaxy growth continues, the S/D shape finally becomes the characteristic 'Diamond' shape with the desired 2.5 GPa channel stress for p-FET to realize N/P current matching. However, this comes with a penalty of a 58% increase of Cgg, which raises a concern for the AC performance. As a result, the engineering between channel stress and parasitic RC would be critical to unlock the full potential of GAA NS-FETs technology.

3. Self-Align S/D Trim Scheme
3.1. S/D Trimming Process Flow

To address the abovementioned trade-off issue between channel stress and parasitic RC, a new self-aligned S/D trimming integration scheme is proposed. This self-aligned trimming process after S/D EPI consists of (1) dielectric fill and pull back recess to expose the top of the S/D, (2) the exposed S/D parts undergo selective TiN deposition to form a self-aligned cap [7], (3) dielectric etch back, and (4) use the TiN as a hard-mask to trim off the side tips of 'Diamond' S/D structure [8]. The proposed integration flow is implemented and demonstrated in Sentaurus SProcess, as shown in Figure 2a, and the impact of trimming on NS-FETs channel stress and Cgg were evaluated as in Figure 2b. In the region of Ymax > 25 nm, the channel stress loss is almost negligible (<3%), even when Ymax~15 nm, the channel stress loss is within 10%. At Ymax~15 nm, which is approximately the same width of as-grown "Square" structure, the stress can retain to 2.18 GPa compared to 1.6 GPa for the as-grown one, while the parasitic capacitance can be reduced by 37%.

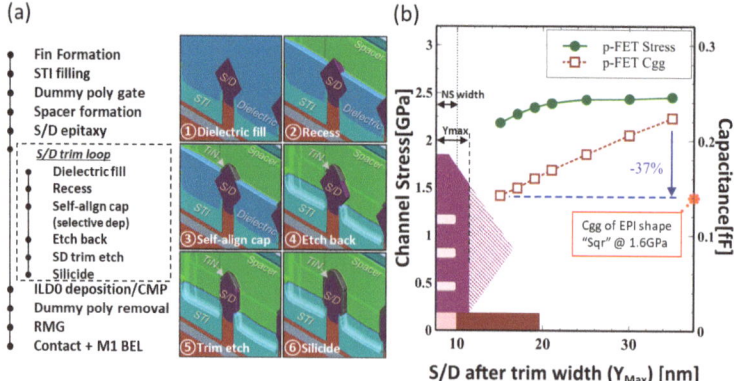

Figure 2. (a) Process details of self-aligned S/D trimming scheme and its relative position in the main flow with step by step process simulation animation; (b) average p-FET nanosheet channel stress and its corresponding capacitance as a function of its max width after trim.

3.2. The Impact of S/D Trimming on Electrical Behavior

Besides the suppression of parasitic capacitance, meanwhile, the S/D trimming process could potentially improve the S/D parasitic resistance. In advanced nodes, S/D access resistance plays an important role in device performance, which accounts for more than 30% of the total resistance [9]. For the 3 nm node, with an estimated cell height of 115.5 nm and N/P separate distance 20.5 nm [10], the maximum width of M0A contact is limited to 47.5 nm. As shown in Figure 3a, limited by the width of M0A contact (W_{ct}), for large diamond shaped S/D, the contact can only land on the top of S/D, which causes large access resistance, especially for the bottom nanosheet channel. With the S/D trimming, wrap-around-contact (WAC) technology would be feasible and can make all the stacked nanosheets electrically equivalent as illustrated in Figure 3a. The quantitative impact has been simulated as in Figure 3b, showing for S/D trimming to Ymax~20 nm, the transistor driving current can be boosted by 31% and 24% for p-FET and n-FET, respectively. With trimming to Ymax~15 nm, n-FET drive current gains an additional ~3% due to further reduction of access resistance. Meanwhile, p-FET drive current almost remains unchanged, mainly due to the small loss of channel stress.

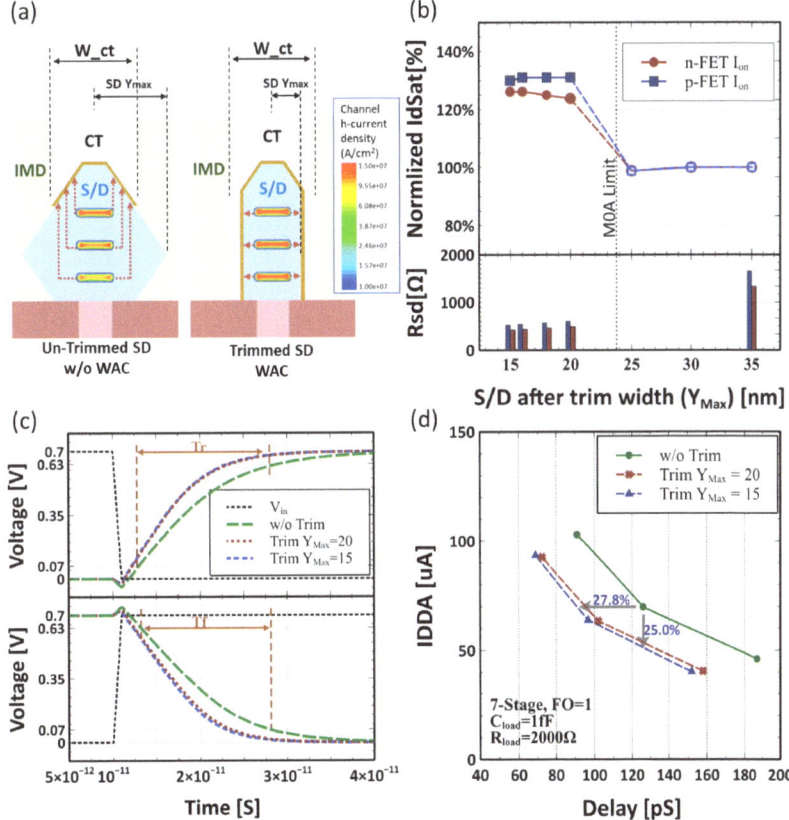

Figure 3. (a) Current from bottom NS of trimmed S/D and WAC has much a shorter distance to contact compared to an un-trimmed case, resulting in a more equally distributed current between all three nanosheets and better efficiency; (b) ion performance and extracted Rsd at varied S/D trimming position; (c) switch delay of an inverter on a fixed load of 1fF. Both Tr and Tf of NS-FETs reduce with S/D trimming; (d) power–delay curve of a constructed 7-stage ring oscillator.

To examine the AC performance with S/D trimming, inverter and 7-stage ring oscillator (RO) [11] are evaluated by plugging in the device's model generated by Primarius BSIMplus [10] . The inverter switch behavior is shown in Figure 3c, showing rising/falling time (Tr/Tf) improved by 49% and 53%, respectively. The IDDA-Delay plot of the 7-stage RO in Figure 3d demonstrates that the S/D trimming scheme could save up to 25 percent of power or 27.8 percent of delay. All these are attributed to the reduction of parasitic S/D RC and preservation of device drive current.

3.3. More Vertical Stacking of Nanosheets

One of the most promising advantages of GAA NS-FETs technology is the vertical scalability [2]. However, as the stacking layers increase, the epitaxy S/D also becomes bigger. This will cause even higher access resistance and bigger parasitic capacitance, degrading the device performance [12]. As shown in Figure 4a, with the proposed S/D trimming scheme, the capacitance increase with more stacking NS becomes much slower, and the S/D after trimming can be adjusted to embrace the WAC, thus further reducing the S/D parasitic Resistance. As demonstrated in Figure 4b, as the number of stacked nanosheets increases, more benefits could be gained from the S/D trimming regarding the inverter delay. This may provide a very promising solution for S/D parasitic suppression and GAA technology extension beyond the 3 nm node.

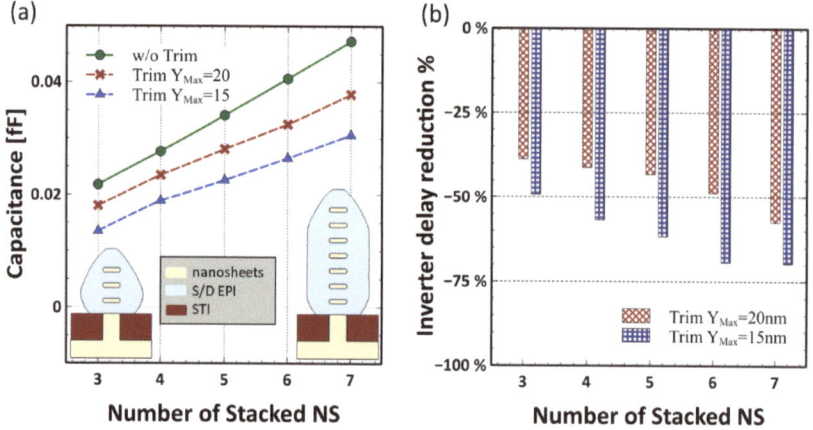

Figure 4. (**a**) NS-FETs vertical scaling challenge: under the same EPI growth condition, the size of S/D increases to accommodate more stacking layers, resulting in worse parasitic RC performance. (**b**) Inverter delay reduction percentage is calculated based on the corresponding un-trimmed case. Higher stacking NS benefits more by implementing the S/D trimming process.

4. Conclusions

By applying the proposed self-aligned S/D trimming process, the parasitic S/D RC has been significantly reduced while retaining the channel stress at most. As a result, the AC performance of inverter and RO is greatly improved. Furthermore, it enables more stacking of nanosheets for higher density and performance, which could be a key factor for GAA NS-FETs technology.

Author Contributions: Conceptualization methodology and analysis, K.C. and J.Y.; software and validation, T.L., D.W. and W.L.; resources, S.X.; data original draft preparation, K.C.; writing—review and editing, M.X., C.W. (Chunlei Wu) and C.W. (Chen Wang); supervision, D.W.Z. All authors have read and agreed to the published version of the manuscript.

Funding: This work is sponsored by the platform for the development of next generation integration circuit technology.

Acknowledgments: The authors acknowledge Wenchao Liu from Primarius Technologies Co., Ltd. for the insightful discussions.

Conflicts of Interest: The authors declare no conflict of interest.

References

1. Loubet, N.; Hook, T.; Montanini, P.; Yeung, C.W.; Kanakasabapathy, S.; Guillom, M.; Yamashita, T.; Zhang, J.; Miao, X.; Wang, J.; et al. Stacked nanosheet gate-all-around transistor to enable scaling beyond FinFET. In Proceedings of the 2017 Symposium on VLSI Technology, Kyoto, Japan, 5–8 June 2017; pp. T230–T231. [CrossRef]
2. Barraud, S.; Previtali, B.; Vizioz, C.; Hartmann, J.M.; Sturm, J.; Lassarre, J.; Perrot, C.; Rodriguez, P.; Loup, V.; Magalhaes-Lucas, A.; et al. 7-Levels-Stacked Nanosheet GAA Transistors for High Performance Computing. In Proceedings of the 2020 IEEE Symposium on VLSI Technology, Honolulu, HI, USA, 16–19 June 2020; pp. 1–2. [CrossRef]
3. Loubet, N.; Kal, S.; Alix, C.; Pancharatnam, S.; Zhou, H.; Durfee, C.; Belyansky, M.; Haller, N.; Watanabe, K.; Devarajan, T.; et al. A Novel Dry Selective Etch of SiGe for the Enablement of High Performance Logic Stacked Gate-All-Around NanoSheet Devices. In Proceedings of the 2019 IEEE International Electron Devices Meeting (IEDM), San Francisco, CA, USA, 7–11 December 2019; pp. 11.4.1–11.4.4. [CrossRef]
4. Barraud, S.; Lapras, V.; Samson, M.; Gaben, L.; Grenouillet, L.; Maffini-Alvaro, V.; Morand, Y.; Daranlot, J.; Rambal, N.; Previtalli, B.; et al. Vertically stacked-NanoWires MOSFETs in a replacement metal gate process with inner spacer and SiGe source/drain. In Proceedings of the 2016 IEEE International Electron Devices Meeting (IEDM), San Francisco, CA, USA, 3–7 December 2016; pp. 17.6.1–17.6.4. [CrossRef]
5. Gendron-Hansen, A.; Korablev, K.; Chakarov, I.; Egley, J.; Cho, J.; Benistant, F. TCAD analysis of FinFET stress engineering for CMOS technology scaling. In Proceedings of the 2015 International Conference on Simulation of Semiconductor Processes and Devices (SISPAD), Washington, DC, USA, 9–11 September 2015; pp. 417–420. [CrossRef]
6. Reboh, S.; Coquand, R.; Loubet, N.; Bernier, N.; Augendre, E.; Chao, R.; Li, J.; Zhang, J.; Muthinti, R.; Boureau, V.; et al. Imaging, Modeling and Engineering of Strain in Gate-All-Around Nanosheet Transitors. In Proceedings of the 2019 IEEE International Electron Devices Meeting (IEDM), San Francisco, CA, USA, 7–11 December 2019; pp. 11.5.1–11.5.4. [CrossRef]
7. Breil, N.; Carr, A.; Kuratomi, T.; Lavoie, C.; Chen, I.C.; Stolfi, M.; Chiu, K.D.; Wang, W.; Van Meer, H.; Sharma, S.; et al. Highly-selective superconformai CVD Ti silicide process enabling area-enhanced contacts for next-generation CMOS architectures. In Proceedings of the 2017 Symposium on VLSI Technology, Kyoto, Japan, 5–8 June 2017; pp. T216–T217. [CrossRef]
8. Sandberg, M.; Vissers, M.R.; Kline, J.S.; Weides, M.; Gao, J.; Wisbey, D.S.; Pappas, D.P. Etch induced microwave losses in titanium nitride superconducting resonators. *Appl. Phys. Lett.* **2012**, *100*, 262605. [CrossRef]
9. Lee, Y.M.; Na, M.H.; Chu, A.; Young, A.; Hook, T.; Liebmann, L.; Nowak, E.J.; Baek, S.H.; Sengupta, R.; Trombley, H.; et al. Accurate performance evaluation for the horizontal nanosheet standard-cell design space beyond 7nm technology. In Proceedings of the 2017 IEEE International Electron Devices Meeting (IEDM), San Francisco, CA, USA, 2–6 December 2017; pp. 29.3.1–29.3.4. [CrossRef]
10. Yakimets, D.; Bardon, M.G.; Jang, D.; Schuddinck, P.; Sherazi, Y.; Weckx, P.; Miyaguchi, K.; Parvais, B.; Raghavan, P.; Spessot, A.; et al. Power aware FinFET and lateral nanosheet FET targeting for 3nm CMOS technology. In Proceedings of the 2017 IEEE International Electron Devices Meeting (IEDM), San Francisco, CA, USA, 2–6 December 2017; pp. 20.4.1–20.4.4. [CrossRef]
11. Jang, D.; Yakimets, D.; Eneman, G.; Schuddinck, P.; Bardon, M.G.; Raghavan, P.; Spessot, A.; Verkest, D.; Mocuta, A. Device Exploration of NanoSheet Transistors for Sub-7-nm Technology Node. *IEEE Trans. Electron Devices* **2017**, *64*, 2707–2713. [CrossRef]
12. Bufler, F.M.; Jang, D.; Hellings, G.; Eneman, G.; Matagne, P.; Spessot, A.; Na, M.H. Monte Carlo Comparison of n-Type and p-Type Nanosheets With FinFETs: Effect of the Number of Sheets. *IEEE Trans. Electron Devices* **2020**, *67*, 4701–4704. [CrossRef]

Communication

A High-Performance MEMS Accelerometer with an Improved TGV Process of Low Cost

Yingchun Fu [1,2], Guowei Han [3], Jiebin Gu [4], Yongmei Zhao [3,5], Jin Ning [3,5], Zhenyu Wei [3,5], Fuhua Yang [3,5] and Chaowei Si [3,*]

1. Beijing Smart-Chip Microelectronics Technology Co., Ltd., Beijing 100192, China; fuyingchun@icrus.cn
2. Zhongguancun Xinhaizeyou Technology Co., Ltd., Beijing 100192, China
3. Institute of Semiconductors, Chinese Academy of Sciences, Beijing 100083, China; hangw1984@semi.ac.cn (G.H.); ymzhao@semi.ac.cn (Y.Z.); ningjin@semi.ac.cn (J.N.); zywei97@semi.ac.cn (Z.W.); fhyang@semi.ac.cn (F.Y.)
4. State Key Laboratory of Transducer Technology, Shanghai Institute of Microsystem and Information Technology, Chinese Academy of Sciences, Shanghai 200050, China; j.gu@mail.sim.ac.cn
5. Center of Materials Science and Optoelectronics Engineering, University of Chinese Academy of Sciences, Beijing 100049, China
* Correspondence: schw@semi.ac.cn; Tel.: +86-10-8230-5147

Citation: Fu, Y.; Han, G.; Gu, J.; Zhao, Y.; Ning, J.; Wei, Z.; Yang, F.; Si, C. A High-Performance MEMS Accelerometer with an Improved TGV Process of Low Cost. *Micromachines* **2022**, *13*, 1071. https://doi.org/10.3390/mi13071071

Academic Editor: Kun Li

Received: 2 June 2022
Accepted: 2 July 2022
Published: 5 July 2022

Publisher's Note: MDPI stays neutral with regard to jurisdictional claims in published maps and institutional affiliations.

Copyright: © 2022 by the authors. Licensee MDPI, Basel, Switzerland. This article is an open access article distributed under the terms and conditions of the Creative Commons Attribution (CC BY) license (https://creativecommons.org/licenses/by/4.0/).

Abstract: High-performance MEMS accelerometers usually use a pendulum structure with a larger mass. Although the performance of the device is guaranteed, the manufacturing cost is high. This paper proposes a method of fabricating high-performance MEMS accelerometers with a TGV process, which can reduce the manufacturing cost and ensure the low-noise characteristics of the device. The TGV processing relies on laser drilling, the metal filling in the hole is based on the casting mold and CMP, and the packaging adopts the three-layer anodic bonding process. Moreover, for the first time, the casting mold process is introduced to the preparation of MEMS devices. In terms of structural design, the stopper uses distributed comb electrodes for overload displacement suppression, and the gas released by the packaging method provides excellent mechanical damping characteristics. The prepared accelerometer has an anti-overload capability of 10,000 g, the noise density is less than $0.001°/\sqrt{Hz}$, and it has ultra-high performance in tilt measurement.

Keywords: MEMS accelerometer; TGV; distributed stoppers; comb structure

1. Introduction

MEMS devices have the advantage of being low cost and small in size, and they can be fabricated in batches. They are also successful sensors for various environmental monitoring uses. MEMS inclination sensors are widely used in buildings, bridges, power transmission networks, automobiles, and other engineering vehicles [1,2]. Accelerometers with high performance and high reliability have great market value.

At present, the accelerometer with the best performance is the sandwich structure from Murata and Safran. The devices of Colibris are based on three-layer silicon bonding and have excellent temperature stability [3], but the preparation cost is high, and it is often used in high-performance fields such as aerospace engineering [4]. The devices from Murata use a silicon glass composite cover plate as the upper and lower electrodes, and the electrodes are grown on the sidewalls through a hard mask, which achieves a good balance between performance and cost, and has a wide range of applications in the automotive and industrial fields [5]. However, since Murata's structure requires electrodes to be grown on the sidewalls, it is not possible to screen the yield of devices on the wafer [6], and the cost is still higher than that of consumer-grade accelerometers.

In this paper, an accelerometer with comb-tooth structures is proposed. It is implemented based on SOG process, and the structure layer adopts 100 μm silicon to ensure the size of the mass block, thereby reducing the mechanical noise of the device. The packaged capacitor electrodes are led out to the surface of the device through vertical leads, which can meet wafer-level tests and reduce production costs. Relying on the distributed stopper set between the comb teeth, the sensitive mass displacement is constrained, depending on the distributed stopper structure set between the comb teeth. Metal filled in through vias on BF33 glasses serves as the vertical lead cover, the anodic bonding process achieves hermetic packaging, and the released gas provides additional damping, making the device excellent in vibration resistance.

The proposed process solution has the capability of low-cost mass production and can guarantee the performance of the device, which is extremely innovative.

2. Structure Design

From the perspective of detection accuracy, the sensitivity of capacitance change per unit acceleration needs to be considered when designing the tilt angle structure, which is determined by the resonant frequency of the structure, the distance between the capacitor plates, and the total capacitance. Under the action of inertial force, the response displacement of the accelerometer can be obtained from Newton's second law, as follows:

$$x = a/\omega_0^2 \tag{1}$$

where x is the displacement under the acceleration a, and ω_0 is the resonant frequency of the accelerometer. If C_0 is the initial capacitor, then d_0 is the initial distance of capacitor plates, and the capacitance change ΔC is

$$\Delta C = xC_0/d_o \tag{2}$$

The accelerometer resolution is determined by its mechanical noise, and the noise $\langle a_n \rangle$ is related to the resonant frequency and the size of the mass m [7]. Where k is the Boltzmann constant, T is the ambient temperature, m is the mass of the movable structure, and Q is the Quality factor.

$$\langle a_n \rangle = \sqrt{4kT\omega_0/mQ} \tag{3}$$

To ensure sensitivity, the typical resonant frequency of an inclination accelerometer is several hundred hertz. Considering the anti-overload capability of the accelerometer, the shock response displacement should not be too large, which is designed to be 1 KHz here. To ensure that the accelerometer has a detection accuracy of $0.001°$, its output electrical noise should be less than $3\ \mu g/\sqrt{Hz}$, and the corresponding mechanical noise should be at the μg level.

Considering the over-damping situation, the quality factor of the accelerometer is less than 1, and the mass of the accelerometer should be greater than 0.12 mg. At present, the detection capability of the accelerometer detection circuit for differential capacitance is at the zF level. Considering the tolerance, it is hoped that $0.001°$ can correspond to a capacitance change rate of 10 zF, then the capacitance sensitivity of an ideal inclination accelerometer is expected to reach 0.57 pF/g.

For a 100 μm-thick silicon structure MEMS device, considering the ability of deep silicon etching, the capacitor plate spacing should be greater than 3 μm. Considering the processing accuracy, the capacitor plate spacing is set to 4 μm, the length of the plate is 150 μm, and the width is 6 μm. The overlap length is 140 μm. One anchor point can support 80 pairs of comb capacitor plates with a capacitance of 1 pF. The capacitor plates are shown in Figure 1.

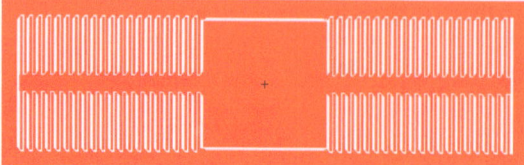

Figure 1. The comb capacitor.

The designed accelerometer has four and a half pairs of comb structures as detection capacitors on one side. The total capacitance is 4.5 pF, the size of the device is 2.4×3.3 mm^2, the size of the mass block is 0.6 mg, the resonant frequency of the device is 1.14 kHz, and the theoretical mechanical noise is 1.43 µg/$\sqrt{\text{Hz}}$. The displacement of the mass block caused by the unit gravitational acceleration is the capacitance sensitivity of 0.19 µm/g, and the theoretical capacitance sensitivity is 0.71 pF/g, all of which meet the design requirements.

The anti-overload design in the detection direction of structure was realized by distributed stopper structure. According to Yoon's conclusions [8], when the stopper collides with the structure, if a flexible collision can be achieved, the stress caused by the collision will be greatly reduced, and the overload resistance of the structure will be improved. In addition to the low resonant frequency of the structure itself, the resonant frequency of an electrode plate is about 100 kHz, which is a good choice as a stopper [9]. Therefore, in this paper, a hypotenuse was designed between the mass block and the electrode plate to reduce the distance between them and the collision distance was reduced, as shown in Figure 2.

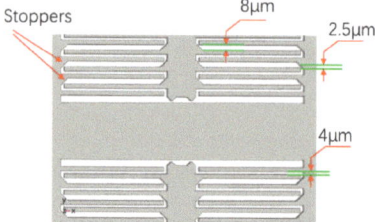

Figure 2. Stoppers located on electrodes.

The stopper has two other advantages. First, the collision between the capacitor plate and the hypotenuse is an edge contact, which does not easily cause adhesion; second, although the collision distance between the hypotenuse and the electrode plate is reduced, the stopper is located in an open area and the etching time will not be increased.

3. Fabrication Process

The designed inclination accelerometer was prepared by the TGV process. The metal filling in the TGV was realized by the electroforming method, the structure layer was prepared by 110 µm-thick (100) crystal silicon, and the resistivity is less than 0.005 Ω cm.

First, the AZ6130 photoresist was patterned on 110 µm-thick monocrystalline silicon, 10 µm-deep silicon was etched, and the bonding area was left. Magnetron-sputtered ITO served as the sacrificial layer, and stripping and cleaning were carried out. Then, the silicon wafer was bonded to a 300 µm-thick BF33 glass.

ITO also served as the mask for structural etching [10]. When ITO is used as a sacrificial layer, the material is transmits light, and the etching of the bottom of the silicon can be observed from the glass layer, avoiding the occurrence of over-etching. In addition, ITO materials with good conductivity can absorb the charged particles of the etching gas, reducing the footprint phenomenon caused by the reflection of F+ particles.

The release of the structure was carried out in sulfuric acid–hydrogen peroxide. After rinsing with pure water, it was boiled in MOS-grade alcohol for 30 min to displace the water in the structure, and dried on a hot plate. The preparation process is shown in Figure 3.

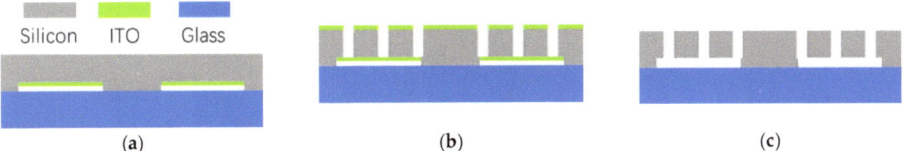

Figure 3. Preparation of the accelerometer structure: (**a**) preparation of structures; (**b**) structures etching; (**c**) structures release.

The advantage of ITO as an etching mask and sacrificial layer material is that the material has different corrosion resistance properties to silicon and glass. When etched in an acid solution, it can be effectively removed without affecting the structure. Compared with the traditional silicon oxide mask, the removal method is more economical and convenient. Using a photoresist as the mask, it is difficult to achieve high aspect ratio pattern etching on the bonding wafer with poor heat dissipation, and during etching, it is also difficult to ensure the etching accuracy of the structure due to the shrinkage of the edge of the photoresist.

The fabricated structures are shown in Figure 4. The designed plate distances are 4 μm and 8 μm, the etched sizes are 3.959 μm and 8.125 μm on the top side of the silicon, and the sizes are 4.306 μm and 8.472 μm on the backside. These values show excellent agreement between design and actual values, and the footprint is effectively suppressed.

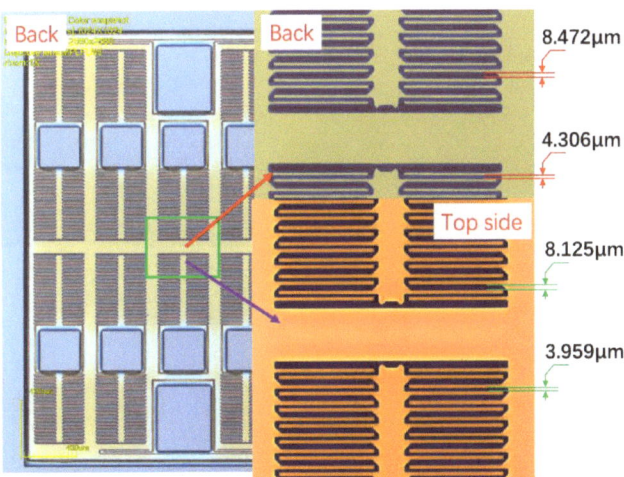

Figure 4. The etched structure.

The TGV cover plate needed to be prepared. First, laser drilling was performed on 300 μm-thick glass [11], the via was designed to the shape of a square, the length of the inlet side was 120 μm, and the length of the outlet side was about 90 μm. The metal filling in the hole was realized by metal casting mold [12], and the excess metal was removed through the CMP process. After CMP, CrAu was used as a mask for glass etching, and the cover plate was prepared. The process is shown in Figure 5.

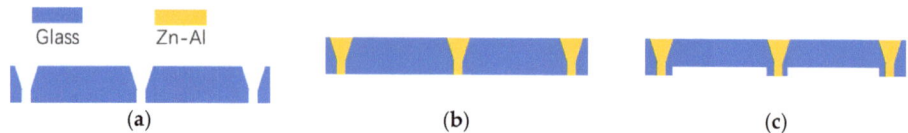

Figure 5. Preparation of the TGV cover: (**a**) laser drilling; (**b**) metal filling and CMP; (**c**) active area corrosion.

Anodic bonding was used to realize the packaging of the tilt accelerometer. When bonding, a three-layer bonding method was used. A positive voltage of 280 V was applied to the silicon structure, and the glass substrate and the TGV cover were grounded, as shown in Figure 6. An excessively high bonding voltage was likely to cause the structure to be adsorbed upward or downward on the glass cover. To ensure bond quality at low voltages, good plasma cleaning of the structure and cover was required. Finally, the interconnection between different electrode groups was achieved through metal traces.

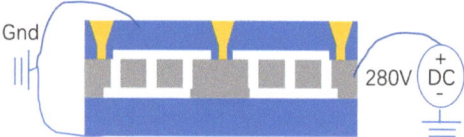

Figure 6. Anodic bonding was used for wafer-level package.

The preparation method used silicon and glass as the main materials, and the cost was low. In addition, the anodic bonding method was used to realize the wafer-level packaging, which also has good air tightness. The oxygen released in the anodic bonding increased the mechanical damping. The anti-vibration capability of the structure was improved. The processed device is shown in Figure 7.

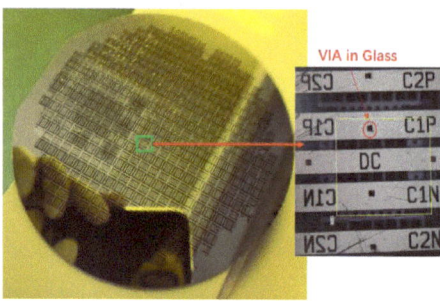

Figure 7. The processed accelerometers.

4. Experiment Results

The dynamic signal analyzer HP35665A was used to test the frequency response characteristics of the accelerometer. It was found that the position of the frequency response curve fluctuated with the bias voltage, but no obvious harmonic response characteristics were measured. So, the released gas in the packaging process provided sufficient damping.

A CV test was carried out on the device using Keysight's B1500, and the test results are shown in Figure 8. The structure of the device on the test surface had a good response to the electrostatic force, the etching and release process of the device was good, and the device was intact and free of adhesion. The capacitances between the mass and the positive and negative electrodes were about 5.52 pF and 5.32 pF, respectively. Considering the parasitic effect, the design value was consistent with the measured value. The differential capacitance deviation caused by the process was only 3.7%.

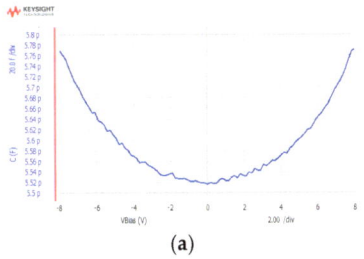 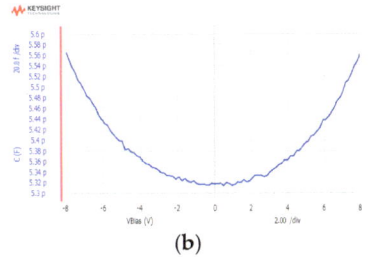

(a)   (b)

Figure 8. CV characteristics of the accelerometer: (**a**) positive direction; (**b**) negative direction.

The capacitance detection was realized through a dedicated ASIC with digital output. The sensitivity of the accelerometer was tested on a centrifuge. The test results showed that the prepared accelerometer had good linearity in the range of plus or minus 2 g. The sensitivity of the accelerometer was 35,459.4 bits/g, and the corresponding sensitivity was about 28.2 μg/bit, which had high detection accuracy, as shown in Figure 9.

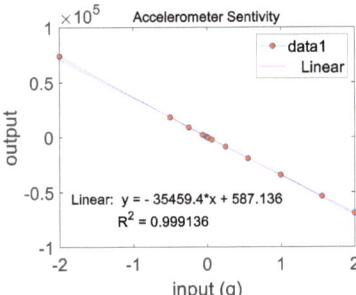

Figure 9. The test result of the scale factor.

The accelerometer output was continuously sampled at 120 Hz for more than two hours when the input was zero, and the result is shown in Figure 10. The corresponding RMS noise was about 163.8 μg, the corresponding tilt noise was 0.0094°, and the noise density near 10 Hz was about 1.2 μg/$\sqrt{Hz}$, the test results are shown in Figure 10.

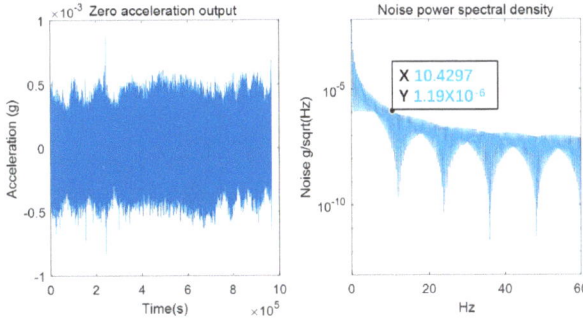

Figure 10. The zero-acceleration output.

The Allan deviation of the accelerometer was calculated, and its zero-bias instability was 9.4 μg, which showed that the device has excellent performance as an inclination detection accelerometer, as shown Figure 11.

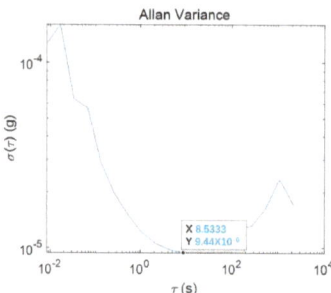

Figure 11. The Allan deviation.

The device was bonded to the ceramic tube shell with DOWSIL™ ME-1030 die-bonding adhesive. The impact test was carried out 6 consecutive times at 10,000 g @ 0.5 ms in the detection direction. The structures of the tested 10 devices were not damaged after the impact, and the device function was good after the power-on test, indicating that the prepared inclination accelerometer had an anti-overload capability of 10,000 g.

5. Discussion

The accelerometer designed in this paper has low mechanical noise due to the use of 100 μm-thick silicon for the structural layer. The method of using ITO as the sacrificial layer reduces the foot effect and ensures the consistency of the device structure and the design value. Glasses with metal-filled through holes are used as the package cover, and the oxygen released by anodic bonding provides sufficient gas damping. Therefore, the accelerometer prepared by this process has extremely low mechanical noise and is suitable for high-precision inclination detection.

Compared with the devices that have been successfully commercialized on the market, the performance of the accelerometer prepared here also shows certain advantages, as shown in Table 1. Colibris's and Murata's pendulum accelerometers have larger masses and thus have lower mechanical noise. The advantage of Colibris's device is that the device is made of all-silicon material, including the upper and lower cover plates and the sensitive structure in the middle. When the ambient temperature changes, the deformation caused by thermal expansion is consistent, so it has excellent temperature stability and repeatability, which is incomparable for other accelerometers, but the cost is too high to be suitable for mass adoption in the industry. Murata's upper and lower cover plates are made of silicon glass composite structure, and the gas-filled method is used to provide over-damping for sensitive structures during bonding, which has excellent reliability and good resistance to vibration and shock. However, the electrodes are grown on the sidewalls and cannot be screened at the wafer level, and the price is much higher than that of consumer-grade MEMS accelerometers.

Table 1. Comparison of parameters about MEMS accelerometers for inclination detection.

Manufacturer	Model	Structure	Package Form	Noise (μg /√Hz)	Bandwidth (Hz)
Colibris [13]	RS9000	Pendulous	Silicon–Silicon	30	30–80
Murata [14]	SCA820	Pendulous	Anodic	220	18
ST [15]	IIS2ICLX	comb	Al-Si	15	12.5
BOSCH [16]	BMA456	Dual-layer comb	Al-Ge	120	1.5–1.6 k
SDI [17]	Model 1521	Teeter–totter		7	0–250
Gatech [18]		Hinge-shaped comb		72	
This Work			Anodic	1.2	10

Bosch has designed multi-layer materials as structural layers to achieve the purpose of increasing the size of sensitive structural mass blocks and reducing the impact of packaging stress on device performance, and is widely used in automobiles. The structure layer should be prepared by the epitaxial polysilicon method, and the thickness of the structure layer is about 30 μm. Compared with the sensitive structure prepared with 100 μm-thick single crystal silicon proposed in this paper, the noise control is slightly insufficient.

In Silicon Designs Inc., the low-noise accelerometer is realized by the double-seesaw process, which has lower noise, but the device is not packaged at the wafer level, which is very unfavorable for wafer-level testing and cost control.

The hinge-shaped accelerometer designed by Galtech has excellent sensitivity, and if the mass size is appropriately increased, then it can also achieve lower mechanical noise; however, this does not solve the packaging problem.

Although the sensitive structure of the high-performance MEMS accelerometer proposed in this paper adopts the traditional comb structure, the mechanical noise of the device is greatly reduced by increasing the thickness of the sensitive structure. Wafer-level packaging in glass with metal-filled through holes also has low-cost mass production capabilities and is a competitive implementation of high-performance MEMS accelerometers.

Author Contributions: Conceived the experiment and wrote the original draft, Y.F.; performed the experiments, G.H.; contributed the metal casting mold process, J.G.; contributed reagents, materials, and key process parameters, Y.Z. and J.N.; performed test and data analysis, Z.W.; technical supervision, F.Y.; writing—review and editing, C.S. All authors have read and agreed to the published version of the manuscript.

Funding: This work is supported by The Laboratory Open Fund of Beijing Smart-chip Microelectronics Technology Co., Ltd., and the Chinese National Science Foundation (Contract No. 52075519 and 61974136).

Institutional Review Board Statement: The study was conducted in accordance with the Declaration of Helsinki, and approved by the Ethics Committee of the Institute of Semiconductors, Chinese Academy of Sciences (1 July 2022).

Conflicts of Interest: The authors declare no conflict of interest.

References

1. Yu, Y.; Ou, J.; Zhang, J.; Zhang, C.; Li, L. Development of Wireless MEMS Inclination Sensor System for Swing Monitoring of Large-Scale Hook Structures. *IEEE Trans. Ind. Electron.* **2009**, *56*, 1072–1078. [CrossRef]
2. Łuczak, S. Erratum to: Guidelines for tilt measurements realized by MEMS accelerometers. *Int. J. Precis. Eng. Manuf.* **2014**, *15*, 2011–2019. [CrossRef]
3. Mejias, L.; Correa, J.F.; Mondragon, I.; Campoy, P. COLIBRI: A vision-Guided UAV for Surveillance and Visual Inspection. In Proceedings of the 2007 IEEE International Conference on Robotics and Automation, Rome, Italy, 10–14 April 2007; pp. 2760–2761.
4. Gonseth, S.; Brisson, R.; Balmain, D.; Di-Gisi, M. Tactical grade MEMS accelerometer. In Proceedings of the 2017 IEEE DGON Inertial Sensors and Systems (ISS), Karlsruhe, Germany, 19–20 September 2017; pp. 1–11. [CrossRef]
5. Vagner, M.; Benes, P. Start-up response improvement for a MEMS inclinometer. In Proceedings of the 2015 IEEE International Instrumentation and Measurement Technology Conference (I2MTC), Pisa, Italy, 11–14 May 2015.
6. Kuisma, H. Glass isolated TSVs for MEMS. In Proceedings of the 5th Electronics System-Integration Technology Conference (ESTC), Helsinki, Finland, 16–18 September 2014; pp. 1–5.
7. Thong, Y.K.; Woolfson, M.S.; Crowe, J.; Hayes-Gill, B.R.; E Challis, R. Dependence of inertial measurements of distance on accelerometer noise. *Meas. Sci. Technol.* **2002**, *13*, 1163. [CrossRef]
8. Yoon, S.W.; Lee, S.; Perkins, N.C.; Najafi, K. Shock-Protection Improvement Using Integrated Novel Shock-Protection Technologies. *J. Microelectromech. Syst.* **2011**, *20*, 1016–1031. [CrossRef]
9. Si, C.; Han, G.; Ning, J. Shock Resistance Design of a High-Performance MEMS Tuning-Fork Gyroscope. *Micronanoelectron. Technol.* **2014**, *51*, 302–307.
10. Cai, M.; Si, C.; Han, G.; Ning, J.; Yang, F. Deep Silicon Etching of Bonded Wafer Based on ITO Mask. *Micronanoelectron. Technol.* **2020**, *57*, 6.
11. Yang, F.; Han, G.; Yang, J.; Zhang, M.; Ning, J.; Yang, F.; Si, C. Research on Wafer-Level MEMS Packaging with Through-Glass vias. *Micromachines* **2018**, *10*, 15. [CrossRef] [PubMed]

12. Gu, J.; Xia, X.; Zhang, W.; Li, X. A Modified MEMS-Casting Based TSV Filling Method with Universal Nozzle Piece That Uses Surface Trenches as Nozzles. In Proceedings of the 2018 19th International Conference on Electronic Packaging Technology (ICEPT), Shanghai, China, 8–11 August 2018; pp. 536–539. [CrossRef]
13. Stauffer, J.M.; Dietrich, O.; Dutoit, B. RS9000, a novel MEMS accelerometer family for Mil/Aerospace and safety critical applications. In Proceedings of the IEEE/ION Position, Location and Navigation Symposium, Indian Wells, CA, USA, 4–6 May 2010; pp. 1–5. [CrossRef]
14. SCA820-D04. Available online: https://www.murata.com/zh-cn/products/productdetail?partno=SCA820-D04 (accessed on 7 June 2022).
15. High-Accuracy, High-Resolution, Low-Power, 2-Axis Digital Inclinometer with Embedded Machine Learning Core. Available online: https://www.st.com/en/mems-and-sensors/iis2iclx.html (accessed on 9 June 2022).
16. High Performance. Thin Package. Low TCO. Acceleration Sensors BMA456. Available online: https://www.bosch-sensortec.com/products/motion-sensors/accelerometers/bma456/ (accessed on 11 June 2022).
17. SDI'S TECHNOLOGY. Available online: https://www.silicondesigns.com/tech (accessed on 13 June 2022).
18. Jeong, Y.; Daruwalla, A.; Wen, H.; Ayazi, F. An out-of-plane hinge-shaped nano-gap accelerometer with high sensitivity and wide bandwidth. In Proceedings of the 2017 19th International Conference on Solid-State Sensors, Actuators and Microsystems (TRANSDUCERS), Kaohsiung, Taiwan, 18–22 June 2017; pp. 2131–2134. [CrossRef]

 *micromachines*

Article

Design and Fabrication of Silicon-Blazed Gratings for Near-Infrared Scanning Grating Micromirror

Sinong Zha [1,2], Dongling Li [1,2,*], Quan Wen [1,2], Ying Zhou [3] and Haomiao Zhang [1,2]

1. Key Laboratory of Optoelectronic Technology and System of the Education Ministry of China, Chongqing University, Chongqing 400030, China; c18722576829@163.com (S.Z.); quan.wen@cqu.edu.cn (Q.W.); 201908131093@cqu.edu.cn (H.Z.)
2. National Key Laboratory of Fundamental Science of Novel Micro/Nano Device and System Technology, Chongqing University, Chongqing 400030, China
3. Chongqing Chuanyi Automation Co., Ltd., Chongqing 401121, China; zhouying314422@163.com
* Correspondence: lidongling@cqu.edu.cn; Tel.: +86-137-5292-5461

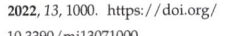

Citation: Zha, S.; Li, D.; Wen, Q.; Zhou, Y.; Zhang, H. Design and Fabrication of Silicon-Blazed Gratings for Near-Infrared Scanning Grating Micromirror. *Micromachines* 2022, 13, 1000. https://doi.org/10.3390/mi13071000

Academic Editor: Ruffino Francesco

Received: 9 May 2022
Accepted: 20 June 2022
Published: 25 June 2022

Publisher's Note: MDPI stays neutral with regard to jurisdictional claims in published maps and institutional affiliations.

Copyright: © 2022 by the authors. Licensee MDPI, Basel, Switzerland. This article is an open access article distributed under the terms and conditions of the Creative Commons Attribution (CC BY) license (https://creativecommons.org/licenses/by/4.0/).

Abstract: Blazed gratings are the critical dispersion elements in spectral analysis instruments, whose performance depends on structural parameters and topography of the grating groove. In this paper, high diffraction efficiency silicon-blazed grating working at 800–2500 nm has been designed and fabricated. By diffraction theory analysis and simulation optimization based on the accurate boundary integral equation method, the blaze angle and grating constant are determined to be 8.8° and 4 μm, respectively. The diffraction efficiency is greater than 33.23% in the spectral range of 800–2500 nm and reach the maximum value of 85.62% at the blaze wavelength of 1180 nm. The effect of platform and fillet on diffraction efficiency is analyzed, and the formation rule and elimination method of the platform are studied. The blazed gratings are fabricated by anisotropic wet etching process using tilted (111) silicon substrate. The platform is minished by controlling etching time and oxidation sharpening process. The fillet radius of the fabricated grating is 50 nm, the blaze angle is 7.4°, and the surface roughness is 0.477 nm. Finally, the blazed grating is integrated in scanning micromirror to form scanning grating micromirror by MEMS fabrication technology, which can realize both optical splitting and scanning. The testing results show that the scanning grating micromirror has high diffraction efficiency in the spectral range of 810–2500 nm for the potential near-infrared spectrometer application.

Keywords: silicon-blazed grating; platform and fillet; anisotropic wet etching; oxidation sharpening; diffraction efficiency

1. Introduction

Diffraction gratings are widely used as core optical splitters in modern optical devices such as ultra-precision measurement systems [1,2], spectrometers [3–6], semiconductor lasers [7,8], display techniques [9,10], and so on. With the development of micro-electromechanical system (MEMS) technology, there are kinds of silicon diffraction gratings developed, including rectangle gratings, sine gratings, and blazed gratings. Particularly, the blazed gratings exhibit excellent optical characteristics since they can focus the major part of the incident energy on a single non-zero diffraction order, which is beneficial for spectral analysis and detection. Generally, the diffraction and spectral characteristics of the blazed gratings are primarily dependent on the shape of the grating grooves, such as blaze angle and the topography of the grating groove. Therefore, it is important to design a proper blaze angle in a certain spectral range and control the grating shape during fabrication. For example, Zamkotsian et al. designed blazed gratings with a blaze angle of 5.04° working at the spectral range of 400–800 nm for high throughput spectrographs in space missions [11]. Sokolov et al. proposed blazed gratings with a blaze angle of 0.6°, 0.8°, and 1.0°, respectively, to achieve high diffraction efficiency at the tender X-ray region [12].

Mechanical ruling technique [12–16], ion beam etching (IBE) technique [17–22], and anisotropic wet etching technique [23–25] are commonly used to fabricate silicon-blazed gratings. Mechanical ruling technique using a specific shape diamond cutting tool extrudes on the grating substrate coated with aluminum or gold film to form grating grooves, which is simple and low-cost. But this method is not suitable for batch manufacturing of large area grating due to its low efficiency. Moreover, it is not compatible with MEMS fabrication process, and it is difficult to fabricate gratings on movable MEMS devices. During the IBE process, the grating structure is shaped through physical sputtering of materials. The blazed gratings with different blaze angle can be fabricated by adjusting the incident angle of ion beam, the duty cycle of the photoresist mask, and the etching rate ratio of the photoresist to the substrate material, which has the advantages of good direction, speed and shape control, high precision and smooth grating surface can be achieved [17,18]. Nevertheless, the fabrication process depends on the IBE equipment, so it is complex and costly. Furthermore, the etching rate of IBE is very low, and it is also not suitable for batch preparation of gratings. Anisotropic wet etching is the most promising technique for fabrication of high-performance blazed grating. It is a simple and convenient method to realize arbitrary blazed angle by using tilted (111) silicon wafers. Moreover, it provides atomically smooth blazed facets since it is formed by (111) lattice plane. However, there is always a small platform or nub left on the top of grating groove after etching, which will lead to amounts of stray light and reduce the diffraction efficiency. Voronov et al. fabricated 6° tilted blazed facets grating grooves by KOH solution, but there were silicon nubs on its top [26]. So, they used piranha solution ($H_2SO_4 + H_2O_2$) to chemically oxidize the silicon nubs, and then sharpened the gratings by hydrofluoric acid (HF) removal of the resulting oxides. However, there were 26 cycles applied to remove the 25-nm wide nubs in total, leading to complex fabrication process and large surface roughness. Frühauf and Krönert prepared triangular grating grooves with 500 nm width platform by anisotropic wet etching process using tetramethylammonium hydroxide (TMAH) solution [27], then the grating grooves were etched for a short time by isotropic etching (mixed solution of HNO_3, HF, and CH_3COOH). Finally, the radius of the convex edge was diminished to 50 nm, but the surface of the triangular gratings was relatively rough. Miles et al. fabricated a master blazed grating with platform width of 30 nm by combining electron beam etching and anisotropic wet etching [28]. To maximize the diffraction efficiency, the master grating was replicated using UV nanoimprinting lithography with the result that the platform became a flat trough which was small enough to be shadowed by the neighboring facet. However, gratings made on the imprint resist are limited in applications compared with those made on the silicon substrate. In our previous work, an 800–1800 nm blazed grating whose blaze angle is 7.4° was prepared by anisotropic wet etching and oxidation sharpening, but the width of the platform was 540 nm and there was a lack of systematic optimization research on etching and sharpening process [29].

The purpose of this paper is to design and fabricate high performance silicon-blazed gratings for near-infrared scanning grating micromirror used in miniature spectrometers. First, the blazed grating working at wavelength range from 800 nm to 2500 nm was designed. Effect of platform and fillet on diffraction efficiency is simulated and optimized. Reactive ion etching (RIE) and wet isotropic etching are investigated to prepare SiO_2 mask and narrow its width. The grating groove is formed by anisotropic wet etching of tilted (111) silicon substrate, and then sharpened by high temperature oxidation sharpening process. The optimal anisotropic wet etching time and oxidation sharpening time was obtained, and the grating grooves were analyzed by optical profiler, scanning electron microscope (SEM), and atomic force microscope (AFM). Finally, the blazed grating is integrated in scanning micromirror by MEMS fabrication technology to form scanning grating micromirror, and the optical properties are tested by diffraction efficiency measurement system and diffracted wavelength range test system.

2. Design of the Grating Structure

Performance of the blazed grating is quite dependent on its groove parameters, including blaze angle and grating constant. In this section, based on diffraction theory and diffraction efficiency simulation, the blazed grating structure is determined.

2.1. Principles

Blazed grating exhibits excellent optical characteristics because it can concentrate light energy at a specific diffraction order through diffraction plane at a specific blaze angle. The blazed grating and the optical path of its diffraction are depicted in Figure 1.

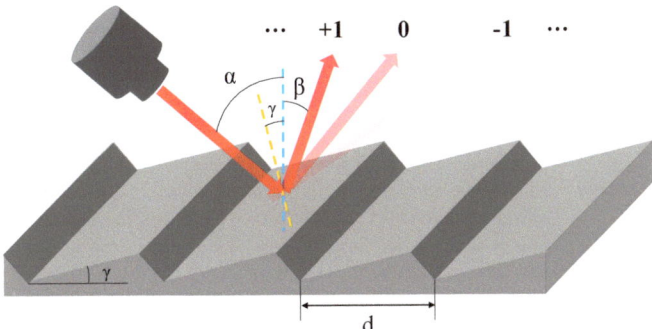

Figure 1. Schematic and optical path diffraction of a blazed grating.

The monochromatic light obliquely incident on the blazed grating, and then it is diffracted into discrete directions. Each grating profile can be considered to be a small, slit-shaped source of diffracted light that combines to form a set of diffracted wave fronts. Because the diffracted lights from different grating surfaces have the same phase, thus resulting in constructive interference. The relationships of incident angle α, diffraction angle β, grating constant d and the incident light wavelength λ are given by the diffraction grating equation, as shown in Equation (1):

$$d(\sin\alpha - \sin\beta) = m\lambda \tag{1}$$

where m is the diffraction order. When the light is blazing at +1 diffraction order, a relationship between the incident, diffraction, and blaze angle γ can be extracted as Equation (2):

$$\alpha - \gamma = \beta + \gamma \tag{2}$$

Then substituting Equation (2) into Equation (1), the relationship between wavelength, grating constant, incident angle, and blaze angle is expressed by Equation (3).

$$\gamma = \frac{1}{2}\left(\alpha - \arcsin\left(\sin\alpha - \frac{\lambda}{d}\right)\right) \tag{3}$$

It is seen that the wavelength of the grating is closely related to blaze angle and grating constant. Moreover, a spectrum will be generated at +1 diffraction order if a polychromatic light incident on the blazed grating in the same way, and any wavelength λ in the spectrum must satisfy the grating equation. Substituting $\sin\beta \geq -1$ into Equation (2), the reasonable range of grating constant is obtained by Equation (4),

$$d \geq \frac{\lambda_{max}}{1 + \sin\alpha} \tag{4}$$

where λ_{max} denotes the maximum wavelength in the spectrum.

As the blazed grating is applied in the miniature near-infrared spectrometer whose spectrum range is from 800 nm to 2500 nm. The incident angle α is fixed at 14.1° as the requirement of optical path in the spectrometer. According to Equation (4), grating constant must satisfy $d \geq 2$ μm. However, a large grating constant will result in a low spectral resolution. Therefore, considering the spectral range, spectral resolution, and lithography resolution, the grating constant is determined to be 4 μm. The blaze wavelength is derived from the thumb rule that the diffraction efficiency drops to 50% of its peak value at $2/3\ \lambda_b$ and at $9/5\ \lambda_b$, where λ_b is the blaze wavelength [30]. The wide spectrum ranges from 800 nm to 2500 nm should contain the range from $2/3\ \lambda_b$ to $9/5\ \lambda_b$ to get high diffraction efficiency. So, the blaze wavelength is calculated as 1200–1389 nm. Substituting it into Equation (3), the desired blaze angle is 8.7°–10.0°.

2.2. Optimization of the Blaze Angle

PCGrate is the software used for the exact calculation of diffraction efficiency by the accurate boundary integral equation method. In this work, the diffraction efficiency curves under different blaze angles are simulated using PCGrate Demo v.6.4 to get optimal blaze angle. Figure 2 shows the grating model for simulation. The gray area is the planar silicon-blazed grating with grating constant of 4 μm. To increase reflectivity, a 100 nm-thick aluminum (Al) film is added to cover the grating surface. Since the grating is fabricated by anisotropic wet etching process, the grating groove is formed by intersecting (111) faces and the apex angle of the grating is 109.47°. So, the border profile of the grating is defined by the blaze angle.

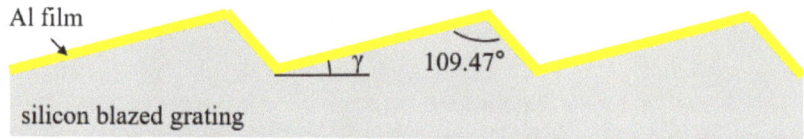

Figure 2. The grating model for diffraction efficiency simulation.

In order to preliminarily determine the range of optimal blaze angel, diffraction efficiency curves under blaze angle of 8.7°, 8.9°, 9.2°, 9.5°, 9.8°, and 10.0° are analyzed, as shown in Figure 3a. It can be seen that the diffraction efficiency increases rapidly and then decreases slowly as the wavelength increases. It gradually goes down in short wavelength and changes in the opposite direction in long wavelength as the increase of the blaze angle. Moreover, some anomalous jumps are clearly observed on these curves, which can be attributed to the exceeding 90° of the diffraction angle at a high order. The diffraction efficiency jumps into the lower orders when higher orders are diffracted into the grating surface and become evanescent. Through comparative analysis of diffraction efficiency at different blazed angles and different wavelengths, it is found that the blazed grating has higher diffraction efficiency when the blaze angle is 8.9°. The diffraction efficiency is higher than 32.84%. It means that the optimal blaze angle is around 8.9°. Then, the blaze angle is further optimized by setting blaze angle as 8.7°, 8.8°, 8.9°, 9.0°, 9.1° respectively. The diffraction efficiency curves are shown in Figure 3b. It shows that the diffraction efficiency is higher than 33.24% when the blaze angle is 8.8°, so the final optimized blaze angle is 8.8°. At this time, the lowest diffraction efficiency is 33.24% at wavelength 800 nm, while the maximum value is 85.62% at blaze wavelength 1180 nm.

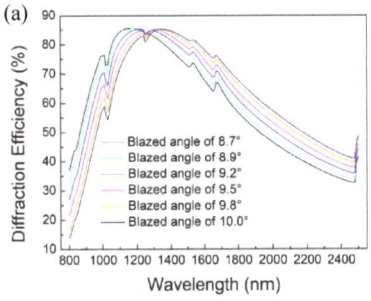
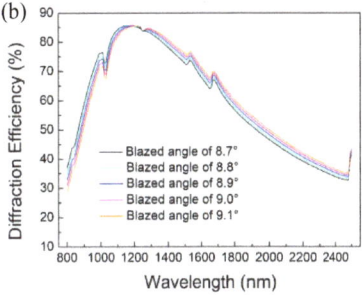

Figure 3. Diffraction efficiency of blazed gratings with blaze angle of (**a**) 8.7°, 8.9°, 9.2°, 9.5°, 9.8°, and 10.0°; (**b**) 8.7°, 8.8°, 8.9°, 9.0°, and 9.1°.

However, a platform or fillet usually remains on the top of the grating groove during the blazed grating fabrication process, which induces stray light and degrades the diffraction efficiency. Therefore, it should be considered in the design process. For a clear view, diffraction efficiency under blaze angle of 8.8° varying with different platform and fillet size are also investigated, as shown in Figure 4. It can be seen that the diffraction efficiency gradually decreases as the platform becomes wider. Assuming the top of the grating is a platform, the diffraction efficiency at blaze wavelength of 1150 nm is 84.03% when the platform width is 400 nm, while it is 85.45% when the platform width is 100 nm. While for fillet, the diffraction efficiency at blaze wavelength is 82.54% and 85.39% respectively, when the radius is 400 nm and 100 nm. The result denotes that the diffraction efficiency indeed decreases with the increase of platform or fillet size, so it needs to be minimized in subsequent fabrication process. Furthermore, the effect of platform or fillet on diffraction efficiency becomes small when the width of the platform or fillet is less than 100 nm. Therefore, the platform width or the fillet radius of the blazed gratings should be sharpened to less than 100 nm to obtain high performance.

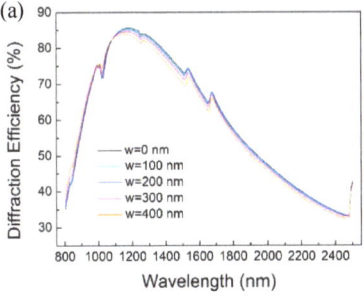
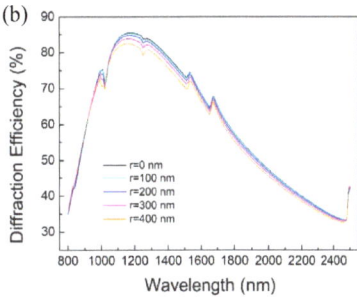

Figure 4. Diffraction efficiency with different (**a**) width of platforms; (**b**) radius of fillets.

3. Fabrication and Characterization of Blazed Gratings

After the determination of the grating structure, it is important to control the groove shape during the next fabrication process. In this section, an improved process is proposed to fabricate the blazed gratings.

3.1. Experimental

To fabricate the sawtooth grating groove, the tilted (111) silicon wafer was used as the substrate. Figure 5a illustrates the schematic diagram of titled silicon wafer cutting. It was cut by (111) silicon ingot, which rotated a certain angle θ from (111) crystal plane to (110) plane along [$\bar{1}$10] orientation. The cutting angle θ is equal to the designed blaze angle γ.

According to the characteristics of anisotropic wet etching, the grating groove was formed by the intersection of slowly etched surface (111) faces, as shown in Figure 5b.

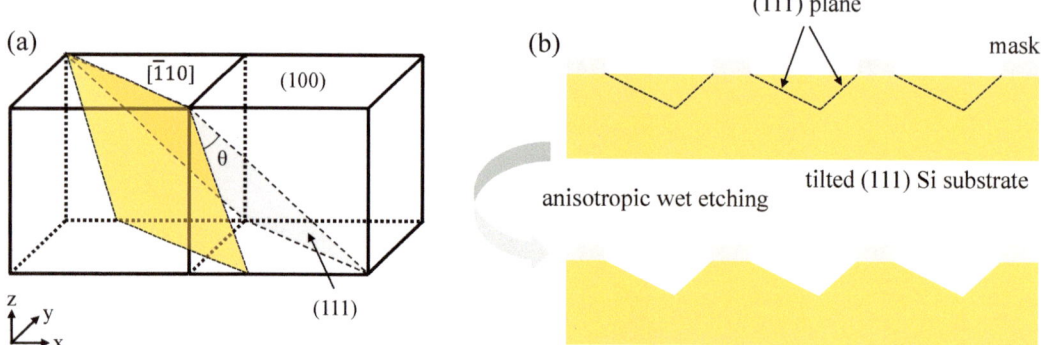

Figure 5. (**a**) The schematic diagram of tilted (111) silicon cutting; (**b**) formation of blazed grating groove by anisotropic wet etching.

The microfabrication process of the blazed grating is depicted in Figure 6. First, the double sides polished, tilted (111) silicon wafer with cutting angle of 8.8° was used as the substrate (Figure 6a), whose thickness was 500 ± 20 μm, resistivity was 70–80 Ω·cm, and error of the cutting angle is ±0.5°. Then, a 300-nm-thick SiO_2 layer was thermally oxidized on the wafer surface (Figure 6b). Next, a photoresist layer was coated on the SiO_2 layer and a series of 500-nm-wide photoresist stripes which had 4 μm period and paralleled to the [$\bar{1}$10] direction was fabricated by stepper lithography (Figure 6c). The SiO_2 layer was etched by reactive ion etching (RIE) technique and buffered oxidate etch (BOE) to form etching mask of gratings (Figure 6d). Subsequently, the grating groove was shaped by anisotropic wet etching (Figure 6e). An additional oxidation of gratings without removing the SiO_2 etching mask was purposely taken to further decrease the platform on grating top (Figure 6f). The desired grating groove was obtained by removing SiO_2 on both sides of the wafer (Figure 6g). Finally, a 100 nm-thick aluminum was deposited on the grating surface using magnetron sputtering to increase the reflectivity for spectral range of 800–2500 nm (Figure 6h).

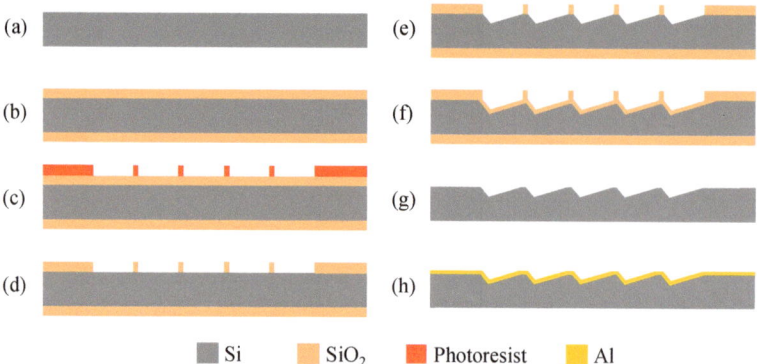

Figure 6. The fabrication flow of the blazed grating. (**a**) substrate preparation; (**b**) thermal oxidation of the substrate; (**c**) photoresist mask preparation; (**d**) SiO_2 mask preparation; (**e**) anisotropic wet etching; (**f**) thermal oxidation of the grating; (**g**) SiO_2 removing; (**h**) aluminum film deposition.

3.2. Fabrication of Grating Etching Mask

SiO$_2$ mask was used as etching mask of gratings in the experiment. In order to reduce the influence of platform on the performance of gratings, the SiO$_2$ mask should be minimized. RIE and BOE isotropic etching was used to fabricate SiO$_2$ mask and reduce its width. Generally, it was difficult to accurately etch 300 nm-thick SiO$_2$ layer and stop on the Si surface only using RIE technique. If the SiO$_2$ layer was overetched, the underlying Si substrate would be also slightly etched, forming a large nub on the top of the grating groove after anisotropic wet etching process, as shown in Figure 7a. The nub was also shaped by (111) plane and hard to be eliminated, so the diffraction efficiency is reduced because of introducing stray light and narrowing the blazed facet. Whereas, long-time wet etching in BOE could damage and cause peeling of the photoresist. Therefore, after RIE etching of SiO$_2$ for 250 nm, BOE was applied to etch the remaining SiO$_2$, and the etching would spontaneously stop at the interface of SiO$_2$ and Si. The grating groove fabricated in this way is shown in Figure 7b, and the nub is successfully avoided.

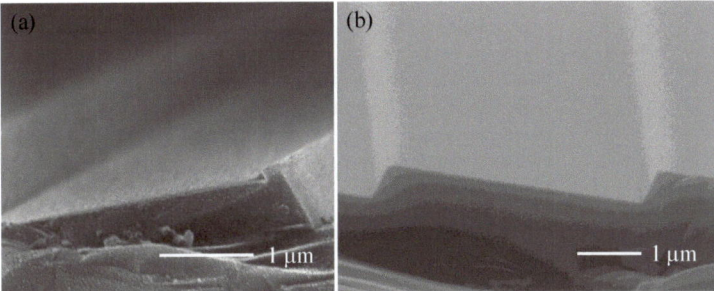

Figure 7. Grating grooves prepared under different grating mask preparation process. (**a**) only using RIE technique; (**b**) combining RIE technique and BOE wet etching technique.

Optimized parameters of RIE and BOE wet etching were investigated for good etching uniformity and small size etching mask. Sulfur hexafluoride (SF$_6$) was used as the etchant gas, and oxygen (O$_2$) was used as protection gas in the RIE process. A high SF$_6$ flow and high etching power resulted in a high etching rate, which was set to be 30 sccm and 150 W respectively. As the O$_2$ flow increased, anisotropy and uniformity of RIE etching were both improved. However, a small width of SiO$_2$ stripes was needed in the experiment, which can be realized by isotropic lateral etching at low O$_2$ flow. Moreover, high O$_2$ flow would easily remove the photoresist, so O$_2$ flow was set to be 5 sccm. After 180 s of etching by RIE, the thickness of the remaining SiO$_2$ was 50nm, then it was overetched by BOE solution for 90 s. Images of photoresist stripes before etching process and SiO$_2$ stripes after photoresist removing are shown in Figure 8. It could be seen that the width of the grating mask was narrowed from 500 nm to 200 nm.

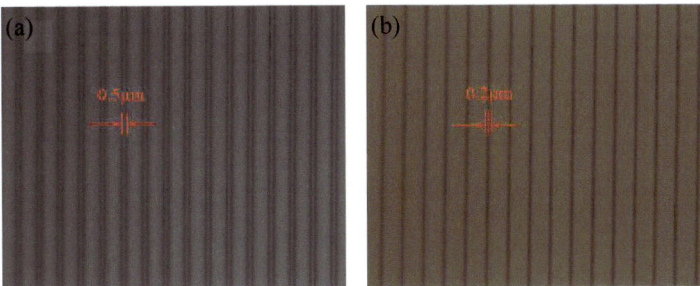

Figure 8. Images of (**a**) photoresist stripes before the etching process; (**b**) SiO$_2$ stripes after photoresist removing.

3.3. Anisotropic Wet Etching of Blazed Gratings

The gratings were prepared by anisotropic wet etching process. The exposed silicon was etched in the anisotropic etching solution at different rates according to the crystal direction. When the two (111) crystal planes intersected, the etching stopped, and the triangle blazed gratings were obtained. The blaze angle of the gratings was the tangent angle of tilted silicon wafer. For smooth grating surface, tetramethylammonium hydroxide (TMAH) solution with mass fraction of 25% was used in the experiment. The grating groove shapes at different etching time of 75 s, 3 min, 5 min, and 7 min were observed by scanning electron microscopy (SEM) after SiO_2 mask removing, and the morphology of the platform on the top of the grating was analyzed. It was seen that a 165-nm-width platform was formed on the top of the grating due to the protection of a narrow SiO_2 mask as the etching time was 75 s, as shown in Figure 9a. The rounded edge was caused by lateral corner cut corrosion. Then, a small nub emerges as the etching time increases, due to the over etching of the (111) planes and shaped by new (111) sidewalls. The height of the nub was 32 nm and the width was 163 nm after 3 min etching, as shown in Figure 9b. The nub became larger in depth and smaller in width with the increasing of etching time because of the constant etching of (111) planes, as shown in Figure 9c,d. When the etching time was 7 min, the width of the nub was 52 nm, and the height was 57 nm. However, the nub was difficult to eliminate further because the SiO_2 stripes fell off as continuous etching. Moreover, the long-term wet etching led to an increase in surface roughness of the grating surface, which ultimately affected the diffraction efficiency. Hence, the suitable etching time was 75 s.

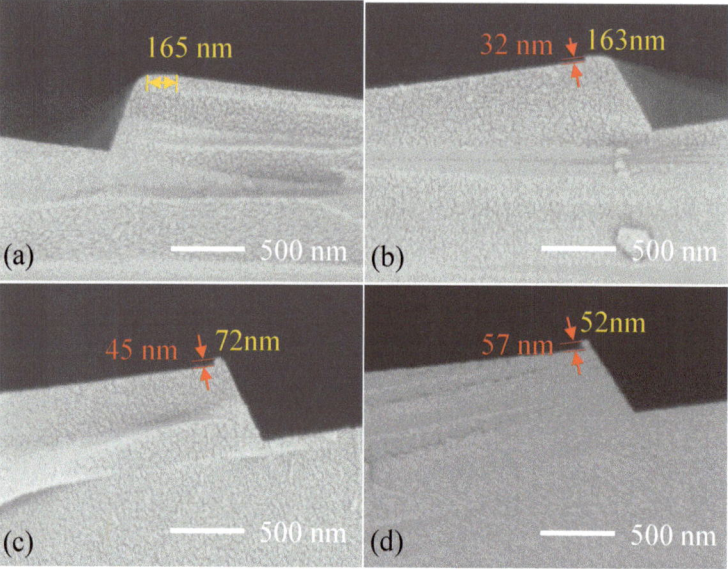

Figure 9. Grating grooves etched by different time. (**a**) 75 s; (**b**) 3 min; (**c**) 5 min; and (**d**) 7 min.

3.4. Oxidation Sharpening

After anisotropic wet etching process, an oxidation sharpening process was subsequently employed to further decrease the platform without removing the SiO_2 etching mask. Figure 10 schematically displays the oxidation sharpening process. The grating samples were thermally oxidized for hours at 1050 °C, oxidation rate at the top of the grating groove was much lower than that of the groove sides since the SiO_2 mask was served as a barrier layer. As a result, the thermal oxide layers are mainly grown at the sides of the groove. Subsequently, the oxidation layer was etched in BOE solution until the Si surface was hydrophobic, and the grating groove was effectively sharpened.

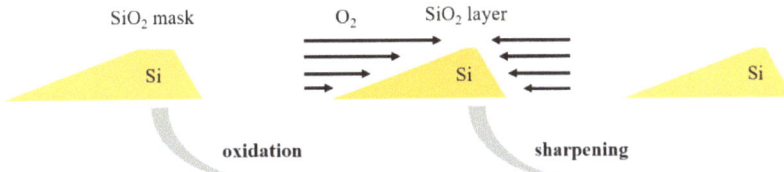

Figure 10. Schematic diagram of oxidation sharpening process.

The grating groove depended on the oxidation time. The grating samples was oxidated at 1050 °C for 1 h, 2 h, and 4 h, respectively, and the SEM images of the gratings after the oxide layer removing are shown in Figure 11. It can be seen that the platforms decreased and finally converted to a fillet. The radius of the fillet first decreased and then increased as the oxidation time increased. There was a 150-nm-wide platform on the grating top when the oxidation time was 1 h. When the oxidation time increased to 2 h, the platform turned to a fillet, whose radius was 50 nm. However, continued oxidation for 4 h caused the fillet to become larger, and the radius was about 130 nm. Therefore, the optimized oxidation sharpening time was 2 h.

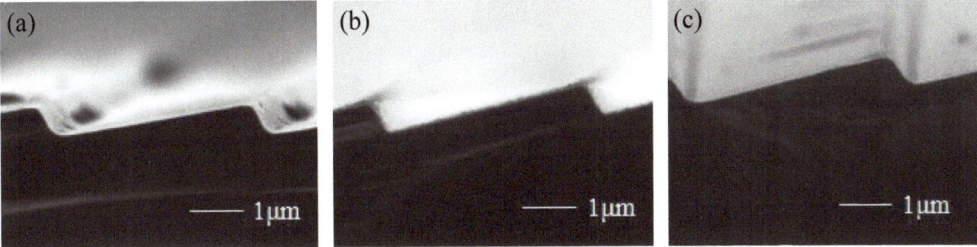

Figure 11. Grating grooves prepared under different oxidation time of (**a**) 1 h; (**b**) 2 h; and (**c**) 4 h.

The grating profile measured by the optical profiler is shown in Figure 12a. The actual blaze angle was 7.4°, which had an offset of 1.4° to the designed value. This deviation mainly came from two sources. One was the cutting angle error of the tilted (111) substrate, another was the etching time difference between upper and lower surfaces. The atomic force microscopy (AFM) image of the blazed grating is also presented in Figure 12b. It could be seen that the measured root mean square (RMS) roughness of the grating surface with aluminum reflective coating was 0.477 nm, that proved the grating surface possessed great smoothness.

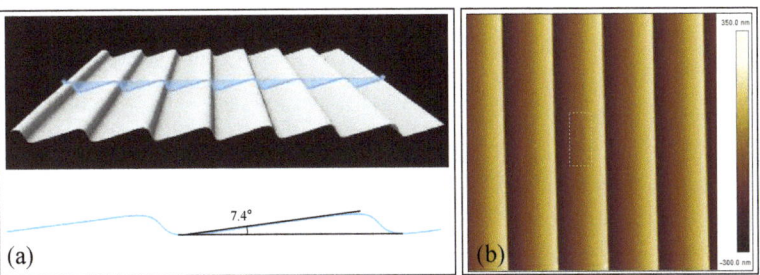

Figure 12. (**a**) Image of blazed grating profile; (**b**) AFM image of the blazed grating.

4. Application in Scanning Grating Micromirror

The fabricated blazed grating was integrated on the movable micromirror to form a scanning grating micromirror applying in the miniature near-infrared spectrometer. It could realize optical splitting and scanning at the same time. The schematic diagram of the scanning grating micromirror is shown in Figure 13. A compound light dispersed into lights arranging in order of wavelength to form a spectrum by the dispersion role of the integrated blazed grating. With the micromirror rotating around torsional springs, the incident angle continuously changed as well as the diffraction angle, resulting in scanning of the diffracted lights. Consequently, a single detector at a fixed position could be able to detect the diffraction spectrum. Applying this device in the near-infrared spectrometer can avoid using expensive near-infrared detection array and reduce the spectrometer size. Using standard MEMS fabrication technology, the blazed grating was integrated on the backside of the scanning micromirror to form scanning grating micromirror. The fabricated device is shown in Figure 13b,c.

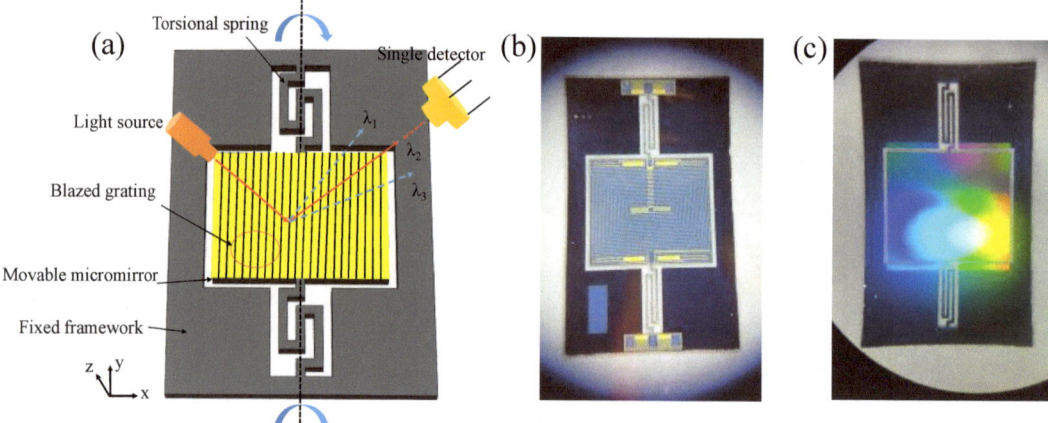

Figure 13. (**a**) Schematic diagram of the scanning grating micromirror; (**b**) frontside of fabricated scanning grating micromirror chip with actuated coils; (**c**) backside of fabricated scanning grating micromirror chip integrated with blazed gratings.

The absolute diffraction efficiency is an important performance parameter for the scanning grating micromirror, which indicates the optical energy transfer capability. It is a measure of how much optical power is diffracted into a designated direction compared to the power incident onto the diffractive element, which is defined as:

$$\eta_{\lambda,m} = \frac{I_{\lambda,m}}{I_\lambda} \quad (5)$$

where $I_{\lambda,m}$ is diffraction light power at diffraction order of m, and I_λ is the incident light power. Figure 14a schematically showed the measurement for absolute diffraction efficiency. The set up mainly consisted of a laser device, an optical power meter, and a scanning grating micromirror. First, the laser was normally incident into the optical power meter to directly detect the incident laser power I_λ. Then, the laser was incident into the center of the scanning grating micromirror with the incident angle of 14.1°, and the optical power meter was moved to detect the light power at +1 diffraction order $I_{\lambda+1}$. According to Equation (5), the absolute diffraction efficiency could be calculated. Figure 14b shows the photographic image of the measurement set up.

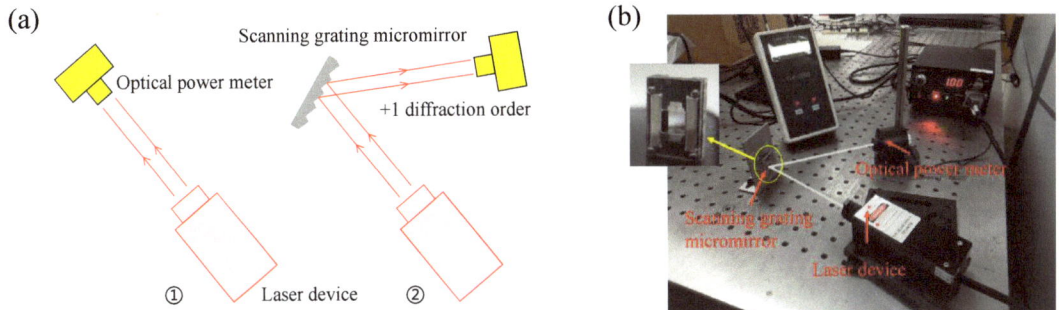

Figure 14. (a) Schematic diagram and (b) photography of the grating diffraction efficiency measurement setup.

Since the laser emitted TM-polarized wave in our test system, the measured results indicated as diffraction efficiency under TM polarization. Here, theoretical diffraction efficiency of blazed grating with blaze angle of 7.4° under TM polarization was obtained by PCGrate, and the theoretical values at given wavelength spots are given in Table 1. The actual diffraction efficiency was tested and compared to its corresponding theoretical value. The results showed high diffraction efficiency of the blazed grating but an error of 4–5% to the ideal value. This error was derived from the oxidation of Al reflective layer and dust adhesion on the grating surface in the test environment, which was acceptable.

Table 1. The measured results of the diffraction efficiency at +1 diffraction order.

λ (nm)	ηTM (experimental)	ηTM (theoretical)	error
808	56.79%	61.52%	4.73%
1064	86.35%	90.87%	4.52%
1550	67.73%	72.22%	4.49%

The diffracted wavelength range of the proposed scanning grating micromirror was tested with our previous NIR spectrometer setup [31]. As Figure 15a demonstrates, the light was emitted through a filter and an entrance slit. Then it was collimated by an off-axis parabolic mirror and reflected toward the scanning grating micromirror by using a flat mirror. The diffracted lights were reflected and focused on the detector's plane through an exit slit. As the scanning grating micromirror tilted, the diffraction lights with varying wavelengths were scanned through the detector which gave the spectrum information for the incoming lights.

During the diffracted wavelength range measurement, a set of narrow band filters and a halogen tungsten lamp were used. The detector's response over scanning process was amplified and recorded with an oscilloscope. The measured data with 810 nm (bandwidth of 10 nm) and 2580 nm (bandwidth of 50 nm) filters are given in Figure 15b and c respectively. The yellow line is the driving voltage for the rotation of scanning grating micromirror, and the green line is the spectral signal received by the detector. Two peaks of detected signal per period of driving voltage are observed since the diffracted light scans through the detector back and forth in a rotation period of scanning grating micromirror. The peaks on the recorded detector's response (the green line) indicates that the proposed scanning grating micromirror has enough diffraction efficiency in the spectral range of 810–2500 nm for potential spectrometer application.

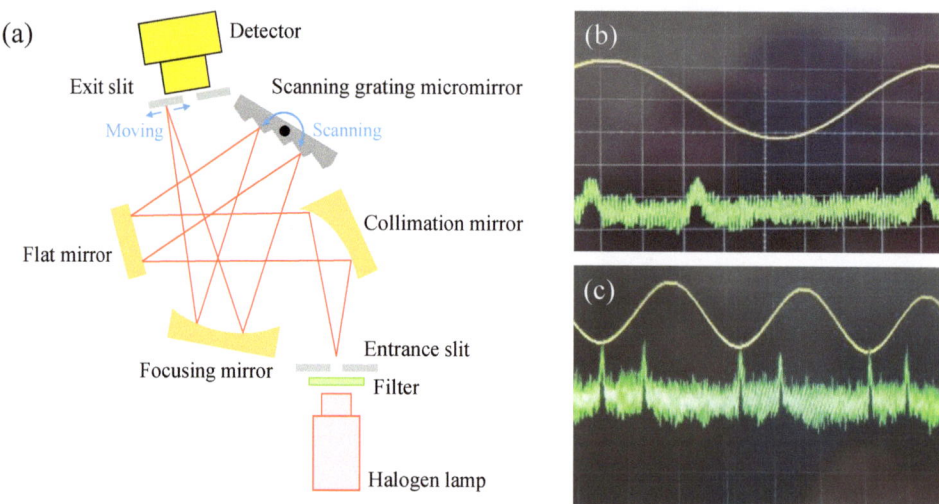

Figure 15. (**a**) Schematic diagram of diffracted wavelength range testing; (**b**) response of the photodetector to monochromatic light (**b**) 810 nm; (**c**) 2580 nm.

5. Conclusions

In this paper, a blazed grating applied in near-infrared scanning grating micromirror was designed, fabricated, and characterized. Based on the diffraction theory and diffraction efficiency simulation, the blaze angle and grating constant were determined as 8.8° and 4 µm, respectively. The effects of the platform and fillet on the diffraction efficiency were also investigated, showing a larger platform and fillet resulted in lower diffraction efficiency. However, the diffraction efficiency hardly decreased when the platform width or the fillet radius was less than 100 nm, which provided a target for controlling the defects on grating shapes in the fabrication process. By optimizing RIE and wet etching process, the SiO_2 mask width was narrowed from 500 nm to 200 nm. The blazed grating was fabricated by anisotropic wet etching combined with high temperature oxidation sharpening process using tilted (111) silicon substrate. The platform was minished to a fillet with radius of 50 nm by optimizing etching time and oxidation sharpening time. The grating grooves were analyzed by optical profiler and AFM, showing the fabricated blaze angle was 7.4° and the RMS roughness of the grating surface with aluminum reflective coating was 0.477 nm. Finally, the blazed grating was integrated in scanning micromirror to form scanning grating micromirror using MEMS fabrication technology, which can realize both optical splitting and scanning. The testing results show that the scanning grating micromirror has high diffraction efficiency in the spectral range of 810–2500 nm. The proposed design and fabrication method of blazed gratings are promising to obtain high-performance blazed gratings applied in spectral analysis instruments.

Author Contributions: S.Z. and D.L. designed and fabricated the blazed grating. Y.Z. gave suggestions on grating design and test. Q.W. designed the scanning grating micromirror and its test system. S.Z. and H.Z. completed the optical performance measurement of the scanning grating micromirror. S.Z. and D.L. wrote and revised the paper. All authors have read and agreed to the published version of the manuscript.

Funding: This work was supported by the National Key R&D Program of China (Grant No. 2018YFF01011200) and the National Natural Science Foundation of China (Grant No. 61804016).

Institutional Review Board Statement: Not applicable.

Informed Consent Statement: Not applicable.

Data Availability Statement: Data available on request due to restrictions, e.g., privacy or ethical. The data and material presented in this study are available on request from the corresponding author.

Conflicts of Interest: The authors declare no conflict of interest.

References

1. Chen, F.; Brown, G.M.; Song, M.M. Overview of 3-D shape measurement using optical methods. *Opt. Eng.* **2000**, *39*, 10–22.
2. Sansoni, G.; Patrioli, A. Noncontact 3D sensing of free-form complex surfaces. *Proc. SPIE* **2001**, *4309*, 232–239.
3. Lin, D.M.; Fan, P.Y.; Hasman, E.; Brongersma, M.L. Dielectric gradient metasurface optical elements. *Science* **2014**, *345*, 298–302. [CrossRef] [PubMed]
4. Zhao, B.S.; Meijer, G.; Schöllkopf, W. Quantum reflection of He$_2$ several nanometers above a grating surface. *Science* **2011**, *331*, 892–894. [CrossRef] [PubMed]
5. Xing, J.Y.; Cui, H.; Hu, P.H.; Jin, S.Q.; Hu, M.Y.; Xia, G.; Hu, H.B. Gratings in dispersion-compensated polarization Sagnac interferometer. *Opt. Commun.* **2020**, *458*, 124806. [CrossRef]
6. Sandfuchs, O.; Kraus, M.; Brunner, R. Structured metal double-blazed dispersion grating for broadband spectral efficiency achromatization. *J. Opt. Soc. Am. A* **2020**, *37*, 1369–1380. [CrossRef] [PubMed]
7. Cheng, F.M.; Zhang, J.C.; Wang, D.B.; Gu, Z.H.; Zhuo, N.; Zhai, S.Q.; Wang, L.J.; Liu, J.Q.; Liu, S.M.; Liu, F.Q.; et al. Demonstration of high-power and stable single-mode in a quantum cascade laser using buried sampled grating. *Nanoscale Res. Lett.* **2019**, *14*, 123. [CrossRef]
8. Yang, K.; Liu, Y.G.; Wang, Z.; Li, G.Y.; Han, Y.; Zhang, H.W.; Yu, J. Five-wavelength-switchable all-fiber erbium-doped laser based on few-mode tilted fiber Bragg grating. *Opt. Laser Technol.* **2018**, *108*, 273–278. [CrossRef]
9. Wang, X.; Wilson, D.; Muller, R.; Maker, P.; Psaltis, D. Liquid-crystal blazed-grating beam deflector. *Appl. Opt.* **2000**, *39*, 6545–6555. [CrossRef]
10. Zhou, F.; Hua, J.Y.; Shi, J.C.; Qiao, W.; Chen, L.S. Pixelated blazed gratings for high brightness multiview holographic 3D display. *IEEE Photonics Technol. Lett.* **2020**, *32*, 283–286. [CrossRef]
11. Zamkotsian, F.; Zhurminsky, I.; Lanzoni, P.; Tchoubaklian, N.; Schneider, C.; Fricke, S.; Schnieper, M.; Lütolf, F.; Luitot, C.; Costes, V. Convex blazed gratings for high throughput spectrographs in space missions. *Proc. SPIE* **2019**, *11180*, 1118051.
12. Sokolov, A.; Huang, Q.S.; Senf, F.; Feng, J.T.; Lemke, S.; Alimov, S.; Knedel, J.; Zeschke, T.; Kutz, O.; Seliger, T.; et al. Optimized highly efficient multilayer-coated blazed gratings for the tender X-ray region. *Opt. Express* **2019**, *27*, 16833–16846. [CrossRef] [PubMed]
13. Wood, R.W. The echelette grating for the infra-red. *Philos. Mag.* **1910**, *20*, 770–778. [CrossRef]
14. Xu, D.; Owen, J.D.; Papa, J.C.; Reimer, J.; Suleski, T.J.; Troutman, J.R.; Davies, M.A.; Thompson, K.P.; Rolland, J.P. Design, fabrication, and testing of convex reflective diffraction gratings. *Opt. Express.* **2017**, *25*, 15252–15268. [CrossRef]
15. Siewert, F.; Lochel, B.; Buchheim, J.; Eggenstein, F.; Firsov, A.; Gwalt, G.; Kutz, O.; Lemke, S.; Nelles, B.; Rudolph, I.; et al. Grating for synchrotron and FEL beamlines: A project for the manufacture of ultra-precise gratings at Helmholtz Zentrum Berlin. *J. Synchrotron. Radiat.* **2018**, *25*, 91–99. [CrossRef]
16. Montesanti, R.C.; Little, S.L.; Kuzmenko, P.J.; Bixler, J.V.; Jackson, J.L.; Lown, J.G.; Priest, R.E.; Yoxall, B.E. Strategies for single-point diamond machining a large format germanium blazed immersion grating. *Proc. SPIE* **2016**, *9912*, 991233.
17. Kowalski, M.P.; Cruddace, R.G.; Heidemann, K.F.; Lenke, R.; Kierey, H.; Barbee, T.W.; Hunter, W.R. Record high extreme-ultraviolet efficiency at near-normal incidence from a multilayer-coated polymer-overcoated blazed ion-etched holographic grating. *Opt. Lett.* **2004**, *29*, 2914–2916. [CrossRef]
18. Kowalski, M.P.; Cruddace, R.G.; Barbee, T.W.; Hunter, W.R.; Heidemann, K.F.; Nelles, B.; Lenke, R.; Kierey, H. High-efficiency multilayer-coated polymer-overcoated blazed ion-etched holographic gratings for high-resolution EUV astronomical spectroscopy. *Proc. SPIE* **2004**, *5488*, 910–921.
19. Fu, Y.; Bryan, N.K.A.; Zhou, W. Self-organized formation of a blazed-grating-like structure on Si (100) induced by focused ion-beam scanning. *Opt. Express* **2004**, *12*, 227–233. [CrossRef]
20. Shen, C.; Tan, X.; Jiao, Q.B.; Zhang, W.; Wu, N.; Bayan, H.; Qi, X.D. Convex blazed grating of high diffraction efficiency fabricated by swing ion-beam etching method. *Opt. Express* **2018**, *26*, 25381–25398. [CrossRef]
21. Lin, H.; Li, L.F. Fabrication of extreme-ultraviolet blazed gratings by use of direct argon-oxygen ion-beam etching through a rectangular photoresist mask. *Appl. Opt.* **2008**, *47*, 6212–6218. [CrossRef] [PubMed]
22. Lin, H.; Zhang, L.C.; Li, L.F.; Jin, C.S.; Zhou, H.J.; Huo, T.L. High-efficiency multilayer-coated ion-beam-etched blazed grating in the extreme-ultraviolet wavelength region. *Opt. Lett.* **2008**, *33*, 485–487. [CrossRef] [PubMed]
23. Fujii, Y.; Aoyama, K.; Minowa, J. Optical demultiplexer using a silicon echelette grating. *IEEE J. Quantum Electron.* **1980**, *16*, 165–169. [CrossRef]
24. Philippe, P.; Valette, S.; Mata Mendez, O.; Maystre, D. Wavelength demultiplexer: Using echelette gratings on silicon substrate. *Appl. Opt.* **1985**, *24*, 1006–1011. [CrossRef] [PubMed]
25. Voronov, D.L.; Lum, P.; Naulleau, P.; Gullikson, E.M.; Fedorov, A.V.; Padmore, H.A. X-ray diffraction gratings: Precise control of ultra-low blaze angle via anisotropic wet etching. *Appl. Phys. Lett.* **2016**, *109*, 43112. [CrossRef]

26. Voronov, D.L.; Ahn, M.; Anderson, E.H.; Cambie, R.; Chang, C.H.; Gullikson, E.M.; Heilmann, R.K.; Salmassi, F.; Schattenburg, M.L.; Warwick, T.; et al. High-efficiency 5000 lines/mm multilayer-coated blazed grating for extreme ultraviolet wavelengths. *Opt. Lett.* **2010**, *35*, 2615–2617. [CrossRef]
27. Frühauf, J.; Krönert, S. Wet etching of silicon gratings with triangular profiles. *Microsyst. Technol.* **2005**, *11*, 1287–1291. [CrossRef]
28. Miles, D.M.; McCoy, J.A.; McEntaffer, R.L.; Eichfeld, C.M.; Lavallee, G.; Labella, M.; Drawl, W.; Liu, B.; DeRoo, C.T.; Steiner, T. Fabrication and diffraction efficiency of a large-format, replicated X-ray reflection grating. *Astrophys. J.* **2018**, *869*, 95. [CrossRef]
29. Nie, Q.Y.; Xie, Y.Y.; Chang, F. MEMS blazed gratings fabricated using anisotropic etching and oxidation sharpening. *AIP Adv.* **2020**, *10*, 65216. [CrossRef]
30. Mouroulis, P.; Wilson, D.W.; Maker, P.D.; Muller, R.E. Convex grating types for concentric imaging spectrometers. *Appl. Opt.* **1998**, *37*, 7200–7208. [CrossRef]
31. Huang, L.K.; Wen, Q.; Huang, J.; Yu, F.; Lei, H.J.; Wen, Z.Y. Miniature broadband NIR spectrometer based on FR4 electromagnetic scanning micro-grating. *Micromachines* **2020**, *11*, 393. [CrossRef]

Article

Development of a Fault Detection Instrument for Fiber Bragg Grating Sensing System on Airplane

Cuicui Du, Deren Kong * and Chundong Xu

School of Mechanical Engineering, Nanjing University of Science & Technology, Nanjing 210094, China; 218101010101@njust.edu.cn (C.D.); dccnjust@126.com (C.X.)
* Correspondence: kongderen218@126.com

Abstract: This study develops a fault detection device for the fiber Bragg grating (FBG) sensing system and a fault detection method to realize the rapid detection of the FBG sensing system on airplanes. According to the distribution of FBG sensors on airplanes, the FBG sensing system is built based on wavelength division multiplexing (WDM) and space division multiplexing (SDM) technologies. Furthermore, the hardware and software of the fault detection device and the relevant FBG demodulator are studied in detail. Additionally, in view of the similar features of the healthy FBG sensor in the same measuring point, a rapid fault diagnosis method based on a synthetical anomaly index is proposed. The features (light intensity I, signal length L, standard deviation of original sample σ and energy value in time-domain P) of FBG sensors are extracted. The aggregation center value of the above feature values is obtained through the loop iteration method. Furthermore, the separation degrees of features are calculated and then form the synthetical anomaly index so as to make an effective diagnosis of the state of the FBG sensor. Finally, the designed fault detection instrument and proposed fault detection method are used to monitor the 25 FBG sensors on the airplane, the results indicated that three faulty and two abnormal FBG sensors on the airplane are identified, showing the effectiveness of the proposed fault detection method.

Keywords: fault detection instrument; fiber Bragg grating sensing system; feature characteristics; synthetical anomaly index; aircraft

Citation: Du, C.; Kong, D.; Xu, C. Development of a Fault Detection Instrument for Fiber Bragg Grating Sensing System on Airplane. *Micromachines* **2022**, *13*, 882. https://doi.org/10.3390/mi13060882

Academic Editor: Kun Li

Received: 5 May 2022
Accepted: 29 May 2022
Published: 31 May 2022

Publisher's Note: MDPI stays neutral with regard to jurisdictional claims in published maps and institutional affiliations.

Copyright: © 2022 by the authors. Licensee MDPI, Basel, Switzerland. This article is an open access article distributed under the terms and conditions of the Creative Commons Attribution (CC BY) license (https:// creativecommons.org/licenses/by/ 4.0/).

1. Introduction

Fiber Bragg grating (FBG) sensors have been widely used in the structural health monitoring (SHM) of aircraft because of their advantages of a wide measuring range, high measuring precision and sensibility, small size, anti-electromagnetic interference, and corrosion resistance, which all make it reliable for high precision, real-time monitoring the strain, temperature and vibration of the aircraft [1–3].

In practical SHM aircraft engineering, a large number of FBG sensors are typically installed on the key components of aircraft (e.g., wings, tails, frame, etc.) to obtain complete and comprehensive physical information (e.g., strain, temperature, etc.) for effective structural component monitoring [4–6]. Furthermore, it is well known that FBG multiplexing technology is commonly used to connect a number of FBG sensors in series, which could share the same light source and demodulation system [7,8]. It also reduces the size of the network circuit and improves space usage [9]. Presently, the common multiplexing technologies such as the wavelength division multiplexing (WDM), time-division multiplexing (TDM), space division multiplexing (SDM), and frequency division multiplexing (FDM) [10–12]. In the WDM technology, FBG sensors with different central wavelengths are connected in series. It reduces work quantity but decreases the broadband light source power, which fails to meet the multipoint and distributed sensing measurements [13]. All FBG sensors are connected to an optical switch used to control and select the transmission channel in the SDM technology. Nevertheless, the optical switch transformed among extensive FBG sensors will affect the sampling frequency [14,15]. The separate WDM, SDM, or TDM cannot meet the demand for the SHM on aircraft.

The accuracy of FBG sensor data is of great significance to the assessments of the measuring structure of the key airplane [16]. However, the FBG sensor's faults or failures become more frequent than the structural damage during the sensor lifetime owing to their own natural aging, line failure, installation failure, harsh operational environment (e.g., temperature, humidity and moisture), and so on. Furthermore, the sensor detection using the one-by-one method will take a long time and increase the maintenance costs, especially in the large sensor network [17,18]. The early identification of faults or failures in FBG sensors is the most important aspect of fault detection to ensure data accuracy and reliability [19,20]. Therefore, it is essential to find a rapid and effective sensor fault or failure detection method to ensure high-precision results in aircraft SHM systems. In some literature [21–23], the available measurements and output of the observer are compared and then the differences between them are exploited to further identify the malfunctions. The effectiveness of this method is dependent on the accuracy of the observer. CAZZULANI G. et al. [24] proposed a sensor fault identification algorithm based on the analysis of the residuals of the measurement estimation, which could identify different typologies of sensor fault or malfunctioning. In [25], a technique to assess the reliability of FBG sensors was proposed. However, it requires an expensive instrument and it is also hard to apply online. Rama Mohan Rao A. et al. [26] presented a sensor fault detection and isolation technique based on the null subspace method. An experimental test was carried out using a scaled-down model of a bridge, which clearly indicated the effectiveness of the proposed algorithm. Huang H. B. et al. [27] presented a sensor-fault detection and isolation method based on a statistical hypothesis test and missing variable approach and its performance was validated and demonstrated on the bridge health monitoring. However, the proposed method is effective in detecting and isolating both bias and drift sensor faults and is not suitable for fault detection on FBG sensors. In [28], a deep learning-based method was proposed for damage identification of the operational functionality of the smart interface device. While the method has shown promising prospects for electro-mechanical impedance (EMI)-based damage detection, it has limitations on other fields. Hence, developing and designing a fault detection device for the FBG sensing system, as well as a fault detection method to detect and maintain FBG sensors and optical fiber demodulator on an airplane is critical. Therefore, in this study, we develop an FBG sensing system based on the WDM and SDM technology, design a fault detection device with the matched FBG demodulator equipment based on the FBG sensing system, propose a wavelength demodulation method based on the volume phase grating (VPG) and linear array photodetector and a Gaussian curve fitting method for finding the peak value of the reflection spectrum. We also propose a fault detection method based on a synthetical anomaly index. The parameters (light intensity I, signal length L, standard deviation of original sample σ and energy value in time-domain P) of FBG sensors are selected as the signal features. The separation degrees of these features are obtained and merged into the synthetical anomaly index for rapidly detecting the state of FBG sensors. Finally, the proposed method is validated and demonstrated for the 25 FBG sensors on an airplane, which proves the effectiveness and good guidance in making quick and effective decisions for the state of FBG sensors.

2. Fiber Bragg Grating Sensing System on the Airplane

FBG strain sensors with advantages of small size, wide measuring range, high measuring precision and sensibility and anti-electromagnetic interference have made them increasingly adopted as an important strain measurement for the aircraft. In Figure 1, 25 FBG sensors (18 FBG strain sensors and 7 FBG temperatures) are installed in different locations of the aircraft. The FBG sensors 1–7, 8–10, 11–12, 13–15, 16–17, 18–20, 21–23, and 24–25 are installed on the Dome, right wing, right tail, left fuselage, left wing, and vertical tail of an aircraft, respectively. The FBG strain sensors are used to measure the strain variation of the relevant measuring points, which could provide an important basis for identifying the status of the aircraft.

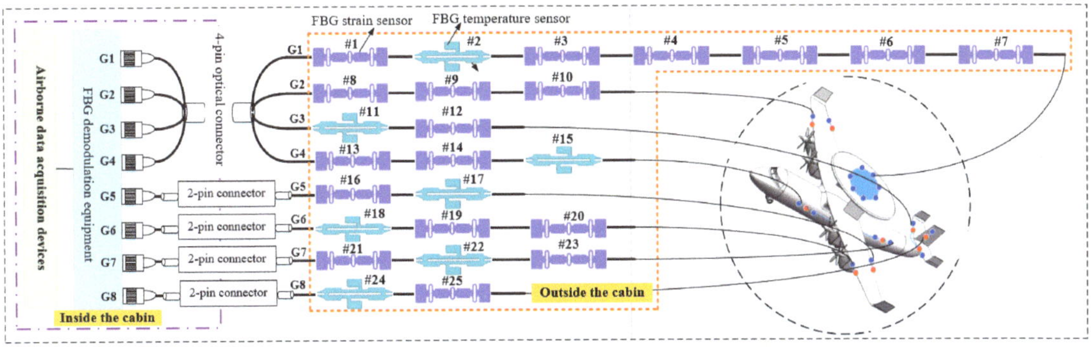

Figure 1. Diagram of the FBG sensing system on the plane.

As shown in Figure 1, the FBG sensing system on the airplane is designed using WDM and SDM technologies. All FBG sensors on an airplane are divided into eight channels, as shown in Figure 1. A certain number of FBG sensors with different central wavelengths are connected in series in one channel through WDM technology. Each FBG sensor is assigned a bandwidth to ensure that the wavelength shift is within the bandwidth. Using SDM technology, the optical switch array in FBG demodulation equipment switches through different channels. Note that the FBG temperature sensors are used to make the temperature compensation for the FBG strain sensors. The FBG sensors, which are installed outside the cabin, are connected to the optical fiber demodulation equipment in the cabin through the cabin optical connector. Furthermore, the airborne data collector is used to collect and save the demodulation information from the FBG demodulation equipment. The combination of WDM and SDM technologies could realize the distributed measurement of multipoint measurement and ensure the FBG sensor system's lightweight and less optical cable.

3. System Design and Implementation

3.1. Hardware Design of Fault Detection Instrument

The fault detection device of the FBG sensing system mainly includes a lithium battery, industrial personal computer (IPC), temperature control module, FBG sensing module, power supply interface, communication interface, display, etc., as shown in Figure 2. A fault detection device is usually used with the FBG demodulator.

The fault detection device of the FBG sensing system is powered by a lithium battery, which also provides power to the IPC, control circuit, and FBG demodulator through the power supply interface. The FBG demodulator's output data are transmitted to the IPC via a communication interface, and the upper computer software processes the data. The output optical cable of the FBG sensors is connected to the flange plate located at the device's front panel, and 16 FBG temperature sensors are pasted on the sensing module. The armored optical cables are equipped with FC/APC optical connectors at both ends. One end of the optical cable is connected to the outside of the flange plate. The other is actualized to connect the external FBG demodulator to realize the conduction of the optical path.

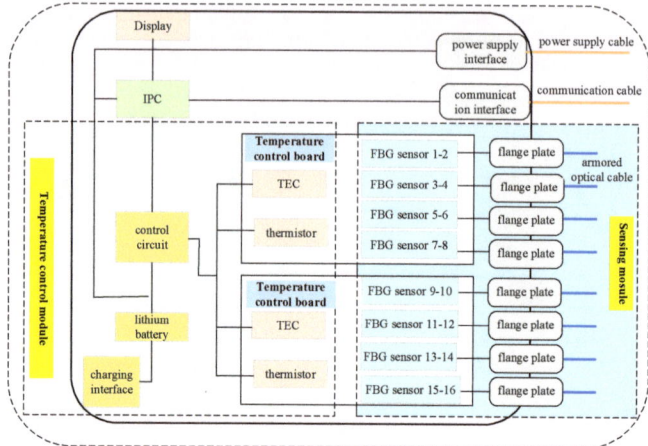

Figure 2. Hardware architecture of the fault detection instrument for FBG sensing system.

A temperature control module comprises a temperature control board (cold board and hot board), thermoelectric cooler (TEC), thermistor, and control circuit. TEC, thermistor, and the FBG temperature sensors are all pasted on the surface of the cold board. A thermistor is used to perceive the temperature change of the cold board, convert the change into an electrical signal, and then transmit the signal to the control circuit. The control circuit algorithm is used to control the TEC or thermistor to achieve a relatively constant temperature of the cold board and further make the FBG temperature sensors enhance a stable output. Note that the strain induced by the external force or other factors will not affect the performance of the FBG temperature owing to its mounting ear structure. When the temperature control module's temperature accuracy reaches 0.1 °C, the installed FBG temperature sensor can achieve a temperature sensitivity of 0.028 nm/°C, and its wavelength shift can be controlled within 3 pm.

The function of the FBG demodulation equipment is used to make the sensing information detected by the FBG sensors demodulated in the form of the wavelength encoding. The external FBG demodulator equipment is typically used with a designed fault detection device when the FBG sensors on an airplane need to be detected. Hence, the matched FBG demodulation equipment is also designed, as shown in Figure 3. The FBG demodulation equipment is mainly composed of a power module, photoelectric processing module, communication interface module and heating module. The heating module provides a heat preservation function for the spectral analysis module and an optical switch circulator.

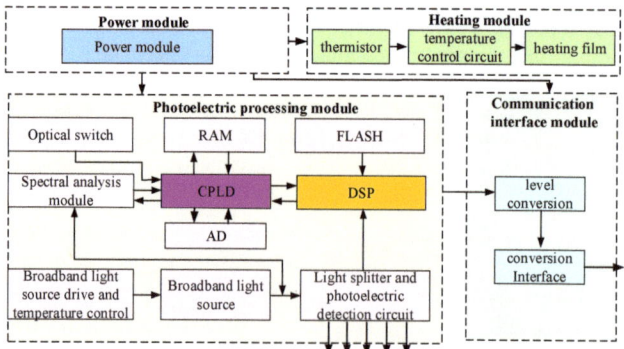

Figure 3. Hardware implementation of the matched FBG demodulation equipment.

As for the photoelectric processing module, the broadband light source that is driven by a constant current circuit could provide the broadband light with a wavelength range from 1510 nm to 1590 nm. The photoelectric detection circuit is mainly used to monitor the light intensity value of the laser and give feedback to the control circuit so that the light intensity value can be adjusted. The output light enters the optical switch through the circulator and the reflected light from the FBG sensor enters the spectral analysis module which is used to convert the light signal into a pixel voltage signal. The voltage signal is converted to digital quantity through an AD converter. A complex programmable logic device (CPLD) realizes the control function and transmits the spectral data to digital signal processing (DSP) through the communication interface. The central wavelength of the FBG sensor was fitted and calculated in DSP. In the spectral analysis module, spectrum detection and demodulation technology based on the volume phase grating (VPG) and linear array photodetector is adopted. The VPG spectrum module with the advantages of small size, without the piezoelectric materials and moving parts, is applicable in the aircraft with large vibration situation.

3.2. Software Design of Fault Detection Device

The program flowchart of the fault detection instrument for the FBG sensing system is shown in Figure 4. Figure 4a shows the detection process of the FBG sensors. As the system has been initialized, the CPLD is in waiting for the subframe synchronization signal sent by the data acquisition unit. It receives the signal and starts the spectral signal acquisition, as well as provides the relevant driving clock in accordance with the spectral analysis module's timing requirements. The collected data include pixel voltage signal and internal temperature signal of the spectral analysis module, and the two types of signals are switched over through an optical switch. The spectral analysis module includes 512-pixel units and all need to be collected and stored in RAM.

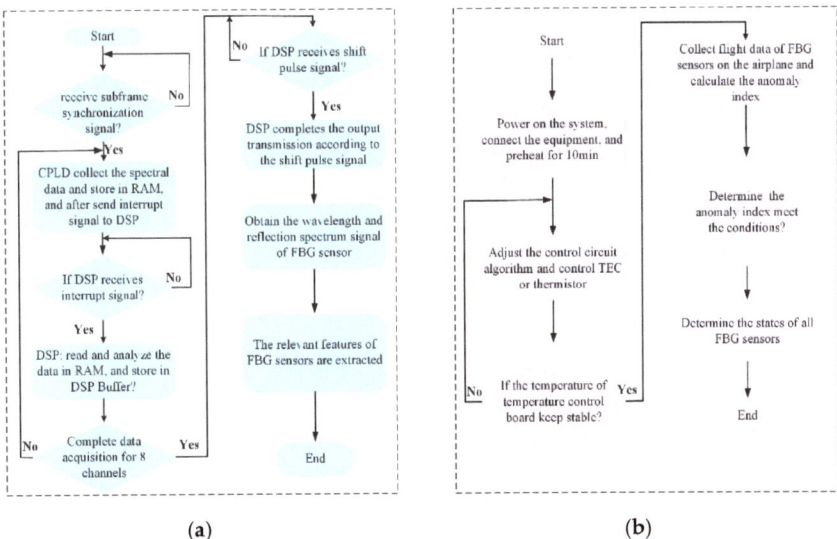

Figure 4. Program flowchart of the fault detection instrument. (**a**) FBG sensors; (**b**) FBG demodulation equipment.

The CPLD then sends the interrupt signal to DSP. If DSP receives the interrupt signal and begins solving the collected data and further stores the results in the DSP buffer. The DSP then sends the command to the CPLD to begin to make data acquisition from the next channel until it completes the data acquisition of eight channels Furthermore, the

DSP could sort out the data of eight channels according to the requirement of the data format and then wait for the shift pulse signal from the data acquisition unit. Once the DSP receives the shift pulse signal, then it completes the data transmission through the communication interface. Till now, the wavelength data and reflection spectrum signals of FBG sensors have been obtained. Finally, the performance of FBG sensors in the normal state could be observed and compared to the central wavelength or light intensity.

Figure 4b shows the FBG demodulation equipment detection process after the system is powered on and connected, the control circuit algorithm is used to control the TEC or thermistor until the temperature of the temperature control board achieves a steady state. Flight data is collected and the anomaly index of FBG sensors on the airplane according to the extracted features is calculated. The performance of the FBG demodulation instrument is reflected by the connected FBG sensors. Hence, the state of the FBG demodulation instrument is validated by the 16 FBG sensors in the fault detection instrument or the states of FBG sensors on the aircraft.

4. Wavelength Demodulation and Fault Detection Methods

4.1. Wavelength Demodulation Method

As shown in Figure 5, when light from the broadband light source (BBS) is incident on the fiber grating sensor network through the circulator, the light that meets the Bragg conditions is reflected and subsequently enters the spectrum detection and demodulation system through the circulator. The optical path switch is used to convert the different channels' optical signal transmission. The wavelength demodulation method is based on a VPG and a linear array photodetector. The reflected light source becomes the parallel light after passing through the collimating lens, which is irradiated on the VPG1. Due to the light splitting effect of the VPG, the light with different wavelengths is separated into the light with different refraction angles. Light beams with different refraction angles are converged on the concave mirror and then transmitted to different positions of the linear array photodetector. The variation in ambient temperatures of the FBG sensor will change the period and the effective refractive index of FBG, resulting in a wavelength shift of the grating signal. Furthermore, the image position on the linear array photodetector will also change accordingly.

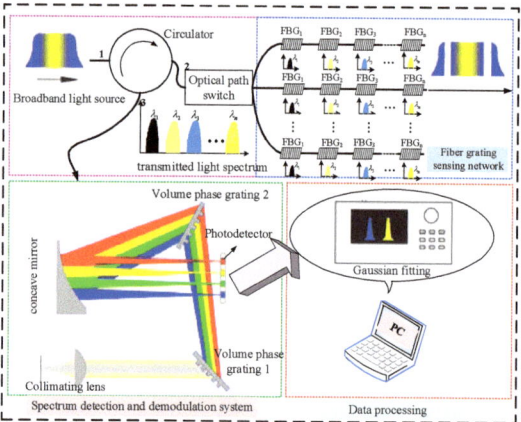

Figure 5. Wavelength demodulation principle and peak searching method.

The pixel voltage signals of the linear array photodetector are collected and processed to obtain the central wavelength shift of the fiber grating sensor. The relationship between the number of pixels and wavelength is expressed as [29]:

$$\lambda = A + B_1 pix + B_2 pix^2 + B_3 pix^3 + B_4 pix^4 + B_5 pix^5 \qquad (1)$$

where, pix is the number of pixels and λ is the wavelength of the FBG sensor. A, B_1, B_2. B_3, B_4 and B_5 are the polynomial fitting coefficients.

Equation (1) is obtained through the least squares method which is used to complete the fifth-order polynomial fitting between λ and pix. Then, the polynomial fitting coefficients are determined: $A = 1.595306 \times 10^3$, $B_1 = -1.355909 \times 10^{-1}$, $B_2 = -6.160845 \times 10^{-5}$, $B_3 = -3.346493 \times 10^{-11}$, $B_4 = -1.224188 \times 10^{-11}$, and $B_5 = 1.133598 \times 10^{-14}$.

The measurement accuracy of the FBG sensor is largely dependent on the finding peak degree of the central wavelength. It is generally known that the shape of the Gaussian curve is extremely similar to the reflection spectrum of FBG [30]. To improve the FBG sensor detection accuracy, the obtained reflection spectrum signal is made denoizing and the Gaussian curve fitting method is used for linear regression to further obtain the central wavelength of the FBG sensor. The Gaussian fitting formula is given by:

$$I(\lambda) = I_0 \exp[= 4 \ln 2 (\frac{\lambda - \lambda_s}{\Delta \lambda_s})] \qquad (2)$$

where, I_0 is the peak value of reflection spectrum intensity, λ_s is the wavelength value when the reflection spectrum intensity is I_0. $\Delta \lambda_s$ is the half-width of reflection spectrum intensity.

4.2. Fault Detection Method

(a) Fault detection principle

The main load affected by FBG sensors is stress during the flight of the aircraft. The variation of loads on the same monitoring structure point is roughly similar. Furthermore, it was found that the parameters of output signals (e.g., the light intensity I, wavelength amplitude et al.) of the healthy FBG sensors have a similarity. Moreover, the multiple statistical features of a healthy FBG sensor are in an aggregation state. When the FBG sensor fails, the features will cause change and tend to be in a separated state. Therefore, the separation degree of the related features of the relevant measuring points could be obtained to determine the state of the FBG sensors. The common fault types and relevant signal characteristics are listed in Table 1.

Table 1. Features values and synthetical abnormal index of five faulty FBG sensors.

Fault Type	Signal Characteristics
Damage or fall off	No change in wavelength and intensity or little change
Excessive shock	Abrupt change of signal amplitude or missing
Break off optical cable	No signal output
Excessive bend optical cable	Weaken the signal or small amplitude variation
Sudden power failure and recover	Data missing and data length reduction

(b) Feature extraction of FBG sensor

Generally, the signal statistical characteristics of the FBG sensor mainly include the light intensity, signal length, standard deviation of the original sample, energy value in the time-domain, amplitude value and variance, etc. According to the actual function of the signal, the first four parameters of the above characteristics are selected as the features for fault detection on FBG sensors. The definition of the four feature parameters is shown in Table 2.

Table 2. Definitions of four feature parameters.

Feature Parameters	Definition
Minimum light intensity I	$I_{min} = \min[I(n)]$
Signal length L	$L = length(\lambda_n)$
Standard deviation of original sample σ	$\sigma = \sqrt{\left\{\sum_{n=1}^{N}[\lambda(n) - \overline{\lambda}]^2\right\}/N}$
Energy value in time-domain P	$P = \sum_{n=1}^{N}[\lambda(n) - \overline{\lambda}]^2$

Here, $\{\lambda_n\}(n = 1 \sim N)$ is the original wavelength data of FBG sensor (N is data points)

The linear array photodetector's measurement range is 1.6 V–3.8 V; thus, if the minimum light intensity I_{min} is close to or equal to 1.6 V, the FBG sensor is in an abnormal or fault state. When the data acquisition device collects the data of FBG sensors on an airplane, the wavelength data that exceeds a certain threshold will be discarded. Thus, the inconsistent length L could indicate an abnormal or faulty state of the FBG sensor. σ could reflect the dispersion degree of the data set, and it is also the basis for judging the signal stability of measuring points. P could reflect the fluctuation intensity of the output signal, and it has obvious indications on the FBG sensor with excessive gain or no obvious change in output amplitude. The feature characteristics of m FBG sensors are sequentially extracted, and the feature vector is obtained as follows:

$$f = \begin{bmatrix} I(1) & L(1) & \sigma(1) & P(1) \\ I(2) & L(2) & \sigma(2) & P(2) \\ \vdots & \vdots & \vdots & \vdots \\ I(m) & L(m) & \sigma(m) & P(m) \end{bmatrix} \quad (3)$$

4.3. Fault Detection Process

The FBG sensor fault detection process is shown in Figure 6. First, to obtain the separation degree of features, the aggregation center values of four features (I, L, σ, P) must be acquired. To avoid the effect of the too large separation values on the aggregation center values, the loop iteration method is used to search the feature point with the largest number of aggregation points nearby. The aforementioned feature point is regarded as the aggregation center value of the relevant feature.

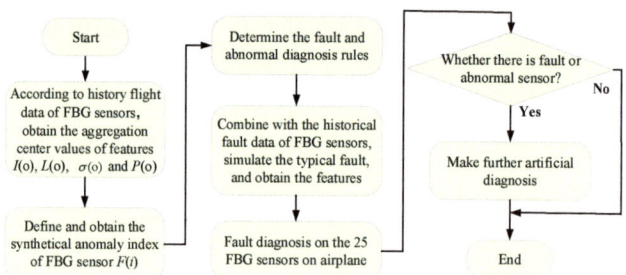

Figure 6. Fault detection process of FBG sensors on the airplane.

The data aggregation coefficient is defined by:

$$C(X) = \frac{a}{l}\sum_{b=1}^{l}|X(b) - \overline{X}|, \ (X = I, L, \sigma, P) \quad (4)$$

where, l is the length of the relevant feature $X(b)$ and $\overline{X}$ represent the four feature data and their own mean value, respectively, and a is a multiple of the data aggregation coefficient.

Furthermore, according to Equation (5), the Euclidean distance between two different feature points is calculated through the loop iteration method to find the aggregation center values of four features.

$$d(i,j) = |X(i) - X(j)| \tag{5}$$

where, i and j are the serial numbers of different measuring points.

Determine the $d(i,j)$ and $C(X)$ whether it meets the condition $d(i,j) < C(X)$; if yes and it shows that the jth value is aggregated near the ith value. Calculate the Euclidean distance between ith point and all other points until finding the point with the largest number of aggregation points near the aggregation center values of four features (I, L, σ, P) are called $I(o), L(o), \sigma(o)$, and $P(o)$, respectively.

Second, the Z-score data standardization method is used to achieve standardization for the offset distance between feature values and their aggregation center value when calculating the separation degree of features. The FBG sensor's synthetical anomaly index $E(i)$ is defined as follows:

$$E(i) = \sum_X \frac{|X(i) - X(o)|}{\sigma_X}, \ (X = I, L, \sigma, P) \tag{6}$$

$$\sigma_X = \sqrt{\left\{\sum_{i=1}^{l} [X(i) - \overline{X}]^2\right\}/l}, \ (X = I, L, \sigma, P) \tag{7}$$

where, σ_X is the standard deviation of the relevant feature vector.

Finally, the fault and abnormal diagnostic rules are described as:

$$R[E(i)] = \begin{cases} E(i) \leq t_1 \cdot \overline{E}, & \text{healthy sensor} \\ t_1 \cdot \overline{E} < E(i) \leq t_2 \cdot \overline{E}, & \text{abnormal sensor} \\ E(i) > t_2 \cdot \overline{E}, & \text{faulty sensor} \end{cases} \tag{8}$$

where, $\overline{E}$ is the mean value of the anomaly index of all measured FBG sensors. t_1 and t_2 are the anomaly coefficient and faulty coefficient, respectively.

5. Experimental Verification and Discussion

To evaluate the performance of the designed fault detection device and the proposed fault detection based on the FBG grating sensing system, a typical fault simulation experiment of FBG sensors and the detecting experiments of FBG sensors on an airplane perform, respectively.

5.1. Typical Fault Simulation Test

The flight data from 25 FBG sensors (the number codes are 1–25) with the healthy state on an airplane are selected as the healthy sample (Sample length of each sensor: 289,081). Additionally, the typical faults of FBG sensors are simulated, respectively. The simulation setup of typical faults on FBG sensors is shown in Figure 7.

According to the installation technology and the working environment of the FBG sensors on the airplane, five FBG sensors (relevant number codes are 26–30) are pasted on the fixture component with the same material as the measuring point structure of the airplane. As shown in Figure 7, the fixture component is installed on the vibration table, which provides the random vibration for the five FBG sensors. Figure 8 shows five typical faults of FBG sensors (26: break off optical cable, 27: excessive bend optical cable, 28: excessive shock, 29: sudden power failure, and 30: fall off FBG sensor) are simulated, and their partial wavelength and light intensity variation curves.

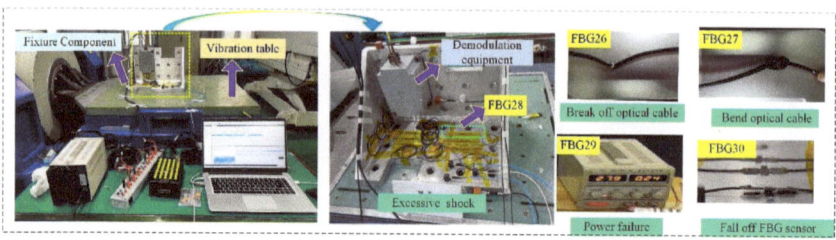

Figure 7. Simulation setup of typical faults on FBG sensors.

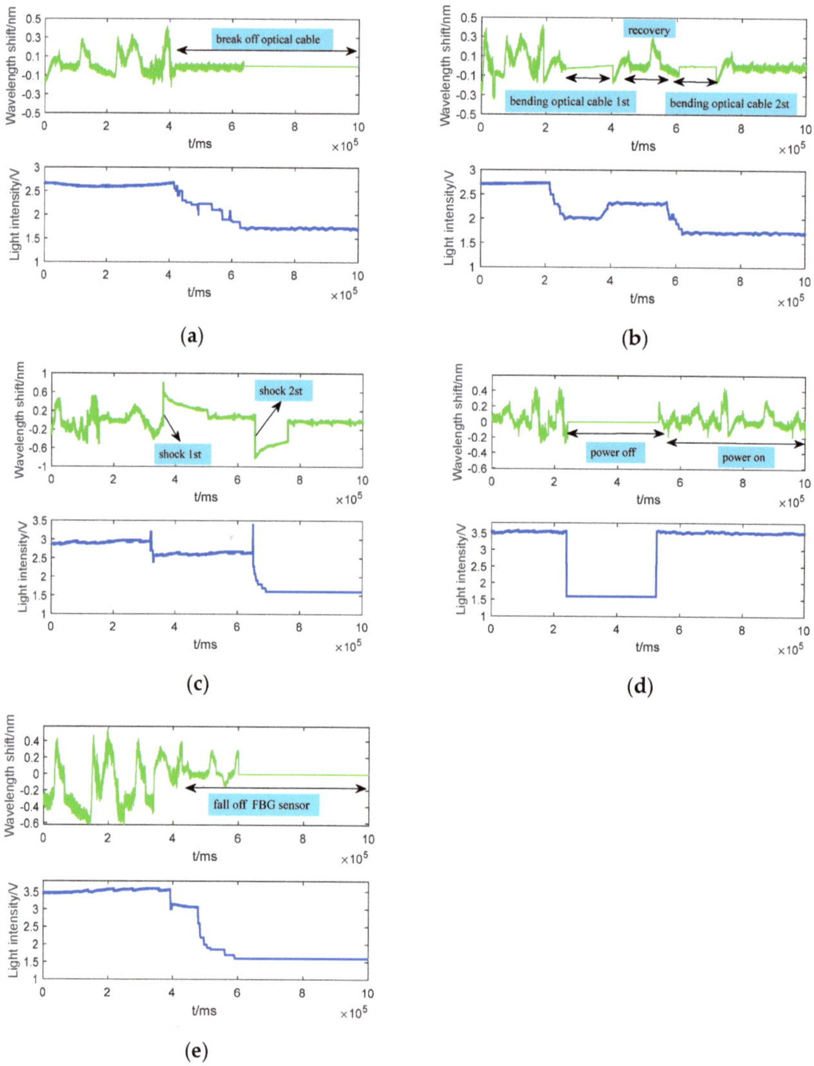

Figure 8. Wavelength and light intensity variation curves of five typical faults. (**a**) Break off the optical cable. (**b**) Bending the optical cable. (**c**) Shock test. (**d**) Sudden power off and power on. (**e**) Fall off FBG sensor.

As shown in Figure 8a–e that the wavelength shifts and light intensities of the FBG sensors cause obvious changes when the faults occur. The data of 25 healthy FBG sensors on an airplane and the above five faulty sensors are regarded as test data. The length of each simulated fault data was equal to the data of each healthy FBG sensor. Faulty sensors using the proposed fault detection method. The features and synthetical anomaly index of faulty FBG sensors are listed in Table 3, and the fault detection results of 30 FBG sensors are clearly shown in Figure 9.

Table 3. Features values and synthetical abnormal index of five faulty FBG sensors.

FBG Sensors Number	Light Intensity I/V	Signal Length L	Standard Deviation σ/nm	Energy Value P/nm^2	Synthetical Anomaly Index
FBG26	1.6	267,341	0.0718	5361.538	3.556338
FBG27	1.6	289,081	0.0808	6781.024	3.881325
FBG28	1.6	289,081	0.2482	34,049.181	10.02711
FBG29	1.6	192,679	0.0874	2741.718	8.97647
FBG30	1.6	227,828	0.1883	36,879.550	9.296857

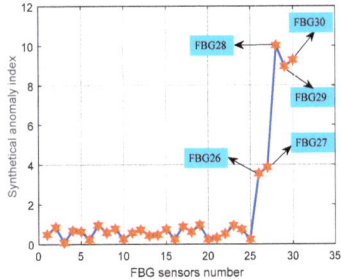

Figure 9. Fault detection results of 30 FBG sensors.

As shown in Figure 9 that the synthetical anomaly index of faulty 26, 27, 28, 29, and 30 FBG sensors is significantly higher than that of the other normal 25 FBG sensors. Furthermore, according to Equation (8) and Figure 9, we can know that too many abnormal or faulty points could cause excessive discrete distribution of features, leading to poor detection results. Then, the anomaly and faulty coefficients t_1 and t_2 are bound to make further adjustments according to the discrete degree of features.

5.2. Fault Detection on 25 FBG Sensors on Airplane

The test sample is flight data from 25 FBG sensors on an airplane (from November 2021 to December 2021). As shown in Figure 10, the designed fault detection instrument and the matching FBG demodulator are placed in the cabin. Through an optical connector, 25 FBG sensors installed on the key structures of an airplane are connected to the fault detection device in the cabin. The measuring data of 25 FBG sensors were extracted from the historical database, and the relevant features of all sensors were also calculated. The fault detection results of 25 FBG sensors are shown in Figure 11, and the features and synthetical anomaly index of faulty and abnormal FBG sensors are shown in Table 4.

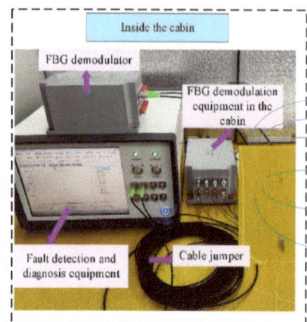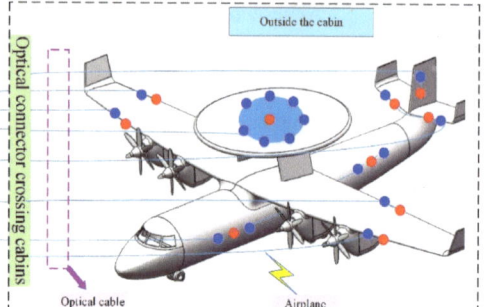

Figure 10. Experiment setup diagram of fault detection instrument for the FBG sensing system on an airplane.

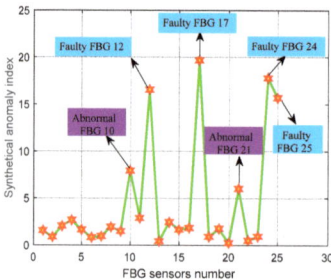

Figure 11. Fault detection results of 25 FBG sensors on airplane.

Table 4. Features values and synthetical abnormal index of faulty and abnormal FBG sensors.

FBG Sensors Number	Light Intensity I/V	Signal Length L	Standard Deviation σ/nm	Energy Value P/nm^2	Synthetical Anomaly Index
FBG 10	2.56	403,581	0.2708	6985.123	7.9565
FBG 12	1.60	257,562	0.1280	4909.545	16.5480
FBG 17	1.60	403,581	0.2035	4293.836	19.6785
FBG 21	2.79	403,581	0.0946	7290.521	6.0214
FBG 24	1.65	327,890	0.1023	5031.056	17.7976
FBG 25	1.71	403,581	0.1245	4986.301	15.6780

As shown in Figure 11, the proposed fault detection method detects four faulty and two abnormal FBG sensors. After making further manual verification on the above faulty and abnormal sensors, it was found that the optical cable near the FBG 12 sensor had broken, FBG 17 sensor had fallen off the measured point, and the optical cable of the FBG 24 sensor was excessively bent, as shown in Figure 12a–c, respectively. The FBG 24 and 25 sensors were connected in one optical cable, and the bent optical cable also caused the abnormal measuring signal. Furthermore, the two abnormal FBG sensors (FBG 10 and 21) were affected by the high-frequency signals, which caused many abrupt changes in signals and then further generated abnormal feature values. The above experimental results show that the designed fault detection device and proposed fault detection method have better guidance in making quick and effective decisions for the states of FBG sensors on an airplane. According to the detection results, the maintenance worker would make a further diagnosis and then troubleshoot the faulty or abnormal sensors immediately and effectively. Furthermore, it is also of great significance to improve measuring data accuracy.

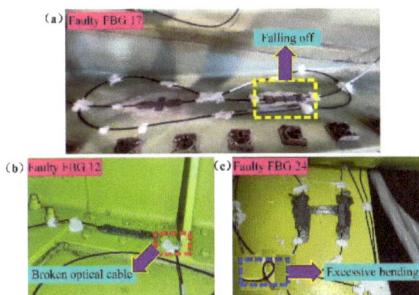

Figure 12. Diagram of faulty FBG sensors on airplane. (**a**) Falling off; (**b**) Broken optical cable; (**c**) Excessive bending.

6. Conclusions

The major contribution of this research is to develop and design a fault detection device and propose a fault diagnosis method based on a synthetical anomaly index. In this study, we built an FBG sensing system based on WDM and SDM technologies based on the FBG sensor network distribution on the aircraft. Furthermore, the hardware and software of the fault detection device and the matched FBG demodulator are illustrated in detail. The simulation test and verification experiments all proved the effectiveness and good guidance in making quick and effective decisions for the states of FBG sensors. The designed fault detection device has the abilities of data acquisition, data storage, data display, and data printed in both text format and image format. It is also easy to operate and could efficiently and quickly detect the state of the FBG sensor and FBG demodulation equipment on an airplane, which could shorten the detection time, reduce the maintenance cost, and provide an important basis for the SHM of the critical structures of the aircraft. Furthermore, the proposed fault detection method could provide reliable detection results for FBG sensors without prior knowledge. However, the major limitation of the fault detection method is that the users determine the concrete fault of the relevant FBG sensor, and it needs further research.

Author Contributions: Conceptualization, Data curation, formal analysis, methodology, writing—original draft preparation, C.D.; software, validation, investigation, resources, supervision, D.K.; writing—review and editing, visualization, project administration, funding acquisition, C.X. All authors have read and agreed to the published version of the manuscript.

Funding: This work was funded by Basic Technology Research Project of Science, Technology and Industry Bureau of National Defense: 995-14021006010401, National Natural Science Foundation of China: No. 11372143, Natural Science Foundation of Jiangsu Province: BK20190464.

Institutional Review Board Statement: Not applicable.

Informed Consent Statement: Not applicable.

Data Availability Statement: The data presented in this study are available in Tables 3 and 4 in this article.

Conflicts of Interest: The authors declare no conflict of interest.

References

1. Arcadius, T.C.; Gao, B.; Tian, G. Structural Health Monitoring Framework Based on Internet of Things: A Survey. *IEEE Internet Things* **2017**, *4*, 619–635. [CrossRef]
2. Schenato, L.; Aguilar-López, J.P.; Galtarossa, A.; Pasuto, A.; Bogaard, T.; Palmieri, L. A Rugged FBG-Based Pressure Sensor for Water Level Monitoring in Dikes. *IEEE Sens. J* **2021**, *21*, 13263–13271. [CrossRef]
3. Sarkar, S.J.; Inupakutika, D.; Banerjee, M.; Tarhan, M.; Shadaram, M. Machine Learning Methods for Discriminating Strain and Temperature Effects on FBG-Based Sensors. *IEEE Photonics Technol. Lett.* **2021**, *33*, 876–879. [CrossRef]

4. Qiu, L.; Fang, F.; Yuan, S.F. Improved density peak clustering-based adaptive Gaussian mixture model for damage monitoring in aircraft structure under time-varying conditions. *Mech. Syst. Signal Process.* **2019**, *126*, 281–304. [CrossRef]
5. Gelman, L.; Petrunin, I.; Parrish, C.; Walters, M. Novel health monitoring technology for in-service diagnostics of intake separation in aircraft engines. *Struct. Control Health Monit.* **2020**, *27*, e2479. [CrossRef]
6. Dutta, C.D.; Kumar, J.; Das, T.K.; Sagar, S.P. Recent Advancements in the Development of Sensors for the Structural Health Monitoring (SHM) at High-Temperature Environment: A Review. *IEEE Sens. J.* **2021**, *21*, 15904–15916. [CrossRef]
7. Yu, X.K.; Song, N.F.; Song, J.M. A novel method for simultaneous measurement of temperature and strain based on EFPI/FBG. *Opt. Commun.* **2020**, *459*, 125020. [CrossRef]
8. Jin, J.; Zhu, Y.H.; Zhang, Y.B.; Zhang, D.W.; Zhang, Z.C. Micrometeoroid and Orbital Debris Impact Detection and Location Based on FBG Sensor Network Using Combined Artificial Neural Network and Mahalanobis Distance Method. *IEEE Trans. Instrum. Meas.* **2021**, *70*, 7005210. [CrossRef]
9. Ab-Rahman, M.S.; Ridzuan, A.M.; Kaharudin, I.H.; Hwang, I.S. Real time FTTH network monitoring using binary coded fiber Bragg grating. *Optik* **2022**, *251*, 168408. [CrossRef]
10. Guan, X.; Shi, W.; Rusch, L.A. Ultra-Dense Wavelength-Division Multiplexing with Microring Modulator. *J. Lightw. Technol.* **2021**, *39*, 4300–4306. [CrossRef]
11. Feng, Y.W.; Chang, J.; Chen, X.H.; Zhang, Q.D.; Wang, Z.L.; Sun, J.C.; Zhang, Z.W. Application of TDM and FDM methods in TDLAS based multi-gas detection. *Opt. Quantum Electron.* **2021**, *53*, 195. [CrossRef]
12. Wu, Q.; Wang, J.D.; Shigeno, M. A novel channel-based model for the problem of routing, space, and spectrum assignment. *Opt. Switch. Netw.* **2022**, *43*, 100636. [CrossRef]
13. Maru, K.C. Two-dimensional spatially encoded cross-sectional velocity distribution measurements based on coherent bias-frequency encoding and wavelength-division multiplexing. *Opt. Commun.* **2021**, *485*, 126740. [CrossRef]
14. Zhang, C.B.; Gao, Y.Y.; Zuo, M.Q.; Lei, P.; Liu, R.W.; He, B.B.; Li, J.H.; Chen, Z.Y. Using ASE sources in remote beamforming system with Space-Division-Multiplex fiber. *Opt. Commun.* **2022**, *504*, 127477. [CrossRef]
15. Yin, S.; Chen, Y.D.; Ding, S.C.; Zhang, Z.D.; Huang, S.G. Crosstalk-aware routing, spectrum, and core assignment based on AoD nodes in SDM-EONs with bidirectional multicore fibers. *Opt. Switch. Netw.* **2022**, *43*, 100647. [CrossRef]
16. Mohapatra, A.G.; Talukdar, J.; Mishra, T.C.; Anand, S.; Jaiswal, A. Fiber Bragg grating sensors driven structural health monitoring by using multimedia-enabled iot and big data technology. *Multimed. Tools Appl.* **2022**. [CrossRef]
17. Soman, R.; Wee, J.; Peters, K. Optical Fiber Sensors for Ultrasonic Structural Health Monitoring: A Review. *Sensors* **2021**, *21*, 7345. [CrossRef]
18. Jinachandran, S.; Rajan, G. Fibre Bragg Grating Based Acoustic Emission Measurement System for Structural Health Monitoring Applications. *Materials* **2021**, *14*, 897. [CrossRef]
19. Tamoghna Ojha, S.M.; Singh, R.N. Wireless sensor networks for agriculture: The state-of-the-art in practice and future challenges. *Comput. Electron. Agric.* **2015**, *118*, 66–84. [CrossRef]
20. Li, D.L.; Wang, Y.; Wang, J.X. Recent advances in sensor fault diagnosis: A review. *Sens. Actuators A Phys.* **2020**, *309*, 111990. [CrossRef]
21. Patton, R.J.; Chen, J. A survey of robustness problems in quantitative model-based fault diagnosis. *Appl. Maths. Comput. Sci.* **1993**, *3*, 339–416.
22. Frank, P.M.; Ding, X. Survey of robust residual generation and evaluation methods in observer-based fault detection systems. *J. Process Control* **1997**, *7*, 403–424. [CrossRef]
23. Kerschen, G.; De, B.P.; Golinval, J.C.; Worden, K. Sensor validation for on-line vibration monitoring. In Proceedings of the European Workshop on Structural Health Monitoring, Munich, Germany, 7–9 July 2004; pp. 819–827.
24. Cazzulani, G.; Cinquemani, S.; Ronchi, M. A fault identification technique for FBG sensors embedded in composite structures. *Smart Mater. Struct.* **2016**, *25*, 055049. [CrossRef]
25. Cazzulani, G.; Cinquemani, S.; Comolli, L. A technique to evaluate the good operation of FBG sensors embedded in a carbon fiber beam. *Proc. SPIE* **2013**, *8794*, 87942S.
26. Rao, A.R.M.; Kasireddy, V.; Gopalakrishnan, N.; Lakshmi, K. Sensor fault detection in structural health monitoring using null subspace–based approach. *J. Intell. Mater. Syst. Struct.* **2015**, *26*, 172–185. [CrossRef]
27. Huang, H.B.; Yi, T.H.; Li, H.N. Sensor Fault Diagnosis for Structural Health Monitoring Based on Statistical Hypothesis Test and Missing Variable Approach. *J. Aerosp. Eng.* **2015**, *B4015003*, 1–14. [CrossRef]
28. Nguyen, T.T.; Kim, J.T.; Ta, Q.B.; Ho, D.D.; Phan, T.T.; Huynh, T.C. Deep learning-based functional assessment of piezoelectric-based smart interface under various degradations. *Smart Struct. Syst.* **2021**, *28*, 69–87.
29. Xu, D.Y.; Tong, J.P.; Gao, J.X.; Wang, F. Fiber Spectrometer Optical Simulation Optimization and Calibration. *Chin. J. Lasers* **2015**, *42*, 1–5.
30. Wen, X.Y. *Research on Principle and Technology of Fiber Grating Sensor*; Wuhan University of Technology Press: Wuhan, China, 2019; pp. 104–106.

Article

High Selectivity, Low Damage ICP Etching of *p*-GaN over AlGaN for Normally-off *p*-GaN HEMTs Application

Penghao Zhang [1], Luyu Wang [1], Kaiyue Zhu [2], Yannan Yang [1], Rong Fan [1], Maolin Pan [1], Saisheng Xu [1], Min Xu [1,*], Chen Wang [1,*], Chunlei Wu [1] and David Wei Zhang [1,*]

[1] State Key Laboratory of ASIC and System, School of Microelectronics, Fudan University, Shanghai 200433, China; phzhang19@fudan.edu.cn (P.Z.); wangly20@fudan.edu.cn (L.W.); yangyn20@fudan.edu.cn (Y.Y.); 20212020129@fudan.edu.cn (R.F.); 21112020100@m.fudan.edu.cn (M.P.); ssxu@fudan.edu.cn (S.X.); wuchunlei@fudan.edu.cn (C.W.)
[2] Department of Electrical and Electronic Engineering, Xi'an Jiaotong-Liverpool University, Suzhou 215123, China; kaiyue.zhu18@student.xjtlu.edu.cn
* Correspondence: xu_min@fudan.edu.cn (M.X.); chen_w@fudan.edu.cn (C.W.); dwzhang@fudan.edu.cn (D.W.Z.)

Abstract: A systematic study of the selective etching of *p*-GaN over AlGaN was carried out using a BCl_3/SF_6 inductively coupled plasma (ICP) process. Compared to similar chemistry, a record high etch selectivity of 41:1 with a *p*-GaN etch rate of 3.4 nm/min was realized by optimizing the SF_6 concentration, chamber pressure, ICP and bias power. The surface morphology after *p*-GaN etching was characterized by AFM for both selective and nonselective processes, showing the exposed AlGaN surface RMS values of 0.43 nm and 0.99 nm, respectively. MIS-capacitor devices fabricated on the AlGaN surface with ALD-Al_2O_3 as the gate dielectric after *p*-GaN etch showed the significant benefit of BCl_3/SF_6 selective etch process.

Keywords: *p*-GaN; selective etching; ICP; surface morphology; MIS capacitor

Citation: Zhang, P.; Wang, L.; Zhu, K.; Yang, Y.; Fan, R.; Pan, M.; Xu, S.; Xu, M.; Wang, C.; Wu, C.; et al. High Selectivity, Low Damage ICP Etching of *p*-GaN over AlGaN for Normally-off *p*-GaN HEMTs Application. *Micromachines* **2022**, *13*, 589. https://doi.org/10.3390/mi13040589

Academic Editor: Kun Li

Received: 8 March 2022
Accepted: 6 April 2022
Published: 9 April 2022

Publisher's Note: MDPI stays neutral with regard to jurisdictional claims in published maps and institutional affiliations.

Copyright: © 2022 by the authors. Licensee MDPI, Basel, Switzerland. This article is an open access article distributed under the terms and conditions of the Creative Commons Attribution (CC BY) license (https://creativecommons.org/licenses/by/4.0/).

1. Introduction

GaN-based high electron mobility transistors (HEMTs) have recently attracted much attention in applications for power switching due to their properties of high-frequency and low on-resistance [1,2]. Two-dimensional-electron-gas (2DEG) is induced by the strong spontaneous and piezoelectric polarization effect in the AlGaN/GaN heterojunction [3], which causes the conventional devices normally be on, i.e., depletion-mode. However, normally-off, i.e., enhanced-mode transistors with a positive threshold voltage are more desirable for simplified circuit design in practice [4,5]. To deplete the 2DEG under the gate area, several approaches have been invented, such as fluorine-implanted treatment [6], gate-recessed [7] and *p*-GaN gate structure [8]. Among these technologies, *p*-GaN gate HEMTs show broad market prospects [9,10].

Precise etch depth control of the *p*-GaN layer with minimum etch damage to the underlying AlGaN barrier is necessary to recover high-density electrons in the access regions, which is the most critical process in the fabrication of *p*-GaN gate HEMTs [11]. Generally, to fully deplete the 2DEG in the channel for normally off operation, a thick *p*-GaN layer with a thin AlGaN layer in epitaxy technique is employed. Further thinning of the AlGaN barrier due to overetching, even a few nanometers, could lead to a dramatic reduction in the conductivity in the access region, which means degradation of the output performance of the devices [12]. On the other hand, an underetched Mg-doped *p*-GaN layer could form a conducting channel contributing to off-state leakage [13]. Therefore, the precise control of *p*-GaN etch depth with minimum damage on AlGaN surface is needed for higher performance E-mode HEMT devices with higher drive current, lower off-leakage and improved dynamic on-resistance [14].

As reported in reference [15], adding SF_6 gas to BCl_3 gas would form an AlF_x nonvolatile layer on the surface of AlGaN layer after GaN removal, thus achieving high selectivity between GaN and AlGaN, as of 23:1. However, the detailed process optimization and the corresponding impact on the etch damage of AlGaN surface have not been studied yet. In this work, a highly selective ICP etching of p-GaN over AlGaN by the BCl_3/SF_6 mixture was systematically investigated. The influence of chamber pressure, SF_6 gas flow, ICP power and bias power on the etch rates and selectivity were studied. The highest selectivity was obtained through process optimization for BCl_3/SF_6 etch ambient. Atomic force microscope (AFM) image of AlGaN layer exposed after p-GaN selective etching showed a very smooth surface. C–V measurements for $Ni/Al_2O_3/AlGaN$ stack MIS structure further confirmed the advantage of this high selective etch process and the minimum etch damage on AlGaN surface.

2. Experimental

In this work, two commercially available p-GaN/AlGaN/GaN and AlGaN/GaN hetero-structures epitaxially grown on 6-inch Si substrate were used. One is p-GaN (80 nm)/$Al_{0.25}Ga_{0.75}N$ (15 nm)/unintentionally doped GaN (300 nm)/buffer (4.2 μm)/Si (1 mm), and the other is $Al_{0.25}Ga_{0.75}N$ (15 nm)/unintentionally doped GaN (300 nm)/buffer (4.2 μm)/Si (1 mm). They are referred to p-GaN sample and AlGaN sample in the rest of this article.

The etch chamber used to develop the selective p-GaN etch process is a customized ICP tool from NAURA (NAURA Technology Group Co., Ltd., Beijing, China) with designed bias power as low as 5 W. For all the processes, a pure BCl_3 plasma pre-etching was carried out to punch through the (Al)GaN native oxide on the exposed surface [16], right before the main BCl_3/SF_6 etch.

The frequencies of power generator and chamber chiller temperature were set as 13.56 MHz and 20 °C, respectively. Etch process conditions were optimized with SF_6 concentration in the range of 0–30% (constant total flow of 150 sccm), chamber pressure in the range of 20–60 mTorr, ICP power in the range of 200–600 W and bias power in the range of 20–80 W. p-GaN and AlGaN samples were etched simultaneously for process evaluation. The etch depth and surface morphology were evaluated using a Park NX10 AFM. Z scanner resolution of this AFM reached 0.015 nm in order that etch depth of patterned samples could be precisely characterized. Etching profiles were inspected by scanning electron microscopy (SEM), and selectivity was calculated as the ratio of the etch rate of p-GaN to AlGaN.

3. Results and Discussion

3.1. Etching Parameter Optimization

3.1.1. SF_6 Concentration

The selective etch process has a strong dependence on the SF_6 concentration in the ambient, as presented in Figure 1. The other etching conditions were fixed as follows: chamber pressure of 40 mTorr, ICP power of 600 W and bias power of 40 W. A significant enhancement in the p-GaN etch rate is observed when the SF_6 concentration increases from 0 to 15% due to the catalyzed generation of the active chlorine [17,18]. However, further increasing SF_6 gas flow leads to a decrease in the p-GaN etch rate due to the formation of involatile GaF_x [18,19]. In summary, adding SF_6 has two-side impacts on the etch of p-GaN and the concentration could be optimized to have the best p-GaN etch. For the AlGaN sample, the etch rate monotonically decreases with increasing SF_6 concentration due to the formation of nonvolatile AlF_x acting as a powerful etch-stop layer [15]. The selectivity reaches a maximum at 15% SF_6.

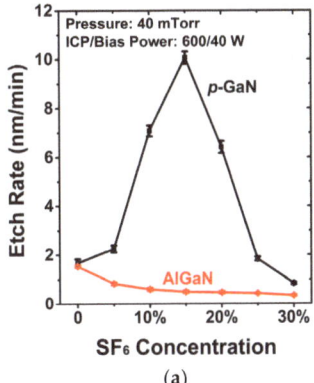
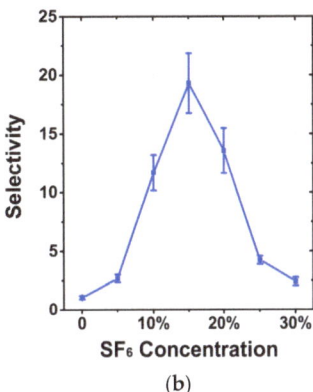

Figure 1. Dependence of (**a**) the etch rates, (**b**) selectivity between *p*-GaN and AlGaN on SF_6 concentration.

3.1.2. Chamber Pressure

The effects of chamber pressure on the etch rates of *p*-GaN and AlGaN layers and the selectivity were examined, as shown in Figure 2. The other etching conditions were fixed as follows: SF_6 concentration of 15%, ICP power of 600 W and bias power of 40 W. At the beginning of increasing chamber pressure, more chemical radicals of chlorine are generated to react with *p*-GaN in order that the etch rate keeps increasing. When the pressure is higher than 40 mTorr, the formed involatile GaF_x becomes the dominant to suppress the etching of *p*-GaN. For AlGaN, more AlF_x will be formed with higher pressure to reduce the etch rate. The etch selectivity increases from 5:1 to 24:1 as the chamber pressure increases from 20 to 60 mTorr. However, there is a trade-off in the pressure range of 40–60 mTorr, considering the slight improvement in selectivity and the sharp drop of *p*-GaN etch rate.

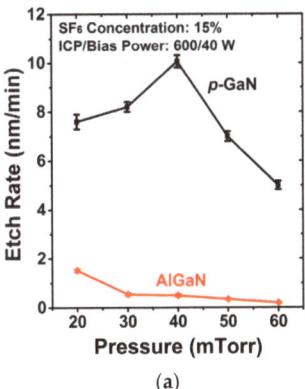
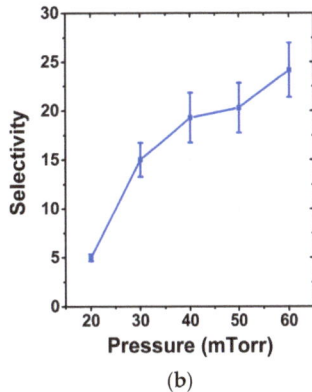

Figure 2. Dependence of (**a**) the etch rates, (**b**) selectivity between *p*-GaN and AlGaN on chamber pressure.

3.1.3. ICP Power

The ICP etching mechanism is a combined process of chemical reaction and ion sputtering. Both plasma density and ion energy can be regulated independently with two RF generators, i.e., ICP and bias power generators. Thus, the variation of source power and bias power can effectively affect the proportion of chemical and physical etching.

The etch rates, selectivity and self-bias voltage as a function of ICP power are shown in Figure 3. The other etching conditions were fixed as follows: SF_6 concentration of 15%, pressure of 40 mTorr and bias power of 40 W. The plasma density and fractional ionization are controlled by the ICP power. The etch rate of the *p*-GaN sample remarkably increases from 1.3 to 10.1 nm/min with increasing ICP power from 200 to 600 W due to more activity and density of the chemical radicals. The declining self-bias voltage is associated with increasing plasma density, indicating that the plasma bombardment is weakened. Increasing chemical reaction proportion and decreasing physical bombardment are exactly desired for high selectivity and low etch damage. However, the AlGaN etch rate slightly increases owing to the competition between chlorine as the etching agent and fluorine as the inhibition agent. AlF_x can be formed more easily than GaF_x, preventing a quick increase of AlGaN etching. As a result, the *p*-GaN/AlGaN etch selectivity increases with ICP power.

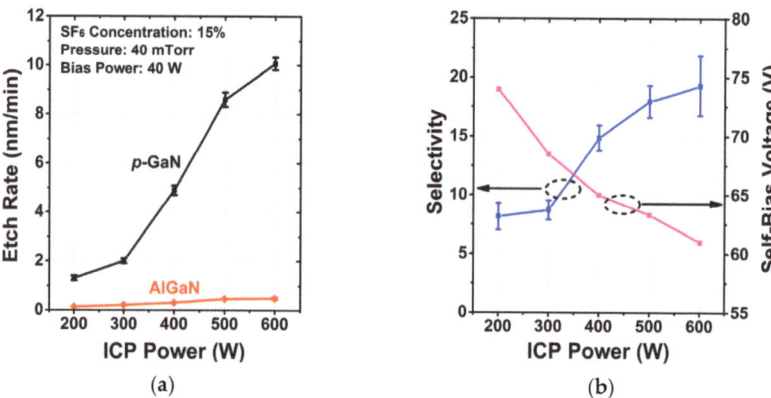

Figure 3. Dependence of (**a**) the etch rates, (**b**) selectivity between *p*-GaN and AlGaN and self-bias voltage on ICP power.

3.1.4. Bias Power

The effects of bias power are shown in Figure 4. The other etching conditions were fixed as follows: SF_6 concentration of 15%, pressure of 40 mTorr and ICP power of 600 W. The bias power is strongly related to physical etching. The self-bias voltage decreases linearly as the bias power decreases, indicating reduced ion bombardment energy. Though both *p*-GaN and AlGaN etch rates decrease proportionally to the decreasing bias power, almost linearly increasing selectivity is obtained. When the bias power drops down to 20 W, the selectivity reaches a maximum of 41:1 at a *p*-GaN etch rate of 3.4 nm/min in this study. The reduction in selectivity at high bias power can be explained in terms of enhanced sputtering of the AlF_x film at the AlGaN surface.

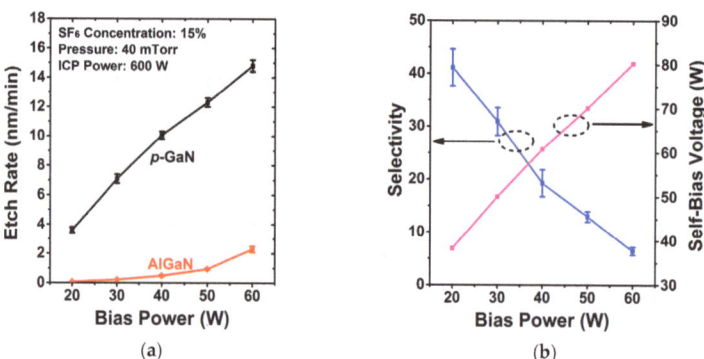

Figure 4. Dependence of (**a**) the etch rates, (**b**) selectivity between *p*-GaN and AlGaN and self-bias voltage on bias power.

To sum up, the final optimized process conditions were determined as SF_6 concentration of 15%, chamber pressure of 40 mTorr, ICP power of 600 W and bias power of 20 W. Table 1 summarizes the results achieved in our work together with other research using the BCl_3/SF_6 mixture. The selectivity in this study is the highest value ever reported, which can be attributed to our systematic optimization and the lowest bias power of our designed etch tool. Additionally, as reported in reference [20], a high selectivity of 33:1 was realized by using a higher frequency bias generator of 40 MHz. The much higher plasma frequency produces lower-energy ions which tends to achieve higher selectivity but with much lower etch rate. This makes the developed process in this work more practical in real device fabrication.

Table 1. Comparison of etch conditions, etch rates and selectivity among BCl_3/SF_6 mixture selective etching recipes.

Reference	[15]	[21]	This Work	[20]
Generator (MHz)	13.56	13.56	13.56	40
SF_6%	20	40	15	40
Pressure (mTorr)	37.5	20	40	10
ICP power (W)	200	200	600	/
Bias power (W)	30	60	20	/
GaN etch rate (nm/min)	12	12	3.4	0.529
AlGaN etch rate (nm/min)	0.52	1.3	0.08	0.016
Max. selectivity	23:1	9:1	41:1	33:1

3.2. Etched Surface and Plasma Damage Analysis

To comprehensively study the practical effects of the developed process on *p*-GaN/AlGaN wafer, firstly the etch depth was measured at different etch times by AFM. As seen in Figure 5, the etch process was quite linear until it reached the AlGaN surface. The X-SEM in the inset clearly shows a very smooth and almost non-recessed AlGaN surface after 2.5 min of overetching under the optimized process, demonstrating a highly selective etch to the AlGaN layer.

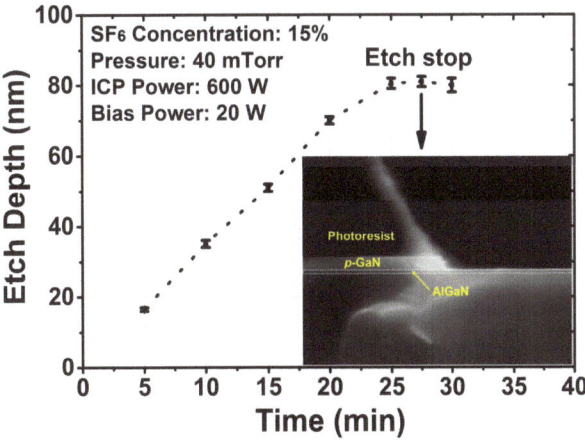

Figure 5. *p*-GaN etch depth as a function of time; the inset X-SEM image of sample with 2.5 min of overetching.

To further evaluate the impact of the developed selective etching process on the AlGaN surface, AFM images (5 μm × 5 μm) of the surface morphology were taken in no-contact mode (NCM) for the abovementioned sample (sample A, 2.5 min over etching under the optimized process) and another etched sample by using the nonselective BCl_3/Ar process (sample B) to etch the 80 nm p-GaN layer. The nonselective process has a p-GaN etch rate of approximately 10 nm/min.

As seen in Figure 6, for sample A the exposed AlGaN surface is quite smooth with the root mean square (RMS) surface roughness of 0.428 nm, which is similar to the as-grown AlGaN surface (0.446 nm in Figure 6a). This is attributed to the advantage of the developed highly selective etching and its low power causing very minimum surface damage. However, with nonselective p-GaN etching for sample B, the exposed AlGaN surface roughness reached as high as 0.987 nm. This is equivalent to the as-grown p-GaN surface, which has 1.053 nm RMS roughness due to the doping of Cp_2Mg. Obviously, the sample B AlGaN surface is much rougher as the morphology is basically inherited from the as-grown p-GaN layer due to the nature of nonselective etching, as illustrated in Figure 6e.

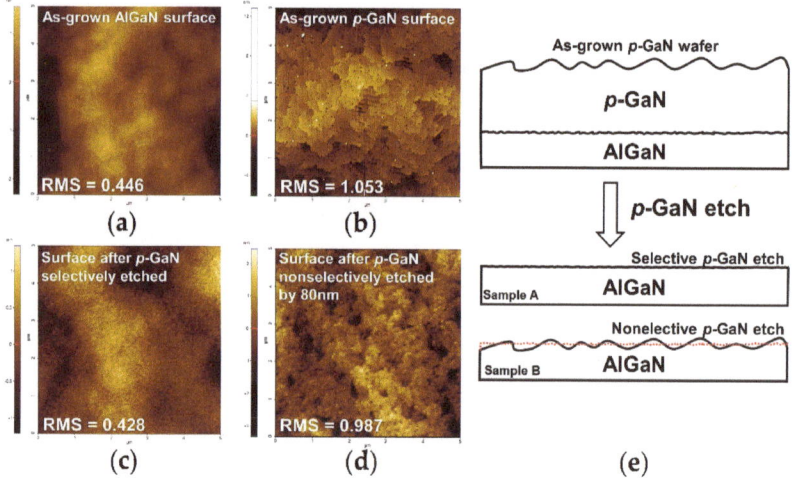

Figure 6. Surface morphology of (**a**) as-grown AlGaN surface, (**b**) as-grown p-GaN surface, (**c**) sample A (p-GaN selectively etched) and (**d**) sample B (p-GaN nonselectively etched), (**e**) diagram to illustrate surface morphology for samples A and B.

MIS capacitors were fabricated to evaluate the possible etch damage on the exposed AlGaN surface for samples A and B. Reference device on as-grown AlGaN wafer was also prepared for comparison. For all the samples, dilute HCl dip was performed to treat the AlGaN surfaces, and 25-nm-thick Al_2O_3 was deposited at 300 °C using trimethylaluminum (TMA) and H_2O as precursors. Ni/Au bilayers were used as electrodes in the inner circle and exterior zone with a ring gap of 50 μm. To avoid the possible repair effect on the etch damage, no anneal was performed in this process.

C–V characterizations using a Keythley 4200A are presented in Figure 7. At a quite negative voltage, the capacitance is close to zero because the 2DEG is depleted for all samples. The nearly flat capacitance C_{2DEG} indicates that 2DEG has been formed at the AlGaN/GaN heterojunction interface. For sample A with selective p-GaN etching, the slope of the C–V curve is quite steep and close to the reference as-grown AlGaN sample, confirming very minimum etch damage on the AlGaN surface after selective p-GaN removal. However, the slope of the C–V curve for sample B shows an obvious stretch-out, indicating that the exposed AlGaN barrier layer was degraded during the nonselective etching of *p*-GaN and thus 2DEG at AlGaN/GaN interface could not be formed efficiently with the gate bias.

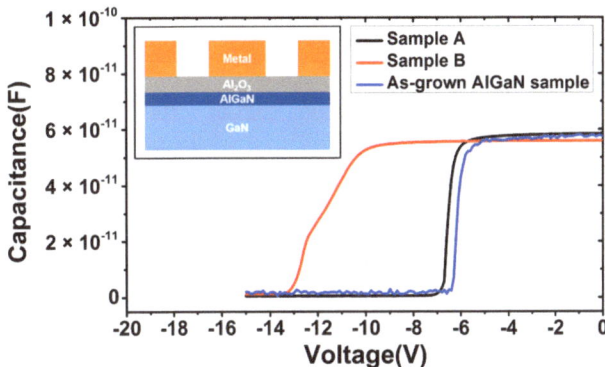

Figure 7. C–V characteristics of Ni/Al$_2$O$_3$/AlGaN MIS capacitors with selective and nonselective *p*-GaN etching. Additionally, as-grown AlGaN sample is referenced here.

4. Conclusions

In this work, a highly selective ICP etch process of *p*-GaN over AlGaN using the BCl$_3$/SF$_6$ mixture was successfully developed, achieving a high selectivity of 41:1. Under the conditions of the optimized SF$_6$ concentration and chamber pressure, as well as high ICP power, and as low as possible bias power benefiting from our dedicated etch tool, a very smooth and high-quality AlGaN surface could be obtained after *p*-GaN etch. On such AlGaN surface, the fabricated Ni/Al$_2$O$_3$/AlGaN MIS capacitor showed comparable C–V characteristics to that on the as-epitaxial AlGaN surface. This phenomenon strongly indicated that there was almost no damage on the AlGaN surface after etching the *p*-GaN layer, making this process very promising to be applied on the fabrication of high-performance *p*-GaN gate HEMTs.

Author Contributions: Conceptualization, P.Z. and L.W.; methodology, P.Z. and L.W.; validation, K.Z. and M.P.; formal analysis, L.W. and K.Z.; investigation, P.Z. and L.W.; resources, S.X. and C.W. (Chunlei Wu); data curation, P.Z. and L.W.; writing—original draft preparation, P.Z.; writing—review and editing, M.X. and C.W. (Chen Wang); visualization, Y.Y. and R.F.; supervision, M.X., C.W. (Chen Wang) and D.W.Z.; project administration, M.X. and C.W. (Chen Wang); funding acquisition, M.X., C.W. (Chen Wang) and D.W.Z. All authors have read and agreed to the published version of the manuscript.

Funding: This research was funded by the National Natural Science Foundation of China, grant number 11904043.

Data Availability Statement: Not applicable.

Conflicts of Interest: The authors declare no conflict of interest.

References

1. Ishida, M.; Ueda, T.; Tanaka, T.; Ueda, D. GaN on Si Technologies for Power Switching Devices. *IEEE Trans. Electron Devices* **2013**, *60*, 3053–3059. [CrossRef]
2. Marcon, D.; Saripalli, Y.N.; Decoutere, S. 200mm GaN-on-Si epitaxy and e-mode device technology. In Proceedings of the Electron Devices Meeting, Washington, DC, USA, 5–9 December 2015. [CrossRef]
3. Sacconi, F.; Carlo, A.D.; Lugli, P.; Morkoc, H. Spontaneous and piezoelectric polarization effects on the output characteristics of AlGaN/GaN heterojunction modulation doped FETs. *IEEE Trans. Electron Devices* **2001**, *48*, 450–457. [CrossRef]
4. Chen, K.J.; Zhou, C. Enhancement-mode AlGaN/GaN HEMT and MIS-HEMT technology. *Phys. Status Solidi* **2011**, *208*, 434–438. [CrossRef]
5. Scott, M.J.; Fu, L.; Zhang, X.; Li, J.; Yao, C.; Sievers, M.; Wang, J. Merits of gallium nitride based power conversion. *Semicond. Sci. Technol.* **2013**, *28*, 74013. [CrossRef]

6. Cai, Y.; Zhou, Y.; Chen, K.J.; Lau, K.M. High-performance enhancement-mode AlGaN/GaN HEMTs using fluoride-based plasma treatment. *IEEE Electron Device Lett.* **2005**, *26*, 435–437. [CrossRef]
7. Saito, W.; Takada, Y.; Kuraguchi, M.; Tsuda, K.; Omura, I. Recessed-gate structure approach toward normally off high-Voltage AlGaN/GaN HEMT for power electronics applications. *IEEE Trans. Electron Devices* **2006**, *53*, 356–362. [CrossRef]
8. Uemoto, Y.; Hikita, M.; Ueno, H.; Matsuo, H.; Ishida, H.; Yanagihara, M.; Ueda, T.; Tanaka, T.; Ueda, D. Gate Injection Transistor (GIT)—A Normally-Off AlGaN/GaN Power Transistor Using Conductivity Modulation. *IEEE Trans. Electron Devices* **2007**, *54*, 3393–3399. [CrossRef]
9. Chen, K.J.; Haberlen, O.; Lidow, A.; Tsai, C.l.; Ueda, T.; Uemoto, Y.; Wu, Y. GaN-on-Si Power Technology: Devices and Applications. *IEEE Trans. Electron Devices* **2017**, *64*, 779–795. [CrossRef]
10. Roccaforte, F.; Greco, G.; Fiorenza, P.; Iucolano, F. An Overview of Normally-Off GaN-Based High Electron Mobility Transistors. *Materials* **2019**, *12*, 1599. [CrossRef] [PubMed]
11. Marcon, D.; Hove, M.V.; Jaeger, B.D.; Posthuma, N.; Wellekens, D.; You, S.; Kang, X.; Wu, T.-L.; Willems, M.; Stoffels, S.; et al. Direct Comparison of GaN-Based E-Mode Architectures (Recessed MISHEMT and p-GaN HEMTs) Processed on 200mm GaN-on-Si with Au-Free Technology. In Proceedings of the SPIE-The International Society for Optical Engineering, San Francisco, CA, USA, 13 March 2015. [CrossRef]
12. Lukens, G.; Hahn, H.; Kalisch, H.; Vescan, A. Self-Aligned Process for Selectively Etched p-GaN-Gated AlGaN/GaN-on-Si HFETs. *IEEE Trans. Electron Devices* **2018**, *65*, 3732–3738. [CrossRef]
13. Su, L.-Y.; Lee, F.; Huang, J.J. Enhancement-Mode GaN-Based High-Electron Mobility Transistors on the Si Substrate With a P-Type GaN Cap Layer. *IEEE Trans. Electron Devices* **2014**, *61*, 460–465. [CrossRef]
14. Greco, G.; Iucolano, F.; Roccaforte, F. Review of technology for normally-off HEMTs with p-GaN gate. *Mater. Sci. Semicond. Process.* **2018**, *78*, 96–106. [CrossRef]
15. Buttari, D.; Chini, A.; Chakraborty, A.; Mccarthy, L.; Xing, H.; Palacios, T.; Shen, L.; Keller, S.; Mishra, U.K. Selective dry etching of GaN over AlGaN in BCl_3/SF_6 mixtures. In Proceedings of the IEEE Lester Eastman Conference on High Performance Devices, Troy, NY, USA, 4–6 August 2004. [CrossRef]
16. Buttari, D.; Chini, A.; Palacios, T.; Coffie, R.; Shen, L.; Xing, H.; Heikman, S.; McCarthy, L.; Chakraborty, A.; Keller, S.; et al. Origin of etch delay time in Cl_2 dry etching of AlGaN/GaN structures. *Appl. Phys. Lett.* **2003**, *83*, 4779–4781. [CrossRef]
17. Lee, Y.S.; Sia, J.F.; Nordheden, K.J. Mass spectrometric characterization of BCl_3/SF_6 plasmas. *J. Appl. Phys.* **2000**, *88*, 4507–4509. [CrossRef]
18. Oh, C.S.; Kim, T.H.; Lim, K.Y.; Yang, J.W. GaN etch enhancement in inductively coupled BCl_3 plasma with the addition of N_2 and SF_6 gas. *Semicond. Sci. Technol.* **2004**, *19*, 172–175. [CrossRef]
19. Feng, M.S.; Guo, J.D.; Lu, Y.M.; Chang, E.Y. Materials Chemistry and Physics. Reactive ion etching of GaN with BCl_3/SF_6 plasmas. *Mater. Chem. Phys.* **1996**, *45*, 80–83. [CrossRef]
20. Beheshti, M.; Westerman, R. Smooth, low rate, selective GaN/AlGaN etch. *AIP Adv.* **2021**, *11*, 25237. [CrossRef]
21. Qian, R.; Cheng, X.; Zheng, L.; Shen, L.; Zhang, D.; Gu, Z.; Yu, Y. Inductively Coupled Plasma Etching of p-GaN Using Different Masks and Etching Gases. *Semicond. Technol.* **2018**, *43*, 449–455. [CrossRef]

Article

Experimental Study on Diesel Engine Emission Characteristics Based on Different Exhaust Pipe Coating Schemes

Keqin Zhao, Diming Lou, Yunhua Zhang *, Liang Fang and Yuanzhi Tang

School of Automobiles Studies, Tongji University, Shanghai 201804, China; zhaokeqin@tongji.edu.cn (K.Z.); loudiming@tongji.edu.cn (D.L.); fangliang@tongji.edu.cn (L.F.); tangyuanzhi@tongji.edu.cn (Y.T.)
* Correspondence: zhangyunhua131@tongji.edu.cn

Citation: Zhao, K.; Lou, D.; Zhang, Y.; Fang, L.; Tang, Y. Experimental Study on Diesel Engine Emission Characteristics Based on Different Exhaust Pipe Coating Schemes. *Micromachines* **2021**, *12*, 1155. https://doi.org/10.3390/mi12101155

Academic Editor: Kun Li

Received: 2 September 2021
Accepted: 24 September 2021
Published: 25 September 2021

Publisher's Note: MDPI stays neutral with regard to jurisdictional claims in published maps and institutional affiliations.

Copyright: © 2021 by the authors. Licensee MDPI, Basel, Switzerland. This article is an open access article distributed under the terms and conditions of the Creative Commons Attribution (CC BY) license (https://creativecommons.org/licenses/by/4.0/).

Abstract: The thermal insulation performance of exhaust pipes coated with various materials (basalt and glass fiber materials) under different braiding forms (sleeve, winding and felt types) and the effects on the emission characteristics of diesel engines were experimentally studied through engine bench tests. The results indicated that the thermal insulation performance of basalt fiber was higher than that of glass fiber, and more notably advantageous at the early stage of the diesel engine idle cold phase. The average temperature drop during the first 600 s of the basalt felt (BF) pipe was 2.6 °C smaller than that of the glass fiber felt (GF) pipe. Comparing the different braiding forms, the temperature decrease in the felt-type braided material was 2.6 °C and 2.9 °C smaller than that in the sleeve- and winding-type braided materials, respectively. The basalt material was better than the glass fiber material regarding the gaseous pollutant emission reduction performance, especially in the idling cold phase of diesel engines. The NO_x conversion rate of the BF pipe was 7.4% higher than that of the GF pipe, and the hydrocarbon (HC) conversion rate was 2.3% higher than that of the GF pipe, while the CO conversion rate during the first 100 s was 24.5% higher than that of the GF pipe. However, the particulate matter emissions were not notably different.

Keywords: diesel engine bench test; basalt fiber; glass fiber; thermal insulation performance; emission characteristics

1. Introduction

Diesel engines are widely applied in the field of commercial vehicles due to their good power features, fuel economy, and reliability. Although they have many advantages, they have a significant impact on the problem of environmental pollution worldwide [1,2]. The nitrogen oxide (NO_x) and particulate matter emissions of diesel engines are considerable, accounting for 70% and 90%, respectively, of the total vehicle emissions, and are also responsible for several health problems [3,4]. To limit the pollutant emissions originating from diesel vehicles, various countries have continuously tightened emission limits, broadened the requirements for low-temperature and low-load emission pollutant control, and greatly reduced the NO_x emission concentration limit. The combination of various after-treatment technologies has become an important means to improve vehicle pollutant emissions and meet the increasingly stringent emission regulations. At present, after-treatment devices largely include optimized combustion + selective catalytic reduction (SCR), and exhaust gas recirculation (EGR) + diesel oxidation catalyst (DOC) + diesel particulate filter (DPF) [5–8]. The exhaust temperature is the main factor influencing the purification effect of the exhaust gas thermal reaction and the after-treatment performance. Within a certain temperature range, the higher the exhaust temperature, the better the purification effect of the thermal reaction [9]. If the exhaust temperature is too low, this may cause problems such as urea solution crystallization in the SCR device [10,11]. Therefore, to improve the catalytic efficiency of after-treatment devices, it is necessary to coat exhaust pipes with an insulating material to reduce heat loss, thereby improving the emission performance of the entire vehicle.

At present, glass fiber and asbestos fiber are mainly used as the insulation materials in the exhaust pipe, but long-term exposure to glass fiber and asbestos fiber will damage the respiratory system and cause health hazards [12,13]. Therefore, it is very important to study new alternative materials. Basalt fiber has increasingly become a substitute for glass fiber in many fields, such as marine and military industries, due to its excellent performance. Additionally, BF is labeled as safe, according to both the USA and the European occupational safety guidelines.

Basalt fiber is a new natural green material. It consists of volcanic extruded rock. Its chemical composition is similar to that of gabbro. The SiO_2 content varies between 45% and 60%. The $K_2O + Na_2O$ content is slightly higher than that in intrusive rocks. The $Fe_2O_3 + FeO$ and MgO contents are slightly lower than those in intrusive rocks. Fiber is produced by putting material into a furnace where it is melted at 1450–1500 °C. Subsequently, the molten material is forced through a platinum/rhodium crucible bushing to create fibers. Compared with glass fiber, basalt fiber is cheaper to produce due to less energy consumed and no additives being required. Additionally, it attains a better high-temperature resistance than glass fiber. Basalt fiber achieves a wider temperature range, from −200 to 800 °C, while the temperature range attained by glass fiber varies between approximately −60 °C and 450 °C [14]. At a working temperature of 400 °C, the breaking strength of basalt fiber can be maintained at 85%, while at a working temperature of 600 °C, its breaking strength can still be maintained at 80%. Moreover, if basalt fiber is pretreated at a temperature ranging from 780 to 820 °C, it can be applied at 860 °C without shrinkage [15]. Due to its excellent high-temperature resistance, basalt fiber has been widely adopted in transportation infrastructure, environmental protection and other fields [16–21].

In recent years, scholars have conducted research on the thermal properties of basalt and glass fiber materials. Their research results indicated that basalt fiber contained a large number of micropores to prevent air convection and heat radiation, and its thermal insulation performance was higher than that of glass fiber [22]; under exposure to the same radiant heat flux, basalt fiber material reached a higher temperature faster than glass fiber material due to its higher thermal emissivity [23]; mass loss occurred in the temperature range of 200 to 350 °C, the thermal decomposition temperature of basalt fiber was 40 °C higher than that of glass fiber, and basalt fiber reached a higher thermal stability [24]. Therefore, basalt fiber provides certain advantages over glass fiber in terms of its thermal performance, and as it is widely used in car mufflers and other parts in the automotive industry [25], basalt fiber exhibits a certain potential in diesel engine thermal management applications. However, there are still few studies on its application in the after-treatment thermal management of diesel engines.

This paper relies on diesel engine bench tests to assess the thermal insulation performance of different exhaust pipe covering schemes considering basalt and glass fiber materials. Moreover, various basalt fiber weaving forms (basalt sleeve (BS), basalt winding (BW), and basalt felt (BF)) and their influence on the emission characteristics of diesel engines are investigated herein.

2. Methodology

2.1. Test Materials

The test exhaust pipe samples adopted basalt sleeve (BS), basalt winding (BW), basalt felt (BF), and glass fiber felt (GF) with the equal thickness and bulk density. The variety of covering samples is presented in Table 1.

Table 1. Test sample.

Serial Number	Prototype	Thickness (mm)	Bulk Density (kg·m^{-3})	Thermal Conductivity (W/m·k)	End-Use Temperature (°C)	Morphology
1	BS	5	120	0.031	780	
2	BW	5	120	0.031	780	
3	BF	5	120	0.031	780	
4	GF	5	120	0.049	400	

2.2. Test Device and Data Processing

The test device mainly included a diesel engine for the test and the bench test system. The system consisted of a Horiba 7200D and an AVL489. The Horiba 7200D was used to measure the CO, THC, NOx, etc., and the AVL489 was used to measure the particle emissions. Table 2 lists the technical parameters of the diesel engine type adopted in the test, and Table 3 summarizes the composition and model of the sampling equipment of the diesel engine bench test system.

Table 2. Parameters of the diesel engine.

Project	Parameter
Diesel engine model	D45
Displacement/L	4.5
Rated power/kW	150
Rated speed/r min^{-1}	2300
After-treatment device	DOC + SCR + DPF

Table 3. Sampling equipment and models.

Sample Item	Device Model	Sample Content
Gaseous substance	Horiba 7200D	NOx, CO, and hydrocarbons (HCs)
Number of particles (PN)	AVL489	PN

Through the bench test, the original emission data of different pollutants before entering the after-treatment system, and the emission data of different pollutants through the after-treatment system with different covering schemes, was obtained. Finally, the conversion rate of different pollutants was determined, and the calculation method is shown in Equation (1) where PE1 is the original emission data, PE2 is the emission data after, and CR is the conversion rate of different pollutants.

$$CR = \frac{PE1 - PE2}{PE1} \times 100\% \quad (1)$$

2.3. Test Conditions and Plan

The cold test of the diesel engine bench adopted the World Harmonized Transient Cycle (WHTC) heavy-duty diesel engine test cycle. The WHTC cycle is a transient operating condition lasting 1800 s that changes at 1 s intervals. The cycle is divided into three stages, namely, 0~600 s as the cold operation stage, 600~1200 s as the transition stage, and 1200~1800 s as the hot operation stage, as shown in Figure 1.

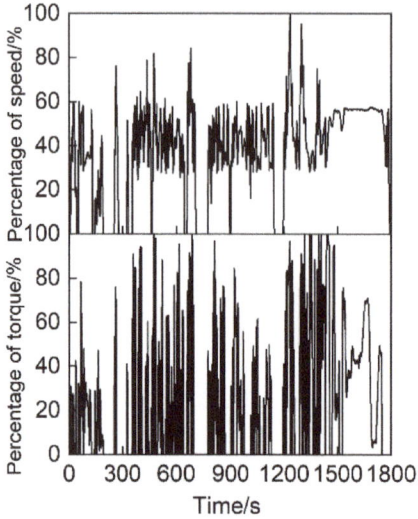

Figure 1. Schematic diagram of the WHTC cycle.

The exhaust pipes were coated with basalt and glass fiber materials with equal thickness and bulk density. The glass fiber material adopted the felt-type weaving method. The basalt fiber material was subjected to three different weaving methods, i.e., sleeve, winding and felt types. The temperature at the outlet end of the diesel engine, and at the inlet of the after-treatment system, was measured under cycling conditions. The heat preservation characteristics of the entire exhaust system were studied, and the emission characteristics (NO_x, CO, HCs, and PN) of the diesel engine containing an exhaust system coated with the above two materials under the same cycle, were analyzed. The test plan is listed in Table 4, and the layout of the test bench and temperature measurement points are shown in Figure 2.

Table 4. Test plan.

Exhaust Gas Temperature	Emission Data	Test Cycle
1. T1 before the pipe 2. T2 after the pipe	Gaseous substances: CO, HCs, and NOx Number of particles: PN	WHTC (cold state)

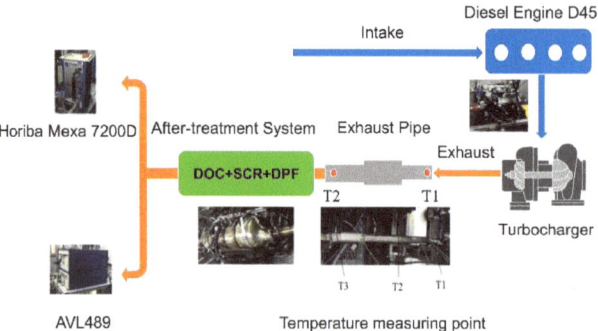

Figure 2. Layout of the test bench and temperature measurement points.

3. Results and Discussion
3.1. Exhaust Temperature Analysis of the Full Cycle

Figure 3 shows the exhaust gas temperature T1 at the rear end of the vortex over time for the different exhaust pipe covering schemes of the tested diesel engine. In essence, the temperature measured at the T1 measurement point of the exhaust pipes with the different materials and covering methods was basically the same, because it was the outlet temperature of the turbocharger and was less affected by the insulation material. The average temperature at each stage during cold, transition and hot operation exhibited a stepwise upward trend, at 201 °C, 258 °C and 328 °C, respectively.

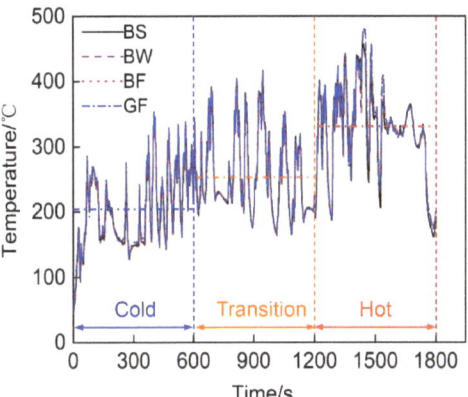

Figure 3. Temperature at the T1 measurement point of the exhaust pipes with different covering schemes over time.

T2 is the temperature at the inlet of the after-treatment system, and its specific changes throughout the cycle are shown in Figure 4. At the T2 measurement point, the exhaust temperature of the exhaust pipes with the different materials and covering methods greatly varied (especially in the cold cycle). The exhaust temperatures of the BS and BW pipes at the cold operation stage were lower than those of the other pipes. Their average temperatures were 3 °C and 2 °C, respectively, lower than the overall average temperature in the cold phase.

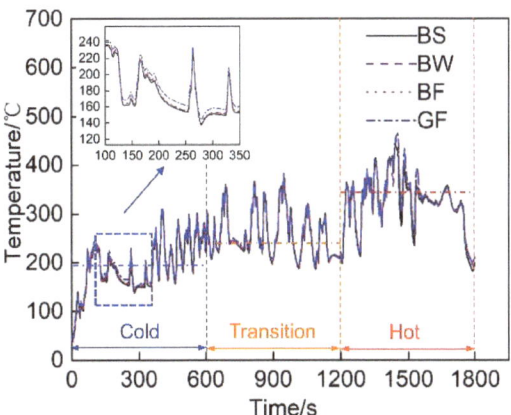

Figure 4. Temperature at the T2 measurement point of the exhaust pipes with different covering schemes over time.

This test considered T1-T2 (Δ T) to intuitively simulate the thermal insulation performance of the vehicle exhaust pipe system. Figure 5 shows the evolution of the temperature decrease in the exhaust pipes with different covering schemes. The overall exhaust pipe temperature experienced a downward trend. This mainly occurred because the exhaust pipe operated in the cold state upon cold start initiation. Some of the heat flux preheated the exhaust pipe, resulting in a notable heat energy loss, therefore the temperature decline was large. With increasing exhaust pipe temperature, the heat loss attributed to preheating gradually decreased, reducing the temperature drop. Figure 6 shows the average temperature drop considering the different covering schemes for the entire circulating exhaust pipe. The results indicated that the average temperature decrease in the BF pipe during the entire cycle was the smallest, which was 0.5 °C smaller than that in the GF pipe. This occurred because basalt fiber is composed of tectosilicates, phyllosilicates, chain silicates, and orthosilicates [26]. The amorphous region in the interior was large, and many grain boundaries, defects and impurities existed. This resulted in a low thermal conductivity. Therefore, the thermal insulation performance of the basalt fiber material was better than that of the glass fiber material. Comparing the different basalt material weaving methods, the average temperature decrease in the exhaust pipe coated with BF was 2.6 °C and 2.9 °C smaller than that in the exhaust pipes coated with BS and BW, respectively, and the heat preservation performance was higher than that of the other two weaving forms. This occurred due to the large size of the pores of the sleeve- and winding-type materials, resulting in serious heat loss and a poor thermal insulation performance.

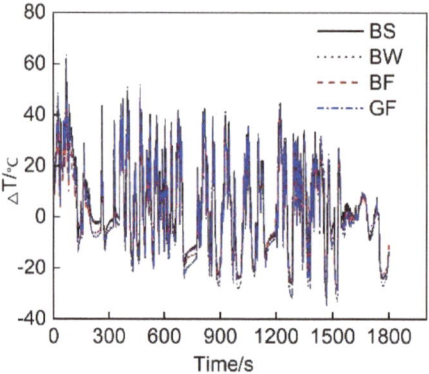

Figure 5. Temperature decrease Δ T in the exhaust pipes with different covering schemes over time.

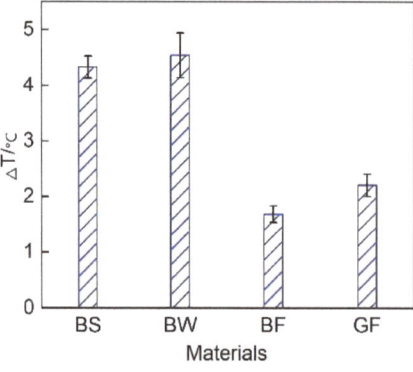

Figure 6. Average temperature decrease in the exhaust pipes with different covering schemes.

3.2. Emission Analysis of the Full Cycle

Figure 7 shows a comparison of the NO_x conversion efficiency of the exhaust pipes between the different covering schemes during each cycle of the diesel engine. The figure reveals that with an increasing load, the NO_x conversion rate gradually increased. Figure 4 shows that the exhaust gas temperature rose, thus promoting SCR. During the first 600 s of the cold-state operation stage, the SCR inlet temperature was low, and the NO_x conversion rate of each pipe did not exceed 50% [27]. At the transition stage (600~1200 s) and hot-state operation stage (1200~1800 s), the NO_x conversion rate of each pipe was greatly improved; this occurred because DOC in the range of 200~400 °C can effectively improve the NO_2/NO_x ratio and improve the NO_x conversion rate [28]. Regardless of the operation stage, the average NO_x conversion rate of the BF pipe was the highest. At the cold operation stage, the average NO_x conversion rate of the BF pipe was 47.6%, 97.6% and 7.4% higher than that of the BS, BW and GF pipes, respectively. At the transition stage (600~1200 s) and hot operation stage (1200~1800 s), the average NO_x conversion rate was 13.6%, 3% and 11% higher, respectively, than that of the other three pipes.

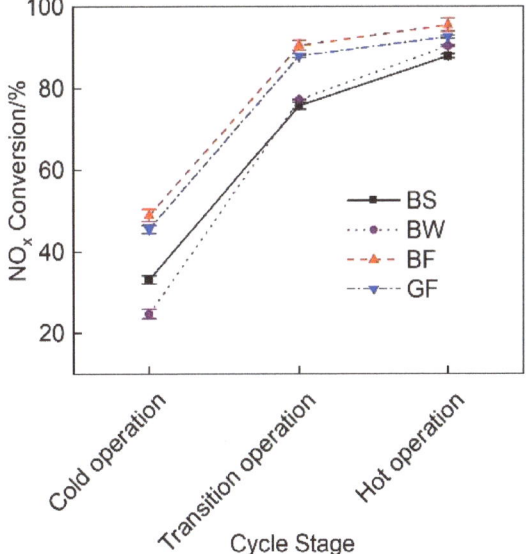

Figure 7. NO_x conversion rate of the different exhaust pipe covering schemes.

Figure 8 shows the average CO conversion rate of the exhaust pipes with the different covering schemes during the WHTC cycle. Under the emission reduction effect of the DOC and DPF device combination, the CO conversion rate of each pipe was maintained at a high level [29], reaching above 85%, and with the increase in exhaust temperature, CO oxidation was promoted and CO conversion rate increased [30]. The average CO conversion rate of the BF pipe was 1.1%, 2.7% and 4.3% higher than that of the BS, BW and GF pipes, respectively, but these differences were small.

Figure 9 shows the HC conversion efficiency of the exhaust pipes with the different covering schemes at each cycle stage. The figure revealed that the overall HC conversion efficiency gradually increased. The average HC conversion rate of the BF pipe reached 96.7%, which was 6.9%, 8.3% and 1.3% higher than that of the BS, BW and GF pipes, respectively.

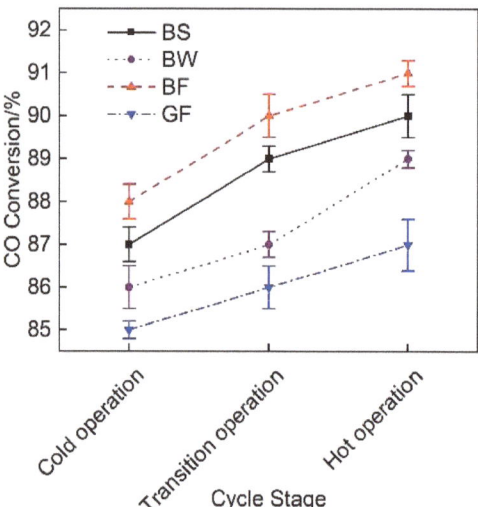

Figure 8. CO conversion rate of the different exhaust pipe covering schemes.

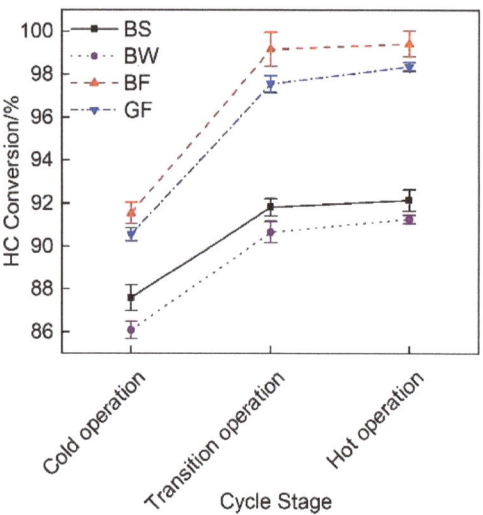

Figure 9. HC conversion rate of the different exhaust pipe covering schemes.

Figure 10 shows the change curve of the particulate matter concentration of the exhaust pipes with the different covering schemes at each cycle stage. Under the effects of the DOC and DPF device combination, the particulate matter emissions of the different insulated pipes were relatively low and exhibited a trend of decreasing first and then slightly increasing. This occurred because at the beginning of the cycle, the exhaust temperature was low and the DOC performance was not high. During hot-state operation, a high exhaust temperature promoted the production of sulfate [31], thus causing a slight increase in particulate matter emissions.

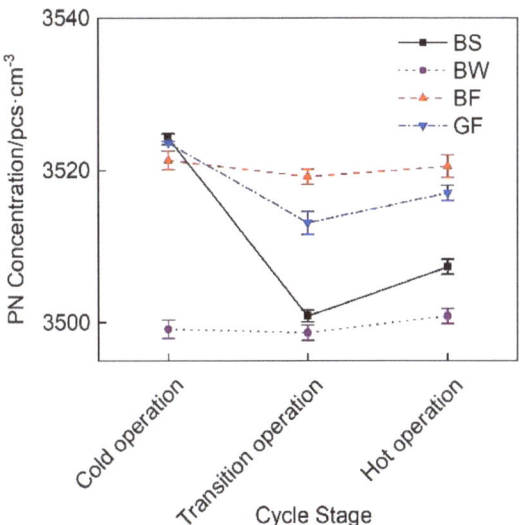

Figure 10. PN concentration change considering the different exhaust pipe covering schemes.

Figure 11 shows the total emissions of gaseous pollutants and particulates during the WHTC cycle of the exhaust pipes with different covering schemes. It is observed that little difference existed in particulate matter emissions. Due to its better thermal insulation performance, the BF pipe achieved a high after-treatment system conversion efficiency, and the gaseous pollutant emissions were the lowest. The total emissions were 67.3%, 49.8% and 23.2% lower than those of the BS, BW and GF pipes, respectively.

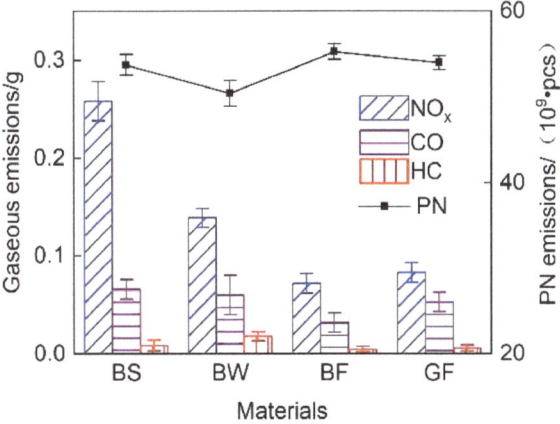

Figure 11. Total emissions during the WHTC cycle.

3.3. Exhaust Temperature Analysis during Cold Operation

Since the WHTC cycle focuses more on the investigation of diesel engine emissions under low-speed and low-load conditions, the diesel engine emission temperature during this cycle is low, and the performance requirements of the SCR after-treatment system are high. Therefore, it is very important to study the temperature and corresponding emission characteristics of cold-state operation. Based on the test results, the cold operation stage (0–600 s) was subdivided, and the change law of the average temperature drop at each substage of the initial cold operation stage was further analyzed.

Figure 12 shows the average temperature drop of the exhaust pipes with the different covering schemes at the various substages of the initial cold-state operation stage. During the cold operation cycle of the diesel engine, the average temperature decrease in the BF pipe over the first 600 s was 2.6 °C smaller than that in the GF pipe.

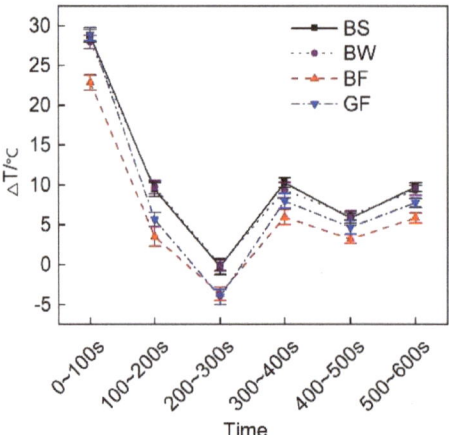

Figure 12. Average temperature drop at each substage of the initial idling stage.

3.4. Analysis of the Emission Results during Cold Operation

Figure 13 shows a comparison of the NO_x conversion efficiency of the exhaust pipes between the different covering schemes at each stage of cold operation. It is observed that the NO_x conversion rate during the first 100 s was lower than 20%. This finding matches the temperature characteristics. Figure 4 shows that during the first 100 s, the average exhaust gas temperature entering the after-treatment system was lower than 200 °C, resulting in a low SCR catalytic activity and a low NO_x conversion rate. The discharge performance of the BF pipe was the highest during cold-state operation, and the average NO_x conversion efficiency was 47.6%, 97.6%, and 7.4% higher than that of the BS, BW and GF pipes, respectively.

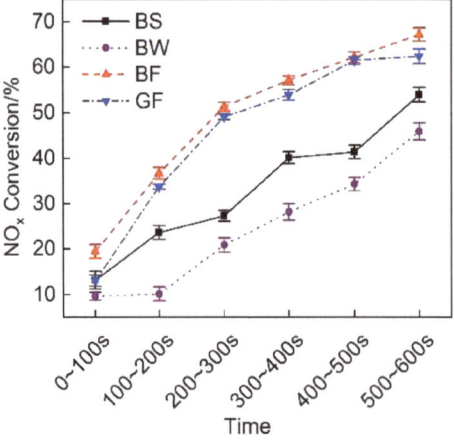

Figure 13. NO_x conversion efficiency at each substage of the initial idling stage of the different exhaust pipe covering schemes.

Figure 14 shows the CO conversion efficiency of the exhaust pipes with the different covering schemes at each stage during cold operation. Figure 14 shows that the CO conversion rate was low during the first 100 s and remained high during the next 500 s. The discharge effect of the BF pipe during the first 100 s was obviously better than that of the GF pipe, and the CO conversion rate of the BF pipe was 24.5% higher than that of the GF pipe.

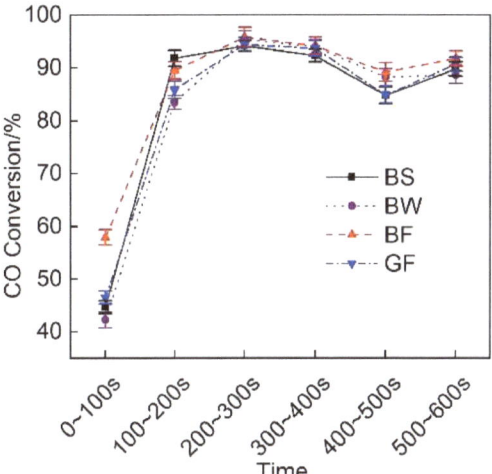

Figure 14. CO conversion efficiency at each substage of the initial idling stage of the different exhaust pipe covering schemes.

Figure 15 shows a comparison of the HC conversion efficiency of the exhaust pipes between the different covering schemes at each stage during cold operation. The figure revealed that the HC conversion rate at each stage was high, being not lower than 75%. The HC conversion rate of the BF pipe was relatively high. The average HC conversion rate at the cold operation stage reached 92%, which was 2.3% higher than that of the GF pipe.

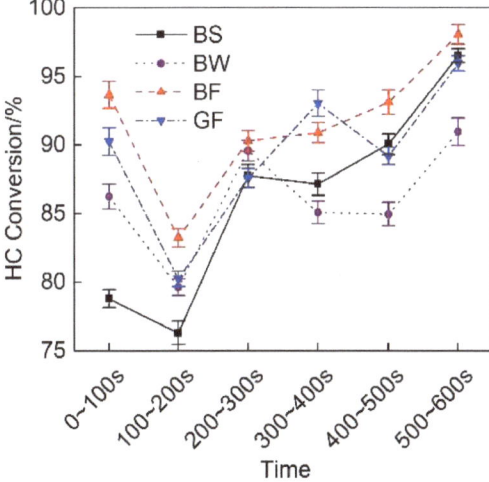

Figure 15. HC conversion efficiency of the different exhaust pipe covering schemes at each substage of the initial idling stage.

Figure 16 shows a comparison of the particulate matter emissions of the exhaust pipes between the different covering schemes at each stage during cold operation. Figure 16 shows that the particulate matter emissions of the BF pipe were the lowest.

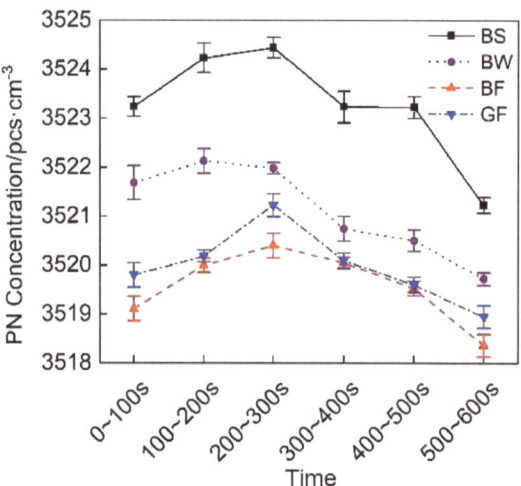

Figure 16. PN concentration of the different exhaust pipe covering schemes at each substage of the initial idling stage.

4. Conclusions

This paper presented an experimental study on the thermal insulation performance of exhaust pipes coated with various materials (basalt and glass fiber materials) under different braiding forms (sleeve, winding and felt types) and the effects on the emission characteristics of diesel engines. The following conclusions can be drawn from this study.

The thermal insulation performance of the basalt fiber material is better than that of the glass fiber material. This occurs because the thermal conductivity of the basalt fiber material is lower than that of the glass fiber material. The average temperature decrease in the BF pipe throughout the entire cold WHTC cycle is the smallest, and its average temperature is 0.5 °C lower than that of the GF pipe, which is a small difference. During the 600 s period before the cold cycle, the thermal insulation performance of the basalt fiber material is obviously better than that of the glass fiber material, and the average temperature drop is 2.6 °C smaller than that of the GF pipe. The basalt fiber material of the felt covering type attains the best thermal insulation performance, and the average temperature drop is 2.6 °C and 2.9 °C smaller than that of the sleeve- and winding-type materials, respectively.

The gaseous pollutant emission performance of the after-treatment system coated with basalt fiber material is better than that of the after-treatment system coated with glass fiber material. Throughout the full WHTC cycle, during the transition phase (600~1200 s) and the thermal operation phase (1200~1800 s), the average conversion rates of NO_x, CO and HCs of each pipe are obviously improved. The average conversion rates of NO_x, CO, and HCs of the BF pipe are all the highest. Among them, the average NO_x conversion rate is 13.6%, 11% and 3% higher than that of the BS, BW and GF pipes, respectively. The BF pipe average CO conversion rate is 1.1%, 2.7%, and 4.3% higher, respectively, and the BF pipe average HC conversion rate is 6.9%, 8.3%, and 1.3% higher, respectively.

During the first 600 s of the cold operation stage, the gaseous pollutant emission performance of the basalt fiber-coated after-treatment system is notably higher than that of the glass fiber-coated after-treatment system. The NOx conversion rate of the BF pipe is 7.4% higher than that of the GF pipe. The CO conversion rate of the BF pipe during the first

100 s is 24.5% higher than that of the GF pipe, and the HC conversion rate of the BF pipe is 2.3% higher than that of the GF pipe. Little difference was observed in diesel particulate matter emissions between the exhaust pipes with the different covering schemes.

Author Contributions: Writing—review and editing, K.Z.; supervision, Y.Z. and L.F.; resources, D.L.; data curation, Y.T. All authors have read and agreed to the published version of the manuscript.

Funding: The authors appreciate the prospective funding support provided by the Nanchang Automotive Innovation Institute and Tongji University (project: Research and Application of Key Thermal Management Technology of Diesel Engine Exhaust System Based on New Basalt Fiber).

Data Availability Statement: The data used to support the findings of this study are included within the article.

Conflicts of Interest: The authors confirm that there are no conflict of interest.

References

1. Yao, H.; Ni, T.; You, Z. Characterizing pollutant loading from point sources to the Tongqi River of China based on water quality modeling. *Int. J. Environ. Sci. Technol.* **2019**, *16*, 6599–6608. [CrossRef]
2. Tangestani, V.; Isfahani, A.H.M. Experimental evaluation of the performance and exhaust emissions of porous medium diesel and Otto engines. *Int. J. Environ. Sci. Technol.* **2020**, *17*, 1463–1474. [CrossRef]
3. Resitoglu, I.A.; Altınısık, K.; Keskin, A. The pollutant emissions from diesel engine vehicles and exhaust after treatment systems. *Clean Technol. Environ. Policy* **2015**, *17*, 15–27. [CrossRef]
4. Prasad, R.; Bella, V.R. A review on diesel soot emission, its effect and control. *Bull. Chem. React. Eng. Catal.* **2010**, *5*, 69–86. [CrossRef]
5. Zhang, Y.H.; Lou, D.M.; Tan, P.Q.; Hu, Z. Experimental study on the particulate matter and nitrogenous compounds from diesel engine retrofitted with DOC+CDPF+SCR. *Atmos. Environ.* **2018**, *177*, 45–53. [CrossRef]
6. Resitoglu, I.A.; Keskin, A.; Özarslan, H.; Bulut, H. Selective catalytic reduction of NO_x emissions by hydrocarbons over Ag-Pt/Al2O3 catalyst in diesel engine. *Int. J. Environ. Sci. Technol.* **2019**, *16*, 6959–6966. [CrossRef]
7. Kang, W.; Choi, B.; Jung, S.; Park, S. PM and NO_x reduction characteristics of LNT/DPF + SCR/DPF hybrid system. *Energy* **2018**, *143*, 439–447. [CrossRef]
8. Zhang, Y.H.; Lou, D.M.; Tan, P.Q.; Hu, Z. Experimental study on the emission characteristics of a non-road diesel engine equipped with different after-treatment devices. *Environ. Sci. Pollut. Res.* **2019**, *26*, 26617–26627. [CrossRef]
9. Yuan, X.M.; Liu, H.Q.; Gao, Y. Diesel Engine SCR Control: Current Development and Future Challenges. *Emiss. Control. Sci. Technol.* **2015**, *1*, 121–133. [CrossRef]
10. Koebel, M.; Elsener, M.; Kleemann, M. Urea-SCR: A promising technique to reduce NO_x emissions from automotive engines. *Catal. Today* **2000**, *59*, 335–345. [CrossRef]
11. Sharariar, G.M.H.; Lim, O.T. Investigation of urea aqueous solution injection, droplet breakup and urea decomposition of selective catalytic reduction systems. *J. Mech. Sci. Technol.* **2018**, *32*, 3473–3481. [CrossRef]
12. Kogan, F.M.; Nikitina, O.V. Solubility of chrysotile asbestos and basalt fibers in relation to their fibrogenic and carcinogenic action. *Environ. Health Perspect.* **1994**, *102*, 205–206.
13. Mcconnell, E.E.; Kamstrup, O.; Musselman, R.; Hesterberg, T.W.; Hesterberg, T.W.; Chevalier, J.; Miiller, W.C.; Thevenaz, P. Chronic inhalation study of size-separated rock and slag wool insulation fibers in Fischer 344/N rats. *Inhal. Toxicol.* **1994**, *6*, 571–614. [CrossRef]
14. Chen, J.; Gu, T.Z.; Yang, Z.J.; Min, L.I.; Wang, S.K.; Zhang, Z.G. Effects of Elevated Temperature Treatment on Compositions and Tensile Properties of Several Kinds of Basalt Fibers. *J. Mater. Eng.* **2017**, *45*, 61–66.
15. Fiore, V.; Scalici, T.; Di, B.G.; Valenza, A. A review on basalt fibre and its composites. *Compos. Part B* **2015**, *74*, 74–94. [CrossRef]
16. Yan, L.; Chu, F.L.; Tuo, W.Y.; Zhao, X.; Wang, Y.; Zhang, P.; Gao, Y. Review of research on basalt fibers and basalt fiber-reinforced composites in China (I): Physicochemical and mechanical properties. *Polym. Polym. Compos.* **2020**. [CrossRef]
17. Militky, J.; Kovaeie, V.; Rubnerova, J. Properties and applications of basalt fibres. *Text. Asia* **2001**, *32*, 29–33.
18. Wolter, N.; Beber, V.C.; Haubold, T.; Sandinge, A.; Blomqvist, P.; Goethals, F.; Van Hove, M.; Jubete, E.; Mayer, B.; Koschek, K. Effects of flame-retardant additives on the manufacturing, mechanical, and fire properties of basalt fiber-reinforced polybenzoxazine. *Polym. Eng. Sci.* **2021**, *61*, 551–561. [CrossRef]
19. Bayraktar, O.Y.; Kaplan, G.; Gencel, O.; Benli, A.; Sutcu, M. Physico-mechanical, durability and thermal properties of basalt fiber reinforced foamed concrete containing waste marble powder and slag. *Constr. Build. Mater.* **2021**, *288*, 123128. [CrossRef]
20. Bulent, O.; Fazll, A.; Sultan, O. Hot wear properties of ceramic and basalt fiber reinforced hybrid friction materials. *Tribol. Int.* **2007**, *40*, 37–48.
21. Novitskii, A.G.; Sudakov, V.V. An unwoven basalt-fiber material for the encasing of fibrous insulation: An alternative to glass cloth. *Refract. Ind. Ceram.* **2004**, *45*, 239–241. [CrossRef]

22. Novitskii, A.G. High-temperature heat-insulating materials based on fibers from basalt-type rock materials. *Refract. Ind. Ceram.* **2003**, *45*, 47–50. [CrossRef]
23. Bhat, T.; Chevali, V.; Liu, X.; Feih, S.; Mouritz, A.P. Fire structural resistance of basalt fiber composite. *Compos. Part A* **2015**, *71*, 107–115. [CrossRef]
24. Hao, L.C.; Yu, W.D. Comparison of the morphological structure and thermal properties of basalt fiber and glass fiber. *J. Xi'an Polytech. Univ.* **2009**, *23*, 327–332.
25. Hafsa, J.; Rajesh, M. A green material from rock: Basalt fiber—A review. *J. Text. Inst.* **2016**, *107*, 923–937.
26. Morozov, M.N.; Bakunov, V.S.; Morozov, E.N.; Aslanova, L.G.; Granovskii, P.A.; Prokshin, V.V.; Zemlyanitsyn, A.A. Materials Based on Basalts from the European North of Russia. *Glass Ceram.* **2001**, *58*, 100–104. [CrossRef]
27. Yang, W.J.; Ren, J.N.; Zhang, H.W.; Li, J.; Wu, C.; Gates, I.D.; Gao, Z. Single-atom iron as a promising low-temperature catalyst for selective catalytic reduction of NO_x with NH_3: A theoretical prediction. *Fuel* **2021**, *302*, 121041. [CrossRef]
28. Bai, S.Z.; Han, J.L.; Liu, M.; Qin, S.; Wang, G.; Li, G.X. Experimental investigation of exhaust thermal management on NO_x emissions of heavy-duty diesel engine under the world Harmonized transient cycle (WHTC). *Appl. Therm. Eng.* **2018**, *142*, 421–432. [CrossRef]
29. Shen, Y.G.; Liao, P.H.; Chen, C.L.; Peng, Y.Y.; Xiang, Y.H.; Chen, G.S. Experimental Study on Performance Evaluation of a Diesel Engine Equipped with Diesel Oxidation Catalytic and Catalytic Diesel Particulate Filter System. *Intern. Combust. Engine Eng.* **2020**, *41*, 17–26.
30. Assanis, D.N.; Wiese, K.; Schwarz, E.; Bryzik, W. The Effects of Ceramic Coatings on Diesel Engine Performance and Exhaust Emissions. *SAE Int. J. Engines* **1991**, *100*, 657–665.
31. Meng, Z.W.; Zhang, C.; Li, L.; Zhang, W.; Chen, C. Experimental Study on the Influence of DOC on Diesel Engine Particulate Matter Emission. *Intern. Combust. Engine Eng.* **2017**, *2*, 67–72.

MDPI AG
Grosspeteranlage 5
4052 Basel
Switzerland
Tel.: +41 61 683 77 34

Micromachines Editorial Office
E-mail: micromachines@mdpi.com
www.mdpi.com/journal/micromachines

Disclaimer/Publisher's Note: The title and front matter of this reprint are at the discretion of the Guest Editor. The publisher is not responsible for their content or any associated concerns. The statements, opinions and data contained in all individual articles are solely those of the individual Editor and contributors and not of MDPI. MDPI disclaims responsibility for any injury to people or property resulting from any ideas, methods, instructions or products referred to in the content.

www.ingramcontent.com/pod-product-compliance
Lightning Source LLC
LaVergne TN
LVHW072318090526
838202LV00019B/2306